ADVANCES IN
HISTORICAL ECOLOGY

The Historical Ecology Series
William Balée and Carole L. Crumley, Editors

This series explores complex links between people and the landscapes that both mold individuals and societies and are fashioned by them. Drawing on scientific and humanistic scholarship, books in the series focus on cognition and on temporal and spatial change. The series offers examples, explores issues, and develops concepts that preserve experience and derive lessons from other places and times.

ADVANCES IN
HISTORICAL ECOLOGY

William Balée,

EDITOR

COLUMBIA UNIVERSITY PRESS / NEW YORK

Columbia University Press
Publishers Since 1893
New York   Chichester, West Sussex
Copyright © 1998 by Columbia University Press

Library of Congress Cataloging-in-Publication Data

Advances in historical ecology   /   William Balée, editor.
        p.   cm.
    Includes bibliographical references and index.
    ISBN 0–231–10632–7 (alk. paper)
    1. Human ecology—History.   2. Nature—Effect of human beings on.
    3. Landscape assessment—History.   4. Landscape changes—History.
    5. Biotic communities.   6. Environmental degradation.   I. Balée,
    William L., 1954—
    GF13.A39   1998
    304.2—dc21   97–34863

∞
Casebound editions of Columbia University Press books are printed on permanent
and durable acid-free paper.

Printed in the United States of America
c 10 9 8 7 6 5 4 3 2 1

CONTENTS

PART II Regional Research and Landscape Analyses in Historical Ecology

FOREWORD

CAROLE L. CRUMLEY

> Or, si presque tout problème humain important appelle ainsi le manie-ment de témoinages de types opposés, c'est, au contraire, de toute nécessité, par type de témoinages que se distinguent les techniques érudites. L'apprentissage de chacune d'elles est long; leur pleine pos-session veut une pratique plus longue encore et quasiment constante. Un bien petit nombre de travailleurs peuvent se vanter d'être également bien préparés à lire et critiquer une charte médiévale; à interpreter cor-rectement les noms de lieux . . . ; à dater, sans erreur, les vestiges de l'habitat préhistorique, celte, gallo-romain; à analyser les associations végétales d'un pré, d'un guéret, d'une lande. Sans tout cela pourtant comment prétendre écrire l'histoire de l'occupation du sol?
> —MARC BLOCH (1993:111)

> (Now, if almost any important human problem thus requires the han-dling of diverse types of evidence, it is then contradictory that fields of technical scholarship are distinguished by type of evidence. The ap-prenticeship for each is long, but full mastery demands a still longer and almost constant practice. Very few scholars can boast that they are equally well equipped to read critically a medieval charter, to interpret correctly the etymology of place-names, to date unerringly the re-mains of prehistoric [and historic] habitations, and to analyze the plants characteristic of a pasture, a field, or a moor. Without all these, however, how could one pretend to write the history of land use? [my translation])

The practice of historical ecology, the twinned necessity and difficulty of which the *Annales* historian Marc Bloch (1993) underscores, requires much of the would-be practitioner. Just to begin, it is necessary to eschew the dualistic history of Western scholarly thought, suspend disciplinary biases, and take up multiple careers. Such an undertaking is even less possible now, in the Information Age, than when Bloch wrote in the early 1940s.

Yet inasmuch as all human activity inevitably takes place *somewhere,* it is em-bedded in a matrix, a context, an environment. Even for a visit to cyberspace, the visitor requires electricity from a power source, air to breathe, and safe surround-ings. It is arguably a hallmark of the twentieth century that during its course the human context has been defined ever more broadly, as the physical and human sci-ences explored the universe, the planet, and the evolution, behavior, and built envi-ronment of our own and other species.

Beginning with the International Geophysical Year in 1958, the quality and quantity of data that document change in the global ecosystem have improved; this has enabled researchers to ask new questions about how humans affect and are affected by their environment. In contrast to the early twentieth century, contemporary research into human/environment relations does not assume a deterministic role for the environment; instead, the assumption is that humans can both instigate and respond to fluctuations in resources or climate.

Although earlier deterministic, mechanistic, and dualistic characterizations have been rejected, there remains a great challenge to contemporary researchers: they must identify and employ an integrated framework, which must accommodate spatially and temporally specific natural and social scientific information and include evidence of changing values, perceptions, and awareness. Construction of this integrated framework has proved particularly difficult; the intellectual architecture of the framework is not yet in place. Scientific understanding of the interconnectivity of the atmosphere, hydrosphere, biosphere, and geosphere in the global system is growing, and provides reasonable background cause-and-effect linkages and cyclicity. Comparable social science theory and methods are needed to monitor and evaluate human activity at all temporal and spatial scales.

One practical issue concerns the scalar incompatibility of human activity with planetary-scale atmospheric phenomena. Patterns of settlement and land use, emissions, and extractive procedures must be investigated at regional and local scales. On the other hand, collective response to global-scale changes (e.g., climate) must be verified at the planetary scale by comparing chronologies of change in widely dispersed regions. Without integrated environmental and cultural information at local and regional scales, no opportunity exists to test and refine global models; without planetary-scale confirmation of the long-term effects of human activity, arguments over values (embedded, among other things, in environmental policy) will continue without resolution until our evolutionary fate is sealed.

Another difficulty has been the changing intellectual fashions. Sometimes scientific explanation has been favored (the Darwinian revolution, modernism), sometimes hermeneutic understanding (religious doctrine, postmodernism). The scientist and author C. P. Snow, in a lecture at Cambridge in 1959, termed this divide the "two cultures" of scientists on one side and humanists on the other. Although Snow's characterization of the values and beliefs of the two camps is not appropriate today (and there is considerable doubt that it ever was; see Collini 1993), his description of the chasm between them is still accurate.

The current attempt to declare a cease-fire among academic disciplines and their factions for the benefit of our species is not the first. The term *ecology* itself (from the Greek *oikos,* dwelling) was meant to draw attention to the reciprocal relationships among living and nonliving elements of our world. The term was first used by natural scientists in the late nineteenth century; by the late 1960s, articulation of the Gaia hypothesis expanded the definition to include all scales (local to global) of relations among living and nonliving elements, including humans.

For the generation that came of age at about the same time our species first set foot off-planet, ecology became a shorthand for the relation of our species to all facets of its *oikos*. The discipline of ecology has since bifurcated and its emphasis has undergone a scalar shift. Today microecology, with ties through cell and molecular biology to schools of medicine and public health, dominates the field; macroecology (e.g., wildlife ecology, global ecology) trains fewer practitioners and garners fewer research dollars than its larger and better-connected twin.

Environmental historians have made an earnest attempt to address the two cultures, and have enjoyed considerable popular success (e.g., Cronon 1983, 1995; Crosby 1986, 1994; Worster 1977, 1994). However, with some notable exceptions (e.g., LeRoy Ladurie 1971; Rotberg and Rabb 1981), historians are often perceived as lacking the appropriate credentials to comment on scientific subjects, and collaborations with natural scientists are few. It is perhaps in the relation between the discipline of history and those of the natural sciences that we can most clearly see the mutual disdain with which their respective methods are considered: historians consider scientific findings trivial and mechanistic; scientists consider historians' methods imprecise and their styles of argumentation histrionic (Ingerson 1994).

Thus while historians rightly concentrate on both intended and unintended consequences of human action, and offer powerful examples of the plastic role of history and culture, they usually have less knowledge of the biophysical systems that also condition human activity. Landscape ecologists (e.g., Naveh and Lieberman 1990; Forman and Godron 1986), for their part, in their comprehensive and authoritative treatment of biophysical systems, either ignore the human presence or assume its effects on "natural" ecosystems to be always negative, even in the case of the research scientists themselves.

Although logically the social sciences should be able to arbitrate the sciences vs. humanities standoff, additional difficulties present themselves. Economists were early recruited to help shape environmental policy, and they went to work forging axioms; like those of microecology, however, rule-based microeconomic prescriptions are criticized as reductionist. The grand economic theories that guided the triumph of capitalism now appear much less certain. For example, attempts to calculate the price of intangibles (values, beliefs) have proved distasteful to humanists (Gore 1992) and empirically questionable to social scientists whose methods are not solely quantitative.

Geographers and anthropologists employ both qualitative (more humanistic) and quantitative (more scientific) methods, and generally understand the relationship between humans and their environments to be reciprocal. However, anthropology and geography enjoy neither the recognition and popular following of history, nor the seductive effects to the policy community of economic dogma, dressed in hard numbers.

Like ecology, the discipline of geography split itself in two for ostensibly economic reasons. After World War II, those trained in the new quantitative and cartographic approaches readily found jobs in a world that was tired of ideology but still enthusiastic about science. Recently their numbers have again increased as remote sensing, geographic information systems, and other computer-based applications shoulder historical and cultural geography aside.

In anthropology, the subdisciplines that routinely investigate the human/environment relation are human ecology (also termed cultural or social ecology), archaeology, and physical anthropology. They have been marginalized within anthropology since the first decades of the century, when Franz Boas defended the role of culture against the determinisms (social, racial, environmental). For primarily historical and political reasons, the anthropological subdisciplines most closely allied with the sciences have been marginalized within anthropology, and even stigmatized in some quarters because of their association with the sciences.

Although the intellectual premise of both geography and anthropology should sustain specialties that serve to bridge the two cultures, scarce resources and old animosities make geographers and anthropologists competitors more often than collaborators. Despite their logical and pivotal position at the center of the human/environment relation, neither geography nor anthropology (nor any other social science discipline) has offered a comprehensive, historically sentient scheme for understanding how humans relate to their immediate, regional, and planetary domicile: *oikos* in its most inclusive sense.

We do not, however, have the unlucky option of concluding that collective effort is impossible. Clearly, humans must respond *both* to global changes that make local differences *and* to local practices that drive global change, employing every means at our disposal. We must search for common ground, in relatively new terrain and on relatively neutral terms.

The term *historical ecology* has been chosen in a renewed effort to foster collaboration in two crucial social science disciplines (anthropology and geography) and among several hybrid fields (e.g., environmental history, environmental sociology, human ecology, landscape ecology) that seek to mend the divide between the two cultures. Don S. Rice (1976) attributes first use of the term to the anthropologist Edward S. Deevey, who directed the Historical Ecology Project at the University of Florida in the early 1970s. Historian Lester J. Bilsky solicited the contributions of anthropologists, a human ecologist, an economist, and fellow historians for *Historical Ecology: Essays on Environment and Social Change* (1980). Anthropologist Alice Ingerson organized a session on historical ecology at the 1984 annual meeting of the American Anthropological Association (AAA) to address the chasm between cultural (e.g., nature as metaphor) and environmental (energy cycles) studies within anthropology, and to explore political economy and social history approaches. Drawing from biology, geography, and history, I employed the term to summarize regional change (Crumley 1987). Archaeologists, anthropologists, and an evolutionary ecologist collaborated in *Historical Ecology: Cultural Knowledge and Changing Landscapes* (Crumley 1994).

The elements that characterize historical ecology are appropriately diverse and drawn from several disciplines and intellectual traditions. The French historical school *Annales* (Burke 1990) plays an important role, reminding American historians, anthropologists, and geographers that it is indeed possible to join forces. Since the first decades of the twentieth century, the humanistic and dialectical *Annales* approach has been employed with considerable success (e.g., the work of Braudel, Duby, LeRoy Ladurie, and many others). Especially useful is the concept of processes operating among temporal scales of varying duration: these broad divisions are termed *événement* (event), *conjoncture* (cultural and historical context), and *longue durée* (long-term history). Broad temporal frameworks such as these also guide the historical sciences. Interpretation, in both history and the sciences, relies on all three: events, whether they are a cleric's account of famine or stratigraphic evidence of flood, must be set in both immediate and far-ranging context.

In the past twenty years American researchers, seeking ways to address time and space simultaneously, have expanded upon the central *Annales* concept of landscape (*paysage*). In landscape both intentional and unintentional acts are recorded; in its study, both humans' role in the modification of the global ecosystem and the importance of past events in shaping human choice and action can be assessed. Broadly defined as the spatial manifestation of the relations between humans and their environment (Marquardt and Crumley 1987:1), "landscape" offers several advantages. It is a common unit of analysis in several diverse fields (geography, archaeology, ecology, geomorphology, architecture, art, regional planning) and thus helps integrate diverse evidence; and it allows changes to be traced through time. The study of changes in the temporal and spatial configurations of landscapes (a traditional pursuit of archaeology), in conjunction with work in cognition, offers a practical means of integrating the natural and social sciences and the humanities.

Because analytic units such as landscape and other concepts (e.g., scale, disturbance, community, region, niche, boundary/ecotone, etc.) are common to several disciplines, diverse fields of study can contribute to a shared grammar. Construction of the grammar is universally instructive and focuses debate on central issues. Geographic information systems (GIS) integrate data at a variety of spatial and temporal scales and support both qualitative and quantitative analysis. Thus can we practice historical ecology in the spirit of Marc Bloch's admonition, by taking advantage of all fields' expertise—whether it is in documentary research, long-term ecological analysis, archaeology, or the liminal theoretical and methodological space between the two cultures.

This volume builds on previous work, but it also advances the effort, inasmuch as the shared concepts, methods, and assumptions that underlie the practice of historical ecology are expanded and applied. A common set of definitions, premises, and postulates begin the volume, and subsequent contributions build a grammar of both practical and theoretical utility. Several concepts from nonhuman ecology are expanded (e.g., fire ecology, chronological ecotones, disturbance, community) to incorporate the history of human populations in environments undergoing change. The contributors focus on traditional societies, situated at the margins of more powerful groups, in which a wide range of shifting strategies are practiced. While geographical coverage is strong for South America, the volume also includes studies from North and Central America, Thailand, India, and Europe. These authors argue that traditional societies are anything but "timeless," with pasts that can be measured in a variety of indigenous and nonindigenous ways. With the perspective of time, it becomes possible to see that subsistence flexibility, as measured by the opportunistic employment of several means of subsistence (e.g., pastoralism, gardening, gathering, hunting), becomes a strategy in itself.

Perhaps the most important characteristic of historical ecology is that it celebrates both the open-mindedness of scientific inquiry and the phenomenological intensity of human experience. The historical ecology of any part of the world is always an unfinished manuscript—passed from hand to hand, critiqued, debated, amended, revised. It is through reflection upon intended and unintended change that all people are moved to action.

References

Bilsky, Lester J. 1980. *Historical Ecology: Essays on Environment and Social Change.* Port Washington, N.Y.: Kennikat Press.

Bloch, Marc. 1993. *Apologie pour l'histoire ou métier d'historien.* Préface de Jacques Le Goff. Edition critique préparée par Etienne Bloch. Paris: Armand Colin. Published in English as *The Historian's Craft.* New York: Vintage, 1953.

Burke, Peter. 1990. *The French Historical Revolution: The Annales School, 1929–1989.* Stanford: Stanford University Press.

Collini, Stefan. 1993. Introduction to C. P. Snow, *The Two Cultures.* Cambridge: Cambridge University Press.

Cronon, William. 1983. *Changes in the Land: Indians, Colonists, and the Ecology of New England.* New York: Hill and Wang.

———. 1995. *Uncommon Ground: Toward Reinventing Nature.* New York: Norton.

Crosby, Alfred W. 1986. *Ecological Imperialism: The Biological Expansion of Europe, 900–1900.* Cambridge: Cambridge University Press.

———. 1994. *Germs, Seeds, and Animals: Studies in Ecological History.* Armonk, N.Y.: Sharpe.

Crumley, Carole L. 1987. Historical Ecology. Pp. 237–264 in Carole L. Crumley and William H. Marquardt (eds.), *Regional Dynamics: Burgundian Landscapes in Historical Perspective.* San Diego: Academic Press.

———, ed. 1994. *Historical Ecology: Cultural Knowledge and Changing Landscapes.* Santa Fe: School of American Research Press.

Forman, Richard T. T. and Michel Godron. 1986. *Landscape Ecology.* New York: Wiley.

Gore, Al. 1992. *Earth in the Balance: Ecology and the Human Spirit.* New York: Houghton-Mifflin.

Ingerson, Alice E. 1994. Tracking and testing the nature/culture dichotomy in practice. Pp. 43–66 in Crumley (ed.), *Historical Ecology.*

LeRoy Ladurie, Emmanuel. 1971. *Times of Feast, Times of Famine: A History of Climate Since the Year 1000.* New York: Doubleday.

Marquardt, William H. and Carole L. Crumley. 1987. Theoretical issues in the analysis of spatial patterning. Pp. 1–18 in Crumley and Marquardt (eds.), *Regional Dynamics.*

Naveh, Zev and Arthur S. Lieberman. 1990. *Landscape Ecology: Theory and Application.* Student ed. New York: Springer-Verlag.

Rice, Don Stephen. 1976. The historical ecology of Lakes Yaxha and Sacnab, El Petén, Guatemala. Ph.D. diss., Pennsylvania State University.

Rotberg, R. I. and T. K. Rabb, eds. 1981. *Climate and History.* Princeton: Princeton University Press.

Worster, Donald. 1977. *Nature's Economy: A History of Ecological Ideas.* Cambridge: Cambridge University Press.

———. 1994. *An Unsettled Country: Changing Landscapes of the American West.* Albuquerque: University of New Mexico Press.

PREFACE

The substance of *Advances in Historical* Ecology is organized into two parts. Part I consists of theoretical treatments of historical ecology and attempts at a very broad coverage of material factors intrinsically important to historical ecology, such as anthropogenic fire, soils, and pathogens. Part II deals with more regionally focused studies of landscape transformations over time, mostly in South America but also in the Mississippi Delta, the Great Basin, Thailand, and Rajasthan, India.

My principal goals as editor will have been achieved if the present work helps students and researchers in human ecology to realize the importance of situating their thinking, research, and writing in relation to the current debates and findings manifested here. It will represent fulfillment for me also if policymakers—whose decisions have effects on peoples and places that possess minimal material wealth and power as measured by Euro-American standards—gain insights into the significance of dialectical relationships between those people and the land as these have unfolded over time in diverse regions of the world without necessarily leading to irreversible degradation, biological extinctions, and cultural impoverishment. Historical ecology holds a key to a more effective (and democratic) management and understanding of biotic and human resources in their inimitably complex, decidedly dualistic relationship over time. In an effort to be true to these goals, any royalties this book may accrue will be donated to the Socioenvironmental Institute (Instituto Socioambiental, formerly the Centro Ecumênico de Documentação e Informação, or CEDI), a nongovernmental organization of São Paulo, Brazil, that has been for a long time intimately connected to the rights and welfare of Brazilian native peoples and their lands.

The idea for this book originated in a conversation over lunch with Brian Ferguson in downtown San Francisco during the 1992 annual meeting of the American Anthropological Association. We discussed the desirability, in the first place, of holding a miniconference on historical ecology that could be partially supported by the Columbia Graduate Anthropology Student Association that Brian was then organizing. I would like to thank Brian and his collaborators in that group of fellow Columbia anthropology graduates—especially Pat Antoniello, Carol Henderson, Janet Chernela, and Barbara Price—for their moral support, for refreshing my memories of Morningside Heights, and for help in publicizing and cosponsoring the miniconference on historical ecology that eventually took place in June 1994 in New Orleans. I am also grateful to the administration at Tulane University for its encouragement and support in holding the conference at Tulane. Thanks are due to discretionary funding by the Provost's Office at Tulane University for partial funding of the conference, and to Tulane graduate students Célia Futemma, Loretta Cormier, and Doug Wells, who supplied much-appreciated assistance with facilitating exchanges among the participants.

Those who presented papers at that conference were Janet Chernela, Brian Ferguson, Ted Gragson, Elizabeth Graham, Thomas Headland, Carol Henderson, Tristram R. Kidder, Darrell Posey, Laura Rival, Anna Roosevelt, Leslie Sponsel, Neil Whitehead, Stanford Zent, and myself. These papers, minus Headland's, became chapters in the present volume. Much useful discussion was added to the conference by the presence of Carole Crumley, Joel Gunn, Fred Damon, Janet Headland, Victoria Bricker, and Eglée Lopes-Zent. This book grew to include chapters by Linda Newson, Stephen Pyne, Robert Bettinger, and Elinor Melville, together with the foreword by Carole Crumley.

I gratefully acknowledge the assistance of all the participants in the conference and the contributors to this volume for their indispensable help in bringing the work into print. I am grateful also to my editor at Columbia University Press, Ed Lugenbeel, for his early interest in the book and his encouragement throughout its composition and editing. Thanks also are due to editorial advice by Julia McVaugh and Roy Thomas. My wife, Conceição, offered her own unique, useful contributions and asides for the sake of maintaining my humor over the course of this project, for which I am, as ever, grateful.

William Balée
New Orleans, October 1997

ADVANCES IN
HISTORICAL ECOLOGY

INTRODUCTION

WILLIAM BALÉE

One senses a generalized discord in the life sciences and the social sciences as the twenty-first century draws near. This discord, like Thomas Kuhn's "essential tension" (1977), may actually represent a necessary impetus toward new ways of thinking about humankind and global environments, ways that may offer better and more humane solutions to some of the critical problems that humans as a species face in the biosphere today. Yet to some degree, the immediate economic and political needs of regional and global systems have taken precedence over science in the search for comprehension of human-ecological conditions, partly because the social and life sciences have lacked a common metalanguage and framework for mutual debate. There is a way out of this impasse by developing an interdisciplinary language that will help connect what Carole Crumley, citing C. P. Snow, refers to in her foreword as the "two cultures."

Historical ecology supplies that metalanguage. Even though it is still nascent, a conceptual centrality focused on the terms *landscape, biosphere, human/environmental dialectics,* and *region* is becoming apparent; such terms occur often in the present book. The concept of *landscape,* above all, seems paramount in historical ecology. These terms and their usages, when comprehended technically, facilitate a more holistic (and therefore more accurate and empirically sound) analysis of human ecology.

A revolution in thinking about humans and the environment is taking place in the life and social sciences simultaneously. The new thinking may bridge disciplines and help reengage the debate between biological and cultural determinism. The models of biological and cultural determinism, by their philosophical opposition, provided the milieu in which much scientific debate was carried on in many fields for more than a hundred years. Now those working models have been largely abandoned by the fields in which they originated, and their demise has left the academy, as a whole, in disarray. The disarray—some call it a crisis—in higher education today is not just related to the information revolution that seems to make much university classroom teaching obsolete, and to the dwindling of funds available for both research and teaching about humans and the environment as traditionally practiced. Rather, it is fundamentally tied to the intellectual void that has followed the

1

evanescence of the two models, however dogmatic, that guided most research in the life and social sciences during this century.

The new framework that can fill that void incorporates the knowledge accumulated by nonstate peoples (i.e., "nonscientists") who are closely related to their local biotic and geophysical resources, knowledge that has been "captured" (Gunn 1994) over the long term, as one guide in understanding the formation and development of *landscapes* (Posey, chapter 5; Rival, chapter 11), while at the same time it maintains scientific standards of method and evidence. This new framework, which has the most promise for reinvigorating research in human ecology, is called historical ecology. It is already beginning to fill the void, especially with the publication of the first treatise on the subject, as I define it below, by Carole Crumley and her colleagues (Crumley, ed., 1994).

I did not invent the term *historical ecology* (see Crumley in the foreword to this book for a useful history of the term), and I am not alone in considering it to be the most important current intellectual advance in the study of human and environmental relationships. A rapidly growing number of scholars from diverse fields also perceive historical ecology as representing a new, powerful, and holistic framework for research and debate on one of the most fundamental problems of our time: the diverse and complex relationships between humans and their environments.

What Is Historical Ecology?

The reader of this book merits a brief overview of the conceptual apparatus that underlies historical ecology and that links the chapters together into an entity, centering on the notion of the *landscape*. At the outset I would mention that not all contributors to this volume agree with me on all specific aspects of my conception of historical ecology. Yet I think we understand each other. I consider that Part I represents a contribution to a healthy debate as to the meaning of historical ecology and its relationship to other—both earlier and contemporary—approaches to human/environmental interaction. Part II consists of substantive findings on the development of past and present landscapes, which are biocultural phenomena, in diverse settings.

Historical ecology is connected to but different from several schools of materialism that were concerned with the relationship between humans and nature, including historical (dialectical) materialism, cultural ecology, cultural materialism, and evolutionary ecology (Balée 1996). With regard to historical materialism, historical ecology picks up where Marx left off. Marx argued that "he [humankind] opposes himself to Nature . . . in order to appropriate Nature's productions in a form adapted to his own wants. By thus acting on the external world and changing it, he at the same time changes his own nature" (1867:177; see Balée, chapter 1). Historical ecologists tend to agree with the premise of the mutual influence of people ("man") and biosphere ("nature") on one another over time as a dialectical phenomenon (Crumley 1996; cf. Whitehead, chapter 2). But Marx did not carry that premise further—he investigated relationships between people, and relationships between people and economic resources in state societies of his time. Historical ecology actually involves the empirical investigation of relationships between humans and the biosphere in specific temporal, regional, cultural, and biotic contexts, regardless of their relationship to (or their incorporation into) nation-states. Historical ecology owes a debt to the pioneers of cultural ecology, especially Julian Steward, insofar as cultural ecology involved a strong emphasis on empirical research into human/environmental relationships (Whitehead, chapter 2; Bettinger, chapter 8). It differs from cultural ecology, however, in at least two important ways. First, cultural ecology focused on how "natural" environments—replete with given geophysical and biotic conditions—affected localized societies and their cultural development over time. In a real sense, cultural ecology effected a separation of nature and culture (cf. Bettinger, chapter 8); by contrast, historical ecology conceives of relationships between nature and culture as a dialogue, not a dichotomy (Ingerson 1994). Second, cultural ecology had a tendency toward environmental determinism (see Zent, chapter 12). Although some disagreement exists on this matter (see Balée, chapter 1; Bettinger, chapter 8), cultural ecology seemingly was never successful in advancing falsifiable theories about the relationships between state societies and their environments; in other words, it was limited in its explanatory framework to egalitarian societies (Balée 1996). Historical ecology, in contrast, is concerned not only with mutual relationships between egalitarian societies and their local environments and regions, but also with all other kinds of human/environmental, human/regional, and human/biosphere relationships over time—since at least the time of *Homo erectus* in the Pleistocene, with its innovations in use of fire, if not since the beginnings of the genus *Homo* before that (Bettinger, chapter 8; Kidder, chapter 7; Pyne, chapter 4; Roosevelt, chapter 9).

Historical ecology is indebted to cultural materialism, and therefore especially to Marvin Harris (1968, 1979), because cultural materialism endeavors to encompass all sociopolitical systems in their relationships with nonhuman environments. Further, both cultural materialism and historical ecology subscribe to the scientific method (I disagree with Neil Whitehead [chapter 2] here, incidentally, because in my view historical ecology itself is not a method, except *sensu lato,* but rather a way of understanding phenomena). The two fields diverge, however, in two important ways. First, cultural materialism holds that developments in the relationship between societies and local environments over time are *evolutionary* rather than historical and ecological (Bettinger, chapter 8; Ferguson, chapter 13), and it also subscribes essentially to neo-Darwinian theory as concerns the biological evolution of species—yet these evolutionary mechanisms are fundamentally unlike analytically. In other words, cultural materialists (like evolutionary ecologists) tend to use the term *evolution* in a polysemous rather than a metalinguistic (more precise) way, with the result that the meaning of cultural evolution is not often clear in specific analyses. The concept of evolution in cultural materialism, moreover, is not applied to reversals in the intensity of land use by specific societies in given regions (Balée, chapter 1; Rival, chapter 11); historical ecologists tend to view such changes as historical or developmental, rather than evolutionary, without denying the mechanisms of evolution (natural selection, genetic drift, and so on) with respect to organisms (Graham, chapter 6). Second, historical ecology differs from cultural materialism, as it does from cultural ecology, insofar as the relationship between humans and the environment in a given regional context is conceived of dialectically: cultural materialism specifically rejects such a synthetic, holistic viewpoint, emphasizing the primacy of infrastructure (modes of production and reproduction taken together) in all cultural developments, regardless of changes in the nonhuman environment. The holistic and dialectical perspective of historical ecology is not antiscientific: it represents objective science, but not an exclusively reductionist science, in the final analysis. In a very specific way, historical ecology is even more dialectical than historical materialism, for it begins with the dialectic of an inalienable link between nature and culture, however defined.

Historical ecology is also related to evolutionary ecology, a point stressed by Winterhalder (1994) and Bettinger (chapter 8). In fact, evolutionary ecology, specifically as concerns humans, is probably a direct descendant of cultural ecology (Bettinger, chapter 8). The difference between historical ecology and evolutionary ecology concerns the ends and means of human/environmental relationships. Evolutionary ecology tends to hold that differing human behavior with regard to habitat and biota results in differences in reproductive fitness of the human actors themselves: some individual human behaviors are selected for, and others are selected against, in the process of differential reproductive fitness. The central problem, however, is whether human behavior as concerns habitat and biota can actually be predicted according to a model based on natural selection (a point wisely raised by Eric Alden Smith, an evolutionary ecologist himself, in the conclusion to *Inujjuamiut Foraging Strategies* [1991]). Such analyses have likewise tended to be restricted to nonstate societies, a limitation also of the earlier cultural ecology. In addition, evolutionary ecology is nondialectical and does not take as axiomatic the dialogue between humans and the nonhuman environment (Ingerson 1994; Balée, chapter 1).

A current debate in historical ecology specifically concerns the problem of how to differentiate history from evolution in understanding the development of relationships between humans and the biosphere—or, at a more finely grained level of analysis, between humans and *landscapes* (Winterhalder 1994; Bettinger, chapter 8; Ferguson, chapter 13; Gragson, chapter 10; Graham, chapter 6; Whitehead, chapter 2). Ted Gragson, in his treatment in chapter 10 of South American foragers of savanna habitats, indicates that the landscape is a dialectical phenomenon. He implies that *landscape ecology* is a synonym for *historical ecology* (although landscape ecologists from the life sciences have tended to exclude the dialectical interrelatedness of people and the landscape, which by their definition is the result of essentially orderly process in nature). Another appropriate synonym would be *dialectical ecology* (Balée, chapter 1).

The debate in anthropology between evolutionism and historicism is not unlike that in paleontology between those who explain changes in the fossil record by reference to the gradual effects of natural selection and those who subscribe to much more rapid, unpredictable changes by reference to punctuated equilibrium, in which extrabiospheric (accidental) phenomena affect the course of evolution of life itself. The sudden disappearance of dinosaurs from the fossil record because of an asteroid impact is empirically analogous to the historical accident of the conquest of the New World by voyagers seeking the Orient who incidentally brought with them diseases from Europe that were unknown in the New World, and to which much of the native population succumbed (Balée 1995). The process of sorting through the similarities and differences between history and evolution (with its attempt to approach both metalinguistically) is one of the most significant contributions to interdisciplinary debate and understanding that this book offers.

A question remains as to whether historical ecology, as a label, embraces unique principles that would pertain to a paradigm in the Kuhnian sense, or to a research strategy in a looser one. Some might argue that historical ecology is a field of inquiry, a theory from another paradigm, or a method of another, already established field of inquiry, such as historical anthropology (Whitehead, chapter 2). This tension in the definition of historical ecology is manifested in the present volume (especially in the chapters by Bettinger, Sponsel, and Whitehead). This book therefore represents an emerging, scientific viewpoint insofar as its chapters' stimulating debate and new insights revolve around key concepts and relationships in human/environmental relationships. I remain content to help in encouraging debate on the status of historical ecology (and Carole Crumley seems to agree with such an aim in her foreword), given that the participants in that debate seem to understand one another partly through their use of key terms and concepts, such as that of the landscape.

Evolutionism vs. Historicism in Historical Ecology: Landscape as the Bridge Between Them

In chapter 2, Neil Whitehead poses the question most forcefully as to whether humans are essentially historical or ecological entities. If historical, then historical ecology must adopt a more rigorous theory of history, rather than merely mapping a simplified temporal dimension (or just a chronology of events) onto an essentially processual, environmental explanation of changes in landscapes over time. A similar *problématique* is clearly present in other chapters. In chapter 13, R. Brian Ferguson argues that "making sense of history requires a sense of evolution" in his documentation of changes in Yanomami tool use and settlement patterns. Likewise, Robert Bettinger (chapter 8), in his insightful analysis of prehistoric settlement patterns in the Great Basin, suggests that "culture and environment interact in complex ways that benefit from an evolutionary analysis in which history is important."

Neil Whitehead, by way of contrast, indicates that history is for him the underlying mechanism of historical ecology. He argues, in incisive discussion of the tension between evolutionism and historicism in modern anthropology, that if persons are essentially historical, then historical ecology can be subsumed under a wider rubric that he calls historical anthropology. Elizabeth Graham (chapter 6) deals with the problem of "history" and "ecology" in anthropology in a more synthetic way. She suggests that neither is related to the disciplinary objectives of anthropology; rather, history, as understood in historical ecology, can now apply to both ecology and anthropology.

On the other hand, Whitehead (chapter 2) contends that we have not yet found a means to bridge the differences between evolutionism and historicism in anthropology, just as the dichotomy between scientism and humanism remains to be eliminated. One can infer from Whitehead's chapter that, in order to comprehend human/environmental relationships, you must be either an evolutionist or a historicist: you cannot be both at the same time. He further implies that historical ecology represents a method, rather than a paradigm. His penetrating questions and critique of ideas in historical ecology are extremely important for debate and cannot go unanswered, if historical ecology is to advance. Many of the chapters do address and try to answer these problems in historical ecology (Balée, chapter 1; Bettinger, chapter 8; Ferguson, chapter 13; and Graham, chapter 6). Tristram Kidder and I treat them in a wider interdisciplinary context in the epilogue.

Regardless of whether humans are to be viewed either as historical or as ecological (evolutionary) entities, Whitehead utilizes the concept of *landscape* in its historical-ecological (holistic) sense. He refers to "human praxis in the landscape," implicitly understood as a biocultural phenomenon. More specifically, in his analysis of anthropogenic black-earth soils in the Amazon, he emphasizes the "synergy" of environment and culture. In other words, he stresses the mutual influence of people and nonhuman nature: the process is really not *either* biological *or* cultural; rather, it is *both*. Whitehead argues that human decision-making in the landscape is the key to understanding the development of the latter over time, as does Graham.

Elizabeth Graham (chapter 6) is also concerned with anthropogenic soils—in this case, of the Maya area in Belize—and notes that soils, both those resulting from ancient geomorphological processes and those affected by human interactions with the *landscape*, are "'solid-phase products of the continuous functioning of the biosphere.'" If termites and leaf-cutter ants can change the composition of soils (Graham, chapter 6; Sponsel, chapter 17), if elephants can create light gaps that increase biodiversity in African tropical forests (Campbell 1991), and if beavers can "build environments" (Sponsel, chapter 17), so can humans—but potentially on a different scale, with a different consciousness, and with a different history (cf. Ingold 1988).

Humans, who are both biological and cultural, are the key mechanism in historical ecology. This is partly how historical ecology bridges the gap between disciplines (specifically between the so-called life sciences and social sciences). As Graham points out in chapter 6, landscapes have a history, rather than merely an evolution behind them. In alluding to Eric Wolf's *Europe and the People Without History* (1982), she notes that earlier scholars without a historical-ecological point of view perceived not only "people without history, but landscapes as well!"

In other words, the mechanistic aspect of historical ecology is simultaneously historical and ecological. In this light, Linda Newson in chapter 3 discusses how chronic and acute infectious diseases affect the historical development of different societies and sociopolitical systems of different kinds in unlike ways. The origins and spread of infectious diseases are both historical and ecological in her analysis. Bacterial and viral pathogens themselves have histories. Infectious disease constitutes a part of the landscape that, depending on historical developments, interacts with people, biota, and regions in distinctive ways.

Anthropogenic soils and infectious diseases represent factors in the landscape. Stephen Pyne in chapter 4 persuasively argues that fire is another such factor, perhaps the most important one. He astutely notes that "fire is a creation of life" (and hence, it is critical to the concept of biosphere in historical ecology—see Balée, chapter 1). Pyne shows how historical-ecological analysis is not limited to egalitarian societies, but rather can be applied to the genus *Homo* (which has a monopoly on fire-making) since the time of *Homo erectus* as a whole—and that includes archaic and modern state societies. He argues convincingly that firefighting in modern Western society has ironically led to more wildfires, whereas low-intensity burning, as practiced by prehistoric and historic native peoples, prevented a buildup of fuels for wildfires. Large, destructive wildfires are artifacts of complex state-level societies, whereas traditional anthropogenic fires (or controlled burning, as with broadcast fires), such as those used in native swidden gardening and in stimulating new growth in the Americas, Africa, Australia, and elsewhere, created landscapes of higher biodiversity (Sponsel, chapter 17).[1]

Sometimes the establishment of national parks and the like in complex societies ironically leads to environmental degradation and biotic impoverishment (Pyne, chapter 4). Carol Henderson (chapter 16) shows how the establishment of a large reserve in the Thar Desert of Rajasthan, India, greatly decreased the availability of scattered, habitable landscapes traditionally used by the agropastoralists of that region—with the result that the utilization of the remaining landscapes was intensified to the point where erosion and decreasing arability, which had not been observed before, became apparent. In chapter 17, Leslie Sponsel also argues convincingly that state conservation efforts are minimal in effectiveness when compared to the long-term degradation of the environment and the decrease in species diversity that, in the specific case of Thailand, they have tended to cause. As Sponsel shows in a detailed historical-ecological analysis, traditional and modern uses of fire in Thailand are strikingly different in their effects on specific landscapes.

Different kinds of fires are associated with different kinds of landscapes and their unlike degrees of biodiversity (Balée, chapter 1; Kidder, chapter 7). Both Stephen Pyne (chapter 4) and Anna Roosevelt (chapter 9) argue for the long-term use of fire involved in the formation of Amazonian landscapes. Pyne points out that the rain forest of the Amazon is an artifact of the Holocene and includes "a biota that has coexisted with humans and thereby with anthropogenic fire." Roosevelt indicates that the antiquity of fire-using hominids in Amazonian tropical forests, which is only now coming to light in her and her colleagues' research, dates back more than 10,000 years.

In addition to fire, anthropogenic soils, and infectious diseases, other factors pertinent to the historical-ecological analysis of landscapes unfold in this volume. Tristram Kidder (chapter 7), Elinor Melville (chapter 15), and Janet Chernela (chapter 14) assess the introduction of alien animals and plants into landscapes where these had not been present before (also see Whitehead, chapter 2, on the history of this type of analysis). Kidder argues that the introduction of South American nutria into southern Louisiana in the 1940s contributed to a profound alteration of landscapes of the Mississippi Delta region. The tendency of nutria to create larger gaps (crevays) in swamp vegetation by outeating and outcompeting the native muskrat has caused some of the increasing erosion and loss of well-drained land noted in the Mississippi Delta. The nutria (which accidentally escaped from the place of introduction on Avery Island to spread throughout southern Louisiana) were introduced by humans associated with complex society. Elinor Melville (chapter 15) argues that the Spaniards' introduction of sheep-grazing to the Otomí people of north-central Mexico, which replaced traditional land use (based mainly on maize) in most areas, has led to a deterioration of the catchment area. The traditional Otomí dictum "do as the land bids" was undermined by the incorporation of their economy into that of the Spanish Empire and of the neo-colonialism that followed it. Janet Chernela (chapter 14) likewise points out how the introduction of intensive sugarcane cultivation by the Portuguese, combined with coerced native labor, led to deforestation and the impoverishment of soils in the Upper Rio Negro region of the Amazon—in contrast to traditional land-use patterns of the same native peoples. In Chernela's words, "landscape and individual were simultaneously 'converted' and 'domesticated'" by the historical accident of the Portuguese conquest.

Finally, landscapes can be formed in other ways. For Kidder (chapter 7) the genus *Homo,* from its very beginnings, is noteworthy for being "untidy." With or without fire, he argues, people have altered formerly natural environments and interacted with these over time in the formation of landscapes (also see Graham, chapter 6). Specifically, Kidder suggests (as do Marquardt [1994] and others) that native peoples of the Mississippi Delta built well-drained environments by the accumulation of *Rangia* clam shells in what were otherwise poorly drained or completely inundated environments. It is remarkable that Kidder's shell-midden sites contrast with well-drained sites formed by natural levees and meanderings of the Lower Mississippi, in terms of species and species diversity: the anthropogenic shell middens exhibit higher species diversity and higher numbers of useful plant species than do the control sites, which are evidently unaffected (at least directly) by human activity. My chapter shows how landscapes (in this case, fallows rather than shell middens) in extreme eastern Amazonia also differ in species composition and are richer in species diversity than control sites (high forests) where there is little or no evidence of past horticulture. Laura Rival (chapter 11) shows how the Huaorani of Amazonian Ecuador distinguish landscapes, replete with many useful species (especially the peach palm), from essentially unaltered forests. Landscapes, which are by definition anthropogenic (or biocultural), have implications for biodiversity, often depending partly on the historical development of society.

It remains to be seen whether nation-states (such as our own) and emergent transnational polities can maintain biodiversity or improve the biosphere more generally for us and other life forms. Much of the research in this volume suggests that traditional, nonstate entities have been more effective in promoting biodiversity (with the possible exception of diversity among bacteria, fungi, parasites, and viruses) and in increasing the arability and habitability of certain world regions, regardless of whether such effects were intended or not. On the other hand, Sponsel (chapter 17) indicates that the environmental ethics of Buddhism in Thailand and elsewhere (a religion associated with state-level society), which encourage the existence of sacred groves and sacred trees—as does Hinduism in Rajasthan and elsewhere in India (Henderson, chapter 16), and as did the Maya religions of Mesoamerica—may have provided a counterweight to otherwise environmentally degrading practices in archaic state-level societies of the past. In historical ecology, the question is *not* whether states or nonstates are inherently better or worse in regard to the sustainability of landscapes; nor does human nature, however understood, have anything to do with it. In each instance, empirical investigation is required in order for people to understand the regional and biospheric contexts of landscapes. The chapters in *Advances in Historical Ecology* suggest possible avenues for crossing disciplinary boundaries and for applying the findings of historical ecology to the modern world by focusing on the central concept of the *landscape.*

Conclusion

The present volume constitutes an effort to effect a more meaningful debate about relationships between humans and the biosphere in specific landscapes and regions. It carries a general message about the dialectical entity conjoined by humans and their landscapes. It may adduce a new literary and scientific span in the bridge of understanding currently being built between the life and social sciences for the mutual benefit of their diverse practitioners. May it fortify that span.

Note

1. I refer to biodiversity in the plant and animal kingdoms. It seems an open question whether sedentary and urban societies may have been associated with higher diversity of bacterial, fungal, parasitic, and viral pathogens (Newson, chapter 3).

References

Balée, W. 1995. Historical ecology of Amazonia. Pp. 97–110 in L. E. Sponsel (ed.), *Indigenous Peoples and the Future of Amazonia: An Ecological Anthropology of an Endangered World.* Tucson: University of Arizona Press.

———. 1996. Anthropology. Pp. 24–49 in J. Collett and S. Karakashian (eds.), *Greening the College Curriculum: A Guide to Environmental Teaching in the Liberal Arts.* Washington, D.C.: Island Press.

Campbell, D. G. 1991. Gap formation in tropical forest canopy by elephants, Oveng, Gabon, Central Africa. *Biotropica* 23(2): 195–196.

Crumley, C. L. 1996. Historical ecology. Pp. 558–560 in D. Levinson and M. Ember (eds.), *Encyclopedia of Cultural Anthropology,* vol. 2. New York: Henry Holt.

————, ed. 1994. *Historical Ecology: Cultural Knowledge and Changing Landscapes.* Santa Fe: School of American Research Press.

Gunn, J. D. 1994. Global climate and regional biocultural diversity. Pp. 67–97 in Crumley (ed.), *Historical Ecology.*

Harris, M. 1968. *The Rise of Anthropological Theory.* New York: Crowell.

————. 1979. *Cultural Materialism: The Struggle for a Science of Culture.* New York: Random House.

Ingerson, Alice E. 1994. Tracking and testing the nature/culture dichotomy in practice. Pp. 43–66 in Crumley (ed.), *Historical Ecology.*

Ingold, T. 1988. The animal in the study of humanity. Pp. 84–99 in T. Ingold (ed.), *What Is an Animal?* London: Unwin Hyman.

Kuhn, T. S. 1977. *The Essential Tension: Selected Studies in Scientific Tradition and Change.* Chicago: University of Chicago Press.

Marquardt, W. H. 1994. The role of archaeology in raising environmental consciousness: An example from southwest Florida. Pp. 203–221 in Crumley (ed.), *Historical Ecology.*

Marx, K. 1977. [1867]. *Capital.* Vol. 1. Trans. S. Moore and E. Aveling. Ed. F. Engels. New York: International Publishers.

Smith, E. A. 1991. *Inujjuamiut Foraging Strategies: Evolutionary Ecology of an Arctic Hunting Economy.* New York: Aldine de Gruyter.

Winterhalder, B. P. 1994. Concepts in historical ecology: The view from evolutionary ecology. Pp. 17–41 in Crumley (ed.), *Historical Ecology.*

Wolf, E. R. 1982. *Europe and the People Without History.* Berkeley and Los Angeles: University of California Press.

PART ONE

Human and Material Factors
in Historical Ecology

CHAPTER 1

Historical Ecology: Premises and Postulates

WILLIAM BALÉE

> Ecological factors never operate in a cultural vacuum nor do the endur-
> ing patterns of language, kinship, and cultural values that every individ-
> ual inherits prevent adaptation to a material environment.
> —R. McC. Netting (1986:101)

Preliminary Definitions and Concepts

Historical ecology concerns itself with interrelationships between human beings
and the biosphere, that part of the earth suffused with life. Historical ecology
clearly requires data drawn from a multitude of disciplines (Crumley 1996), even
though it is centered on humans. It is unlike environmental history (*Journal of
American History* 1990), ecological anthropology (Balée 1996), cultural geography, and other fields that exhibit contrasting positions and orientations. Historical
ecology takes a distinctive perspective on human societies and their interactions
with other life forms and the land, just as do cultural ecology,[1] cultural materialism,
structuralism, and other theoretical orientations. Historical ecology reflects a materialist viewpoint, but cannot be equated with cultural materialism, which is an explanatory device in both environmental history (Worster 1990) and ecological anthropology (Winterhalder 1994:21–22). Because historical ecology emerges from a
dialectical point of view (Crumley 1996), it bears a stronger resemblance to dialectical materialism than to cultural materialism (Crumley 1994b; Ingerson 1994; Patterson 1994). *Dialectical ecology* (Levins and Lewontin 1985) may be an apposite
synonym, therefore, for historical ecology.

Unlike environmental determinism, cultural ecology, cultural materialism, and
cultural evolutionism, historical ecology begins with the premise that historical, not
evolutionary, events are responsible for the principal changes in relationships between human societies and their immediate environments. Historical ecology reveals a dialectical process in the unfolding of these changes. Like the concept of
"punctuated equilibrium," it assumes that historical events may affect biocultural
developments. Where environmental disturbance occurs sporadically or even

chaotically over time such that equilibrium never seems to be reached, biological and cultural impoverishment is unusual.

Historical ecology focuses on the interpenetration of culture and the environment, rather than on the adaptation *of* human beings *to* the environment. In other words, a relationship between nature and culture is conceived, in principle, as a dialogue, not a dichotomy (Ingerson 1994:65). Beyond that premise, I propose that historical ecology explains human/biosphere interrelationships by a core of interdependent postulates. These postulates may help explicate historical ecology as a viewpoint, rather than as a field or method per se.

The postulates are: (1) Much, if not all, of the nonhuman biosphere has been affected by human activity. (2) Human activity does not necessarily lead to degradation of the nonhuman biosphere and the extinction of species, nor does it necessarily create a more habitable biosphere for humans and other life forms by increasing the abundance and speciosity of these. (3) Different kinds of sociopolitical and economic systems (or political economies) in particular regional contexts tend to result in qualitatively unlike effects on the biosphere, on the abundance and speciosity of nonhuman life forms, and on the historical trajectory of subsequent human sociopolitical and economic systems (or political economies) in the same regions. (4) Human communities and cultures together with the landscapes and regions with which they interact over time can be understood as total phenomena. The remainder of this chapter assesses these postulates in the light of known data.

Postulate 1: Much, if not all, of the nonhuman biosphere has been affected by human activity.

In the most important work on the subject of historical ecology to date, Carole Crumley (1994a:240) stated that "no spot on the earth is unaffected by humans." That comment is at least intuitively obvious today—taking into consideration chlorofluorocarbons (CFCs) in the atmosphere, ozone depletion, acid rain, tropical deforestation, global warming, and the like. Yet other researchers would claim that it applies to the prehistoric world as well, in regions as widely separated as Australia (Allan and Baker 1990; Hynes and Chase 1983; Gould 1971; Walsh 1990), Africa (Bailey and Headland 1991; Vansina 1990; Wilmsen 1989), North America (Cronon 1983; Denevan 1992; Kidder, chapter 7, this volume; Lewis 1982; Marquardt 1992; Nicholas 1988; Patterson and Sassaman 1988; Pyne 1982; Stewart 1956), and especially South America (Denevan 1966, 1992; Steven 1993), where human influence is mostly noted through anthropogenic fire and agriculture, but which is most often considered to harbor among the most pristine terrestrial conditions on Earth (short of Antarctica).

Whereas historical ecology assumes that wherever humans are or have been present an interrelationship exists between them and their biotic and abiotic regional environs, this postulate sheds no light on the uniqueness of humans as a species. If humans have affected the entire biosphere, including essentially uninhabited parts such as the Southern Ocean and the Antarctic Continent (Campbell 1992), one can still argue that such pervasive influence, generally of a deleterious sort for biodiversity, has arisen with the appearance of modern nation-state societies. CFCs, global warming, declines in sperm counts, and so forth seem to be mainly twentieth-century phenomena.

If humans routinely affect the biosphere, this does not necessarily distinguish humanity as a life form. According to the Gaia hypothesis, part of the reason why the surface of Earth is habitable yet the surfaces of its sister planets Venus and Mars (which were also formed nearby from the same giant cloud of gas) are not, is that photosynthetic organisms about 3.5 billion years ago witlessly began imbuing Earth with a large supply of atmospheric oxygen and a profound reduction in atmospheric carbon dioxide; for every molecule of oxygen released by photosynthesis, a corresponding molecule of carbon became interred as fossil carbon in sediment (Lovelock 1979, 1992; Barghoorn 1992; McElroy 1992). The atmosphere is partly an artifact of the unfolding of life. In this context, one may consent to the view that Earth and its Latin equivalent, Terra, are misnomers: our planet should have been called Vita (Campbell 1992; Campbell and Durkee 1996)—for it is life itself, rather than any single life form or species (even the human one), that distinguishes it from the other planetary bodies of our solar system, at present. Life as a total phenomenon may even have affected plate tectonics and other supposedly inorganic processes (Margulis and Olendzenski 1992), just as elephants have changed structure and perhaps species composition in African tropical forests by creating major light gaps (Campbell 1991).

Elephants, like humans, are gap-producing species in tropical forests. Unlike humans, though, elephants lack extrasomatic tools such as broadcast fires (Pyne, chapter 4). Humans are unique perhaps only in terms of the scale by which they have modified the planet, as well as in the degree of intentional planning (what is "available to the discursive awareness of the actor" [Giddens 1987:63]), based in the higher functions of the cerebral cortex, that preceded their actions. As Tim Ingold (1988:97) has pointed out, "though humans differ but little from other animal species . . . that difference has mighty consequences for the world we inhabit, since it is a world that, to an ever greater extent, we have made for ourselves, and that confronts us as the artificial product of human activity."

If the human species is somehow unique in its relationship to the biosphere, this may not necessarily be because it has always and everywhere influenced other life forms on Earth. Rather, humans may demonstrate historically a *greater potential* than any other species to affect biodiversity and the biosphere generally. In addition, they have evinced high *adaptability* to a wider range of habitats, as well as *technologies* that are distinctive properties with historical implications. Historical ecological research indicates that the environment (with the possible exception of certain maritime, mountainous, and circumpolar zones of the earth) and society (without exception) are essentially historical constructs, not immutable givens (Gunn 1994). Research in historical ecology has sought neither to deconstruct the role of nature nor to deny the role of evolutionary mechanisms, such as natural selection, in human social life. Rather, it is based on the premise that putatively natural environments that have been subjected to management have progressively become landscapes (Crumley 1993; Crumley and Marquardt 1990)—that is, cultur-ally and historically determined physical environments. Other life forms that lack domesticates (and here one may distinguish between anthropogenic fire and wild-fire, the former a kind of "domesticate"; see Pyne, chapter 4) and agriculture have not created landscapes, however much they may have affected the surface of the earth in other ways.

This human potential to affect biodiversity and the biosphere in qualitatively dif-ferent ways from other life forms and also from the sum total of other life forms does not, however, support the assertion that humanity is biologically programmed, or in some other way overdetermined, to reduce biodiversity and make Earth less habitable generally for other life forms. Nor does it support the opposite view, that human beings are biologically programmed, or in some other way overdetermined, to live in harmony with the other life forms of the biosphere, even behaving so as to increase their abundance and diversity.

Some of the recent criteria employed in characterizing human nature have been implicitly related to biodiversity and environmental conservation. With regard to biodiversity, it may be argued that two doctrines concerning human nature seem to be in constant competition within sociocultural anthropology (and, for that matter, within the bioecological sciences): the Ecologically Noble Savage (Redford 1991; Alvard 1993, 1994), on the one hand, and what I call *Homo devastans* (Balée 1996), its opposite, on the other. The Ecologically Noble Savage implies that in-digenous (especially foraging) peoples tend not to diminish biodiversity, may in fact deliberately act to increase it (e.g., Anderson and Posey 1989; Hynes and Chase 1983; Orr 1992:32; Posey 1985), and exhibit a wisdom and knowledge of local na-ture greater than the potential of Western science to know that nature (e.g., Hughes 1983; C. L. Martin 1992; Nelson 1993; Orr 1992; Reichel-Dolmatoff 1976). Those who subscribe (if only implicitly) to *Homo devastans* argue that indigenous peoples, presumably like human beings everywhere, contribute to lowered biodiversity (Al-vard 1994, 1995; Diamond 1986, 1992), are naturally destructive or polluting of local environments (Rambo 1985; Redford 1991), and do not manage other life forms or increase environmental diversity (e.g., Parker 1992, 1993).

The Ecologically Noble Savage doctrine holds that it is human nature to be cus-todial of the environment, the relationship becoming corrupted only after the rise or intrusion of civilization. The doctrine of *Homo devastans,* in contrast, holds hu-mankind itself accountable for the destruction of natural habitats and of other species. These opposed dogmas have been applied to non-state-level societies, from which some researchers have sought to promote or deconstruct specific view-points on human nature (see critiques by Sponsel 1992; Nabhan and St. Antoine 1993). Both views require the demonstration of sociocultural universals; either would become a mere shibboleth with proof of a single counterexample. It is clear that both views have converts in the scientific community today. Yet research in his-torical ecology seems to support neither view, just as sociocultural universals based on the juxtaposition of biology, language, and culture have been continuously proven to be erroneous since the time of Franz Boas (Sussman 1995).

Many modern environmentalists seem most likely to assume the existence of *Homo devastans.* When referring to some "panhistorical, cross-cultural, and ultimately destructive human 'nature,'" according to Alice Ingerson (1994:52), her environmentalist students really meant the world capitalist system. Many environmentalists, evidently, do not consider the peoples of nonstate, egalitarian societies to make up part of that abstraction formerly called "Man."

The American conservationist George Perkins Marsh showed the internal contradictions of this view in his famous work *The Earth as Modified by Human Action,* originally published in 1864. That work helped inspire many twentieth-century students of human/environmental relationships (Thomas 1956; Turner et al. 1990). Marsh claimed that "the action of man upon the organic world tends to derange its original balances" (1885:vii; also see Graham 1956:688). In a section entitled "Destructiveness of Man," Marsh (1885:33, quoted in Graham 1956:688) declared that "Man is everywhere a disturbing agent; wherever he plants his foot, the harmonies of nature are turned to discords." Further, "Man pursues his [nonhuman living] victims with reckless destructiveness" (Marsh 1885:34). That by "Man" Marsh generally meant the industrializing society of his time seems evident: "Purely untutored humanity, it is true, interferes comparatively little with the arrangements of nature, and the destructive agency of man becomes more and more energetic and unsparing as he advances in civilization" (1885:38–39). Marsh (1885:121) did not exonerate nonstate peoples categorically, however, pointing out that native peoples had extirpated the large flightless land birds, moas, from New Zealand. He seems to have shared in the nineteenth-century evolutionist thought of many of his contemporaries—namely, that the progression to civilization underwrote human history (see Whitehead, chapter 2).

Marsh's view was counterbalanced by that of another nineteenth-century conservationist, Henry David Thoreau, who contrasted his own, environmentally destructive society with that of the American Indian. The Indians possessed true "wisdom" about nature (Worster 1977:96). Since non-Indians could emulate this wisdom, Thoreau promoted a precursor to Kent Redford's (1991) "Ecologically Noble Savage" (see Kidder, chapter 7). The superior state of humankind, in Thoreau's view, was evident in other societies. In this context he wrote (1985:712):

> The kings of England formerly had their forests "to hold the king's game," for sport or food, sometimes destroying villages to create or extend them; and I think that they were impelled by a true instinct. Why should not we, who have renounced the king's authority, have our national preserves, where no villages need be destroyed, in which the bear and panther, and some even of the hunter race [i.e., American Indians], may still exist, and not be "civilized off the face of the earth,"—our forests, not to hold the king's game merely, but to hold and preserve the king himself also, the lord of creation,—not for idle sport or food, but for inspiration and our own true re-creation?

Human nature sought peace with nonhuman nature. The biophilia hypothesis seems to be a more recent incarnation of this view (Kellert 1993).

The environmentally incorrect rival of the Ecologically Noble Savage, *Homo dev-astans,* gained strength in the twentieth century by its expansion of reference. In addition to modern civilization, it came to include hunters and gatherers of the late Pleistocene, who were blamed for the extinction of hundreds of species and genera of mammals worldwide. Certain nineteenth-century scientists, such as the comparative anatomist Jean Baptiste Lamarck and the zoologist John Fleming, earlier believed that the human species had been involved in Pleistocene extinctions, partly because humans were thought to have coexisted with Pleistocene animals and were simultaneously considered to have been capable of extirpating other species—even though empirical evidence for this was scant (Grayson 1984b:23–24). Once archaeological research in France and England during the 1850s proved that human beings and extinct animals of the Pleistocene had been contemporaneous, the famous geologist Charles Lyell declared that the antiquity of man "throws great light on extermination of animals, and in Denmark, of trees" (quoted in Grayson 1984b:24).

This view gained support in the twentieth century with the Pleistocene overkill hypothesis of paleontologist Paul S. Martin, who found that very few late Pleistocene extinctions of megafauna in the Americas, Australia, and Oceania occurred before the arrival of human beings (Martin 1966, 1967, 1973; see Alvard 1994, 1995, for possible recent applications of this view to Amazonia, and the critique in Alvard 1995 by Janis Alcorn). In other words, Martin's argument is based on the *timing* of extinctions—specifically, their coincidence with humankind in given regions. In the Americas, the major extinctions of megafauna occurred from about 11,000 B.P. on, or shortly after the presumed time of the arrival of human beings across Beringia. Because human remains were rarely found with extinct animals of the Pleistocene, there being very few kill sites, Martin (1973, 1984) suggested that there was a "blitzkrieg" of the fauna by Paleo-Indians, whose Clovis projectile points and fully modern human anatomy would have had a deadly efficiency in extirpating species and genera of prey. The evidence from Australia, however, is more equivocal; there, major extinctions of megafauna seem to postdate human occupation by many thousands of years (Grayson 1984a). Martin's model of Pleistocene overkill cannot be falsified because it is a simulation, not an empirically demonstrable hypothesis (Grayson 1984a). The same can be said of the competing dogmas of the Ecologically Noble Savage and *Homo devastans.*

Nevertheless, the concept of Pleistocene overkill gained a number of adherents ready to believe that it was human nature to make other life forms vanish. In light of the evidence for Pleistocene overkill, the Pulitzer Prize winner René Dubos (1974:44) declared that "like the tendency to kill, the tendency to waste and to foul the nest seems to be inscribed in the genetic code of the human species." On the other hand, whereas human hunters certainly killed individuals of the Pleistocene megafauna, it remains to be proved (and the issue is still quite controversial) whether they alone *caused* the extinction of any or all of the taxonomic groups to which these individuals belonged. In this light, *Homo devastans* seems to be as dogmatic and unempirical as its earlier sociobiological counterparts, the creature of microeconomics *Homo economicus* and the Hobbesian, territorial *Homo bellicosus* (see Sahlins 1976:53; also Nabhan and St. Antoine 1993).

This counterargument means that it may not be human nature to be destructive of biodiversity and the environment, however defined. But it may also not be human nature to encourage the growth of biodiversity and increase the habitability of Earth for other life forms—which leads to the next postulate of historical ecology, as I see it.

Postulate 2: Human activity does not necessarily lead to degradation of the non-human biosphere and the extinction of species, nor does it necessarily create a more habitable biosphere for humans and other life forms and increase the abundance and speciosity of these.

Some evidence suggests that the creation of certain landscapes by human beings did not result in irreversible damage to regional biodiversity—thus undermining the doctrine of *Homo devastans.* The evidence is principally associated with the origins of domesticated and semidomesticated plants and animals. In some regions of the Neolithic world, the domestication of plant and animal species may have entailed a net increase in the total number of species present, assuming that sometimes the wild progenitors did not become extinct. The New World contributed more than one hundred species of plants to the world's inventory of domesticated plants (Brücher 1989); in the absence of evidence for local extinctions of ancestral and related species of these plants, this contribution represents an increase in plant biodiversity. Early agrarian societies, and modern ones that have either retained or been forced into an essentially egalitarian political system and reciprocal economy, may have been frequently associated with net regional increases in bioecological diversity.

I would reiterate in this context a definition of resource management: *"the human manipulation of inorganic and organic components of the environment that brings about a net environmental diversity greater than that of so-called pristine conditions, with no human presence"* (Balée 1994:116; emphasis in the original). For Australia and North America, it has been argued (Pyne 1982; Patterson and Sassaman 1988) that the use of broadcast fires by indigenous peoples led to an increase in the abundance of game animals by encouraging new growth of grasses and legumes (also see Cronon 1983; Lewis 1982; Walsh 1990). These fires may have also decreased the risk of large *wildfires,* which tend to be more destructive than constructive with respect to new habitats. In belated recognition of the environmentally enriching effects of certain indigenous activities, the fire-management strategies of Australian Aborigines (as interpreted by government planners) are now being employed by National Park personnel of Australia to control the incidence of wildfires (Allan and Baker 1990). These strategies essentially reduce fuels in the fire environment, and therefore reduce the likelihood of large wildfires (Pyne 1982; and see chapter 4, this volume). In nineteenth-century North America, "light burning" (controlled broadcast fires) was derogatorily referred to as "Paiute forestry." Yet this practice came to be adopted by the U.S. Forest Service during the early years of the twentieth century as a land-management and fire-protection strategy in areas of the West that are prone to wildfires. "Paiute forestry" has continued to form part of the program of conflagration control by the National Park Service, in spite of problems with escape fires (Pyne 1982:100–104).

Other alterations of the landscape by Native Americans that did not necessarily lead to species extinctions, and that may even have enhanced habitats for nonhuman life forms, include the islands of shell mounds created by prehistoric hunter-gatherers on the southwest coast of Florida (Marquardt 1992) and coastal Louisiana (Kidder, chapter 7); and the parkland environment of New England, which encouraged plants adapted to fire regimes and much sunlight (including strawberries and other edible fruits) as well as game animals, such as white-tailed deer (Cronon 1983; Patterson and Sassaman 1988; Denevan 1992).

In Africa, it is becoming increasingly clear that many equatorial forests once thought to be pristine are in fact anthropogenic forests (Bailey 1996; Bailey and Headland 1991; Vansina 1990). Robert Bailey (1996:325) makes the significant observation that "biodiversity exists in central Africa today, not despite human habitation but because of it." This observation also applies to a regional analysis of the forests inhabited by the Ka'apor Indians of eastern Amazonia.

Amazonia evinces human-induced landscape changes since prehistory (Balée 1989; Denevan 1992; Posey 1985; Posey and Balée 1989; Moran 1993). These landscapes occur in the Llanos de Mojos of Bolivia, with its prehistoric raised fields and mounds (Denevan 1966, 1992); the *apêtê* (forest islands) of the cerrado country in north-central Brazil (Posey 1985; Anderson and Posey 1985, 1989); the mounds of Marajó Island (Roosevelt 1991); and widely distributed forest types such as liana forests, Brazil nut forests, babaçu and other palm forests, and forests dominated by the dicotyledonous trees bacuri (*Platonia insignis,* in the clusia family), *Jacaratia spinosa* (in the papaya family), hog plum (*Spondias mombin,* in the cashew family), copal (*Hymenaea parvifolia,* in the caesalpinia family), *Gustavia augusta* (in the Brazil nut family), *Trichilia quadrijuga* (in the mahogany family), *Neea* (in the four-o'clock family), *Simaba cedron* (in the simaruba family), and *Theobroma speciosum* (in the cacao family) (Balée 1989, 1993, 1994).

Eugene Parker's (1992, 1993) theory that the *apêtê* (forest islands on the high savanna) of the Kayapó are completely natural, nonanthropogenic phenomena seems unconvincing. As I have noted elsewhere (Balée 1993, 1994), many species found on *apêtê* of the Kayapó (as reported in Anderson and Posey 1985, 1989; and elsewhere) are also found in fallows of the Ka'apor about three hundred miles to the north and east. The fallows of the Ka'apor, called *taper,* are anthropogenic forest formations brought about by indigenous forest-management practices (Balée 1994). Some of the plant species that fallows share with *apêtê* include *Tapirira guianensis* (cashew family), *Himatanthus sucuuba* (dogbane family), *Schefflera* sp. 1 (ginseng family), *Tabebuia serratifolia* (bignonia family), *Tetragastris altissima* (bursera family), *Maytenus* sp. 1 (staff-tree family), *Casearia* spp. (flacourtia family), *Sacoglottis* spp. (humiria family), *Mascagnia* spp. (malpighia family), *Cecropia palmata* (cecropia family), *Neea* spp. (four-o'clock family), *Coccoloba paniculata* (buckwheat family), *Simaruba amara* (quassia family), and *Vitex flavens* (verbena family). Several of these species occur *only* in areas that have been disturbed by indigenous forest-management practices in the region of the Ka'apor—indirectly suggesting that *apêtê* may be secondary forests in the region of the Gorotire Kayapó.[2] In other words, the Kayapó term *apêtê* would probably best be glossed in Ka'apor as *taper* ("old fallow") or *taper-ran* ("old fallow-similar"), and vice versa.

The question remains, however, whether indigenous forestry conforms to the definition of resource management given above. William Denevan (1966, 1992) implies that the construction in Bolivia of mounds, raised fields, and other transformations of the lowlands ultimately lowered biodiversity, but no direct evidence supports this claim. Many if not all of the forested landscapes of the Llanos de Mojos may be anthropogenic (Denevan 1966, 1992; Erickson 1995; Erickson et al. 1991; Roosevelt 1992; Stearman 1989)—but that does not mean, a priori, a net loss in biodiversity. Clark Erickson (1995) refers to "landscape accumulation" in the Llanos de Mojos by the building of mounds and raised fields, which in many cases today are covered with forest vegetation that otherwise would be absent in the flooded savanna (Balée 1995; for a North American analogue, see Kidder, chapter 7). Some 1,300 miles to the northeast, Ka'apor native forestry practices may have actually enhanced biodiversity in the region by the creation (via disturbance) of the distinctive landscapes that they refer to as *taper*.

In defining the concept of landscape, Carole Crumley and William Marquardt (1990:73) pointed out that, "in interacting with their physical environment, people project culture onto nature"—often unconsciously. Such a dialectical view of interrelationships between humans and the biosphere (or, for earlier writers, Nature) has a clear forerunner in Karl Marx (1867:177; also see Sahlins 1976:126–129). Historical ecology draws on dialectical materialism, even if it goes far beyond that earlier viewpoint by actually investigating human/biosphere interrelationships empirically. It also shows just how dialectical those interrelationships really are.

The landscapes that I call fallows represent a projection of culture onto nature through time. These are living landscapes, even if they have traditionally (and erroneously) been understood to be primary forests by foresters, ecologists, and phytogeographers alike (Balée 1989). Fallows exhibit many species, including some of those mentioned above in the comparison with *apêtê*, that occur nowhere else in the *terra firme*. They are as biologically rich as the high forests in the same region (Balée 1993, 1994; see figure 1.1), but they harbor many species unique to them, and many more that only gain ecological importance in areas disturbed by indigenous agroforestry. These species may be collectively considered as semidomesticates (also see Posey and Balée 1989).[3] Insofar as fallows and their constituent species would not exist without indigenous forestry, it may be concluded that indigenous forestry has actually enhanced the environmental and biological diversity of the region of the Ka'apor, given that there is no evidence for extinctions of plants and animals within that region. Similar findings with respect to the extremely rich habitat of the Huaorani of Amazonian Ecuador have been expressed by Laura Rival (chapter 11). In other words, traditional Ka'apor agroforestry (and no doubt that of other indigenous peoples in Amazonia) undermines the doctrine of *Homo devastans,* even if it does not provide proof of the Ecologically Noble Savage either. The point is that no evidence exists to show that human beings are biologically programmed, or in some other way overdetermined, to be either stewards or destroyers of the diversity of nonhuman life forms.

Postulate 3: Different kinds of sociopolitical and economic systems (or political economies) in particular regional contexts tend to result in qualitatively unlike effects on the biosphere, on the abundance and speciosity of nonhuman life forms, and on the historical trajectory of subsequent human sociopolitical and economic systems (or political economies) in the same regions.

If some Amazonian peoples have by their activities enhanced environmental diversity and increased regional biodiversity, clearly not all nonindustrial political economies have had similar results. The evidence for altered landscapes in Amazonia has led some researchers to believe that this represents an automatic decline in biodiversity and habitability for other life forms (Alvard 1993, 1994; Redford 1991)—but no long-term empirical evidence supports that.

While these statements derive in part from the doctrine of *Homo devastans,* convincing evidence exists that Polynesia, for its part, did suffer severely lowered biodiversity, partly as a result of human occupation. But this increased poverty of the flora and fauna did not result solely from human *nature,* for it can be demonstrated that humans have not everywhere been associated with diminished diversity of other life forms (as in the Ka'apor example given above). Rather, Polynesia suffered lowered biodiversity partly because of the peculiarities of island environments (such as high biological endemism), and partly because of human occupation. Individual islands, unlike regions and continents, tend to be high in endemic species over small expanses of land (Balée 1995). In Polynesia, many of these species evolved over millions of years before human arrival within the last 2,500 years; therefore, many species were unusually susceptible to extirpation by perturbations of the environment caused by humans and perhaps other animals (especially introduced animals) (Meilleur 1996). In addition, prehistoric and modern indigenous peoples of Melanesia may have caused extinctions of numerous bird species (Diamond 1984, 1992), and the indigenous Semang people of Malaysia have been seen as "primitive polluters" (Rambo 1985), no different in kind from civilized societies in their supposed propensity (if I may borrow from Dubos) to "foul the nest."

Yet evidence for increased agricultural biodiversity as a result of indigenous agroforestry complexes in South America is not limited to Amazonia. In the Andes, prehistoric peoples developed many new landraces of potatoes, *oxala* tubers, maize, coca, and other domesticated species; however, after the rise of the state, which put an emphasis on surplus food and monoculture for taxation, a decrease in infraspecific crop biodiversity and a reduction of soil fertility seem to have occurred (Zimmerer 1993, cited in Futemma 1994). The rise of the classic Maya in the Copan val-

SPECIES/AREA CURVES FOR FALLOW AND HIGH FOREST
in the Gurupi, Turiaçu, and Pindaré basins

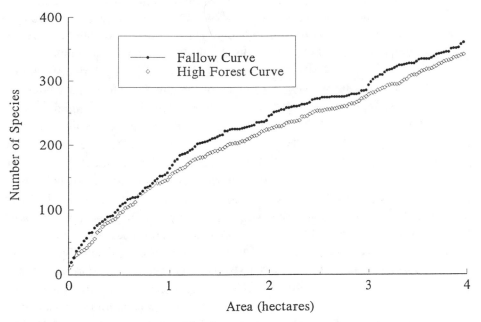

Fig. 1.1 *Species/Area Curves from pre-Amazonia.* Four hectares of fallow forest (*taper*) and four hectares of high (primary) forest (*ka'a-te*) were inventoried in the region. Mapping the increase in species diversity per unit area of each forest type shows that fallows and primary forests accumulate diversity at a similar rate, and that the total numbers of species found in the two forest types do not differ significantly.

ley has also been associated with deforestation and increased pauperization of the environment (Abrams et al. 1996).

With the exception of certain island societies, therefore, the only solid evidence for a human association in certain regions with reduced biodiversity and decreased habitability for other life forms comes from state societies, old and new. It would be counterproductive to abandon the social and politicoeconomic criteria that have distinguished hunter-gatherers, village horticulturalists, chiefdoms, and states, because differences, even if continuous, exist in terms of their mutually distinctive effects on the biota of landscapes and regions. This hypothesis is, in principle, measurable, and hence falsifiable. Although these politicoeconomic types are not linked in an evolutionary sequence (given the obvious fact that states and nonstates still coexist—although perhaps for not much longer), nor is one type morally superior to the other, the differences are significant. These differences, in terms of their demonstrable effects on the biosphere, show that human nature is ultimately not the culprit in today's massive depletion of nonhuman life forms; rather, the political economies of states and multistate organizations (Capistrano and Kiker 1990, cited in Schmidt 1994:99) only are to blame. L. S. B. Leakey was right when he wrote that "we are too apt to blame man and man alone for some of the things that happened"

(1964:26). Whether states are intrinsically destructive of bioenvironmental diversity, however, is an ethnographic and historical, not a biological, question.

Postulate 4: Human communities and cultures together with the landscapes and regions with which they interact over time can be understood as total phenomena.

In an attempt to view society (or culture) and nature as a single phenomenon, Thomas Patterson (1994:230) defines totality as "a dialectically structured and historically determined unity that exists in and through the diverse interpenetrations, connections, and contradictions that join its constituent parts regardless of whether the components are observable or unobservable." In the context of the proposed postulates discussed above, this definition also encompasses, in principle, the landscapes and regions of given political-economic entities over time (Crumley 1994b:9). It is in the visible manifestation of landscapes, such as fallows or cultivated forests (Rival, chapter 11), that culture and nonhuman life forms can be comprehended as one—that is, as a totality. Just as potsherds and other nonliving artifacts may be considered cultural, so too may certain living infrahuman organisms, such as domesticated and semidomesticated plants and animals. From this perspective, culture can be said to reside in, or be expressed by, certain trees as well as people's minds (cf. Roberts 1964:439).

Culture and the environment—together with their many permutations, such as culture and nature, society and nature, people and the biosphere, and so forth—represent a single phenomenon amenable to regional analysis using the paradigmatic concepts and tools of historical ecology. Although the human species remains central to historical ecology, this approach is perhaps less anthropocentric than some others, for whereas humans have conditioned the biosphere through their activities in regions and on landscapes, these same activities have constrained other potential developments. If it is not human nature to be either the nemesis or the steward of nonhuman life forms, the human species as a whole cannot be considered as wholly independent of those other life forms in given regional contexts. Rather, specific types of social and political-economic systems have historically interacted in finite and comprehensible ways with portions of the biosphere. These interactions constitute different totalities in the human experience of the biosphere, and vice versa.

Acknowledgments

I thank all the other contributors to this volume, especially Leslie Sponsel and Stanford Zent, for insightful comments and criticisms. I take full responsibility for any errors herein and for the point of view expressed.

Notes

1. Several writers (e.g., Moran 1990; Murphy 1970; Netting 1986) seem to hold *cultural ecology,* as a term, to mean the field of inquiry into the mechanical links between society and

the environment. I prefer the terms *human ecology* and *ecological anthropology* to refer to that field. Cultural ecology, as originally conceived by Julian Steward, was not a field, but a theoretical construct (see discussion in Balée 1996; Butzer 1990).

2. Several of these species, such as *Tabebuia serratifolia* and *Tapirira guianensis,* occur widely in Amazonia, may be physiologically plastic or ecologically insensitive (Bush 1994; Gentry 1988), but do not seem to be species of areas quite recently disturbed.

3. Charles Clement (1995) divides my use here of semidomesticates into three categories based on degree of human selection.

References

Abrams, F. M., A. Freter, D. J. Rue, and J. D. Wingard. 1996. The role of deforestation in the collapse of the Late Classic Copan Mayan state. Pp. 55–75 in Sponsel, Headland, and Bailey (eds.), *Tropical Deforestation.*

Allan, G. and L. Baker. 1990. Uluru (Ayers Rock–Mt. Olga) National Park: An assessment of a fire management programme. *Proceedings of the Ecological Society of Australia* 16:215–220.

Alvard, M. S. 1993. Testing the "ecologically noble savage" hypothesis. *Human Ecology* 21(4): 355–387.

———. 1994. Conservation by native peoples: Prey choice in a depleted habitat. *Human Nature* 5:127–154.

———. 1995. Infraspecific prey choice by Amazonian hunters. *Current Anthropology* 36(5): 789–818.

Anderson, A. B. and D. A. Posey. 1985. Manejo de cerrados pelos índios Kayapó. *Boletim do Museu Paraense Emílio Goeldi* (ser. Botânica) 2(1): 77–98.

———. 1989. Management of a tropical scrub savanna by the Gorotire Kayapó of Brazil. Pp. 159–173 in Posey and Balée (eds.), *Resource Management in Amazonia.*

Bailey, R. C. 1996. Promoting biodiversity and empowering local people in central African forests. Pp. 316–341 in Sponsel, Headland, and Bailey (eds.), *Tropical Deforestation.*

Bailey, R. C. and T. N. Headland. 1991. The tropical rain forest: Is it a productive environment for human foragers? *Human Ecology* 19(2): 261–285.

Balée, W. 1989. The culture of Amazonian forests. Pp. 1–21 in Posey and Balée (eds.), *Resource Management in Amazonia.*

———. 1993. Indigenous transformation of Amazonian forests: An example from Maranhão, Brazil. *L'Homme* 33(2–4): 231–254.

———. 1994. *Footprints of the Forest: Ka'apor Ethnobotany—The Historical Ecology of Plant Utilization by an Amazonian People.* New York: Columbia University Press.

———. 1995. Indigenous agroforestry and tropical biodiversity: Reconsiderations of refuge theory. Paper presented at the 94th Annual Meeting of the American Anthropological Association, 17 November, Washington, D.C.

———. 1996. Anthropology. Pp. 24–49 in Collett and Karakashian (eds.), *Greening the College Curriculum.*

Barghoorn, E. S. 1992. The antiquity of life. Pp. 71–84 in Margulis and Olendzenski (eds.), *Environmental Evolution.*

Brücher, H. 1989. *Useful Plants of Neotropical Origin and Their Wild Relatives.* Berlin: Springer-Verlag.

Bush, M. B. 1994. Amazonian speciation: A necessarily complex model. *Journal of Biogeography* 21(1): 5–17.

Butzer, K. W. 1990. The realm of cultural-human ecology: Adaptation and change in historical perspective. Pp. 685–701 in Turner et al. (eds.), *The Earth as Transformed by Human Action.*

Campbell, D. G. 1991. Gap formation in tropical forest canopy by elephants, Oveng, Gabon, Central Africa. *Biotropica* 23(2): 195–196.

———. 1992. *The Crystal Desert: Summers in Antarctica.* Boston: Houghton Mifflin.

Campbell, D. G. and V. Durkee. 1996. Biology. Pp. 50–71 in Collett and Karakashian (eds.), *Greening the College Curriculum.*

Capistrano, A. D. and C. F. Kiker. 1990. Global economic influences on tropical broadleaved forest depletion, 1967–1985. Paper presented at the International Society for Ecological Economics, 12–13 May, Washington, D.C.

Clement, C. R. 1995. 1492 and the loss of Amazonian crop genetic resources. Unpublished MS.

Collett, J. and S. Karakashian, eds. 1996. *Greening the College Curriculum: A Guide to Environmental Teaching in the Liberal Arts.* Washington, D.C.: Island Press.

Cronon, W. 1983. *Changes in the Land: Indians, Colonists, and the Ecology of New England.* New York: Hill and Wang.

Crumley, C. L. 1993. Analyzing historic ecotonal shifts. *Ecological Applications* 3(3): 377–384.

———. 1994a. Epilogue. Pp. 239–240 in C. L. Crumley (ed.), *Historical Ecology: Cultural Knowledge and Changing Landscapes.* Santa Fe: School of American Research Press.

———. 1994b. Historical ecology: A multidimensional ecological orientation. Pp. 1–16 in Crumley (ed.), *Historical Ecology.*

———. 1996. Historical ecology. Pp. 558–560 in D. Levinson and M. Ember (eds.), *Encyclopedia of Cultural Anthropology,* vol. 2. New York: Henry Holt.

Crumley, C. L. and W. H. Marquardt. 1990. Landscape: A unifying concept in regional analysis. Pp. 73–79 in K. M. Allen, S. W. Green, and E. B. W. Zubrow (eds.), *Interpreting Space: GIS and Archaeology.* London: Taylor and Francis.

Denevan, W. M. 1966. *The Aboriginal Cultural Geography of the Llanos de Mojos of Bolivia.* Ibero-Americana 48. Berkeley: University of California Press.

———. 1992. The pristine myth: The landscape of the Americas in 1492. *Annals of the Association of American Geographers* 82(3): 369–385.

Diamond, J. M. 1984. Historic extinctions. Pp. 824–862 in Martin and Klein (eds.), *Quaternary Extinctions.*

———. 1986. The environmentalist myth. *Nature* 324:19–20.

———. 1992. The Golden Age that never was. Pp. 317–338 in J. M. Diamond, *The Third Chimpanzee: The Evolution and Future of the Human Animal.* New York: HarperCollins.

Dubos, R. 1974. *Beast or Angel?* New York: Scribner.

Erickson, C. L. 1995. Archaeological methods for the study of ancient landscapes of the Llanos de Mojos in the Bolivian Amazon. Pp. 66–95 in P. W. Stahl (ed.), *Archaeology in the Lowland American Tropics.* Cambridge: Cambridge University Press.

Erickson, C. L., J. Esteves, W. Winkler, and M. Michel. 1991. Estudio preliminar de los sistemas agrícolas precolombinos en el Departamento del Beni. Unpublished MS, University of Pennsylvania and Instituto Nacional de Arqueología.

Futemma, C. 1994. Management of the environment in the Inca Empire. Unpublished MS.

Gentry, A. 1988. Changes in plant community diversity and floristic composition on environmental and geographical gradients. *Annals of the Missouri Botanical Garden* 75:1–34.

Giddens, A. 1987. *Social Theory and Modern Sociology.* Stanford: Stanford University Press.

Gould, R. A. 1971. The archaeologist as ethnographer: A case from the Western desert of Australia. *World Archaeology* 3(2): 143–177.

Graham, E. H. 1956. The re-creative power of plant communities. Pp. 677–691 in Thomas (ed.), *Man's Role.*

Grayson, D. K. 1984a. Explaining Pleistocene extinctions: Thoughts on the structure of a debate. Pp. 807–823 in Martin and Klein (eds.), *Quaternary Extinctions.*

———. 1984b. Nineteenth-century explanations of Pleistocene extinctions: A review and analysis. Pp. 5–39 in Martin and Klein (eds.), *Quaternary Extinctions.*

Gunn, J. D. 1994. Introduction: A perspective from the humanities-science boundary. *Human Ecology* 22(1): 1–22.

Hughes, J. D. 1983. *American Indian Ecology.* El Paso: Texas Western Press.

Hynes, R. A. and A. K. Chase. 1983. Plants, sites, and domiculture: Aboriginal influence upon plant communities in Cape York Peninsula. *Archaeology in Oceania* 18:38–45.

Ingerson, Alice E. 1994. Tracking and testing the nature/culture dichotomy in practice. Pp. 43–66 in Crumley (ed.), *Historical Ecology.*

Ingold, T. 1988. The animal in the study of humanity. Pp. 84–99 in T. Ingold (ed.), *What Is an Animal?* London: Unwin Hyman.

Journal of American History. 1990. *Round Table: Environmental History.* Vol. 76 (March).

Kellert, S. R. 1993. Introduction. Pp. 20–27 in S. R. Kellert and E. O. Wilson (eds.), *The Biophilia Hypothesis.* Washington, D.C.: Island Press.

Leakey, L. S. B. 1964. Prehistoric man in the tropical environment. Pp. 24–29 in *The Ecology of Man in the Tropical Environment.* IUCN Publications, new series, vol. 4. Morges, Switzerland.

Levins, R. and R. Lewontin. 1985. *The Dialectical Biologist.* Cambridge: Harvard University Press.

Lewis, H. T. 1982. Fire technology and resource management in aboriginal North America and Australia. Pp. 45–67 in N. M. Williams and E. S. Hunn (eds.), *Resource Managers: North American and Australian Hunter-Gatherers.* Boulder, Colo.: Westview Press.

Lovelock, J. E. 1979. *Gaia: A New Look at Life on Earth.* Oxford: Oxford University Press.

———. 1992. The Gaia hypothesis. Pp. 295–315 in Margulis and Olendzenski (eds.), *Environmental Evolution.*

McElroy, M. 1992. Comparison of planetary atmospheres: Mars, Venus, and Earth. Pp. 1–14 in Margulis and Olendzenski (eds.), *Environmental Evolution.*

Margulis, L. and L. Olendzenski, eds. 1992. *Environmental Evolution: Effects of the Origin and Evolution of Life on Planet Earth.* Cambridge: MIT Press.

Marquardt, W. H., ed. 1992. *Culture and Environment in the Domain of the Calusa.* Institute of Archaeology and Paleoenvironmental Studies, Monograph 1. Gainesville: University of Florida.

Marsh, G. P. 1885. [1864]. *The Earth as Modified by Human Action.* 2d rev. ed. of *Man and Nature.* New York: Scribner.

Martin, C. L. 1992. *In the Spirit of the Earth: Rethinking History and Time.* Baltimore: Johns Hopkins University Press.

Martin, P. S. 1966. Africa and Pleistocene overkill. *Nature* 212:339–342.

———. 1967. Prehistoric overkill. Pp. 75–120 in P. S. Martin and H. E. Wright, Jr. (eds.), *Prehistoric Overkill: The Search for a Cause.* New Haven: Yale University Press.

————. 1973. The discovery of America. *Science* 179:969–974.

————. 1984. Prehistoric overkill: The global model. Pp. 354–403 in Martin and Klein (eds.), *Quaternary Extinctions.*

Martin, P. S. and R. G. Klein, eds. 1984. *Quaternary Extinctions: A Prehistoric Revolution.* Tucson: University of Arizona Press.

Marx, K. 1977. [1867]. *Capital.* Vol. 1. Trans. S. Moore and E. Aveling. Ed. F. Engels. New York: International Publishers.

Meilleur, B. A. 1996. Forests and Polynesian adaptations. Pp. 76–94 in Sponsel, Headland, and Bailey (eds.), *Tropical Deforestation.*

Moran, E. F. 1990. Ecosystem ecology in biology and anthropology: A critical assessment. Pp. 3–40 in E. F. Moran (ed.), *The Ecosystem Approach in Anthropology: From Concept to Practice.* Ann Arbor: University of Michigan Press.

————. 1993. *Through Amazonian Eyes: The Human Ecology of Amazonian Populations.* Iowa City: University of Iowa Press.

Murphy, R. F. 1970. Boasian ethnography and ecological theory. In E. H. Swanson (ed.), *Languages and Cultures of Western North America.* Pocatello: Idaho State University Press.

Nabhan, G. P. and S. St. Antoine. 1993. The loss of flora and faunal story: The extinction of experience. Pp. 229–250 in Kellert and Wilson (eds.), *The Biophilia Hypothesis.*

Nelson, R. 1993. Searching for the lost arrow. Pp. 201–228 in Kellert and Wilson (eds.), *The Biophilia Hypothesis.*

Netting, R. M. 1986. *Cultural Ecology.* 2d ed. Prospect Heights, Ill.: Waveland Press.

Nicholas, G. P., ed. 1988. *Holocene Human Ecology in Northeastern North America.* New York: Plenum Press.

Orr, D. W. 1992. *Ecological Literacy: Education and the Transition to a Postmodern World.* Albany: SUNY Press.

Parker, E. 1992. Forest islands and Kayapó resource management in Amazonia: A reappraisal of the *apêtê. American Anthropologist* 94(2): 406–428.

————. 1993. Fact and fiction in Amazonia: The case of the *apêtê. American Anthropologist* 95(3): 715–723.

Patterson, T. C. 1994. Toward a properly historical ecology. Pp. 223–237 in Crumley (ed.), *Historical Ecology.*

Patterson, W. A. III and K. E. Sassaman. 1988. Indian fires in the prehistory of New England. Pp. 107–135 in Nicholas (ed.), *Holocene Human Ecology.*

Posey, D. A. 1985. Indigenous management of tropical forest ecosystems: The case of the Kayapó Indians of the Brazilian Amazon. *Agroforestry Systems* 3:139–158.

Posey, D. A. and W. Balée, eds. 1989. *Resource Management in Amazonia: Indigenous and Folk Strategies.* Advances in Economic Botany, vol. 7. Bronx: New York Botanical Garden.

Pyne, S. J. 1982. *Fire in America: A Cultural History of Wildland and Rural Fire.* Princeton: Princeton University Press.

Rambo, A. T. 1985. *Primitive Polluters.* Ann Arbor, Mich.: Museum of Anthropology.

Redford, K. H. 1991. The ecologically noble savage. *Cultural Survival Quarterly* 15(1): 46–48 (reprinted from *Orion Nature Quarterly,* 1990, 9[3]: 24–29).

Reice, S. R. 1994. Nonequilibrium determinants of biological community structure. *American Scientist* 82:424–435.

Reichel-Dolmatoff, G. 1976. Cosmology as ecological analysis: A view from the rain forest. *Man* 11:307–318.

Roberts, J. M. 1964. The self-management of cultures. Pp. 433–454 in W. H. Goodenough (ed.), *Explorations in Cultural Anthropology.* New York: McGraw-Hill.

Roosevelt, A. C. 1991. *Moundbuilders of the Amazon: Geophysical Archaeology on Marajó Island, Brazil.* San Diego: Academic Press.

———. 1992. Secrets of the forest. *The Sciences* 32(6): 22–28.

Sahlins, M. 1976. *Culture and Practical Reason.* Chicago: University of Chicago Press.

Schmidt, P. R. 1994. Historical ecology and landscape transformation in eastern equatorial Africa. Pp. 99–125 in Crumley (ed.), *Historical Ecology.*

Sponsel, L. E. 1992. Myths of ecology and ecology of myths: Were indigenes noble conservationists or savage destroyers of nature? Pp. 27–37 in *The Second Annual Conference on Issues of Culture and Communication in the Asia/Pacific Region (Conference Proceedings).* Honolulu: East-West Center.

Sponsel, L., T. N. Headland, and R. C. Bailey, eds. 1996. *Tropical Deforestation: The Human Dimension.* New York: Columbia University Press.

Stearman, A. M. 1989. *Yuqui, Forest Nomads in a Changing World.* New York: Holt, Rinehart and Winston.

Steven, W. K. 1993. Garden of Eden in ancient America? It's only a myth. *New York Times,* 30 March.

Stewart, O. 1956. Fire as the first great force employed by man. Pp. 115–133 in Thomas (ed.), *Man's Role.*

Sussman, R. W. 1995. The nature of human universals. *Reviews in Anthropology* 24(1): 1–11.

Thomas, W. Jr., ed. 1956. *Man's Role in Changing the Face of the Earth.* Chicago: University of Chicago Press.

Thoreau, H. D. 1985. [1864]. *The Maine Woods.* Pp. 593–845 in R. F. Sayre (ed.), *Henry David Thoreau: Notes and Selections.* New York: Library of America.

Turner, B. L. II, W. C. Clark, R. W. Kates, J. F. Richards, J. T. Mathews, and W. B. Meyer, eds. 1990. *The Earth as Transformed by Human Action: Global and Regional Changes in the Biosphere over the Past Three Hundred Years.* New York: Cambridge University Press with Clark University.

Vansina, J. 1990. *Paths in the Rainforests: Toward a History of Political Tradition in Equatorial Africa.* Madison: University of Wisconsin Press.

Walsh, F. J. 1990. An ecological study of traditional Aboriginal use of "country": Martu in the Great and Little Sandy Deserts, Western Australia. *Proceedings of the Ecological Society of Australia* 16:23–37.

Wilmsen, E. N. 1989. *Land Filled with Flies: A Political Economy of the Kalahari.* Chicago: University of Chicago Press.

Winterhalder, B. P. 1994. Concepts in historical ecology: The view from evolutionary ecology. Pp. 17–41 in Crumley (ed.), *Historical Ecology.*

Worster, D. 1977. *Nature's Economy: The Roots of Ecology.* San Francisco: Sierra Club Books.

———. 1990. Transformations of the earth: Toward an agroecological perspective in history. *Journal of American History* 76(4): 1087–1106.

Zimmerer, K. S. 1993. Agricultural biodiversity and peasant rights to subsistence in the Central Andes during Inca rule. *Journal of Historical Geography* 19(1): 15–32.

CHAPTER 2

Ecological History and Historical Ecology: Diachronic Modeling Versus Historical Explanation

NEIL L. WHITEHEAD

Anthropology finds itself at a moment when it is being challenged on two fronts: first, as to the epistemological basis of its representation of others in ethnographic, historical, and archaeological interpretation; and second, as to how those representations are historically contingent upon particular categories of cultural representation, such as *nature-culture-society, evolution-process-structure,* and *ecology-economy-environment.* Although this challenge is substantive, and not to be lightly dismissed (Sangren 1988; Reyna 1994), the knowing of other times, persons, and places can yet be achieved—but only through a more rigorous *philosophical* analysis of the basis of anthropological understanding (see Whitehead 1995). The historical and ecological praxis of persons is therefore an important part of anthropological knowing, since persons are always situated in time and place, the latter being key to a historicized ecology. The idea of *praxis* is used here to refer to persistent features in the sociocultural repertoire of human physical and mental behavior that is overtly oriented to the achievement of an identifiable goal. *Ecological praxis*, then, refers to those forms of human activity that structure usages of the environment.

So "ecology," understood in the widest conceptual sense of the term—and despite its popular perception as a holistic alternative to positivist and instrumentalist representations of people and their environment—faces the same kind of critique that is currently general throughout the social sciences. Dissatisfaction with the way in which some kinds of "energetic" and "adaptability" analyses reduce human behaviors to caloric measures, and the perception that those "objective" and quantifiable measures vary "subjectively" according to a *culturally conditioned* cost-benefit analysis (Beckerman 1983), have led to a growing emphasis on human meanings as a source for understanding "natural" ecological systems (Balée 1994; Descola 1994; Moran 1993). At the same time, the rapid changes to which various kinds of "ecologies" have been subject, especially as the result of European colonial occupation, have led to new kinds of studies on the ways in which human activity shapes ecological outcomes through time.

Together, these trends in ecological analysis make it seem opportune to propose a fully "historical" ecology—the implication being that such a style of reasoning about ecology should put persons, not organic systems, at the center of explanations of changing ecological relationships through time. Moreover, these recent analyses—which emphasize the cultural contingency of individual perceptions of the costs and benefits of varying subsistence strategies, as well as the variability of individual success, especially in activities like hunting—suggest that *individual* praxis, rather than an aggregated or abstracted social or cultural praxis, should be the initial level of our analysis. The problem of situating that individual praxis within the overall repertoire of observed behaviors then forces a consideration of the wider anthropological issue of representation (as was indicated above). However, just as the historicizing of anthropology is often proclaimed but rarely fulfilled, so too the conjunction of the terms *historical* and *ecology* does not mean that any such intellectual activity is necessarily possible, or that it represents a true "paradigm" (Kuhn 1970) rather than merely an interesting methodology. Indeed, there are compelling reasons to think that the whole idea is premature, since the historical ecological literature to date has barely even acknowledged such issues, let alone offered any distinctive solutions to them.

Accordingly, this chapter outlines how the debate on history and anthropology can illustrate the theoretical problems that a "historical ecology" must address if it is to become a paradigm, rather than just an innovative methodology. Certainly it is laudable that ideas of "interpenetration" replace those of "adaptation" as a methodological aim in the description of human ecological praxis through time, but this theoretical activity can hardly be called "paradigmatic." Notwithstanding this I illustrate how a *methodology* of historical ecology could be applied for the purposes of field study. I then conclude by reconsidering the question whether we really need to invent a historical ecology: for one can argue that a *historical anthropology* already comprises the issues and topics that a "historical ecology" might address.

This question arises because "historical ecology" has no theory of history or of historiography, and so remains dependent on historical anthropology to develop the conceptual tools required for the interpretation of behaviors, texts, and artifacts, even where the latter are also biological entities like plants or trees (see Balée, chapter 1). Historical anthropology analyzes the interplay of *all* structures of human activity—such as polity, economy-ecology, society, culture, and so forth—with historical events, understood as the human praxis that innovatively evinces and so imperfectly reproduces these structures through time. As such, historical anthropology is quite distinct from ethnohistory, with which it is sometimes confused or conflated. *Ethnohistory*, within this framework, refers to either the history of a given ethnological group, or (following cognate usages of the prefix *ethno* in other fields) the autorepresentations of the past by a particular ethnic group. In either case, these intellectual activities are part of the analytical materials for historical anthropology but do not delimit its range of interests, which nonetheless include the comparative study of ethnohistories and ethnologies.

Concerning Amazonia in particular, issues of the relationship between the historical and ecological praxis of native peoples are particularly acute. In the first place, it is precisely in terms of a culturally restrictive and overgeneralized picture of Amazonian ecology as a "counterfeit paradise" (Meggers 1971) that many denials of native historical praxis have been couched (see discussions in Moran 1993; Whitehead 1993, 1996). By the same token, even the careful study of ecological praxis, as a product of the socialization of nature and naturalization of society (Descola 1994), must refer to historical praxis if the representation of this ecological praxis as "homeostatic" is to be improved—as it must be.

In anthropology more generally a theoretical ambiguity over the usage of the terms *economy* and *ecology* (even though the words derive from the single Greek term for "household/habitat," οικος), reflects the way this debate has been carried on in other regions. Thus, *economy* and *ecology* have tended to designate, respectively, cultural and natural systems of activity. Moreover, the unacknowledged force of this usage is to suggest that "primitive" societies (as in Amazonia) are dominated by their ecological setting, and "complex" ones by their economy. In either case, however, the issue of how human consciousness arises from, represents, and acts on its οικος is to the fore, which then reminds us that all human behavioral structures and events are an interwoven field of human activity (praxis) for which the complementary opposition of Nature/Culture is an insufficient analytical tool.

To the extent, then, that this dichotomy is present in "historical ecological" theories—whether or not the opposition is mediated by relationships of "interpenetration," "dialectic," "dialogue," and so forth—such theories will be inadequate for the representation of human praxis through time. This is quite simply because that praxis has *already* been analytically bifurcated into the "natural" and the "cultural": a contingent complementary opposition, but one to which ecological method, as quantification, measurement, and prediction, is necessarily committed (see Winterhalder 1994).

History in Anthropology and Ecology

Ever since the inception of modern anthropology in the middle of the last century there has been a tension between two competing paradigms for the explanation of the past: *evolutionism* and *historicism*. Essentially the difference here is between, respectively, the diachronic modeling of processes in the distribution of phenomena, and the explanation of human thought and behavior by reference to avowed meanings and reconstructed events in the past.

It is well known that such early figures in the emergence of professional anthropology as Edward Tylor (1871) and Lewis Henry Morgan (1877) were evolutionists in their approach to understanding the past of non-Western societies. They both proposed schemes of evolutionary progression through various stages (or "ethnical periods," in Morgan's terminology). Human culture was a unitary product of the species, and so a people's relative position in the evolutionary sequence implicitly assumed that this was also their achieved historical circumstance. There is no need to rehearse now the standard objections to these ideas (Evans-Pritchard 1962), but it is important to emphasize the extent to which evolutionist ideas necessarily underlie all nonhistorical reasoning about the human past—whether that reasoning proceeds from biology, ecology, or archaeology.

The work of Franz Boas (1936), Bronislaw Malinowski (1945), and their intellectual heirs, by contrast, emphasized the particularity of the cultures they studied and the ways in which those cultures therefore had to be understood *in their own terms*. This ethnographic particularism was useful in eventually showing that human groups often failed to match meaningfully the criteria generated by many evolutionists. "Simple" hunter-gatherers, as in Australia, were found to be "complex" in sociocultural terms, and some were found to have been in a dynamic symbiosis with both agriculturalists and the state (see Shott 1992).

The intellectual and cultural disenchantment with evolutionist theory, especially with its implicit idea of an inevitable social progression through time, also led some early ethnographers to insist on a functionalist paradigm of explanation. Since history was unknowable in the case of pre-textual tribal peoples, it could only be conjectural; as A. R. Radcliffe-Brown put it, "the view taken here is that such speculations are not merely useless but are worse than useless" (1958:3). Later ethnography, in an effort to recognize change and process in forms of culture and society, if not history per se, in turn was driven by varieties of "structuralism."

Such approaches emphasized either the role of cognitive structures in producing systematic variation, as with Claude Lévi-Strauss (1963), or the role of social structure in differentiating cultural plurality, as with Edmund Leach (1965). In either case, these abstractions from specific human contexts gave space for the theoretical discussion of the evolution and typology of these structures. So, by dealing with sociocultural phenomena rather than with the persons who embodied those phenomena, these approaches were very receptive to evolutionist ideas, especially the notion of *systemic process*, as an explanation for social and cultural patterns. The advent of poststructuralism therefore actually represented an intellectual retreat from these issues, at least ethnographically, into a kind of hyperparticularism. This trend is necessarily less evident in historical anthropology, since there is a continuing interest in the theoretical issue of relating the long-term structures of social and cultural activity to the particular practices of individuals or groups (see Obeyesekere 1992; Ohnuki-Tierney 1990; Sahlins 1995).

The "history of history" within anthropology is therefore strongly imprinted on the development of ecological anthropology as a whole. In the United States, Julian Steward played a very influential role—not least for Amazonia, in his capacity as editor of the *Handbook of South American Indians*. Cultural ecology, as envisaged by Steward (1963), was an extension and refinement of the old evolutionist paradigm, but with the all-important caveat that such evolution could result in *multilinear* trajectories for future development. Again we should note that it is "cultures," not persons, that represent the unit of evolutionary selection; and so this mode of analysis, as developed by Robert Netting (1977) and others, did not meet the historicist objections that cultural typologies tell us little because ethnographically (as mentioned above) it is rare that persons perfectly instantiate the relevant category of classification. Hunter-gatherers and pastoral nomads are found to be symbiotically linked with settled agriculturalists; and all the while, in Amazonia at least, characterizations of horticulturalism had failed to address the importance of fish capture, despite the intensity of debate on "protein scarcity" (Beckerman 1994; Gragson 1992).

This breakdown in the evolutionist paradigm, which in Amazonia became a kind of *environmental determinism* in the hands of Betty Meggers (1971) and Steward (1948), was also heralded within ecological anthropology by a form of neofunctionalism that examined the use of ritual or cosmology in the regulation of ecological relationships. The extent to which this was seen as a more or less *conscious* activity on the part of native peoples differed from author to author, ranging from the divine authority of Desana shamans (Reichel-Dolmatoff 1976), through Roy Rappaport's (1968) operationalized vs. cognized models in Tsembaga land, to the extreme functionalism of Harris (1977) in his solutions to the "riddles" of culture, such as food taboos and anthropophagy.

Subsequently the ecological paradigm has seen evolutionist models reemerge in various forms of biobehavioral ecology and evolutionary culture theory. In these works the role of human meanings is all but absent, and the agency of change is seen to be the structural properties of evolutionary process, whether that is seen as genetic or as cultural. The sociobiology of Napoléon Chagnon (1990) would be a relevant example for Amazonianists. Most recently, the advent of ethnoecology promises to reintegrate human meaning and understanding into models of ecological systems, stressing the need to gain access to these understandings if adequate scientific models of contemporary ecology are to be built (see Posey and Balée 1989).

It seems evident, however, that even this ethnologically sensitive kind of contemporary understanding remains incomplete, particularly in view of the nature of European global expansions during the last five hundred years. In short, one can now appreciate that such native understandings themselves are a product of *mutual* historical forces, and that the character of these historical forces is such as to have often produced radical disjunctures between past and present practices. This creates a presumption against the projection of the ethnographic present back into the past (Whitehead 1994). Moreover, in Amazonia it is already clear that making any absolute distinctions between indigenous and nonindigenous ecological understanding is highly problematical, since many non-Amerindians have occupied a range of Amazonian environments since the early seventeenth century.

Into this intellectual situation bravely marches the idea of "historical ecology." I will address the implications of a recent volume of that title (Crumley 1994) later; for the moment, I wish to draw attention to a number of studies that anticipate the topic, but are given limited treatment in that work. Chief among these are the works of Carl Sauer and Alfred Crosby. Crosby in *The Columbian Exchange* (1972) and Sauer in *The Early Spanish Main* (1966) both anticipated the themes that are central to the methods of historical ecology, even if they failed to overtly theorize about the issues that this chapter has so far raised. Sauer is actually more concerned with the immediate and local impacts of Europe on America, rather than with the systemic and extended consequences of a series of such impacts that preoccupies Crosby (especially in his later work, *Ecological Imperialism* [1986]).

In Crosby's first volume there is an attempt to discuss how, not just individual biota, but complexes of biota had a systemic impact on the environments extant in the Americas in 1492. This theme is extended in the later work to include non-American contexts, such as the Canaries or Australia, and has been expanded beyond the historical moment of European colonialism to include earlier interactions, such as the Norse in Greenland. Since Crosby has paid attention both to the biologically systemic interactions that produce environments and to the contingent structures of human decision that initiated particular forms of systemic biological interaction, the more recent volume (1986) is therefore aptly entitled *Ecological Imperialism*.

In this form of historical study there is now a "second generation" of research that has resulted in the detailed study of particular cases of ecological imperialism, such as William Cronon's *Changes in the Land* (1983) and Carolyn Merchant's *Ecological Revolutions* (1989). Sauer and Crosby were thus among the first to see the intellectual force of this way of discussing the ecological consequences of human decisions, and the extent to which those human decisions were fundamental in producing what were apparently pristine landscapes (see also Geertz 1963; Hughes 1975). Subsequent work on the "pristine myth" (see Denevan 1992), then, has the potential to tell us as much about humans and their history as it does about landscapes and their development, which must surely be the defining element of any historical ecological methodology. As Donald Worster notes (1988:289), this involves no longer *reducing* human agency to statistical measures, but *enhancing* the physical agency of environment.

It is therefore useful to comment on whether the recent volume edited by Carole Crumley (1994) does or does not live up to this potential. Because the volume proceeds not from anthropology but from landscape science, few of the issues and problems raised so far are dealt with therein as a whole. The most cogent interpretation that can be given to the contributors' intent in this work is that some *processual* form of ecology should be favored over a systemic or functionalist kind. In this way, the question addressed seems to be no different from that which arose between evolutionism and functionalism—namely, what is the best explanation given either by considering the way in which a system maintains itself, or by knowing by what processes it may have been produced? Certainly this is an interesting question, but it is not one that has not already been thoroughly explored in anthropology, as was explained above. Even if we were to pursue a historicist understanding of a particular "ecosystem," as Bruce Winterhalder suggests (1994:19), this would still leave open the critical issue of what causal variables would be addressed in the explanation given.

Thus the problem is *not* one of inventing some new paradigm, as Crumley (1994) argues, where the opposition of culture and nature is broken down; for that would be to lose precisely the human-centered explanation that the use of the epithet "historical" implies. Systems, distributions, and phenomena show processual or evolutionary change; humans show historical change. To try and mix these metaphors is to confuse types of analysis. In any case, it seems doubtful that the scientific representation of "natural systems" requires in any sense an appreciation of their particular historical characteristics, for this would completely contradict the nomothetic ambitions of a *generalizing* endeavor.

Rather, the point of a methodology of historical ecology must be to make human decision-making, and the consciousness that drives it (see Posey, chapter 5), the independent variable in our analysis of environmental dynamics. The varying "ecologies" (understood as a *praxis*—see also Descola 1994) that drive those decisions and their associated environmental impacts then become a special subset of study for cultural history, as reflected by works such as Keith Thomas's *Man and the Natural World* (1983) or Robert Brightman's study of Cree hunting, *Grateful Prey* (1993). But to equate a historical ecology with a landscape history, however expansive that notion of landscape might be, is to miss the point that an ecology represents human praxis *in* the landscape. The fact that human practices never take place in a "pristine" environment, but always in a landscape shaped by the past ecological praxis of others, only serves to emphasize the importance of understanding the historical dynamics of human usages; it does not entail that such usages are equated only with particular data sets that might be used to partially reconstruct them, such as landscape forms. Simply to chart changes in landscape through time once again places phenomena, rather than persons, at the center of explanation.

The history of an ecology is more complex than this, and must include an account of the synergetic impacts of changing human ideas. That such a project is only intermittently realizable, due to the limits of historical text and archaeological data, thus threatens to produce a situation where sequential functionalist explanations of different "landscapes" are proffered as historical ecology. This situation might be partly remedied by means of a more sophisticated notion of "history." The many forms in which the past is culturally presented as "history," as well as the variety of purposes that such constructions serve, mean that "history" is not necessarily descriptive, or even chronological—yet Crumley and her fellow contributors seem content to define it in this way, even as this particular cultural idiom of historiography is being challenged by professional historians themselves. While all may agree that the historical aspect of human affairs is fundamental to their meaningful analyses, in the case of the *Historical Ecology* volume, the contributors seem largely content with grafting a temporal dimension onto the chronological study of systems. For historical anthropologists this will be an insufficient notion of "history," since the nonconflation of the evolutionary change of phenomena (including human phenomena) with the history of persons (including their decisions that affect their landscapes) is central to the whole historical paradigm.

This analysis is borne out by the way that the issues in landscape history are reproduced in historiography at a more general level. For example, one asks: Do we read landscape retrospectively or regressively? That is, do we intend to understand past ecologies and their processual responses to change, as in the early works of Clifford Geertz (1963) and J. Donald Hughes (1975); *or*, how past ecologies produce present ones, as in the work of Crosby, Cronon, and Merchant mentioned above? Obviously it is the latter kind of research agenda that is potentially distinct enough to form at least a methodology for historical ecology. However, the fact that all these authors must engage a series of ethnographic and historiographic issues indicates that their ecological themes are necessarily secondary to the writing of the past more generally. How, then, might field study be designed to be epistemologically valid within the parameters outlined above?

Studying Historical-Ecological Praxis in Amazonia

The soils of the Amazon have traditionally been viewed as "poor." However, the existence of a limited class of high-fertility soils called *terra preta do Indio* (Indian black earth) has been known informally for a long time in the literature of the region (Nimuendajú 1952; H. H. Smith 1879).

Terra preta (latosol amarelo húmido antropogênico) is a prevalent soil classification in Amazonia today—yet whether it actually forms a unified soil type remains to be investigated, and the distribution, composition, and formation of such soils are only beginning to be understood. *Terra preta* exhibits circumscribed distributions in a wide variety of environmental contexts, such as *terra firme* bluff edge, interior *terra firme*, and *várzea*; however, the relative densities of *terra preta* in each of these settings are poorly known due to the lack of systematic survey coverage. In spite of pioneering chemical and physical characterization studies, the issue of the exact composition of such soils, as it relates to the genesis and evolution of the *terra preta* category, is still unresolved. For example, it is still unclear even why these soils are "black" (Eden et al. 1984; Kern and Kampf 1989; N. Smith 1980).

Such soils are often closely identified with human habitation and activity, being replete with cultural remains. In this context, the determination of the variation in soil composition would establish a dependent variable for the evaluation of the cultural context of the *terra preta* sites. It seems likely that the contribution from human activities to the formation of these soils differs among various sites. Indeed, there is also the possibility that localized subsurface geological characteristics may have led to the "natural" formation of *terra preta*, which was subsequently occupied and augmented by human activities. Equally, the nonpedological characteristics of a given site, such as a location that straddled existing active trade routes or gave access to particular fluvial resources, could have led to a sufficiently stable occupation for the direct formation of a *terra preta* deposit (see review in White-head 1996). Whatever the site-specific explanation, either situation produces a synergistic effect between environment and culture.

It is this *synergy* that is really the analytical object of a humanistic ecology—and most particularly that of a historical ecology. Thus, the usage of *terra preta* sites is known from the historic record (Whitehead 1994) but, because of the character of colonial occupation (Whitehead 1993), little of ancient Amerindian technique has been transmitted to the current occupants of Amazonia (Denevan 1992). Researching that context through archaeology and history then allows the interpretation of soil characteristics in the context of other scientific measures of human usage, and the opportunity to relate these findings to the historical record. Together such research strategies provide an active cultural and historical context for the interpretation of the soil data.

In short, this will permit a characterization of the soil compositional differences in ethnological and historical terms, as well as providing ecological referents for social and historical difference within this cultural pattern. Initial work on the historical and archaeological literature, as well as on soil science, can be used as an element for definition of the study area. In turn, the emerging results of historical and archaeological reconstruction can be progressively applied to the data from specific site investigations.

By such means it will be possible to characterize the various classes of soil formations mentioned above, suggest the human context of their development, and provide a link to broader questions of current anthropological debate in Amazonia concerning human impacts on the environment and their implications for long-term cultural development. Indigenous usages are thus a key element in the understanding of Amazon ecology—but more important is the history of those usages *in* the successive contexts of changing landscapes.

Conclusion

In the practical and intellectual contexts outlined in this chapter, it seems relevant to ask: Is a new *paradigm*, as opposed to a methodology, of historical ecology achievable, or even desirable?

The achievability of a historical ecology would actually seem to rest on the same grounds as historical anthropology generally—with the critical proviso that a historical ecology is absolutely unable to resolve the paradigmatic issues because it does not theoretically address the philosophical problems posed by a conjunction of a positivist "science" and a relativist "humanism." However, even if these contrasting approaches are philosophically incommensurable, as Thomas Kuhn (1970) and Paul Feyerabend (1975) imply, this does not entail that they are hermeneutically so (Rorty 1979; Whitehead 1995). Therefore a historical conception of human ecological relationships can form a powerful idiom for the explanation of people and place. Yet methodologically this deals only intermittently with the phenomena (informants, texts, artifacts) that are the central source of the theoretical imponderabilia in anthropology, and so it is unclear how such a methodology would address these epistemological questions paradigmatically. Unless ecologically oriented anthropologists are prepared to invest the same theoretical and research effort to these issues as historical anthropologists already do, then at the very least a historical ecology will have to wait on the progress of historical anthropology in general.

In turn, and perhaps in a more parochial sense, I am tempted to wonder whether a historical ecology is even desirable, since the ecological paradigm in Amazonia has developed in a way that precluded both economy and history. I have already referred in passing to the great "protein debate" in Amazonia and have mentioned how the ethnographic data were simply incomplete as regards modes of protein capture. But worse than this was the impression created that indigenous decision-making was so constrained by the ecological imperative that no economic solutions to resource differentials, such as trade and/or the intensification of agricultural production, were ever contemplated—despite the presence of such economic mechanisms as both ethnographic and historic fact.

Clearly the integration of data sets, the expansion of causal variables, and an appreciation of systemic interactions are some of the positive methodologies that an ecological approach brings to anthropology—but these are part of anthropology anyway, and without the integration of such methodologies no independent historical ecological theory can emerge. Similarly, ideas about succession, persistence, and change are integral to ecological discussions of the past—but this, too, is part of anthropology already. In short, historical and ecological explanations are most vital to any adequate study of humanity, but the question for the would-be historical ecologist must be: *What critique (if any) can historical ecology offer not just of a functionalist, ecosystems theory but, more significantly, also of anthropological conceptions of history?*

References

Balée, W. 1994. *Footprints of the Forest: Ka'apor Ethnobotany—The Historical Ecology of Plant Utilization by an Amazonian People.* New York: Columbia University Press.

Beckerman, S. 1983. Carpe diem: An optimal foraging approach to Bari fishing. Pp. 269–300 in R. Hames and W. Vickers, *Adaptive Responses of Native Amazonians.* New York: Academic Press.

———. 1994. Hunting and fishing in Amazonia: Hold the answers, what are the questions? Pp. 177–200 in Roosevelt (ed.), *Amazonian Indians from Prehistory to the Present.*

Boas, F. 1936. History and science in anthropology. *American Anthropologist* 38:137–141.

Brightman, R. 1993. *Grateful Prey.* Berkeley: University of California Press.

Chagnon, N. A. 1990. Reproductive and somatic conflicts of interest in the genesis of violence and warfare among tribesmen. Pp. 77–104 in J. Haas (ed.), *The Anthropology of War.* Cambridge: Cambridge University Press.

Cronon, W. 1983. *Changes in the Land: Indians, Colonists, and the Ecology of New England.* New York: Hill and Wang.

Crosby, A. W. 1972. *The Columbian Exchange: Biological and Cultural Consequences of 1492*. Westport: Greenwood Press.

———. 1986. *Ecological Imperialism: The Biological Expansion of Europe, 900–1900*. Cambridge: Cambridge University Press.

Crumley, C. L. ed. 1994. *Historical Ecology: Cultural Knowledge and Changing Landscapes*. Santa Fe: School of American Research Press.

Denevan, W. M. 1992. The pristine myth: The landscape of the Americas in 1492. *Annals of the Association of American Geographers* 82(3): 369–385.

Descola, P. 1994. *In the Society of Nature: A Native Ecology in Amazonia*. Cambridge: Cambridge University Press.

Eden, M., W. Bray, L. Herrera, and C. McEwan. 1984. *Terra preta* soils and their archaeological context in the Caquetá Basin of southeast Colombia. *American Antiquity* 49(1): 125–140.

Evans-Pritchard, E. E. 1962. *Essays in Social Anthropology*. London: Faber.

Feyerabend, P. 1975. *Against Method*. London: New Left Books.

Geertz, C. 1963. *Agricultural Involution: The Processes of Ecological Change in Indonesia*. Berkeley: University of California Press.

Gragson, T. L. 1992. Strategic procurement of fish by the Pumé: A South American "fishing culture." *Human Ecology* 20(1): 109–130.

Harris, M. 1977. *Cows, Pigs, Wars, and Witches: The Riddles of Culture*. London: Fontana.

Hughes, J. D. 1975. *Ecology in Ancient Civilizations*. Albuquerque: University of New Mexico Press.

Kern, D. C. and N. Kampf. 1989. Comissão V—Gênese, morfologia, e classificação do solo. *Revista Brasileira de Ciências do Solo* 13:219–225.

Kuhn, T. 1970. *The Structure of Scientific Revolutions*. Chicago: University of Chicago Press.

Leach, E. 1965. *Political Systems of Highland Burma*. London: Beacon.

Lévi-Strauss, C. 1963. *Structural Anthropology*. Harmondsworth: Penguin.

Malinowski, B. 1945. *The Dynamics of Culture Change*. New Haven: Yale University Press.

Meggers, B. 1971. *Amazonia: Man and Culture in a Counterfeit Paradise*. Chicago: Aldine.

Merchant, C. 1989. *Ecological Revolutions: Nature, Gender, and Science in New England*. Chapel Hill: University of North Carolina Press.

Moran, E. 1993. *Through Amazonian Eyes: The Human Ecology of Amazonian Populations*. Iowa City: University of Iowa Press.

Morgan, L. H. 1877. *Ancient Society*. London.

Netting, R. McC. 1977. *Cultural Ecology*. Menlo Park, Calif.: Cummings.

Nimuendajú, C. 1952. The Tapajó. *Kroeber Anthropological Society Papers* 6:1–25.

Obeyesekere, G. 1992. *The Apotheosis of Captain Cook*. Princeton: Princeton University Press.

Ohnuki-Tierney, E., ed. 1990. *Culture Through Time: Anthropological Approaches*. Stanford: Stanford University Press.

Posey, D. and W. Balée, eds. 1989. *Resource Management in Amazonia: Indigenous and Folk Strategies*. Advances in Economic Botany, vol. 7. Bronx: New York Botanical Garden.

Radcliffe-Brown, A. R. 1958. Structure and function in primitive society. Pp. 1–14 in M. N. Srivinas (ed.), *Method in Social Anthropology*. Chicago: University of Chicago Press.

Rappaport, R. A. 1968. *Pigs for the Ancestors: Ritual in the Ecology of a New Guinea People*. New Haven: Yale University Press.

Reichel-Dolmatoff, G. 1976. Cosmology as ecological analysis: A view from the rain forest. *Man* 11:307–318.

Reyna, S. P. 1994. Literary anthropology and the case against science. *Man* 29(3): 555–582.

Roosevelt, A. C., ed. 1994. *Amazonian Indians from Prehistory to the Present: Anthropological Perspectives*. Tucson: University of Arizona Press.

Rorty, R. 1979. *Philosophy and the Mirror of Nature*. Princeton: Princeton University Press.

Sahlins, M. 1995. *How Natives Think*. Chicago: University of Chicago Press.

Sangren, P. S. 1988. Rhetoric and the authority of ethnography: "Postmodernism" and the social reproduction of texts. *Current Anthropology* 29(3): 277–307.

Sauer, C. O. 1966. *The Early Spanish Main*. Berkeley: University of California Press.

Shott, M. J. 1992. On recent trends in the anthropology of foragers: Kalahari revisionism and its archaeological implications. *Man* 27(4): 843–871.

Smith, H. H. 1879. *Brazil, the Amazons and the Coast*. New York: Scribner.

Smith, N. J. H. 1980. Anthrosols and human carrying capacity in Amazonia. *Annals of the Association of American Geographers* 70:553–566.

Steward, J. 1963. *Theory of Culture Change*. Urbana: University of Illinois Press.

———, ed. 1948. *Handbook of South American Indians*. Vol. 3, *The Tropical Forest Tribes*. Washington, D.C.: Smithsonian Institution.

Thomas, K. 1983. *Man and the Natural World: Changing Attitudes in England, 1500–1800*. Harmondsworth: Penguin.

Tylor, E. B. 1871. *Primitive Culture*. London.

Whitehead, N. L. 1993. Historical discontinuity and ethnic transformation in native Amazonia and Guayana, 1500–1900. *L'Homme* 28:289–309.

———. 1994. The ancient Amerindian polities of the Amazon, Orinoco, and Atlantic Coast: A preliminary analysis of their passage from antiquity to extinction. Pp. 33–54 in Roosevelt (ed.), *Amazonian Indians from Prehistory to the Present*.

———. 1995. The historical anthropology of text. *Current Anthropology* 36(1): 190–232.

———. 1996. Amazonian archaeology: Searching for paradise? *Journal of Archaeological Research* 4(3): 241–264.

Winterhalder, B. P. 1994. Concepts in historical ecology: The view from evolutionary ecology. Pp. 17–41 in Crumley (ed.), *Historical Ecology*.

Worster, D. 1988. Doing environmental history. Pp. 289–307 in D. Worster (ed.), *The Ends of the Earth: Perspectives on Modern Environmental History*. Cambridge: Cambridge University Press.

CHAPTER 3

A Historical-Ecological Perspective on Epidemic Disease

LINDA A. NEWSON

From the perspective of historical ecology, with its distinct focus on the human species, this chapter examines the history of epidemics by exploring the interrelationships between parasites, human hosts, and their environments. In this process, human societies are considered to be an integral part of the environment. While the emphasis of the chapter is on the ways in which differences in human social organization have affected patterns of epidemic disease through time, the value of a holistic approach is stressed through noting that parasites, their human hosts, and their physical environments have also changed, both singly and together, due to interactions between them and also to their own internal dynamics. The aim of this chapter is not to deny that there are general and long-term correspondences in historical processes, but rather to argue that an understanding of the trajectories of biological and social change and the complexities of the processes at work can be best achieved through viewing epidemics in their specific historical and geographical contexts.

The intimate relationship between humans and the natural world finds perhaps its clearest expression in human disease. Human diseases emerge from interactions between parasites, hosts, and their environments: no understanding of their origins, spread, and impact can be achieved if any of these three elements is excluded. It is necessary to be aware of the biology of parasites and hosts, the manner in which diseases are transmitted, the environmental conditions necessary for their survival, and also the characteristics of the human society affected. The need for a holistic approach is evident.

Infectious disease may be characterized as either chronic or acute. Chronic diseases—such as herpes simplex, tuberculosis, and treponemal infections—are those that persist for long periods, exhibiting little fluctuation in occurrence, and generally without conferring immunity on their hosts. Acute infections, on the other hand—such as measles, rubella, and smallpox—are characterized by short periods of infection that result in high morbidity and potential mortality while conferring lifelong immunity on survivors. The persistence of acute infections depends either on the existence of large populations of previously uninfected persons, or on non-human hosts, such as rodents or insects, in which the parasites may survive or reproduce. Historically, acute infections took the form of epidemics that resulted in short periods of high mortality—becoming endemic only where human populations were of sufficient size to generate enough susceptibles in the form of children to maintain the disease indefinitely. Chronic infections, on the other hand, are generally endemic, taking a regular toll of the population. Although the distinction between chronic-endemic and acute-epidemic diseases is important, it is not fixed: as already indicated, acute infections may become endemic in large populations, while minor changes in a parasite, host, or environment may occasionally cause a normally endemic disease to erupt as an epidemic (Ramenofsky 1987:140). The primary concern of this chapter is with acute infections that historically took the form of epidemics.

Epidemic diseases appear relatively late in the historical record and are commonly associated with the emergence of agriculture and sedentary communities (Cockburn 1971:50; Fenner 1970:64–65; 1980:14–16). Unidentified epidemics afflicted the Mediterranean, and probably India and China, in the first millennium B.C., but the first epidemics to be linked to measles and smallpox occurred in the second and third centuries A.D. (McNeill 1976:102–113). Within the last two hundred years, however, the incidence of epidemics has generally declined. This has been attributed to acquired immunity; to the isolation of disease organisms and the introduction of preventive measures, both medical and environmental; to improvements in living conditions and nutrition; and to changes in the parasites themselves. The relative importance of these factors, both in general and in the case of specific diseases, has been hotly debated (Cohen 1989:54; McKeown 1976:42–72; 1988:77–87). The fact that epidemics have declined in occurrence does not necessarily mean that humans are more healthy—for they continue to be afflicted by chronic infections, while the incidence of noncommunicable disease such as heart disease, diabetes, and cancer has increased (Boyden 1970:204–207; Mascie-Taylor 1993:14). Furthermore—as attested by Legionnaires' disease, Ebola, Marburg, Bolivian hemorrhagic fever, and AIDS—new transfers from other animal species, or changes in human parasites or the environment, mean that hitherto-unknown diseases continue to emerge.

Viewed from the perspective of the present, epidemic diseases appear to have already had their heyday in history, which makes them a fitting focus of historical ecology. Unfortunately, much of the history of human disease remains obscure, since it predates written records. Apart from the Black Death that afflicted Europe in the fourteenth century, the best-documented epidemics relate to the New World and to isolated communities, often located on islands—especially in the Pacific, where the impact of Old World diseases was late and sporadic (see, e.g., Black 1966; Cliff et al. 1981; Kunitz 1993; Stannard 1989). Because of the shortage of evidence for earlier periods, the discussion below draws on studies of the epidemiological and demographic consequences of European expansion, particularly in the Americas, and on observations of historically more recent epidemics on which the mathematical models of the spread and impact of disease have generally been based.

It is worth noting at this stage that although European expansion was particularly significant in the spread of human disease, it was not a historically unique event; nor did it affect non-European peoples equally. Prior to European expansion, human populations in large parts of Eurasia and Africa had already been afflicted by newly emerging diseases to which they had not been previously exposed and had suffered levels of disease mortality comparable to those experienced in the Americas after 1492 (Patterson and Hartwig 1978:8); they had also, through centuries of contact, developed various levels of immunity to epidemic diseases, so that their contact with Europeans did not precipitate demographic disasters on a scale comparable to that experienced in the Americas or the Pacific islands (Crosby 1976; Dobyns 1966; Jacobs 1974). This example highlights the significance of the specific epidemic history of a region for understanding the subsequent occurrence and impact of disease at even the most extended geographical and temporal scale.

Present understanding of the spread and impact of human diseases derives in large part from attempts to control them. For such purposes sophisticated mathematical models, generally disease-specific, have been developed that are based largely on levels of susceptibility and patterns of contact. Such models have sometimes taken the age-structure, immunity, and distribution of populations into account, but differences in environmental and social conditions have not figured significantly, nor have biological differences in populations (human or nonhuman) been considered, particularly over a long timescale (Sattenspiel 1990:246–249, 270). The limitations of deterministic models, particularly the simple ones, have become more apparent with recent developments in the science of chaos that have stressed the complexity of systems and demonstrated how minor differences in initial conditions can produce major and unpredictable effects (Gleick 1987). In ecology, population biology, and epidemiology chaotic models have been developed to account for spatial and temporal discontinuities (Anderson and May 1979; Olsen and Schaffer 1990; Schaffer 1987). While these models have attempted to explain unpredictability in the incidence of infections, the range of factors considered has not expanded. In part this reflects the difficulty of handling complex systems mathematically, for in many cases the potential effect of minor differences in host, parasite, or environment on the origin, spread, and impact of a disease has been recognized, if only in passing.

In the qualitative literature there has been a growing recognition of the influence of environmental and social conditions on geographical and temporal variations in the progress of epidemics (Cohen 1989:7; Kunitz 1994:5–7; Milner 1980:46–47; Newson 1985:48). It cannot be assumed that a newly introduced disease will spread unhindered, and that its impact will be uniform; rather, each epidemic has its own history. In this chapter I will examine the significance of the local social context in understanding the origins, spread, and impact of epidemic diseases; but in order to stress the importance of a holistic perspective for an understanding of the historical

ecology of epidemics, I will consider briefly the types of changes that have occurred over time to disease organisms, their hosts, and their environments, that
have also affected their histories.

Bioenvironmental Conditions: Variations and Change

Studies of the history of human disease (Cockburn 1971; McKeown 1988) have
often recognized the reciprocal relationship between humans and parasites. Humans coexist with parasites as part of the animal world. The human body, especially the lower intestine, is home to hundreds of parasites, and the diversity of
species it houses reflects the diversity of the environment in which it exists. Thus
African and Malaysian hunter-gatherers living in the tropical rain forest contain
twice as many intestinal parasites as Bushmen and Australian Aborigines, who live
in a species-poor environment (Dunn 1972:227–228). Since human activity has
often resulted in a simplification of the environment, over time the number of
species harbored by humans has generally declined (Polunin 1977:16).

Many parasites coexist with humans without causing them any harm. Other parasites are unable to establish themselves in human hosts because when they invade
the body either they kill the host, or else they induce an immune response that results in the parasite's death. Occasionally, however, a parasite is transmitted that is
able to survive without killing its host, and in this way a new human pathogen
emerges. Even then, if it has a short period of communicability it may be unable to
reach another human host before it becomes extinct. The establishment of human
diseases must have been characterized by many false starts. It is worth noting that
most human diseases are thought to have originated in animals, and particularly in
domesticated species with which humans have more intense contact; for example,
smallpox is related to cowpox, and measles probably to canine distemper or rinderpest (Cohen 1989:7; McNeill 1976:54–56).

Once established as human diseases, parasites continue to interact with humans,
sometimes resulting in genetic change. The life cycles of parasites are generally extremely short, being only a fraction of that of humans, so that the chance that a mutation will occur is far greater for them than for humans (Black 1980:42). Since
most human diseases did not emerge until sedentary farming communities had become established, too few generations have passed for appreciable genetic changes
to have occurred in humans (Boyden 1970:193)—with the obvious exceptions of
the sickle-cell and the Duffy negative traits found among African populations, which
provide some immunity to different forms of malaria (Dunn 1993:858–859). The
former has been associated with *Plasmodium falciparum*, which may have emerged
with intensive agriculture about 2,000 years ago; the latter trait, which is linked to
the more benign *Plasmodium vivax*, is more widely dispersed, and this probably reflects its greater antiquity (Black 1980:50; Wiesenfeld 1967:1134–1139). Although
these traits are associated with African populations, some resistance to malaria has
been acquired by New World groups through racial mixing. Most often, however,

the most immediate and effective responses limiting human diseases have been social rather than biological. Vaccination programs, modifications to the environment,
and the development of pesticides and insecticides have all been attempts to control
or mitigate the impact of disease. However, as in the case of malaria, such actions
have sometimes encouraged the emergence of new resistant strains.

While humans have developed little genetic resistance to disease, recent research has suggested that the degree of genetic diversity of human groups may influence disease mortality (Black 1992b). The significance of genetic homogeneity
for disease mortality is that pathogens become preadapted to successive hosts who
possess similar immune systems, resulting in increased virulence and host mortality. Native South Americans lost genetic traits as they passed through the environmental bottlenecks of the Bering Strait and the Panamanian isthmus, becoming genetically more homogeneous and therefore unusually susceptible to newly
introduced diseases (Black 1990:65–68). Francis Black (1992a:4–5) has estimated
that in the New World, which he defines as the Americas, Australasia, and Oceania,
the minimum probability of a pathogen's encountering a host with a similar immune system is 28 percent, compared to only 2 percent in the Old World. In small
isolated groups, however, the tendency toward genetic homogeneity is countered
by random genetic drift—though operating against this process may be the seizure
of captive women and children, or the incorporation of small groups, if their populations are too low to maintain a supply of suitable spouses (Bodmer and Cavalli-
Sforza 1976:392–398). Therefore Black (1990:65) suggests that genetic distances
among apparently unrelated tribes are not much greater than among tribes of one
cultural group. Even if genetic diversity can be shown to affect disease mortality, its
significance relative to environmental and cultural factors in influencing the pattern
of infection remains difficult to determine (Jenkins et al. 1989:28–30; Svanborg-
Eden and Levin 1990:33–34).

Discussions of the spread and impact of particular diseases in history often draw
on the experience of recent epidemics. Yet even when a disease can be identified in
the past—and this in itself is often problematic, given the vague descriptions of
symptoms in the historical record, which in addition are sometimes based on classification systems that are not easy to interpret—it should not be assumed that the
characteristics of the parasite have remained constant. Over a long time period
probably no pathogen has gone unchanged, and some have changed considerably
(McKeown 1988:77). Many viral diseases—especially influenza, but also smallpox
and syphilis—evolve rapidly through mutation, recombination, and reassortment.
These processes may produce different strains of the disease, with different latent
and infectious periods and levels of virulence, to which previous exposure may not
have conferred immunity (Ramenofsky 1987:137–138; Sattenspiel 1990:254).

Until recently it was generally considered that the coevolution of hosts and parasites resulted in reduced virulence, since this was to their mutual benefit (Allison
1982:245; Anderson and May 1991:249–250; Pimental 1968:1437). Indeed, it has
been suggested that a number of human diseases, including smallpox and bubonic
plague, have through repeated infections lost some of their virulence. In recent his-

tory scarlet fever seems to have declined because of changes in the character of the disease (McKeown 1976:82–85). It has also been suggested that the sudden disappearance of plague from western Europe in the seventeenth century may have been due to changes in the virus or in the resistance of rat vectors (Appleby 1980:169–173; Slack 1981:471–472). However, on theoretical grounds it has been argued that the coevolution of parasites and hosts may not necessarily trend toward avirulence but may instead follow many paths, depending on the degree to which parasite transmissibility and recovery rates are linked to rates of host mortality (Anderson and May 1991:649–653; Levin et al. 1982:214). The key point is that each disease organism has its history. It cannot be assumed that the smallpox virus that devastated New World populations in the sixteenth century was the same as that which had afflicted China in the fourth century A.D., or was to strike Hawaii in 1853. Variations in the virulence of particular strains of disease may be relatively small, but in particular contexts they might be of considerable significance.

Before embarking on a discussion of societal and cultural influences on the origin, spread, and impact of human diseases, it is worth commenting on the significance of variations in the physical environment, even though this is often recognized, at least at a general level. Each parasite has certain environmental limits beyond which it cannot survive, of which perhaps the most obvious is climate. The influence of climate appears to be greater for diseases spread by aerial transmission—such as colds, influenza, measles, and smallpox—and for vector-borne diseases such as malaria, yellow fever, and sleeping sickness, which require certain temperatures for the reproduction of the parasite. Many diseases are confined to tropical climates. Seasonal variations in temperature and humidity, most notably in temperate climates, may also result in marked seasonal variations in the incidence of disease (Fenner 1982:112–114; Mascie-Taylor 1993:9–11; Upham 1986:119–122). In some cases the seasonal effects are not direct, but are mediated by human activities that often reflect climatic variations. For example, the regular congregation of people at particular times of the year for wider communal purposes, such as harvests or religious festivals, may facilitate the spread of disease, while seasonal food shortages may affect disease susceptibility.

Other environmental variables affecting the habitat of parasites are more closely related to human activity. Any change in habitat will encourage the expansion of some parasites at the expense of others, and the greater the simplification of habitats the greater the opportunity for particular species to multiply (Polunin 1977:15–16). Hence forest clearance may encourage the reproduction of mosquitoes that propagate malaria, but the specific parasites may also vary according to particular conditions—for example, on the coast of Ecuador the more virulent *Plasmodium falciparum* is found in more open environments, whereas the more benign *P. vivax* is prevalent where the vegetation cover has been less disturbed (Newson 1993:1190). Changes in the level of grazing may also influence the incidence of disease. Cattle may constitute alternative blood meals for mosquitoes, thereby encouraging the spread of malaria, but reduced levels of grazing by altering the environment may bring other disease hazards. In the 1890s rinderpest caused a decline in cattle populations in Africa, which enabled the bush vegetation to recover and thereby encouraged the spread of tsetse flies carrying sleeping sickness (Curtin et al. 1978:553–554). Other forms of human environmental change associated with the increased incidence of disease are irrigation and dam building. These activities create stagnant water bodies that constitute favored habitats for the reproduction of mosquitoes and snails involved in the transmission of malaria and schistosomiasis, respectively (Boyden 1992:204; Cockburn 1971:49–50; Cohen 1989:42–43; Grove 1980:196–198; Kunitz 1994:11; McKeown 1988:190; Wirsing 1985:313). However, the process of human-induced environmental change is not unidirectional or always irreversible. Particular habitats may be abandoned in favor of others as a result of population decline, wars, or technological change (Denevan 1992:381).

Given the significance of environmental conditions for the incidence of epidemic disease, it is important to note that even when the impact of human activities is excluded, conditions have not remained constant. The influence of long-term climatic change has often been recognized, but environments have also experienced short-term perturbations and disturbances, sometimes caused by fire or drought, that have occurred as part of ecological processes (Cronon 1983:11; Demeritt 1994:23–26; Worster 1990:8–11). Ecologists now feel less comfortable with the concept of an ecosystem following a succession toward some sort of "climax" or natural equilibrium; they prefer to view nature as a landscape of patches that are continually changing through time and space as a result of their own dynamic processes. Nature is not only acted upon, but has its own history. Temporal and spatial variations in environmental conditions may not only affect the distribution of parasites, but through influencing the availability of resources also affect subsistence patterns, nutrition, and disease susceptibility. Studies of the subsistence economies of traditional societies show that they are affected, albeit often unconsciously, by relatively small shifts in environmental conditions (Dean et al. 1985:550; Larson, Johnson, and Michaelsen 1994:271, 288).

It has been shown that bioenvironmental conditions have not remained constant, but have changed through time, both individually and through interactions between them. While certain broad generalizations may be made about the nature of those changes, the changes themselves have often been unpredictable. Against this background of variability in hosts, parasites, and environments, the second part of this paper will examine how differences in social conditions affect their interaction and hence influence the origins, incidence, spread, and impact of epidemic disease.

The Origins and Incidence of Human Disease

When human populations subsisted on wild food resources, they were too small and too mobile to enable most parasites to become established as specifically human pathogens. Acute infections such as smallpox, measles, or influenza, which are characterized by short periods of infectivity, could not have survived, because they would have died out before reaching new hosts. The most successful diseases were those that had longer infectious periods and did not kill their hosts, and those that could survive on a nonhuman vector or intermediate host. Diseases that can persist for long periods in small populations include herpes simplex, chicken pox, typhoid, dysentery, hepatitis, leprosy, and treponemal infections; and, of those with nonhuman vectors, leishmaniasis, malaria, filaria, and schistosomiasis (Black 1975: 515–518; 1980:45–49; Cockburn 1971:50; Fenner 1980:14–15; Garruto 1981: 560–564; McKeown 1988:38, 49). As human populations increased and began to settle in permanent nucleated settlements, the opportunities expanded for parasites to develop as human pathogens. Thus with the beginnings of agriculture and sedentary life, the major killer diseases emerged (Cockburn 1971:48–51; Fenner 1970:48–68; McKeown 1988:48–56; McNeill 1976:54–57).

Population size is critical for understanding the incidence of infection. Since endemic infections are characterized by latency and recurrence, they can persist in small societies. Herpes simplex and chicken pox can survive in populations of less than 1,000, and even in isolated family units (Fenner 1970:58, 64). Acute infections require much larger populations. In classic papers, Maurice Bartlett (1957) argued that for measles to become endemic in U.S. cities, 7,000 susceptibles are required in an urban population of 250,000–300,000; Black (1966:210), using evidence from island communities, suggested that a threshold exceeding 500,000 may be required for a densely settled population. Since smallpox spreads less rapidly than measles, its threshold population has been estimated at 200,000 (Fenner et al. 1988:118).

Where populations are small and dispersed, such as among tribal groups, the shortage of new susceptibles to infect means that the spread of a disease is slow and "fade-outs" are common (Cliff and Haggett 1988:245–246; Haggett 1994:10–11; Neel 1977:160). Small communities may therefore remain relatively disease-free for long periods, but their lack of exposure to infection leads to a build-up of susceptibles so that when a disease is reintroduced through contact with a larger population it is associated with a higher level of mortality that affects adults as well as children. Whether diseases can become endemic or not is significant, because, as will be elaborated below, adult losses may undermine the functioning of the group and have particularly adverse effects on demographic trends, especially where populations are small. A further characteristic of the pattern of infection among small dispersed populations is that it is highly irregular in space and time (Dobson 1989:280; Haggett 1994:12). This is because contacts are fewer and there are frequent "fade-outs," so that in an epidemic some communities are likely to escape infection. If the same disease is reintroduced within a short period its spatial impact will be significantly affected by the irregular pattern of mortality and immunity produced by the previous epidemic. Within any region of dispersed population, therefore, there are likely to be considerable differences between neighboring communities in terms of their epidemic histories and their demographic trajectories.

Interest in the influence of population size on the spread of disease has focused on the population threshold needed to maintain a pool of susceptibles, but it is increasingly recognized that the latter is dependent not only on the size of the population, but on a range of other factors, including the birth rate, migration, and the loss of immunity (Cohen 1989:49). Increasingly these variables have been incorporated into epidemic models. The birth rate has often been regarded as central to understanding both endemicity (Anderson and May 1979:366; Black 1966) and levels of recovery from epidemics (Thornton, Miller, and Warren 1991:30–39). In noncontracepting societies the birth rate is largely determined by social practices such as age at marriage, breast-feeding, child spacing, and sexual taboos (Marcy 1981:309–323). Hence a change from polygamous to monogamous practices, often associated with Christian conversion, by encouraging an increase in birth rate (Hern 1992:53–64; 1994:127, 137; Krzywicki 1934:201–202; Reid 1987:39–41) might have implications for the persistence of a disease as well as for demographic recovery. Thus while population size relates to culture in general, the size of the pool of susceptibles may vary widely from community to community according to social custom.

The Spread of Human Diseases

Human diseases are transmitted in different ways. Some are spread by direct face-to-face contact, while others are spread by a vector such as a rodent or insect. In the latter case, human location near the nonhuman reservoirs is essential for disease transmission, and, as already indicated, disease occurrence is often highly dependent on environmental conditions. The following discussion will concentrate on directly transmitted diseases.

Epidemics may break out when changes in the parasite or the environment create more favorable conditions for a pathogen's spread, but historically many were introduced from other regions. Indeed, the incidence of infection is often highly dependent on the intensity of outside contacts or the movement of peoples to new disease environments. In examining the history of epidemics, William McNeill (1976) has identified three major periods of disease outbreaks that followed significant population movements. These movements included the establishment of contact between Europe and Asia in the early Christian era, the expansion of the Mongol empire in the thirteenth century, and the beginning of European overseas expansion in the fifteenth century. Over time the improvements in methods of transport have enabled epidemic diseases to spread more quickly and widely. The development of ocean navigation and European colonial expansion from the fifteenth century is particularly significant in the history of disease, since it vastly expanded contacts and introduced infections to previously unexposed populations (Fenner 1970:65–66; 1980: 19–20; McNeill 1976:185–216; 1979:97). Colonial expansion did not only bring epidemic diseases, it also established new trading networks and patterns of migration that facilitated their spread, while it transformed native economies and societies in ways that affected their impact (Dobson 1989:287–294; Jackson 1994:161; Kunitz 1993:130, 132, 157; Newson 1985:49–66; 1993:1188, 1193–1194). Because colonial empires often sought to capture sources of labor, their efforts were concentrated where there were dense native populations; since epidemics could spread more easily among these groups, it was they who bore the initial impact of disease. Subsequently, more remote (and often smaller) populations were affected—but, as will be shown later, here the effects differed.

Epidemic models developed to explain the spread of infections have tended to calculate the probability of contacts based on population size, without considering the geographical distribution and mobility of the population and the nature and intensity of human contacts. In small, mobile, dispersed populations, contacts are generally fewer, irregular, and unsustained, and the build-up of parasites is more limited (Cohen 1989:39–41; Coimbra 1988:90–91; Polunin 1977:8–13; Wirsing 1985:311). This contrasts with large, permanent, nucleated settlements, which engender unhealthy conditions by encouraging the build-up of wastes and parasites, and also facilitate the rapid spread of directly transmitted diseases. Historically, towns and cities could not sustain their populations independent of migration from the countryside, which replenished urban losses from disease and famine (McNeill 1976:65; 1979:96). More recent historical processes involving the congregation of formerly dispersed peoples in nucleated settlements—to achieve greater administrative control or Christian conversion, or to facilitate major development projects such as dam building—have encouraged the spread of disease (Cook 1943:30–34; Curtin et al. 1978:554; Wirsing 1985:313). The significance of population concentration may be extended to the scale of individual households. In sixteenth-century Ecuador, Spanish attempts to control promiscuity by replacing extended-family or multifamily households with nuclear family residences were recognized at the time as a factor reducing the levels of disease mortality (Newson 1992:109).

It is not only population size and distribution per se that are important in understanding the spread of disease, but also the geographical location of communities and the character and intensity of contacts between them (Milner 1980:47; 1992:110–111). Rugged terrain or adverse climatic conditions are likely to hinder the spread of disease, while easy communications are likely to facilitate it (Shea 1992:160–161). Ann Ramenofsky (1990:41–42) has proposed a model relating the impact of disease to settlement types—specifically, to their location, duration (sedentary or mobile), and form (nucleated or dispersed). She hypothesizes that nucleated settlements (whether sedentary or mobile) located along primary drainage systems have the lowest probability of persistence because of more frequent and regular contacts, both between their members and with outside groups. In general, friendly contacts are likely to be more frequent and sustained. Indeed, trading contacts have generally been recognized as important channels for the spread of disease, and in the African context it has even been suggested that there might be variations in patterns of infection among ethnic and occupational groups according to their participation in trade (Hartwig 1978:25–43; Patterson and Hartwig 1978:12). In addition, where friendly relations exist, native responses to health crises may involve visiting the sick, the convening of communal gatherings, or the provision of hospitality for those fleeing from epidemics—all of which might actively promote their spread (Crosby 1976:297; Dobyns 1983:16; Krech 1978:715–716; Wirsing

1985:311). Hostile relations, on the other hand, generally discourage contacts and the spread of acute infections, even though these might be transmitted during a brief raid. In some cases uninhabited buffer zones between hostile groups, such as probably existed in the Amazon Basin and parts of North America, may have acted as effective disease barriers (De Boer 1981:365; Myers 1976; Snow and Lanphear 1988:17; Thornton, Warren, and Miller 1992:192–193).

The Impact of Epidemic Disease

Patterns of disease mortality will vary in the first instance with the particular pathogen responsible and the level of immunity acquired by a population, the latter depending in part on the size of the population and whether the disease has become endemic. Endemic diseases persist among human populations, taking a small but regular toll of infants as they are weaned and lose the immunity acquired from their mothers. (At this stage the nutritional status of an individual may play a significant role in determining susceptibility. The relationship between nutrition and suscepti-bility to disease will be considered further below.) Where diseases fail to become endemic, communities may remain disease-free for long periods, but this leads to a build-up of susceptibles, so that when an infection is reintroduced from the outside it results in high levels of mortality among adults as well as children.

Mathematically it can be shown that high death rates from infrequent epidemics have a less-significant effect on the general level of mortality and rate of population growth than do endemic infections that kill a significant proportion of children within ten years of birth (Dobson 1989:288; McKeown 1976:69). However, this generalization is not particularly helpful in predicting the overall impact of epi-demics, for it takes no account of the indirect effects of child and adult losses on the continued functioning of the society and on demographic trends, which would vary according to specific social and environmental conditions. Where losses occur among children they may retard demographic recovery, but they do not threaten the functioning or survival of the group. Adult losses, however, not only result in an im-mediate loss of reproductive capacity, but may also undermine a wide range of ac-tivities necessary for the community's maintenance and recovery (Kunitz 1984:560; McNeill 1979:96; Whitmore 1991:479–480). This pattern is most common among small populations, which commonly possess other features (to be elaborated upon below) that can increase the effect of adult losses and raise mortality levels even higher. In order to understand the impact of an epidemic it is therefore necessary to be aware of the particular individuals or social groups affected by the disease.

Adult losses may undermine, or even destroy, subsistence patterns, create food shortages or famines, and thereby enhance disease susceptibility. Falling numbers may result in decreased production, weaken reciprocal social obligations, and cause a shift in emphasis on different activities, sometimes closing off subsistence options altogether and rendering communities more vulnerable in the face of environmental perturbations (Larson, Johnson, and Michaelsen 1994:276). The devastating impact of Old World epidemics in the Americas led to the collapse of many forms of intensive production—such as irrigation canals, terracing, and raised fields—that could not be maintained with declining populations (Denevan 1992:375–381). In some cases it led to the abandonment of agriculture altogether (Balée 1992:51).

Significant differences appear to exist between state and nonstate societies in their ability to sustain food supplies and economic production during epidemics. State societies are generally more productive, have forms of storage and food distribution systems that may overcome temporary shortages, and are characterized by political organizations that can mobilize labor to maintain production in times of crisis. Even though, in general, state societies may be better able to cope with epidemics, food supplies will vary between social groups (class, race, gender, and age) according to their status and access to resources. The economies of nonstate societies, especially those dependent on wild food resources, are often highly adapted to specific environments. Food production prospects are generally dependent on whether subsistence strategies or cooperative activities can remain viable with a reduced population. In small societies the loss of only a small number of those with special skills, such as hunting, may be a serious threat to food supplies—particularly in regions of marked seasonality, or where groups are dependent on a limited range of resources (Dobyns 1983:16, 332; Hill 1989:12; Krech 1978:717; McGrath 1991). The vulnerability of these groups is often enhanced by the limited surpluses produced, and by their lack of familiarity with methods of food storage. Furthermore, their leaders may lack the authority to organize practical responses to epidemic crises that might aid survival (Stannard 1991:531; Zeitlin 1989:57–60). During the measles epidemic that afflicted the Yanomama in the 1960s, James Neel and others observed that village life collapsed completely: only a few members remained capable of providing food and water or tending the sick, and the concern for well-being seldom extended beyond the immediate family (Neel et al. 1970:427).

In assessing the impact of epidemics on food production, nutritional levels, and disease susceptibility, it is important to recognize that the nutritional status of a community is not determined solely by food supplies: it is also dependent on the members' energy requirements (Walter and Schofield 1989:17–21). These are greater in colder environments and where individuals are involved in arduous labor or are constantly fighting infection. The harsh labor conditions often introduced by colonial regimes not only affected the health of native peoples directly but also contributed to their declining nutritional status, which in many cases was already threatened by diminished food supplies resulting from new external demands on their lands, labor, and production (Newson 1985:62–66). According to Stephen Kunitz (1994:51), a significant factor explaining the greater depopulation of Hawaii and New Zealand compared to western Polynesia (Samoa and Tonga) was the more extensive expropriation of land, which destroyed native subsistence production and increased disease susceptibility. Clearly the study of variations in environmental and social conditions is essential to understanding nutritional status.

Epidemics may bring famine, starvation, and death, but in many cases the greatest impact of reduced levels of production is through declining nutritional levels that may enhance disease susceptibility (McKeown 1988:52–55). However, the direct relationship between malnutrition and infection is difficult to substantiate, since malnourished individuals are also likely to experience poor living conditions, where crowded accommodation and inadequate sanitation may favor the spread of disease. Furthermore, deficiencies in protein or in particular vitamins and minerals may actually provide some resistance to infection (Cohen 1989:167); for example, many disease organisms need iron to thrive, so that short-term iron deficiency may assist the body to fight infection. In general, the relationship between disease and nutrition is now thought to be less clear, and certainly more complex, than previously envisaged.

In regard to disease mortality, the link with malnutrition appears stronger for some diseases than for others (Cohen 1989:167; Livi-Bacci 1991:35–39; McKeown 1988:52–53; Rotberg and Rabb 1983:305–308). In the case of measles and most respiratory and intestinal infections, levels of morbidity and mortality appear to increase with poor nutrition; whereas smallpox, plague, yellow fever, and malaria seem to be relatively unaffected by nutritional status, and have in the past killed both affluent and poor (McKeown 1988:52). Also, nutritional status is probably insignificant in the case of particularly virulent strains, and where individuals have not been previously exposed to the disease. However, where microorganisms become endemic, malnutrition may exert a greater influence on disease mortality, particularly infant and child mortality (Harpending, Draper, and Pennington 1990:257–258). Class and gender differences also may be of significance at this stage.

The importance of local conditions for understanding the impact of epidemic disease becomes particularly evident when social reactions are considered. Epidemics may result in social disorganization, as marriages break down due to death or flight and as political authority and religious beliefs are questioned for their failure to explain or cope with disaster (McGrath 1991:417; Stannard 1991:531; Zubrow 1990:761). These processes of social disintegration may lead to the biological or cultural extinction of the group, but they may also result in the emergence of new forms of social organization that may aid recovery. The particular demographic path followed by a group is closely linked to existing social practices and demographic regimes, while the degree of any recovery achieved will affect the size of the population, the numbers of susceptibles, and the future pattern of disease spread.

Epidemics that result in high adult mortality result in greater social disorganization. However, the impact of the loss of equal numbers of adults will differ in different societies. Adult losses are more difficult to sustain in small communities, where the formation of new unions may be more difficult to achieve because the availability of partners is by definition more limited and sex ratios are often more volatile (Early and Peters 1990:137, 140; Kunitz 1994:9). The problem of unbalanced sex ratios may be aggravated in the case of particular diseases—such as smallpox, influenza, malaria, and dysentery—that cause high levels of mortality among pregnant women (McFalls and McFalls 1984:60–61, 130, 533–534; Stannard 1990:336–347). The ability to form new unions will also depend on marriage practices and social structures. Cultural restrictions on the suitability of spouses and on remarriage, including restrictions on crossing class boundaries, may constitute significant obstacles to the formation of new unions and will thus limit reproductive capacity. In preindustrial societies where high fertility levels are required in order to maintain the population in the face of high infant mortality and low life expectancy, even a small reduction in the fertility rate can be significant (Harvey 1967:195–196). In small societies it may limit any demographic recovery.

In conditions of declining numbers and limited marriage pools, in order to ensure biological survival groups may have to modify their population policies and social attitudes. As Charles Wagley (1951) showed for two Tupi-Guaranian groups, even within the same cultural-linguistic group differences in social practice may mean that some populations are preconditioned to cope more effectively with epidemic stress. Groups with rigid social practices that function to control population numbers may be significantly disadvantaged in demographic crises, whereas those that are more flexible in their population policies or in their attitudes toward outsiders may be able to survive. Francis Johnston and his coworkers (Johnston et al. 1969:33) observed that the Peruvian Cashinahua were able to recover from a severe epidemic in 1951 by abandoning birth control practices. It is also noteworthy that many surviving indigenous societies are exogamous and in past periods have been prepared to absorb or be absorbed by outsiders (Dobyns 1983:306, 310–311; Mc-Grath 1991:414; Milner 1980:47; Thornton 1986:128–129). Why some groups opt for change and others do not is unclear. Henry Dobyns (1983:303–306, 310–311) has suggested that amalgamation with other cultural groups occurs where native populations have fallen below their conception of the ideal size of a community. In other cases, decisions appear to be more pragmatic. For example, Black (1975:516) notes how two small societies in northern Brazil joined the Tiriyó when faced with the problem of finding marital partners not forbidden by incest taboos. Anthony F. C. Wallace (1956:269–272) doubts that such decisions emerged from community deliberations; rather, he thinks they were directed by individuals—particularly native leaders and shamans, perhaps guided by vision experiences brought on by physical and psychological stress. In the case of the Cashinahua noted above it was the shaman who proscribed contraceptives and abortifacients (Johnston et al. 1969:33). Just as the spread of epidemics may be highly variable in time and space, so social practice and action may enhance that variability, resulting in some communities' becoming extinct while their neighbors survive.

Social structures and belief systems can also influence a group's response to an epidemic crisis. Individuals' ability to cope with stress, which in itself can enhance disease susceptibility (Cassell 1976:107–123), may be affected by the actions of their leaders and by the way the epidemic is explained. As previously noted, leaders may play critical roles in shaping community reactions, instilling a "will to survive," and organizing practical responses to epidemic crises, such as ensuring continued production, introducing public health measures, or providing social support (McGrath 1991:417; Stannard 1991:531; Zeitlin 1989:57–60). Even though the medical treatments employed may be of little intrinsic value, the provision of basic needs is clearly essential to survival; and it is less commonly recognized that nursing care can reduce mortality levels significantly (Carmichael 1983:59–60; Crosby 1976:294; McCaa 1995:420–422). Indeed, Mark Cohen (1989:39) argues that one of the few advantages of sedentism in regard to the impact of disease is the ability to care for the sick.

The limited political power exercised in nonstate societies might render them less able to cope with epidemic crises than those with strong leadership. Any authority the leaders may possess could be weakened further by their apparent inability to control an epidemic, especially where leadership has a functional rather than a hereditary basis. Yet while, in general, state societies may be better able to cope with disasters (and it might even be suggested that highly centralized polities would be most effective in crisis situations), when leaders succumb to the epidemic or the political organization fails to contain the crisis, dissension over policies and procedures can lead to the rejection of authority—leading to social disorder, and sometimes violent conflict. This was the case in the Andes between 1524 and 1527 when the death of the Inca ruler, Huayna Capac, probably from smallpox, precipitated civil wars (Newson 1992:88). Conflicts between groups and within groups may not only enhance mortality levels directly, but by disrupting subsistence production and social functioning may aggravate the impact of epidemics (McGrath 1991:412, 417–418; Milner 1992:111).

A successful response depends not only on effective leadership but also on a correct identification of the cause of the epidemic and the means of its transmission. Quarantining might effectively isolate a disease spread by face-to-face contact, such as smallpox and measles, but it would have little effect on reducing the incidence of a disease spread by a vector, such as malaria or typhus; in the latter case, changes to environmental conditions would be more appropriate (McGrath 1991: 410–417). Only within the last two hundred years have the causes of most diseases been accurately identified. Previous success, particularly with unfamiliar infections, was largely a matter of chance, trial and error, and perceived effectiveness, without an understanding of the process of transmission or the biology of parasites (Boyden 1970:197–200; 1992:207–208). Where the cause of an epidemic is not understood it may be ascribed to a supernatural agent. This may result in a fatalistic attitude toward death, or in the placing of blame on a particular individual, group, or class, which may lead to social conflict (Jenkins et al. 1989:37, 44–45; McGrath 1991:411, 417). Increased intergroup warfare following epidemics has been commonly noted in Amazonia, where sickness and death are often attributed to sorcery—which requires revenge (Early and Peters 1990:80; Ferguson 1990:241–242; Hill 1989:12–13; Newson 1995:318). Misidentification of the cause of a sickness and inappropriate responses may thus enhance the levels of mortality that can be attributed directly to disease.

Although specific environmental and social contexts are important in understanding the effect of epidemic disease on individual communities, the preceding discussion suggests that a disease's incidence and impact are significantly affected by the size of the population and its political economy. In state societies, the spread of disease may be facilitated by the relatively large size of the population and its nucleated settlement patterns; but in the longer term, the effect of epidemics on these societies may be moderated as the relatively large size of their populations enables diseases to become endemic and ensures an adequate supply of spouses to maintain the fertility rate. Furthermore, their production systems and sociopolitical organizations generally enable them to cope more effectively with epidemic crises and achieve a level of demographic recovery. Conversely, a settlement pattern of small dispersed populations may retard the spread of diseases and moderate their direct impact; but in these circumstances diseases fail to become endemic, so that when infections are reintroduced from outside they cause high levels of adult mortality that may have particularly adverse effects on fertility levels and food production. Furthermore, such groups often lack the strong leadership needed to foster a "will to survive" and mobilize community efforts to cope with disaster. Thus while disease mortality might be moderated (though this would depend on the frequency of reinfection), the indirect impact of epidemics might threaten any demographic recovery and raise mortality rates to levels greater than those suffered by state societies.

These broad differences in the pattern of infection may partially account for the continued decline of native peoples in lowland areas of Latin America during the colonial period, while most former chiefdom and state societies experienced a degree of demographic recovery, the timing of which varied from region to region but began in some areas in the early seventeenth century (Newson 1985:43–45). This contrast in demographic trends was characteristic of colonial Ecuador, where there are suggestions that in the highlands some diseases were becoming endemic in the early seventeenth century when the Indian population began to recover (Alchon 1991:57–58, 76–77), whereas in the eastern lowlands and on the Pacific coast epidemics and population decline continued throughout the colonial period. Although it may be hypothesized that such differences in demographic trends may be related to differences in the pattern of infection, it is difficult to generalize about differences in their aggregate effects—for clearly the impact of epidemics, particularly on small populations, would depend, among other things, on the frequency of reinfection.

This chapter has focused on epidemics, which historically have exerted a powerful influence on demographic trends. Nevertheless, even in these circumstances disease mortality can provide only a partial explanation for demographic changes, which also reflect changes in fertility rates as well as the effect of other factors on mortality levels. Epidemics have often occurred during periods of economic crisis and political change that in many cases have been characterized by conflict, major social upheavals, economic hardship, the questioning of belief systems, and even environmental change. These processes, perhaps most evident in the expansion of colonial rule, may not only affect demographic trends directly, but also, through interacting with biological processes, may influence patterns of infection and mortality. The coincidence of epidemics with periods of change is often critical in understanding their demographic impact (Kunitz 1993:135; 1994:13; Patterson and Hartwig 1978:10–13). The complexity of the interaction is such that one cannot be understood without the other.

Conclusion

Historical studies of epidemics often assume that, once introduced, diseases have spread unhindered, and that their impact has been uniform. In this chapter I have stressed that the origins, spread, and effect of epidemic disease cannot be understood without acknowledging the significance of differences in the character of the parasite, the host, and the physical and social environment in which they interact over time. In particular, I have shown how the impact of epidemic disease is influenced by the size, distribution, and character of human populations—especially their settlement patterns, subsistence systems, sociopolitical organization, and ideology. This means that the effect of even a single disease is likely to be highly variable in time and space. Although I have here stressed the importance of particular environmental and social circumstances, my aim has not been to argue for a relativistic approach, but rather to demonstrate the value of adopting a holistic framework that acknowledges human variability and the dynamic quality of interactions. In essence, I am advocating a fuller recognition of the complexity of the processes at work—for it is only from this perspective that the highly variable temporal and spatial patterns in the spread and impact of epidemic disease, which are unexplained by simple deterministic models, can be understood, and that long-term correspondences can be identified.

References

Alchon, S. A. 1991. *Native Society and Disease in Colonial Ecuador*. Cambridge: Cambridge University Press.

Allison, A. C. 1982. Coevolution between hosts and infectious disease agents and its effects on virulence. Pp. 245–267 in Anderson and May (eds.), *Population Biology of Infectious Diseases*.

Anderson, R. M. and R. M. May. 1979. Population biology of infectious diseases: Part I. *Nature* 280:361–367.

———. 1991. *Infectious Diseases of Humans: Dynamic and Control*. Oxford: Oxford University Press.

———, eds. 1982. *Population Biology of Infectious Diseases*. Heidelberg: Springer-Verlag.

Appleby, A. B. 1980. The disappearance of plague: A continuing puzzle. *Economic History Review* 33(2): 161–173.

Balée, W. 1992. People of the fallow: A historical ecology of foraging in lowland South America. Pp. 35–57 in K. H. Redford and C. Padoch (eds.), *Conservation of Neotropical Forests: Working from Traditional Resource Use*. New York: Columbia University Press.

Bartlett, M. S. 1957. Measles periodicity and community size. *Journal of the Royal Statistical Society*, ser. A, 120:48–70.

Black, F. L. 1966. Measles endemicity in insular populations: Critical community size and its evolutionary implications. *Journal of Theoretical Biology* 11:207–211.

———. 1975. Infectious diseases in primitive societies. *Science* 187:515–518.

———. 1980. Modern isolated pre-agricultural populations as a source of information on prehistoric epidemic patterns. Pp. 37–54 in Stanley and Joske (eds.), *Changing Disease Patterns and Human Behaviour*.

————. 1990. Infectious disease and evolution of human populations: The example of South American forest tribes. Pp. 57–74 in Swedlund and Armelagos (eds.), *Disease in Populations in Transition*.

————. 1992a. Low polymorphism in the immune system puts New World populations at risk from variant pathogens. Unpublished MS.

————. 1992b. Why did they die? *Science* 258:1739–1740.

Bodmer, W. F. and L. L. Cavalli-Sforza. 1976. *Genetics, Evolution, and Man*. San Francisco: Freeman.

Boyden, S. 1970. Cultural adaptation to biological maladjustment. Pp. 190–218 in S. Boyden (ed.), *The Impact of Civilisation on the Biology of Man*. Canberra: Australian National University Press.

————. 1992. *Biohistory: The Interplay Between Human Society and the Biosphere: Past and Present*. Paris: UNESCO and Parthenon.

Carmichael, A. G. 1983. Infection, hidden hunger, and history. Pp. 51–66 in Rotberg and Rabb (eds.), *Hunger and History*.

Cassell, J. 1976. The contribution of the social environment to host resistance. *American Journal of Epidemiology* 104(2): 107–123.

Cliff, A. and P. Haggett. 1988. *Atlas of Disease Distributions*. Oxford: Blackwell.

Cliff, A. D., P. Haggett, J. K. Ord, and G. R. Versey. 1981. *Spatial Diffusion: An Historical Geography of Epidemics in an Island Community*. Cambridge: Cambridge University Press.

Cockburn, A. T. 1971. Infectious diseases in ancient populations. *Current Anthropology* 12:45–62.

Cohen, M. N. 1989. *Health and the Rise of Civilization*. New Haven: Yale University Press.

Coimbra, C. E. A. Jr. 1988. Human settlements, demographic patterns, and epidemiology in lowland Amazonia: The case of Chagas's disease. *American Anthropologist* 90:82–97.

Cook, S. F. 1943. *The Conflict Between the Californian Indian and White Civilization: 1. The Indian Versus the Spanish Mission*. Ibero-Americana, vol. 21. Berkeley and Los Angeles: University of California Press.

Cronon, W. 1983. *Changes in the Land: Indians, Colonists, and the Ecology of New England*. New York: Hill and Wang.

Crosby, A. W. 1976. Virgin soil epidemics as a factor in the aboriginal depopulation in America. *William and Mary Quarterly* 33:289–299.

Curtin, P., S. Feierman, L. Thompson, and J. Vansina. 1978. *African History*. Harlow: Longman.

Dean, J. S., R. C. Euler, G. J. Gumerman, F. Plog, R. H. Hevly, and T. N. V. Karlstrom. 1985. Human behavior, demography, and paleoenvironment on the Colorado plateaus. *American Antiquity* 50(3): 537–554.

De Boer, W. 1981. Buffer zones in the cultural ecology of aboriginal Amazonia: An ethnohistorical approach. *American Antiquity* 46:364–377.

Demeritt, D. 1994. Ecology, objectivity, and critique in writings on nature and human societies. *Journal of Historical Geography* 20(1): 22–37.

Denevan, W. M. 1992. The pristine myth: The landscape of the Americas in 1492. *Annals of the Association of American Geographers* 82(3): 369–385.

Dobson, M. J. 1989. Mortality gradients and disease exchanges: Comparisons from Old England and colonial America. *Social History of Medicine* 2: 259-295.

Dobyns, H. F. 1966. Estimating aboriginal American population: An appraisal of techniques with a new hemispheric estimate. *Current Anthropology* 7:395–416.

————. 1983. *Their Number Become Thinned: Native American Population Dynamics in Eastern North America*. Knoxville: University of Tennessee Press.

Dunn, F. L. 1972. Epidemiological factors: Health and disease in hunter-gatherers. Pp. 221–228 in R. B. Lee and I. DeVore (eds.), *Man the Hunter*. Chicago: Aldine-Atherton.

———. 1993. Malaria. Pp. 855–862 in K. F. Kiple (ed.), *The Cambridge World History of Human Disease*. Cambridge: Cambridge University Press.

Early, J. D. and J. F. Peters. 1990. *The Population Dynamics of the Mucajai Yanomama*. New York: Academic Press.

Fenner, F. L. 1970. The effects of changing social organization on the infectious diseases of man. Pp. 48–76 in Boyden (ed.), *The Impact of Civilisation on the Biology of Man*.

———. 1980. Sociocultural change and environmental diseases. Pp. 7–26 in Stanley and Joske (eds.), *Changing Disease Patterns and Human Behaviour*.

———. 1982. Transmission cycles and broad patterns of observed epidemiological behaviour in human and other animals. Pp. 103–119 in Anderson and May (eds.), *Population Biology of Infectious Diseases*.

Fenner, F. L., D. A. Henderson, I. Arita, Z. Jezek, and I. D. Ladnyi. 1988. *Smallpox and Its Eradication*. Geneva: World Health Organization.

Ferguson, R. B. 1990. Blood of the Leviathan: Western contact and warfare in Amazonia. *American Ethnologist* 17:237–257.

Garruto, R. M. 1981. Disease patterns of isolated groups. Pp. 557–597 in H. Rothschild (ed.), *Biocultural Aspects of Disease*. New York: Academic Press.

Gleick, J. 1987. *Chaos*. London: Heinemann.

Grove, D. L. 1980. Schistosomes, snails, and man. Pp. 187–204 in Stanley and Joske (eds.), *Changing Disease Patterns and Human Behaviour*.

Haggett, P. 1994. Prediction and predictability in geographic systems. *Transactions of the Institute of British Geographers*, n.s., 19:6–20.

Harpending, H. C., P. Draper, and R. Pennington. 1990. Cultural evolution, parental care, and mortality. Pp. 251–265 in Swedlund and Armelagos (eds.), *Disease in Populations in Transition*.

Hartwig, G. W. 1978. Social consequences of epidemic diseases: The nineteenth century in eastern Africa. Pp. 25–45 in G. W. Hartwig and K. D. Patterson (eds.), *Disease in African History*. Durham, N.C.: Duke University Press.

Harvey, H. R. 1967. Population of the Cahuilla Indians: Decline and causes. *Eugenics Quarterly* 14:185–198.

Hern, W. M. 1992. Polygyny and fertility among the Shipibo of the Peruvian Amazon. *Population Studies* 46(1): 53–64.

———. 1994. Health and demography of native Amazonians: Historical perspective and current status. Pp. 123–149 in A. Roosevelt (ed.), *Amazonian Indians from Prehistory to the Present: Anthropological Perspectives*. Tucson: University of Arizona Press.

Hill, J. D. 1989. Ritual production of environmental history among the Arawakan Wakuénai of Venezuela. *Human Ecology* 17(1): 1–25.

Jackson, R. H. 1994. *Indian Population Decline: The Missions of Northwestern New Spain, 1687–1840*. Albuquerque: University of New Mexico Press.

Jacobs, W. R. 1974. The tip of the iceberg: Pre-Columbian Indian demography and some implications for revisionism. *William and Mary Quarterly* 31:123–132.

Jenkins, C., M. Dimitrakakis, I. Cook, R. Sanders, and N. Stallman. 1989. Culture change and epidemiological patterns among the Hagahai, Papua New Guinea. *Human Ecology* 17(1): 27–57.

Johnston, F. E., K. M. Kensinger, R. L. Jantz, and G. F. Walker. 1969. The population structure of the Peruvian Cashinahua: Demographic, genetic, and cultural relationships. *Human Biology* 41:29–41.

Krech, S. III. 1978. Disease, starvation, and northern Athapaskan social organization. *American Ethnologist* 5:710–732.

Krzywicki, L. 1934. *Primitive Society and Its Vital Statistics*. London: Macmillan.

Kunitz, S. J. 1984. Mortality change in America, 1620–1920. *Human Biology* 56(3): 559–582.

———. 1993. Historical and contemporary mortality patterns in Polynesia. Pp. 125–166 in Mascie-Taylor (ed.), *The Anthropology of Disease*.

———. 1994. *Disease and Social Diversity: The European Impact on the Health of Non-Europeans*. Oxford: Oxford University Press.

Larson, D. O, J. R. Johnson, and J. C. Michaclscn. 1994. Missionization among the coastal Chumash of central California: A study in risk minimization strategies. *American Anthropologist* 96(2): 263–299.

Levin, B. R., A. C. Allison, H. J. Bremermann, L. L. Cavalli-Sforza, S. A. Levin, R. M. May, and H. R. Thieme. 1982. Evolution of parasites and hosts. Pp. 213–243 in Anderson and May (eds.), *Population Biology of Infectious Diseases*.

Livi-Bacci, M. 1991. *Population and Nutrition: An Essay on European Demographic History*. Cambridge: Cambridge University Press.

McCaa, R. 1995. Spanish and Nahuatl views on smallpox and demographic catastrophe in Mexico. *Journal of Interdisciplinary History* 25(3): 397–431.

McFalls, J. A. and M. H. McFalls. 1984. *Disease and Fertility*. New York: Academic Press.

McGrath, J. W. 1991. Biological impact of social disruption resulting from epidemic disease. *American Journal of Physical Anthropology* 84:407–419.

McKeown, T. 1976. *The Modern Rise of Population*. London: Edward Arnold.

———. 1988. *The Origins of Human Disease*. Oxford: Basil Blackwell.

McNeill, W. M. 1976. *Plagues and Peoples*. Oxford: Oxford University Press.

———. 1979. Historical patterns of migration. *Current Anthropology* 20(1): 95–98.

Marcy, P. T. 1981. Factors affecting the fecundity and fertility of historical populations. *Journal of Family History* 6:309–326.

Mascie-Taylor, C. G. N. 1993. The biological anthropology of disease. Pp. 1–25 in C. G. N. Mascie-Taylor (ed.), *The Anthropology of Disease*. Oxford: Oxford University Press.

Milner, G. R. 1980. Epidemic disease in the postcontact Southeast. *Mid-Continental Journal of Archaeology* 5(1): 39–56.

———. 1992. Disease and sociopolitical systems in late prehistoric Illinois. Pp. 103–116 in J. W. Verano and D. H. Ubelaker (eds.), *Disease and Demography in the Americas*. Washington, D.C.: Smithsonian Institution Press.

Myers, T. P. 1976. Defended territories and no-man's-lands. *American Anthropologist* 78:354–355.

Neel, J. V. 1977. Health and disease in unacculturated Amerindian populations. Pp. 155–177 in K. Elliott and J. Whelan (eds.), *Health and Disease in Tribal Societies*. CIBA Foundation Symposium 49. Amsterdam: Elsevier.

Neel, J. V., W. R. Centerwall, N. A. Chagnon, and H. L. Casey. 1970. Notes on the effects of measles and measles vaccine in a virgin-soil population of South American Indians. *American Journal of Epidemiology* 91:418–429.

Newson, L. A. 1985. Indian population patterns in colonial Spanish America. *Latin American Research Review* 20(3): 41–74.

———. 1992. Old World epidemics in early colonial Ecuador. Pp. 84–112 in N. D. Cook and W. G. Lovell (eds.), *Secret Judgments of God: Old World Disease in Colonial Spanish America*. Norman: University of Oklahoma Press.

———. 1993. Highland-lowland contrasts in the impact of Old World diseases in early colonial Ecuador. *Social Science and Medicine* 36(9): 1187–1195.

———. 1995. *Life and Death in Early Colonial Ecuador*. Norman: University of Oklahoma Press.

Olsen, L. F. and Schaffer, W. M. 1990. Chaos versus noisy periodicity: Alternative hypotheses for childhood epidemics. *Science* 249:499–504.

Patterson, K. D. and G. W. Hartwig 1978. The disease factor: An overview. Pp. 3–24 in G. W. Hartwig and K. D. Patterson (eds.), *Disease in African History*. Durham, N.C.: Duke University Press.

Pimental, D. 1968. Population regulation and genetic feedback. *Science* 159:1432–1437.

Polunin, I. 1977. Some characteristics of tribal peoples. Pp. 5–20 in K. Elliott and J. Whelan (eds.), *Health and Disease in Tribal Societies*. CIBA Foundation Symposium 49. Amsterdam: Elsevier.

Ramenofsky, A. F. 1987. *Vectors of Death: The Archaeology of European Contact*. Albuquerque: University of New Mexico Press.

———. 1990. Loss of innocence: Explanations of differential persistence in the sixteenth-century southeast. Pp. 31–48 in D. H. Thomas (ed.), *Columbian Consequences*, vol. 2. Washington, D.C.: Smithsonian Institution Press.

Reid, A. 1987. Low population growth and its causes in pre-colonial Southeast Asia. Pp. 33–47 in N. G. Owen (ed.), *Death and Disease in Southeast Asia: Explorations in Social, Medical, and Demographic History*. Oxford: Oxford University Press.

Rotberg, R. I. and T. K. Rabb, eds. 1983. *Hunger and History*. Cambridge: Cambridge University Press.

Sattenspiel, L. 1990. Modeling the spread of infectious disease in human populations. *Yearbook of Physical Anthropology* 33:245–276.

Schaffer, W. M. 1987. Chaos in ecology and epidemiology. Pp. 233–248 in H. Degn, A. V. Holden, and L. F. Olsen (eds.), *Chaos in Biological Systems*. New York: Plenum.

Shea, D. E. 1992. A defense of small population estimates for the central Andes in 1520. Pp. 157–180 in W. M. Denevan (ed.), *The Native Population of the Americas in 1492*. 2d ed. Madison: University of Wisconsin Press.

Slack, P. 1981. The disappearance of plague: An alternative view. *Economic History Review* 34(3): 469–476.

Snow, D. R. and K. M. Lanphear. 1988. European contact and Indian depopulation in the Northeast: The timing of the first epidemics. *Ethnohistory* 35(1): 15–33.

Stanley, N. F. and R. A. Joske, eds. 1980. *Changing Disease Patterns and Human Behaviour*. London: Academic Press.

Stannard, D. E. 1989. *Before the Horror: The Population of Hawaii on the Eve of Western Contact*. Honolulu: University of Hawaii.

———. 1990. Disease and fertility: A new look at the demographic collapse of native populations in the wake of Western contact. *Journal of American Studies* 24(3): 325–350.

———. 1991. The consequences of contact: Toward an interdisciplinary theory of native responses to biological and cultural invasion. Pp. 519–539 in D. H. Thomas (ed.), *Columbian Consequences*, vol. 3. Washington, D.C.: Smithsonian Institution Press.

Svanborg-Eden, C. and B. R. Levin. 1990. Infectious disease and natural selection in human populations: A critical reexamination. Pp. 31–46 in Swedlund and Armelagos (eds.), *Disease in Populations in Transition*.

Swedlund, A. C. and G. J. Armelagos, eds. 1990. *Disease in Populations in Transition: Anthropological and Epidemiological Perspectives*. New York: Bergin and Garvey.

Thornton, R. 1986. History, structure, and survival: A comparison of the Yuki (Ukomno'm) and Tolowa (Hush) Indians of northern California. *Ethnology* 25:119–130.

Thornton, R., T. Miller, and J. Warren. 1991. American Indian population recovery following smallpox epidemics. *American Anthropologist* 93(3): 28–45.

Thornton, R., J. Warren, and T. Miller. 1992. Depopulation in the Southeast after 1492. Pp. 187–196 in J. W. and D. H. Ubelaker (eds.), *Disease and Demography in the Americas*. Washington, D.C.: Smithsonian Institution Press.

Upham, S. 1986. Smallpox and climate in the American southwest. *American Anthropologist* 88:115–128.

Wagley, C. 1951. Cultural influences on population. *Revista do Museu Paulista* 5:95–104.

Wallace, A. F. C. 1956. Revitalization movements. *American Anthropologist* 58:264–281.

Walter, J. and R. Schofield. 1989. Famine, disease, and crisis mortality in early modern society. Pp. 1–73 in J. Walter and R. Schofield (eds.), *Famine, Disease, and the Social Order in Early Modern Society*. Cambridge: Cambridge University Press.

Whitmore, T. M. 1991. A simulation of the sixteenth-century population collapse in the basin of Mexico. *Annals of the Association of American Geographers* 81(3): 464–487.

Wiesenfeld, S. L. 1967. Sickle-cell trait in human biological and cultural evolution. *Science* 157:1134–1140.

Wirsing, R. L. 1985. The health of traditional societies and the effects of acculturation. *Current Anthropology* 26(3): 304–322.

Worster, D. 1990. The ecology of order and chaos. *Environmental History Review* 14(1–2): 1–18.

Zeitlin, J. F. 1989. Ranchers and Indians on the southern isthmus of Tehuantepec: Economic change and indigenous survival in colonial Mexico. *Hispanic American Historical Review* 69(1): 23–60.

Zubrow, E. 1990. The depopulation of native America. *Antiquity* 64:754–765.

CHAPTER 4

Forged in Fire: History, Land, and Anthropogenic Fire

STEPHEN J. PYNE

Quest for Fire: The Competition for Combustion

The capture of fire by the genus *Homo* marks a divide in the natural history of the earth. Fire of course long predated hominids, and many organisms display adaptations, often highly specific, to fire. But from the time of *Homo erectus* one species acquired the capacity to start and stop fires, a niche filled by no other organism. A uniquely fire creature became bonded to a uniquely fire planet.[1]

The earth would burn. Anthropogenic fire would have to "compete" with other ignition sources, other means of consuming biomass, other ecological cycles of chemicals, and other successions of species and habitats. With or without people, the earth will burn, though its human firebrands preferentially consume biomass by means of combustion and burn it in particular ways. They use their fire power to re-shape the planet, to render it more suitable to their needs. In effect, humans began to cook the earth. They reworked landscapes in their ecological forges, assisted by the anvils and hammers of their technological repertoires. Anthropogenic fire was different from other regimens of combustion. Anthropogenic environmental change dates from that moment.

Fire, Fuel, and First Principles: A Primer on Fire Ecology

Fire is a creation of life: terrestrial life provides the fuel, and life everywhere furnishes the oxygen required for combustion. While ignition comes from various sources, all except lightning were trivial, at least until the Pleistocene. Evidence of fire is apparent in all the coal-bearing strata of the geologic record. For eons, then, fire has shaped ecosystems, informed evolutionary trends, and shared in the complexity of living systems.

The patterning of fire—the fire regime—resembles a two-cycle engine, oscillating between wet and dry conditions, with lightning supplying the spark. There has to be enough moisture to produce fuel, and enough dormancy or drought to ready that fuel for burning. This wet-dry cycle can crack open a biota the way a frost-thaw cycle can split open rock. In many environments this informing cycle occurs annually as a well-recognized fire season; in other places it appears over the course of decades or even centuries, as climatic tides of wet and dry weather wash over the biota. In desert regions, outbursts of exceptional rains create the fuels that otherwise do not exist in sufficient mass for fire. In rain forests, periods of exceptional drought prepare ever-abundant biomass for burning. The sharper the contrasts between wet and dry—the more intense the gradient, and the faster the transition—the more vigorous the fire regime. Hence Mediterranean climates, monsoonal forests, and landscapes subject to El Niño–Southern Oscillation–style bouts of drought and deluge are particularly prone to fire. But fire can be found nearly everywhere, and it appears more profusely during times of rapid and extreme climatic change, or when, through their technologies, humans can force conversions of equal magnitude.

Adaptations to fire are many, complex, and subtle. Rarely are they fire-specific or unique to a particular type of fire. More typically, organisms show a suite of traits that adapt to a suite of conditions; and they adapt not to fire in the abstract, but to a particular regimen of fire. Thus drought, grazing, and fire, for example, form a complex of pressures that result in similar kinds of traits—such as storing a larger proportion of biomass in root stocks, or the capacity to resprout from top-killed stems. The ability to adapt to disturbances of many kinds also helps an organism adapt to fire.

What matters is not whether fire exists or not—the prospect of fire exclusion is an illusion of industrial societies—but the character of the fire regime. A change in regimes will cause a biotic shift within an ecosystem, favoring some organisms over others. Although fire is present in both instances, the changes in intensity, frequency, seasonal timing, size, and so on will affect an ecosystem differently. The choice is not between adapting to fire or not, but rather between adapting to various fire regimes. Most of the planet's terrestrial biotas—excepting perhaps tundra, stony deserts, and spring-fed grottos—have in fact adapted to anthropogenic fire, or have had their biotic ore smelted and hammered in its forge.

Firepower: Fire Creature, Fire Planet

The human capture of fire was thus a profound event for the earth. It forced biotas to adjust to new fire regimes and new fuel complexes as these were shaped, deliberately or accidentally, by human societies. Equally, it forced humans to compete against natural sources of fire. If fire is the primary technology for rendering their world habitable, then it is essential that humans, not lightning, control the fire regime. It is no less imperative that humanity not allow any other creature to possess fire. None does, and it is unthinkable that humans would allow any other species to break the monopoly.

The exercise of this power takes many forms. The most obvious expression is through direct control over ignition itself. Humans can start fires at will, although the environment will not always accept them; and they can stop fires, although this is an even more troublesome task. Historically, the most common way to contain wildfire is to replace it with domesticated fire, to base fire control upon fire use— which is to say, that fire is best controlled through controlled burning.

Only with industrialization have societies attempted outright fire suppression— in effect, substituting for open burning the controlled combustion of fossil fuels embedded in firefighting technologies. In most regions this strategy, by itself, cannot be sustained. The long-term effect is to substitute infrequent, large fires for frequent, low-intensity fires. This shift in fire regimes has serious consequences for fuel complexes (and wildfire) as well as for biodiversity. Figure 4.1 shows the impact of suppression on the distribution of fires, as manifest in northern Ontario. Figure 4.2 documents the shift in fire regimes in the southern Sierra Nevadas of California and in northern Arizona as a result of attempted fire exclusion.

No less importantly, human societies can also influence a fire environment in indirect ways by restructuring its fuels. For this purpose fire is also a vital catalyst, making possible many other mechanical and biological technologies, from smelting

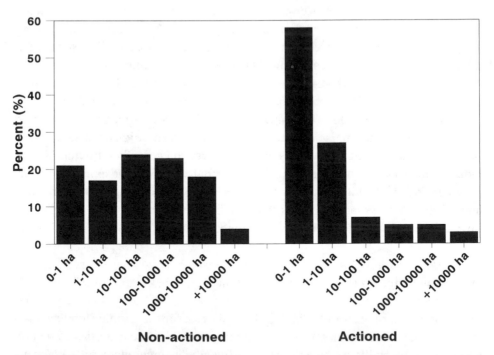

Figure. 4.1 The Effects of Fire Suppression as Revealed by the Size of Fires for Which Fire Control Is Attempted ("Actioned") or Not ("Non-actioned"). The data derive from the removal of firefighting forces from northern Ontario in the years 1976–1988. With suppression the average size of fires decreases, not only because early attack is effective but because smoldering fires no longer reside on the land and experience repeated runs. (Data from Stocks 1991.)

Figure. 4.2 *The Ecology of Fire Removal.* The two photographs, identically positioned but taken seventy years apart (*top*: 1890; *bottom:* 1960), document the changes in forest structure from fire exclusion. The indigenes routinely burned the forest floor; European settlers removed such fires and swatted out any that occurred in the mistaken belief that they could thereby protect the sequoia groves. (Source: Yosemite National Park.)

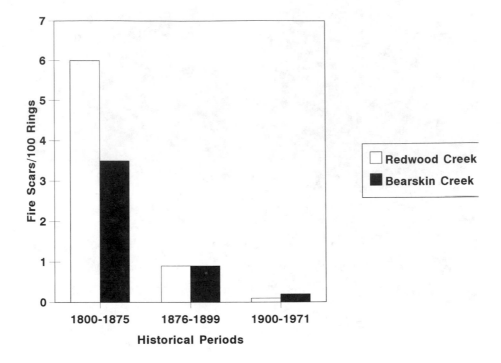

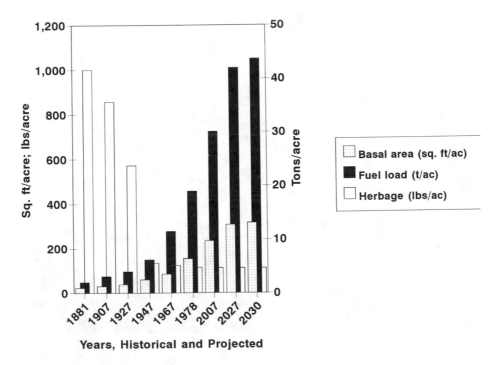

Figure. 4.2 *(continued)* The two graphs show the same shift in fire regimes. The *top graph* traces the reduction in fire-scarring with the reservation of sites as public land. (Redrawn from Kilgore and Taylor 1979.) The *bottom graph,* from ponderosa pine forests in northern Arizona, calculates the actual and projected changes in grasses, trees, and fuel loads. (Data from Covington and Moore 1992.)

to grazing. Farms, cities, metallurgy, and domestic herds may indirectly compete with lightning fire for a landscape's available biomass, and thus shape the fire regime. Until very recently most of these technologies required fire, so even in these secondary forms anthropogenic fire practices still competed with natural fire. Anthropogenic fire is the quintessential interactive technology. Almost never does it occur by itself; it almost always occurs in association with other practices, each leveraging the power of the other.

Figure 4.3 sketches the boundary conditions for anthropogenic fire as a "competitor" within the earth's combustion environment. Initially humans competed with lightning for the available biomass, and the amount of burning was limited by the source—the fuels—accessible in common to both. Increasingly, however, anthropogenic fire is being forced into competition with another human combustion technology, namely, the burning of fossil fuels. In this case anthropogenic fire competes not over a common source, biomass, but over a common sink, the atmosphere—that is, the amount of emissions that regional or global airsheds can absorb without major distortion. In neither case is the combustion frontier well understood.

Tending the Fire: Understanding Anthropogenic Fire

The fire-mediated relationship between humans and the earth is fantastically complex. Not only can humans, within limits, control ignition and manipulate the available fuels, but they exercise even greater range through fire's power as a catalyst—

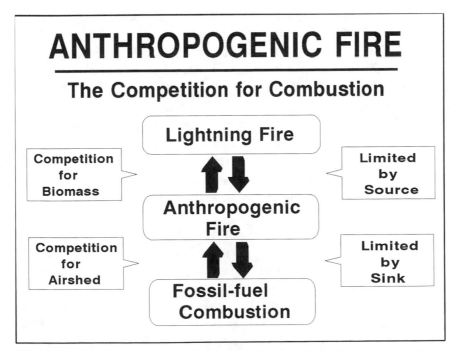

Fig. 4.3 *The Competition for Combustion.* A simplified schema for identifying the major boundaries of anthropogenic fire.

an enabling device for hunting, foraging, farming, pastoralism, heat engines, and fire-dependent technologies from fire-tempered spears and stones to ceramics and metallurgy. Accordingly, the wholesale anthropogenic modification of the biosphere did not begin with the industrial revolution or with the Neolithic revolution, but with the hominid revolution announced with Promethean splendor by the capture of fire.

Nearly all fire origin myths identify the acquisition of fire as the means of passage from life among the beasts into special status as a human being. In ecological terms, that mythology contains more than a kernel of truth. Anthropogenic fire is as much a cultural artifact as are chopping stones and skyscrapers, and landscapes forged in those fires are as much a creation of human societies as are marble sculptures and parking lots. If fire takes on the character of the landscapes within which it burns, so also it assumes the traits of the people who oversee it. Thus fire is at once cause, consequence, and catalyst.[2]

The pervasiveness of fire makes it a near-universal instrument for analyzing how humans have interacted with their environment, and a handy index for synthesizing the outcome of that engagement. These benefits come, however, at an intellectual cost. What is true of everything may say nothing special about anything. What is universal may become diffuse to the point of obscurity. The interdisciplinary methodology demanded for fire history, even as a subset of historical ecology, may resemble a conceptual breccia, the explosive fusing of fragmented disciplines. The study of fire may lack precisely what fire has traditionally given to human society, which is to say, a focus. (The Latin *focus* means "hearth.")

These concerns become inescapable when one considers the spectrum of meanings attached to "fire history." At one end stand the natural sciences; at the other, the humanities. The first study fire, fire regimes, and fire ecology: the physical record of fire, and the ways in which fire, of whatever source and purpose, interacts with the natural environment around it. The second uses fire as a tool by which to analyze human experience. Just as people apply fire to remake their environment, to transform it into forms more habitable and usable, so it is possible to exploit fire to reexamine the historical record—to smelt data-sets into ingots of new insight, to drive the character of humanity out of archival scrub. The ultimate ambition is the construction of moral universes: to address issues of character and identity, of what it means to be human and of how humans ought to behave. Because of humanity's species monopoly over fire, fire history is wonderfully positioned to inquire into these matters.

Syntheses are troubling, however. Fire makes a dynamic, unstable weld. Forcing the sciences and humanities together often results in epistemological schizophrenia, although the gibberish is sometimes punctuated by striking insights. The social sciences propose a third point, not along this conceptual spectrum so much as outside it, making a kind of fire triangle. Perhaps a better trope is to imagine the disciplines seated around a campfire sharing stories. What, in the end, they all share is the fire between them. The beginning of any mutual understanding requires an understanding of that common fire.

First Contacts, Lost Contacts: The Competition with Lightning

In most of the earth the boundary between anthropogenic and lightning fires became obscured long ago. For thousands (and in some cases, for hundreds of thousands) of years anthropogenic fire practices have defined the matrix within which combustion must proceed. It is likely that the initial encounter came as a shock wave, a flaming front of rekindled hominid migration as *Homo sapiens* added to the hominid tool kit and propagated itself around the globe.

This front involved burning on a major scale, as humans sought through fire to restructure the new lands into forms more suitable to them. In many environments—Australia for example—colonization seems to be recorded in the form of a charcoal horizon (figure 4.4). In recent years Australian Aborigines have rekindled that tradition on islands in the Gulf of Carpentaria (Lewis 1994). After the shock of first contact the record shows a different presence for fire, typically one of routine, low-intensity burning characteristic of aboriginal societies. Similar records are found in more recent colonizations of previously uninhabited islands—the Maoris in New Zealand, the Norse in Iceland and Greenland, the Portuguese in Madeira, the Malagasy in Madagascar, and so on.[3]

So thoroughly has anthropogenic fire replaced lightning fire that early explorers often dismissed the possibility of lightning as an ignition source in many landscapes

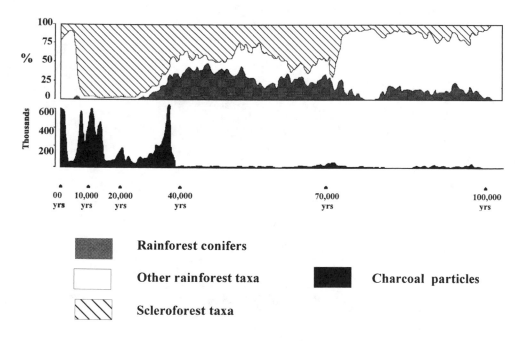

Fig. 4.4 *Fire Front:* The Hominid Colonization of Australia. A summary pollen diagram from Lynch's Crater, northeast Queensland, Australia. The sudden eruption of large charcoal corresponds closely to the earliest dated remains of humans. (Redrawn from Clark 1981.)

for which it now exists. In the early 1950s Americans denied lightning fires in central Alaska; Europeans today continue to dismiss lightning fire as other than a freak of nature. Partly this reflects a failure to observe, partly it represents an unwarranted extrapolation that the magnitude of native burning witnessed by the observer must apply everywhere, and partly it testifies to the success with which human societies had replaced wildfire with a more domesticated fire. By farming, grazing, and burning, human societies simply controlled the overall combustion environment (see figure 4.5).

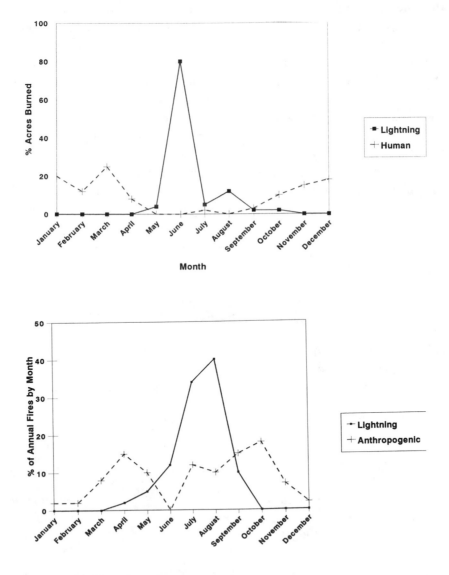

Fig. 4.5 *Combustion Competitors:* Biomass. Examples of how anthropogenic fire practices redefine the dimensions of fire seasons, and of how, by offset seasonal burning, humans contain lightning fire. (*a: top*) The Everglades. (Data from Everglades National Park.) (*b: bottom*) The Northern Great Plains. (Data from Higgins 1984.)

Where humans have withdrawn, however, lightning fire has often returned. This has happened historically where wars, diseases, and migrations have temporarily depopulated a region, but there are a number of interesting contemporary examples as well in which people have voluntarily withdrawn from sites they once occupied. The most common are nature reserves where "natural" processes are left untrammeled. Thus lightning fire has reasserted itself in Australia, North America, and even Europe, not in its historical forms but in a new regime for which management philosophy (and environmental ideology) have dictated the acceptable regime (figure 4.6).[4]

A dramatic illustration of lost contact is the revival of fire in the region around the Chernobyl nuclear plant (Dusha-Gudym 1992). After the catastrophic meltdown of 1986, radioactive contamination forced the abandonment of the surrounding region. With no one to tend the vegetation, fuels rapidly built up, and in the summer of 1992 fires broke out for which control was virtually impossible. The fires in turn entrained and redistributed some of the radioactive soils and plants in a horrifying cycle of positive feedback. It is not clear whether lightning kindled these fires, but in the absence of a local population it is likely; and regardless, the revealed pattern of feral fire reclaiming once-domesticated land is a scenario that is becoming increasingly common.

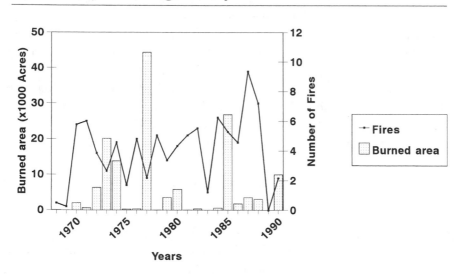

Lightning Fire Returns
Prescribed Natural Fires
Sequoia-Kings Canyon National Park

Fig. 4.6 *Lost Contacts, Restored Contacts: The Deliberate "Restoration" of Lightning Fire.* It would be fatuous to assert that humans can stop and start lightning fires at will. But it is equally absurd to claim that calculated attempts to reintroduce lightning fires (and only lightning fires) are not an artifact of human society. Compare this graph to the overall reduction in fire load shown in fig. 4.2. (Data from Sequoia–Kings Canyon National Park.)

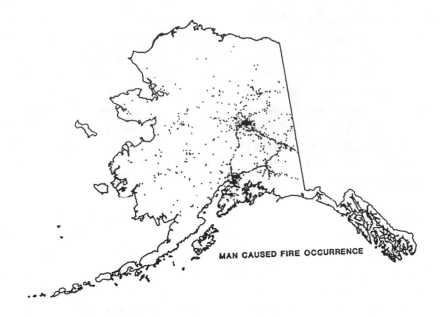

Fig. 4.7 *The Modern Fire Geography of Alaska.* The different domains of anthropogenic (*left map*) and lightning fire (*right map*) reflect decisions about land use. (From Gabriel and Tande 1983.)

Figure 4.7 shows the recent geography of fire in Alaska. In this instance, land legislation has determined where humans may settle and what kinds of technologies they may use. The division of the land into various categories has effectively defined separate domains for lightning and anthropogenic fire. Lightning fire is confined to the interior valley; anthropogenic fire, to fields of settlement and corridors of traffic. Only in a few areas do they overlap. The lightning fire regime is far from "natural," however. Gone are most traditional anthropogenic fires. Gone, too, are the older patterns of fire. Through firefighting institutions like the Alaska Fire Service, the National Park Service, and the Fish and Wildlife Service, humans "manage" many lightning fires through decisions about what fires to control and in what forms. The larger fire regime remains very much an expression of human will. Humans have deliberately sought to segregate their presence (and their fires) from that of nature (and its fires); the fire geography of Alaska expresses exactly this cultural decision.

Another illustration of how landscape ecology reflects anthropogenic fire (and vice versa) is the story of twentieth-century Greece. Figure 4.8 traces the historical record of fire. Note, particularly, the acceleration in burning over the past fifty years. In this case the disintegration of traditional agriculture, even outright land abandonment, is encouraging a steady buildup of fuels that sustains an increase in fire. The spikes in burning correspond to periods of social unrest when ignition increases and the capacity for fire suppression decreases. Similar scenarios, the stockpiling of fuels due to the breakdown of intensive farming, characterize other northern Mediterranean landscapes.[5]

LIGHTNING FIRE OCCURRENCE

Fig. 4.7   *(Continued)*

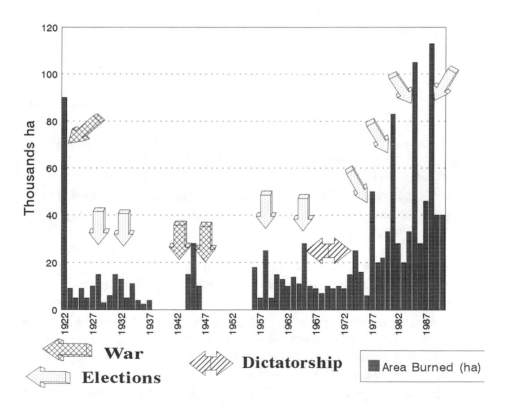

Fig. 4.8   *Greek Fire: Area of Forests and Scrublands Burned.* The overall increase in area burned relates to changing land use, particularly the decay of traditional agriculture; the spiked ignitions correspond to periods of social unrest. (Data from Kailidis 1992.)

While lightning has not apparently contributed to this revival—lightning fire is believed to be rare in the Mediterranean Basin, although a high fraction of fire causes are officially recorded as "unknown"—the potential exists, and it may be that lightning will find an environment more suitable than any it has known since the Pleistocene. The absence of lightning fire may have been another artifact of very long-term human presence, like terraced olive orchards and hard-browsed maquis. Whether it is a prelude or a freak, it is worth noting that lightning kindled one of the largest wildfires in recent years in Spain, an event that would have been unthinkable before large quantities of fuel became generally available.

When Torch Meets Torch: The Competition Among Anthropogenic Fires

The exchange of a lightning fire regime for one based on anthropogenic fire occurred long ago in most of the world. But this is not to say that regimes ceased to change beyond that. Torch met torch. Societies evolved, clashed, merged, and otherwise engaged in ways that have affected how fire has been applied or withheld from the environment. Various anthropogenic fires began to compete among themselves.

A full narrative of this competition is nothing less than the history of humanity and its interaction with the environment. This involves, of course, many peoples and many lands, in many narratives played out through many climates. Moreover, the parts themselves change through time. Human societies evolve as well as migrate; ecosystems adjust to species gained or lost, and to processes added or subtracted; climate fluctuates both locally and globally. Empires assemble and then disintegrate; epidemics strip societies bare, like the Black Death in fourteenth-century Europe and the pandemics that swept the Americas after the European Encounter; cattle and horses replace bison on the Great Plains of North America, and metastasize exponentially throughout the Argentinean pampas; the spread of metallurgy deforests local landscapes for charcoal; a fashion for beaver hats upsets the ecology of Northern Hemisphere forests—the list is endless. Each of these changes affects the local fire regime and the global fire budget. Equally, fire serves as an indicator, and integrator, of these other environmental reformations.

The major divisions of anthropogenic fire history thus follow the prominent divisions of human history—the initial colonization of continents; the Neolithic revolution; the expansion of Europe; fossil-fuel combustion associated with industrialization. For each of these processes there are unique historical examples. Aboriginal fire shaped Australia. The Americas mixed aboriginal and agricultural fire practices, but without domesticated livestock. Portions of central Eurasia and Africa relied more or less exclusively on pastoral burning. To varied degrees, Africa, Europe, and Asia experienced mixed economies and displayed fire regimes appropriate to farming, herding, and, in more remote sites, hunting and foraging. Antarctica exhibits a continent without fire—and, for that reason, is unlivable without imported fire. (If you think that fire is difficult to cope with, consider the alternatives.)

The magnitude of human impact varied widely, of course. Where environmental conditions were favorable, as in Australia and California, aboriginal peoples could reshape landscapes wholesale. The firestick became a lever that, suitably placed, allowed them to move continents. In the Great Plains of North America, climate, landforms, and fire promoted vast grasslands; but through anthropogenic burning that biome could be perpetuated even after climatic changes nominally favored trees. The power of aboriginal fire is nicely illustrated by its removal. Everywhere the abolition of aboriginal burning has inspired dramatic ecological changes. For the most part, these involve shifts from grass to woods, from prairies or forest-steppes to scrub and forest.[6]

Humans made other regions more favorable through allied technologies, notably agriculture. Even simple swiddeners outfitted with stone axes—and particularly graziers with small flocks of goats, sheep, cattle, and swine—could induce dramatic reformations in the landscape. For the first, it was not necessary to fell all (especially large) trees, only the small ones. For both, it was not necessary to fell, only to kill the shade-producing species, and this they could easily do by girdling and stripping. Once the forest surface was exposed to sun and wind, it rapidly dried and became available for burning, as did standing dead trees. With livestock and human foragers to select among the species that sought to reclaim the site, it was possible to restructure landscapes extensively. When this process was extended over centuries (or millennia), whole landscapes became, in effect, anthropogenic artifacts.

In this way shifting agriculturalists could break up putatively stable, old-growth forests. A consistent outcome was an increase in biodiversity: there were more niches, a tighter mosaic of biotic patches, more dynamism to the cavalcade of species successions. The classic description, of course, comes from Denmark through the work of the palynographer Johannes Iversen (1973). The entombing lime (*Tulia*) forest of Mesolithic Europe splintered before the diffuse *landnam* (literally, "land taking") of Neolithic tribes (figure 4.9). Species driven into sun-refugia along riverbanks and lakes now percolated throughout the landscape. Under the impress of clearing, grazing, and fire, heath and moor also proliferated. Interestingly, where those pressures have been removed, as in Sweden's Dalby Söderskog National Park, biodiversity has collapsed. Between 1925 and 1970 the park lost 41 percent of its species, largely through inattention, largely because the park excluded rather than promoted the chronic disturbances that had nurtured its nascent variety.[7]

Of course, during any period of change, some species thrive and some falter. Some species cannot tolerate fire. But increasingly ecologists have discovered that fire exclusion (or the exclusion of fire-catalyzed practices) is a powerful destroyer of biodiversity, and more and more ethnographers are describing fire-forged landscapes for which, once the fire was extinguished, the biota turned cold, immobile, and monolithic. In the Brazilian *cerrado*, South African *fynbos*, Mediterranean *maquis*, North American tallgrass prairie, Swedish *heden*, southern pineywoods,

Everglades sawgrass, Australian spinifex—in an endless register, the power of aboriginal and agricultural fire has been demonstrated and reconfirmed, not only for individual species but for habitats and landscapes. Earlier critics had assumed that fire, so visible and dramatic, was inevitably a force of destruction. The real issue, however, was not the presence of fire but its regime, its scale, and the fire-catalyzed technologies with which it was allied.[8]

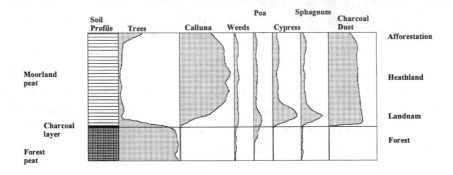

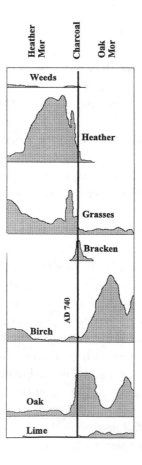

Fig. 4.9 *Landnam: Neolithic Fire.* With other tools, and especially with domesticated animals, Neolithic tribes in Europe could exploit fire to greater effect than could Mesolithic peoples. A first-contact swidden was the outcome, often resulting in significant floral shifts. *Top: Landnam* in Denmark. In the paradigmatic case study, as reconstructed by Iversen, the shock of contact broke open the old shade forest like a dropped pumpkin. Where a more regular cultivation did not subsequently claim the land, agriculturalists had to reinitiate *landnam* at a later date. (Redrawn from pollen diagrams in Iversen 1973.) *Left: Landnam* in Norway. Here the shock converted stable forest to stable heath, not unlike the way swidden has often transformed tropical forests into grasslands. The critical element for Europe, however, was the presence of livestock, which interacted synergistically with burning. Norse colonists carried the practice across the Atlantic. Their *landnam*ed charcoal dates their varied arrivals. (Redrawn from Behre 1988.)

Fire and Fallow: Agricultural Burning

Sedentary agriculture, too, relied on some regimen of burning. After all, until very recently agriculture was rotational, either because it moved the farm through the landscape or because, in effect, it moved the landscape through the farm; both forms demanded fire. Agriculturalists looked to fire to perform its twin duties, to purify and to fertilize. (The major exception was irrigation agriculture, which substituted water for fire—although even here it was common practice to burn the stubble.) But in order to burn it was necessary to have fuel, or, in agronomic language, fallow.

"Swidden" is a poor description for the robust spectrum of such fire-fallow regimes. (In its origin the term derives from old Norse, meaning "to burn"; it came to England with Norse invaders, and referred specifically to the decadal burning of ling, or *Calluna* heather.) One reason why the term fails is that most landscapes were compounds and mixtures of habitats, not just fields, and they experienced a medley of fire practices. There was slash-and-burn cultivation in outfields; stubble burning in infields; pastoral burning in abandoned swidden plots, wetlands, and rough fields; underburning to promote native tubers and berries for foraging; broadcast burning to sustain hunting grounds; smoking to harvest bees; patch burning to produce roofing thatch and branches for basketry. Landscapes encountered not one fire but many, and found those burns remarkably specific, subtle, and, in their sum, pervasive.

Especially among European agronomists, the fallows were considered a badge of shame. This was land out of production, land that with proper management could grow food and fodder to sustain Europe's spiraling population. Ideally, agricultural savants sought to substitute a calculated succession of useful crops to replace those species that flourished opportunistically after abandonment. They hoped that ultimately it would be possible to dispense with fallows altogether—which is to say, with fire, for a primary purpose of the fallows was to produce the fuel without which fire was impossible. Agricultural systems raised fuel for field burning just as they cultivated oats for draft horses.[9]

The fallows served other purposes, and were often rich in semicultivated (or at least selected) species that could be put to various uses. That they were the scenes of chronic disturbances by grubbing, grazing, cutting, harvesting, planting, and burning made them notoriously profuse in species diversity. But for agronomic critics biomass, not biodiversity, was the measure of success. Political theory, too, preferred sedentary to shifting cultivators. As Europeans expanded their dominion across the globe, those notions of proper farming, good societies, and healthy landscapes dominated their understanding of the agricultural systems they encountered.

Vestal Fire: The European Exemplar

The dominant story over the past half-millennium has been the expansion of Europe. Mariners reconnected in decades what geologic processes had separated over tens of millions of years. Plants, animals, microbes, diseases, and peoples mixed—

invading, transplanting, eradicating, mingling—in what Alfred Crosby (1986) has proposed as the greatest biological event since Pleistocene glaciation. What resulted was not merely a new global economy but a reconstituted global ecology and, after the consolidation of modern science, a global scholarship by which to understand these events. European science would replace indigenous lore as the interpreter of fire, and Europe would declare itself the standard by which fire practices should be judged.

The expansion of Europe and the subsequent spread of industrialization based on fossil-fuel combustion thus offers a range of encounters. In Australia, Europe met peoples who practiced neither farming nor herding. In the Americas it confronted hunters, foragers, and farmers. In Africa and Asia it interacted with mixed economies much like its own. The fire histories of these regions thus differ, not only because of the intensity of colonial rule but also because of the shock wave of encounter itself. The greater the difference, the greater the fragmentation. In general, however, the amount of open burning has declined, woody biomes have replaced grasslands, and fire, where it persists, has moved from the field to the furnace.

Because of Europe's influence, the fire history of Europe holds particular interest. Virtually every fire practice found in the world finds expression in European history. Europeans practiced fire hunting and fire fishing. They burned to improve berries and mushrooms. They practiced slash-and-burn cultivation in nearly every environment; the agricultural revolution that preceded the industrial revolution was based, in part, on the extensive conversion of heath, moor, and other organic soils through a regimen of slashing and burning (see figures 4.10 and 4.11). Swidden was still practiced in Scandinavia and the Midi region of France in the 1920s, and continued in Finland through the refugee resettlements after World War II. The agricultural geography of Tsarist Russia resembled a series of nested dolls, the scale of each sized by the length of its fallows (figure 4.12). There was no absence of fire (see Steensberg 1993; Kuhnholtz-Lordat 1938, 1958; Sigaut 1975; Pallot and Shaw 1990; Pyne 1997).

Until the mid-eighteenth century Europe was little different from other parts of the world subjected to similar economies. Its fire regimes were only marginally distinctive, a reflection of local environments and histories; its encounter with fire regimes outside Europe reflected more the indirect effects of its expansion in terms of plants and animals introduced and extirpated, and of diseases and the demographics of contact, than of any deliberate policy aimed at restructuring those lands through fire. But officials and intellectuals, almost all of them urban, detested fire and denounced it with edict and epistle.

Increasingly, European intellectuals of the Greater Enlightenment sought to constrain fire. They saw abused landscapes—organic soils pared and burned down to rock, fallowing cycles shortened dangerously, expanding heaths, felled woods, marauding shepherds, eroded hillsides, mountain torrents. Reason demanded some way to protect humus and eliminate fallows. They saw agriculture, under the crush of increasing population, trying to square the circle instead of close it. They identified fire, ever present, as a cause of destruction. Even the legendary Linnaeus was

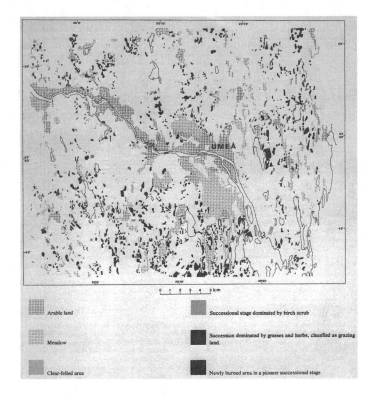

Fig. 4.10 *Swidden in Sweden.* Map of Umeå showing the mosaic of managed lands. The pattern is similar to traditional slash-and-burn cultivation elsewhere in the world. (From Zackrisson 1977.)

forced to delete from his 1752 Skåne journal favorable comments regarding swidden in Småland and to substitute a passage on manure.[10]

Intellectuals unambiguously separated primitive from rational agriculture according to whether or not practitioners used fire. Intensification demanded an end to fallows; increased production required that biomass go to the slow combustion of metabolizing livestock and humans, rather than to the fast combustion of open flame. Agronomists had their reasons. National wealth depended on agriculture, agriculture on the state of the humus, the humus on whether or it not it was burned. Besides, broadcast burning of almost any kind, from hunting to pastoralism, works best where populations cycle through a landscape; as enclosure and fixed ownership spread, fires were less desirable because they were less possible. Free-burning fires belonged with the flaming front of colonizers and pioneers—but most of Europe was beyond this. So too were Europe's intellectuals, almost all of whom were urban. Only those of rural backgrounds saw much value to fire—Virgil and Linnaeus being among the most prominent. Others knew fire only through the terrors of urban conflagration, and that, as often as not, was the outcome of hostile siege and sack.

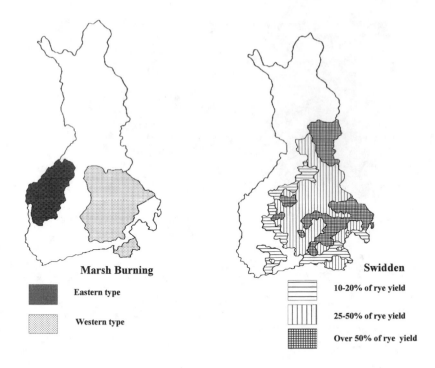

Fig. 4.11 *Finnish Fire.* Two patterns of slash-and-burn cultivation still apparent in 1830s Finland. In the better-drained sites, forests were swiddened (*right map*); in the mires, the organic soils (*left map*)—a practice that characterized much of northern Europe during the agricultural revolution of the eighteenth and early nineteenth centuries. (Redrawn from Soininen 1974.)

The swelling population itself brought such pressures on the land that there was simply less and less free biomass available for burning. Peasants raked up pine needles for stable bedding, seized every reed and blade, lopped off tree limbs for fuel and fodder; in response, agronomists became obsessed with restoring the organic component of the soil, primarily by manuring. The degradation of Mediterranean landscapes by promiscuous browsing and burning complemented Edward Gibbon's *Decline and Fall of the Roman Empire* as an allegory of political economy and moral malfeasance. Encounters with tropical islands such as St. Helena, Barbados, St. Vincent, and Mauritius became semicontrolled experiments on the disastrous consequences of abusing lands—microcosmic warnings of macrocosmic doom. Cutting, grazing, and burning became the unholy trinity of environmental critics.[11]

Counter Fire: The Engineering of Free-Burning Fire

The European hearth became a norm, a skewed vestal fire for the planet. Europe's experiences within its own borders were sharpened by the encounter with the "strange fire" of distant colonies and eventually became the basis for a more sys-

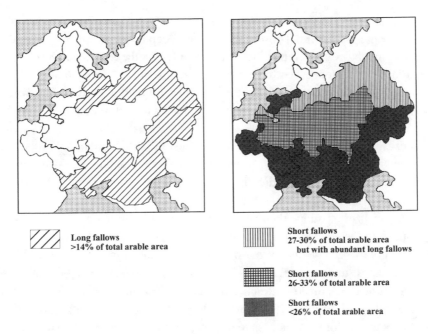

Long fallows
>14% of total arable area

Short fallows
27-30% of total arable area
but with abundant long fallows

Short fallows
26-33% of total arable area

Short fallows
<26% of total arable area

Fig. 4.12 *Fire and Fallows: The Russian Example.* The pattern corresponds both to climatic gradients (the domain of the mixed European forest) and to the east-trending spread of Russian settlement. Even longer fallows characterize Siberian agriculture. (Redrawn from Pallot and Shaw 1990.)

tematic understanding of biomass and burning. The fiery exchange with overseas colonies, however, went only one way. Europe exported its science and fire control apparatus, but never imported indigenous knowledge about fire ecology and practices. These experiences were institutionalized as the eighteenth-century Enlightenment remade classical arboriculture into modern forestry.

Academic forestry grew out of European agronomic thinking, and thus accepted humus as a universal index of environmental health. Since fire burned humus, it was intrinsically evil. Besides, swiddeners burned forests before they could mature into sawtimber, and pastoralists routinely fired woodlands to encourage browse and grass and in so doing directly attacked trees. Foresters freely identified the weapon, fire, with the assailant, herders and shifting cultivators, and sought to control the users by controlling the use. Out of all the disciplines that might have addressed fire, the task fell to foresters—with consequences that are still very much with us.

Foresters struggled to create institutions to suppress traditional burning and to explain the damages wrought by fire. In general, indigenous lore, like indigenous fire practices, was extinguished; this was as true in Europe as elsewhere. Fire suppression became the norm, and the relatively fire-free forests of central Europe, an aberration, became a standard. Germany's forests stood to landscapes as Max Weber's Prussian bureaucracy did to social institutions: a Platonic type. Foresters in turn joined those other great corps of engineers—civil, hydraulic, and mining—that were so influential in projecting European influence throughout the world.[12]

The presence or absence of forestry became in fact a defining criterion in the evolving geography of fire. Those nations that accepted it, or had it imposed, removed too much fire but created an infrastructure for the management of forests and wildlands that later served for the reintroduction of fire. Those that did not develop forestry continued traditional burning (but within an altered and much more densely populated world) and today generally lack an institutional means to contain those fires.

This exercise in technology transfer was easiest in North America. It was messiest in the Southern Hemisphere, where biotas were dramatically different and where, as in Asia and Africa, indigenous peoples and livestock outnumbered European imports. There traditional burning persisted in defiance of European desires. In fact, colonists often adopted native fire practices or hybridized with them in mutual hostility to the edicts of imperial proconsuls and the theories of European savants. Fire policy became an expression of colonial rule, a means of reordering the ruled landscapes. In response, natives burned illicitly—at once a political protest and an attempt to restore traditional lands to traditional purposes. Regardless of whether local peoples retained legal access to state-reserved lands, without the use of fire they lacked biological access to the potential resources those lands held. The character of the land changed, often irrevocably.

All this did not pass unnoticed. Typically a debate ensued, often formal and even published, that pitted European standards against local practices, fire control against fire use. The most dramatic confrontations flared in British and French colonies (or former colonies), where foresters nurtured in Franco-German traditions were most aggressive at imposing policies that aspired to fire exclusion. In North America, fire control triumphed; in India, Australia, and South Africa, awkward compromises resulted that first denounced controlled burning, only to recant later and ultimately to absorb them into official doctrine. In time, controlled fire returned to North America as well, although in a language ("prescribed burning") that denied the legitimacy of its earlier incarnation ("light burning").[13]

Fire Ecology Reconsidered: Recycling a Concept

All this has reformed the meaning of fire ecology. The range of anthropogenic fire matches that of human society, no longer confined to particular sites or limited to a penumbral shadow of off-site effects. Of course, ecosystems had long accommodated anthropogenic fire practices. An ecosystem's energy flows, its nutrient cycles, its structure, its pattern of disturbances, its fire regimes—all had been influenced, if not dominated, by humans. As human societies expanded and evolved, so did the nature of fire ecology. Increasingly Europe's widened contacts rewired ecosystems, recalibrated their scales, and modified their tempo.

A global economy cycled nutrients on a global scale. It could, for example, mix guano from Pacific islands with forests in France, replace native eucalypts in Australia with American pines or substitute eucalypts for Indian deodar, trade Ukrainian wheat for British steel. More-or-less-bounded ecosystems broke up and were reconstituted, at least in part, on different scales through the trophic chain of a money economy. Human institutions rapidly connected lands previously bound only through the longest and most tenuous of biogeochemical cycles.

The organization of distant colonies followed from the knowledge (or ignorance, or misinformation) of imperial powers. Science and engineering, no less than economic imperatives, created institutional valences that bonded India to Cape Colony, and Corsica to the Ivory Coast. What foresters (for example) learned in one part of the British Empire would be felt in other parts: imperial foresters could transfer fire practices developed for Madhya Pradesh to Burma and Kenya and the Yukon.

In the United States, fire effects similarly cycled through national institutions. The 1910 fires so traumatized the U.S. Forest Service that it would not consider controlled burning until its founding generation had passed from the scene, thirty years later. By killing fifteen firefighters, the otherwise obscure 1937 Blackwater fire in Wyoming stimulated both the smokejumper and the forty-man-crew programs, each of which profoundly altered the fire regimes of America's public lands. By forcing the National Park Service and the U.S. Forest Service to reevaluate their programs, the Yellowstone fires of 1988 had ecological consequences for cognate lands in Oregon and Minnesota and New Mexico.[14]

While fire scientists (especially Americans) sought to reconstruct the "natural" ecology of fire, landscapes in fact reflected their human histories and the dynamics of social institutions. The cycling of information through bureaucracies was as critical as the flow of carbon through lodgepole pine forests. Budget cycles became as important as the rhythms of the seasons; policy determinations, as potent as evolutionary selection; the flow of money, as vital as that of sunshine. The study of fire ecology without reference to the human presence was meaningless— like physicists studying perpetual motion machines because they have decided to disregard entropy.

Into the Furnace: The Competition
with Fossil-Fuel Combustion

Not least of all, industrial Europe began to sublimate the power of fire into machines. Controlled combustion began to replace controlled burning. By the end of the eighteenth century industrialization had compounded imperialism into a new order. The encounter between Europe and the remainder of the planet broadened, and its tempo quickened. For fire history the critical events involved the exhumation of fossil biomass, the invention of new fire engines that could combust it, and the substitution of artificial herbicides, pesticides, and fertilizers for tasks once done by fire. Those engines redefined the character of pyrotechnologies, while coal and petroleum became, in effect, fossil fallows that allowed agriculture to transcend its closing ecological circles.

Fire Engines: From Flame to Combustion

In the beginning industrial combustion, and the industrial economy, depended on biomass burning or other sources of power such as watermills. But fossil hydrocarbons soon altered that, and imposed a change in the character of the earth as a fire planet (figure 4.13). Fossil-fuel combustion burned without regard to the living environment. It stood outside the ancient ecology of fire, and outside the traditional social mores that had guided anthropogenic fire use. Humanity metamorphosed from the keeper of the flame to the custodian of the combustion chamber.

Industrialization's reach exceeded its grasp. Fossil fuels burn beyond the parameters of traditional fire ecology. Their combustion proceeds with savage indifference to season, time of day, and the biotic rhythms of nutrient capture, growth, and decay. They restructure fire regimes—both directly, by redefining and controlling ignition, and indirectly, by reorganizing the biosphere and its fuels. They have redefined what are natural resources, how they might be exploited, and what lands will be used in what ways. Converted to petrochemicals and combustibles, fed into tractors, diesel trucks, and the turbines of commercial ships, fossil fuels have allowed agriculture to overcome, at least momentarily, the closed ecology of fallow farming.

What had long exploited—in fact, required—fallows now had an alternative. Fossil fallow could substitute for living fallow. Output was not restricted to available inputs. European agriculture, in particular, had long been obsessed with recy-

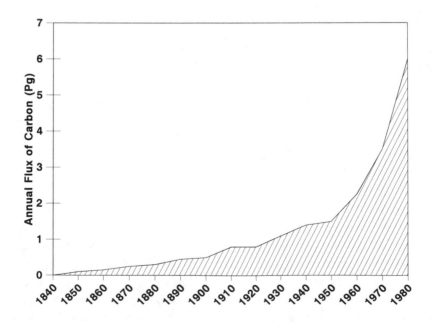

Fig. 4.13 *Combustion Competitor:* Fossil Biomass. The escalation of global fossil-fuel combustion, as measured by the annual flux of carbon (Pg = pentagram, one billion metric tons). (Data from Houghton and Skole 1990.)

cling waste, whether dung or fallow fields; only by increasing the biomass fed into a system could farmers and pastoralists hope to extract more out of it. That was the agronomic logic behind the European critique of folk burning. But when coal and oil could substitute for fuelwood, when diesel-powered tractors, trucks, and trains could replace draft animals and their fodder requirements, when petroleum-derived pesticides, herbicides, and fertilizers could do what only fallow-fueled fire had done in the past, then it was possible to break out of the limits of fire-fallow agriculture—or, more properly, to transcend its biotic geometry. Agriculture could stop chasing its tail around the field and leap over the fences that had once confined it.

No longer, or at least not for several centuries, does available biomass limit productivity. Fossil fallow is matter and energy added to, not extracted from, landscapes of cultivation. The new combustion regime is no longer restricted by what the land can grow. Instead, fossil hydrocarbons are a kind of biotic bullion, plundered from the past. Unfortunately, fossil biomass is acting on nature's economy as the sixteenth-century flood of Inca gold and Mexican silver did on Spain's: it is inflationary, often ecologically corrupting, and shaped less by productivity than by consumption, by the capacity of the system to absorb the excess effluent. Increasingly the earth's ecology traces the nutrient cycles and energy flows of a global economy. Increasingly the limits to productivity are set by available sinks rather than sources.

The full range of interactions is not well understood. There are not even formal terms to discriminate between traditional burning and industrial burning. Call the new regime "combustion," since it has disaggregated burning from flame. Reserve "fire" for traditional open burning. It is clear, however, that if lightning fire defines one border to anthropogenic fire, then fossil-fuel combustion defines another (see figure 4.3). This pyric revolution has demanded not only new combustion technologies for furnace and forge, but new fire practices for field and forest. Perhaps the most critical difference ecologically is that fossil-fuel combustion competes with anthropogenic fire not over a common source, biomass, but over their common sinks, principally the atmosphere.

Again, the nature of this competitive boundary is not well understood. While industrial combustion can substitute for many human pyrotechnologies, it cannot supplant the full range of ecological effects that come with free-burning fire. It can replace older fire technologies, but not fire ecologies; fire as a tool has surrogates, fire as a process may not. It is one thing to substitute an electric wall heater for a fireplace, another to propose that chain saws and bulldozers can mimic the disturbance and fire regimes of a forest. But if the ecological processes are blurred, the historical border is remarkably precise. Fire follows fuel. Fire's modern age tracks exactly the exploitation of fossil hydrocarbons. Had industry merely burned biomass, the ancient fire-fallow cycle would have held. But industrial technology discovered, instead, a new fuel, outside the nagging limits of its grown biomes. In doing so, it inaugurated a new era of fire history.

The range of encounters between new and old is large, confusing, and often incommensurable. More and more, the earth divides between two geographies: one of industrial combustion, another of open fire; with the exception of India (here, as in other matters, seeming to absorb new elements into its fire castes), lands belong to one or the other. The steady abolition of fire from industrialized landscapes reflects not only the removal of agriculture and its human tenders, but also the application of fire suppression technologies, all dependent on such combustion-powered machinery as portable pumps, wildland fire engines, and slurry-dropping aircraft; even fire-retarding chemicals derive from fossil hydrocarbons. Previously, the best—often the only—protection against wildfire was to burn under controlled conditions, which had the effect of keeping fire, some fire, domesticated fire, on the land. Paradoxically, the dramatic diminution of controlled fire has allowed biomass to bloom, and in many exurban landscapes, from Los Angeles to Sydney, these proliferating fuels have powered devastating wildfires. Even more curiously, industrial societies have cultivated an alternative to the biodiversity-rich fallow of agricultural landscapes. Nature reserves have become the fallow-surrogates of the industrial Earth.

Making and Unmaking Agriculture:
The Amazon and the Mediterranean

Since the late 1980s Amazonia has become an international symbol of combustion run amok. Slashing and burning on a colossal scale have, it is charged, degraded landscapes, impoverished biotas, and unhinged the climatic order of the planet. Yet, in fact, fire is everywhere in Amazonia's natural history. Charcoal lenses underlie the soil; small-scale swidden accounts for perhaps 35–40 percent of the biodiversity of many sites; the grassy *cerrado* is fire-dependent, soon overwhelmed by species-poor woods if not regenerated routinely by burning. The rain forest appears to be an artifact of the Holocene, a biota that has coexisted with humans and thereby with anthropogenic fire. How much of this forest is a secondary product of depopulation following a violent, disease-brokered encounter with Europe is unclear. In Mesoamerica the evidence suggests that burning has not yet returned to preconquest levels (figure 4.14). Undoubtedly this is also true for most of South America, and almost certainly it is true for North America (see Balée 1993 and Suman 1991).

Not enough is known to say whether this also characterizes Amazonia. Clearly, however, the reckless deforestation of the past two decades is a product of industrial combustion, either directly through internal transportation grids, or indirectly through linkages to a global economy that determines the value of Amazonian land and makes profitable its production of commodities. Industrialization, after all, is the announced goal of the Brazilian state and stands behind its subsidies of development. Similar scenarios characterize Indonesia, Malaysia, Thailand, and parts of India.

The Mediterranean Basin presents the obverse scenario. Here a traditional agriculture that in one form or another has existed for several thousand years is disintegrating. The region has a fire-prone climate, a pyrophytic biota, and a very long history of anthropogenic burning for fuel, pastoralism, and farming. The intensity of traditional usage is such that wildfires are rare—there is simply not enough wild

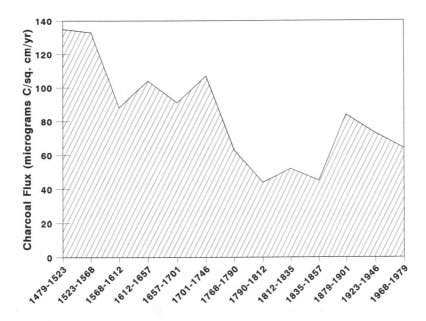

Fig. 4.14 *Lost Contacts: The Removal of Anthropogenic Fire.* Charcoal fluxes embedded in Pacific sediments from offshore Mesoamerica. (Data from Suman 1991.)

fuel to sustain them.[15] Industrialization and integration into the European Community are forcing changes, however. Without massive subsidies subsistence agriculture cannot endure. Accordingly cities are growing, and farm land is being abandoned or converted to second homes or other uses. The liberation of fuels has liberated fire. Figure 4.15 shows the relationship between the intensity of usage and the propensity to uncontrolled fire. This change, of course, underlies the record of renewed fire for Greece expressed in figure 4.8. Similar situations characterize North America and Australia, where exurbanites are reclaiming land from rural economies but, because they do not engage the land—harvesting, grazing, gathering, burning— are allowing fuels to build up. This "intermix" fire scene is generally recognized as the dominant problem of the American fire community. The Southern California conflagrations of 1993 demonstrate its volatility and destructive potential.[16]

Hostile Hearth: The Energy Exchange

A related process characterizes the change in domestic fuels. The substitution of fossil fuels for biomass is, in many places, allowing for a buildup of woody vegetation. Figure 4.16 shows this exchange for Korea and the United States, a transformation typical of industrialization.

Many observers consider this change desirable. Biomass burning releases greenhouse gases, fouls the air near human inhabitants, wastes energy, and subverts surrounding ecosystems. The explosive growth of human populations has made the

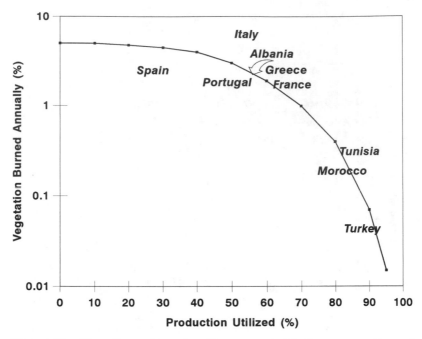

Fig. 4.15 *The Competition for Biomass: A Mediterranean Example (1980–1985).* The greater the utilization of biomass, the less the area burned. (Redrawn from Rego 1992.)

hearth, long an emblem of human solidarity, into a symbol of human degradation of the environment. Larger and urban populations demand a shift in domestic energy from biomass to some more efficient combustion based on fossil fuels. In principle, the amount of carbon released can be absorbed by the vegetation that is spared.

Of course the issue is more complicated than this, and among its unexpected side-effects is the prospect for wildfire. If nothing else consumes that renewed growth, fire typically will. This in turn upsets the fire regime of the ecosystem in ways that are often as damaging as the direct harvesting of its flora. In this sense the substitution of fossil fuels for biomass does not abolish fire, but instead argues the need for a new regimen of anthropogenic burning in which controlled burning can substitute for wildfire.

Friendly Fire: Wildland Management

Among the curious consequences of industrialization is the attempt to suppress fire of all kinds both in commercial forests and in nature reserves—landscapes of special concern to the kind of urban societies that industrialization has made possible. Breaking the fire cycle was deemed necessary to protect such sites for large-scale logging and, after the enthusiasm for wild nature became fashionable, for parks, wildlife sanctuaries, and nature preserves. In both cases traditional fire practices were prohibited; fire management was based on fire control, not on fire use. Thus

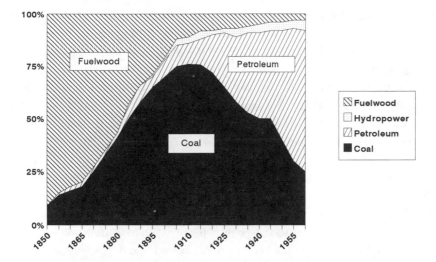

Industrial Combustion Sources
Korea

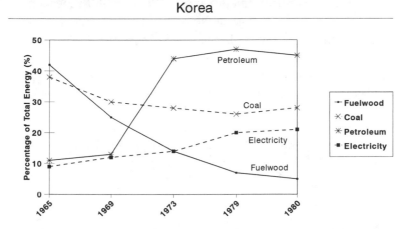

Fig. 4.16 *Hostile Hearth: The Substitution of Fossil Fuels for Fuelwood Consumption in the United States (top) and Korea (bottom).* The exploitation of fossil biomass to supplement (or supplant) living biofuels traces the spread of industrialization. The conversion, with fossil fuels burned in proper combustion chambers or at some distance from the home, has reduced domestic air pollution drastically—an improvement in public health. But the power plants in turn have only displaced, not abolished, the air quality issue: instead of polluting the kitchen or house, the industrial hearth threatens to contaminate the regional, and even the global, atmosphere. (Redrawn from Williams 1989 [*top*] and Smith 1987 [*bottom*].)

forestry revealed its origins in European agronomic thought, and park administrators their belief that a prelapsarian state of nature represented the natural order. For the first use, administrators had to intervene to reshape a quasi-natural system to more suitable purposes; for the second, they had (so they believed) only to prevent human practices of any kind other than low-impact tourism and scientific research.

The experiment has been particularly vigorous in the Northern Hemisphere, where it has progressed for almost a century. Figure 4.17 shows some examples of how dramatically the fire load has dropped. One explanation for this putative success is that state forests and nature reserves have eliminated traditional burning. Comparative histories suggest that very large returns result from initially small investments in fire control. Even a first-order infrastructure can lead to dramatic reductions in burned area. The real dilemma comes from trying to sustain this condition through ever-escalating inputs of money, machinery, and manpower. The ecological crisis was not that wildfires were suppressed, but that controlled fires were no longer set. A change in fire regimes, not fire exclusion, should have been the goal.

There were other reasons for the delayed appreciation of what industrial fire control meant, reasons peculiar to the geography of industrial firepowers. Almost all these nations are situated in the north, where, indeed, the bulk of the world's landmass and forests also lie. It is characteristic of the boreal forest that a small number of large fires during drought years accounts for most of the acreage burned; rapid attack can control those critical ignitions at great savings. Still another explanation is that the ponderous rate of change in the northern landscapes disguised the full consequences of fire exclusion for decades, whereas in the Southern Hemisphere the negative impact was felt within a handful of years.

But felt it was, even in the north. Landscapes changed, fuels built up, species responded. It became apparent that to withhold fire was as powerful an ecological act

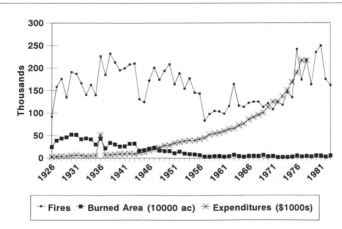

Fig. 4.17 *The Expulsion of Fire.* (*a*) United States. There are two logistic curves, one tracing the removal of free-burning fire, the other the costs of these acts; neither is sustainable. Reforms have sought, with limited success, to control costs and reintroduce fire. (Data from *Historical Statistics of the United States.*)

Fire in Sweden

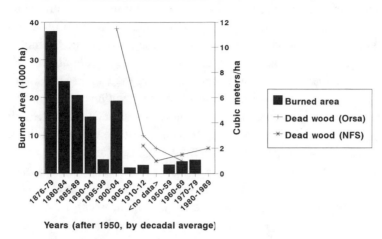

rsa=Orsa Forest Commons; NFS=national forest survey

Fig. 4.17 *(continued)* (*b*) Sweden. In converting farm and woods to commercial forest, Sweden treated fire as it would other agricultural pests, and intensive use (note the decline in dead wood) has virtually abolished fire. No coherent statistics exist for many decades, although large fires did break out in 1934. (From Högbom 1934; Stocks 1991.)

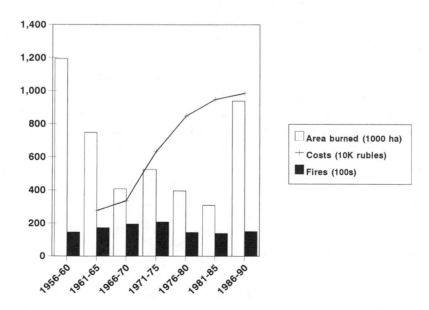

Fig. 4.17 *(continued)* (*c*) USSR. The figures are almost certainly false—they should be increased by a factor of 2–10. But the trend is probably correct, and shows remarkable similarities to that in the United States. (Data from Avialesookhrana, Russian Forest Service.)

as to introduce it. Systems degenerated, biodiversity decayed, and lands became
unstable with respect to fire. Fire control had succeeded in maintaining an incre-
mental improvement by investing in elaborate technologies—by ultimately fighting
fire with fire, but through internal combustion engines rather than through torches.
But this soon reached another point of diminishing returns. The triumphant profile
of suppressed fire was equally a profile of an accumulating fire deficit.[17]

For the past two decades the fire organizations of the United States, for example,
have struggled to reinstate fire into public forests and wildlands. A grand experiment
in letting nature determine the fire regime more or less ended with the debacle surrounding the Yellowstone fires of 1988. It is clear that the historic environment had
adapted to anthropogenic fire, not to fire in the abstract, and that an anthropogenic
fire regime must be reinstated if that landscape is to be preserved in something like
its traditional state. Thus, even in lands from which people have been excluded,
there is a compelling case for active fire management, which is to say controlled
burning. There is a fire tithe owed most ecosystems, and it has been a human duty to
pay it on behalf of the biota. If humans fail to do so, then lightning, arson, or accident will extract it in ways that will not always agree with human goals.

Scorched Sky: The Iconography
of Uncontrolled Combustion

The boundary between anthropogenic fire and industrial combustion differs from
those between lightning fire and human fire, and between various regimes of anthropogenic burning. Of course, their domains do express themselves geographically (figure 4.18). But they do not compete directly for biomass so much as for air-
shed. Air, not humus, has become the universal index of environmental health,
supplemented by measures of biodiversity; and fire has become part of an iconog-
raphy of international environmentalism.

Nuclear winter, greenhouse summer, slashed-and-burned biodiversity in Brazil
and Indonesia, Kuwait oil fires, the meltdown of the Chernobyl nuclear plant, wildfires in Manchuria, Borneo, and suburban California, even the parbroiled extinction
of the dinosaurs—concerns for all of these have been animated through images of
fire as destroyer. One effect has been to make fire itself seem a cause of disaster
rather than a catalyst for it, or even a consequence of mismanagement.

Trial by Fire: Fire and the Future

The capture of fire announced a unique power for early hominids, and established a
special relationship between humanity and the earth. Since that time humans have
used fire in every conceivable place for every conceivable purpose. With the acquisition of fire, humans stamped their personality onto the planet and began to alter the
global carbon cycle. With the acquisition of its firepower, humanity confronted the
need for an environmental ethos as well as for suitable environmental engineering.

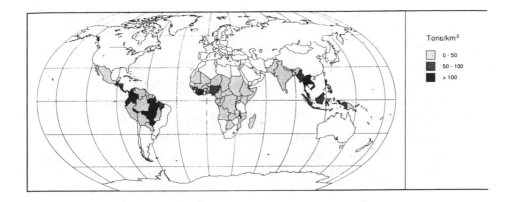

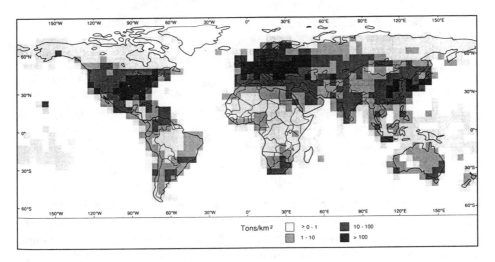

Fig. 4.18 *The Two Geographies of Contemporary Combustion: Fossil Fuel and Biomass. Top map*: The geographic distribution of the net flux of carbon from terrestrial ecosystems to the atmosphere in 1980. (From Marland, Rotty, and Treat 1985). *Bottom map*: The geographic distribution of emissions of carbon from fossil-fuel combustion. The one area of significant overlap is monsoonal Asia. (From Houghton et al. 1987.)

Fire Report

It is probable that there is presently too much combustion and not enough fire. A disequilibrium exists between fossil-fuel and biomass combustion. The fire that does persist is maldistributed—too much wildfire in the wrong places at the wrong times; too much burning in furnaces, and not enough fire in the fields. Biomass burned is not equivalent to biota burned. Free-burning and confined fires have different biological consequences, and thus different atmospheric effects. Fossil fallows function differently than contemporary ones; eventually their added chemicals must go somewhere, and this typically involves some form of industrial combustion. How an index of "fire" translates into carbon cycles and greenhouse gas emissions and so on, however, is unclear.

Over the last five hundred years the earth has probably experienced a reduction in fire—how much is impossible to say (my own guess is that perhaps 20 percent as much area is burned today as when Columbus sailed; in the United States, perhaps 5 percent). This has been felt at different places at different times to different degrees. There seems to be no good measure, no explicit effort (as yet) to quantify the process. Anecdotal evidence gleaned from the historic record has its equivalent in anecdotal numbers generated in the absence of any valid historic baseline. The world did not begin with CO_2 measurements in East Anglia.

In particular, it is far too easy to dismiss aboriginal or agricultural burning, especially in pre-Encounter landscapes. The Australian example is neither trivial nor exceptional. Perhaps a majority of the world's grasslands were created through a convergence of climate and fire, but they were subsequently maintained through routine anthropogenic burning. When those pressures were removed, the landscape spontaneously regenerated to forest or scrub. For much of the world, the European Encounter has meant an increase in woody vegetation at the expense of grasslands.

The terrestrial biotas of the planet had more or less adapted to anthropogenic fire—or at least they had until the expansion of Europe scrambled ecosystems, and industrialization through fossil-fuel combustion began refiguring the combustion calculus of the earth. The reduction and redistribution of biomass burning has thus created for many ecosystems a fire deficit. Fossil-fuel combustion may substitute for some lost processes—may establish new energy flows, fashion novel nutrient cycles, substitute money and information for phosphorus and nitrogen, and so on (see figure 4.19). But for systems retained in quasi-natural conditions there is no real alternative except the restoration of anthropogenic fire according to something like its historical regime. Burning fuelwood in ovens does not have the same ecological effect as burning it in the woods.

In this context "pollution" and "disruption" are relative terms, like "weed" or "socially maladjusted." Granted the longevity, power, and ubiquity of anthropogenic fire, it is difficult to establish a baseline for change. There is no meaningful "natural" condition, only an existential earth that has known many pasts and can evolve into many futures.

Existential Earth

All this has meanings both theoretical and practical for historical ecology. They begin with the search for a template. If the changes inspired by industrial combustion are to be assessed, then they must be measured against some standard. Without a meaningful historical baseline, numbers are only numbers and the scenarios that computer models generate from them are only a technological fairy tale. A standard is essential for scientific understanding; it is also essential for political and ethical judgments about what humans should do. But anthropogenic fire has for so long and with such complexity interacted with the earth's landscapes that a norm is difficult to discover.

Europe's Agricultural Fire Triangle

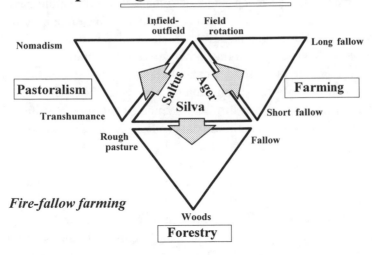

Fire-fallow farming

Europe's Industrial Fire Triangle

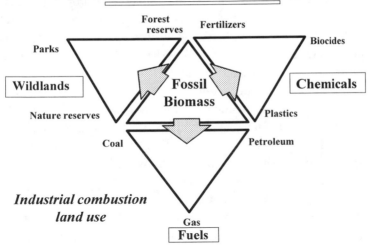

*Industrial combustion
land use*

Fig. 4.19 *The Two Fire Triangles of Europe. Top*: Since ancient time agriculture has been subdivided around three land uses: *ager* (arable), *saltus* (pasture), and *silva* (woods). Each has forms of fallow, and both saltus and silva often serve as long-term fallow for ager. Since fire follows fuel, the fire regimes of agricultural Europe have obeyed the rhythms of the fallowed fields. *Bottom*: The industrial conversion: fossil biomass has substituted for fallow, and industrial pyrotechnologies for traditional burning fueled by living biomass. Coal, gas, and oil supply fuel, and allow for the mechanical replacement of draft animals by vehicles; they furnish substitutes for the chemical by-products of open burning, from petroleum distillates to herbicides, pesticides, and plastics; and they replace agricultural fallow with nature reserves, an uncultivated landscape better suited to the values and practices of an industrial and urban society. There is little interest in traditional fire. Yet fire remains. Revanchist fuels feed a new era of wildfire unless surrogate forms of controlled burning (like prescribed fire) preempt its role.

There is in Western thought a long tradition of belief in something called a balance of nature, a natural order independent of humans that reflects, ultimately, the willed design of a Creator. That natural order in turn sustains a moral order, such that human law should build upon natural law. Literary and philosophical conventions like the Noble Savage and Virgin Forest derive their cultural strength from just such suppositions: they testify to a natural standard against which European civilization can judge itself. This supposition clearly underwrites part of the arguments for nature reserves that can thrive untrammeled by human artifice. Their purer natural order preserves a purer moral order, a less corrupted text of revelation.

It is difficult, however, to detect a clear baseline in fire history. Too much has changed—too many landscapes, too many climates, too long a history of combustion, and especially too ancient an alliance with anthropogenic fire. No one time or place or biota can claim privileged status as primary nature. Worse, the longevity of anthropogenic fire makes the task of disentangling a purely natural fire from the symbiotic regimes created by millennia of human activity all but impossible. And what is true for fire history is true for the history of landscape ecology at large.

The European premise that a still-Edenic nature existed on earth was based on the belief that native peoples in the Americas did not share in the Fall, and that they lived in a state of prelapsarian innocence with the landscape around them. That is not true—not, at least, in the sense that savants willed it. The wilderness that European explorers discovered had been shaped by its human inhabitants. Wilderness is not an ecological condition so much as it is the interplay between a constantly evolving state of nature and a constantly changing state of mind.

What, then, is the order of nature, and where the natural order of fire? This is no merely academic question. It has practical consequences for the management of nature reserves and for understanding the consequences underscoring the choices between biomass burning and fossil-fuel combustion. What makes fire particularly compelling is that there is no way to avoid choice. What humans choose *not* to do can be as powerful as what they choose to do. Removing anthropogenic fire from ecosystems, even those managed as nature preserves, has not restored a prelapsarian nature, but has probably created an environment that has never before existed and is in fact an artifact of human judgment, however ignorant, and of human will, however incomplete.

Nor are all the values that humans instill into the landscape mutually compatible. Reducing fire in some landscapes can, for example, enhance carbon sequestration, but at the cost of biodiversity. Promoting biodiversity by selective burning can overload airsheds, at least locally. Accepting that anthropogenic fire is necessary does not determine just which fire regime is best. Why prefer one regime to another? Why promote one landscape or habitat over another? Why is biodiversity a more important measure than others?

But these are secondary issues. The primary conundrum is to understand better the nature of the competition between the burning of fossil hydrocarbons and biomass; to accept that the earth is a fire planet and that the choice of combustion and fuels has profound ecological consequences; to acknowledge that humans are, for better or worse, the keepers of the earth's flame. Denying the fire does not explain the artifacts forged within it, and extinguishing the flame does not allow those artifactual landscapes to be tended or remade. The trying fire continues, as it has from the beginning, to assay the human presence.

Notes

1. On the human monopoly of fire, see Goudsblom 1992; Komarek 1967; Sauer 1975; Hough 1926; and Bellomo 1990.

2. The best compendium is still Frazer 1974; and see Frazer 1923a:269–300. A useful (and competent) abbreviation of European fire ceremonies is available in Frazer 1923b:638–641.

3. For Australia, see Singh 1982; Lewis 1994. For Iceland, see Thorarinsson 1944. For New Zealand, see Cumberland 1962. For Madagascar see Burney 1986. The chronicles claim that when the Portuguese discovered Madeira in 1420 they set a fire that managed to burn for seven years. Whether this refers to one remorseless burn or to seven years of recurrent burning, the episode rings true to patterns of human colonization.

4. A good source for the reintroduction of fire are the many symposia held in the United States since the mid-1960s. A good summary is Lotan et al. 1985.

5. The literature on the Mediterranean is very large. For general overviews see Conrad and Oechel 1982; Mooney and Conrad 1977. For recent reviews, see U.N. ECE/FAO/ILO 1990, 1992.

6. These topics form another large literature. A recent, representative summary drawn from the Southwest is Krammes 1990. For a distillation of the classic fire-starved forest, consult Mutch et al. 1993. An interesting record of changes are comparative photos, of which a score of studies have been published. The classic is Progulske with Shideler n.d.

7. Information about Dalby Söderskog comes from displays at the park. See also Steensberg 1979, 1993.

8. Again, a vast literature. In addition to the Mediterranean fire symposia listed in n. 5, see Goldammer 1990; Booysen and Tainton 1984; Cowling 1992; Burbidge 1985; Balée 1992, 1993; Anderson and Posey 1991; and Blackburn and Anderson 1993.

9. Probably the best published summary is Steensberg 1993, although he does not organize fire regimes around fallow regimes. A study that does is Pyne 1997.

10. A good digest of progressive nineteenth-century thinking, with a European focus, is Marsh 1965. See also Glacken 1967. An account of the Linnaeus story in English is available in Weimarck 1968.

11. Again, see Marsh 1965. For the transition, however, see Grove 1994.

12. The standard, still-useful history of forestry's development is Fernow 1907. For examples of how it affected fire practices, see Pyne 1982, 1991, 1995.

13. The fullest account is Pyne 1982:100–122. An updated, abbreviated version is found in Pyne 1995.

14. See Stephen Pyne, "The summer we let wild fire loose" and "Vestal fires and virgin lands," in Pyne 1995.

15. See the references under n. 5. French foresters have been particularly active in fire problems in Provence; see *Les incendies de forêts* 1975; *Espèces forestiers et incendies* 1990; and *Forêt Méditerranéenne* 1993.

16. For the intermix scene, see Stephen Pyne, "The fire this time," in Pyne 1995.

17. An interesting discussion of this problem in Sweden is available in *Tänd eld på skogen!* 1991; and in Olsson 1992.

References

Anderson, Anthony B. and Darrell A. Posey. 1991. Reflorestamento indígena. *Ciência Hoje*, special issue on Amazonia, pp. 6–12.

Balée, William. 1992. Indigenous history and Amazonian biodiversity. Pp. 185–197 in Harold K. Steen and Richard Tucker (eds.), *Changing Tropical Forests: Historical Perspective on Today's Challenges in Central and South America*. Durham, N.C.: Forest History Society.

———. 1993. Indigenous transformation of Amazonian forests. *L'Homme* 33(2–4): 231–254.

Behre, Karl-Ernest. 1988. The role of man in European vegetation history. Pp. 633–671 in B. Huntley and T. Webb III (eds.), *Vegetation History*. Dordrecht: Kluwer.

Bellomo, Randy. 1990. Methods for documenting unequivocal evidence of humanly controlled fire at early Pleistocene archaeological sites in East Africa: The role of actualistic studies. Ph.D. diss., University of Wisconsin-Milwaukee.

Blackburn, T. C. and M. K. Anderson, eds. 1993. *Before the Wilderness: Native Californians as Environmental Managers*. Menlo Park, Calif.: Ballena Press.

Booysen, P. de V. and N. M. Tainton, eds. 1984. *Ecological Effects of Fire in South African Ecosystems*. Berlin: Springer-Verlag.

Burbidge, A. 1985. Fire and mammals in hummock grasslands of the arid zone. Pp. 91–94 in Julian Ford (ed.), *Fire Ecology and Management in Western Australian Ecosystems*. WAIT Environmental Studies Group Report No. 14. Perth: Western Australia Institute of Technology.

Burney, David Allen. 1986. Late Quaternary environmental dynamics of Madagascar. Ph.D. diss., Duke University.

Clark, R. L. 1981. Bushfires and vegetation before European settlement. Pp. 61–74 in Peter Stansbury (ed.), *Bushfires: Their Effect on Australian Life and Landscape*. Sydney: University of Sydney.

Conrad, C. Eugene and Walter C. Oechel, tech. coords. 1982. *Proceedings of the Symposium on Dynamics and Management of Mediterranean-Type Ecosystems*. General Technical Report PSW-58. U.S. Forest Service.

Covington, W. W. and M. M. Moore. 1992. Postsettlement changes in natural fire regimes: Implications for restoration of old-growth ponderosa pine forests. Pp. 81–99 in Merrill R. Kaufmann et al. (tech. coords.), *Old-Growth Forests in the Southwest and Rocky Mountain Regions*. General Technical Report RM-213. U.S. Forest Service.

Cowling, Richard, ed. 1992. *The Ecology of Fynbos*. Oxford: Oxford University Press.

Crosby, Alfred. 1986. *Ecological Imperialism: The Biological Expansion of Europe, 900–1900*. Cambridge: Cambridge University Press.

Cumberland, Kenneth B. 1962. "Climatic change" or cultural interference? New Zealand in Moahunter times. Pp. 88–142 in Marry McCaskill (ed.), *Land and Livelihood*. Christchurch: New Zealand Geographical Society.

Dusha-Gudym, Sergei I. 1992. Forest fires on the areas contaminated by radionuclides from the Chernobyl nuclear power plant accident. *International Forest Fire News* 7 (August): 4–6.

Espèces forestiers et incendies. 1990. *Revue Forestière Française*, special issue.

Fernow, Bernhard. 1907. *A Brief History of Forestry.* Toronto: Toronto University Press.

Forêt Méditerranéenne. 1993. Vol. 14(2).

Frazer, James. 1923a. *Balder the Beautiful: The Fire-Festivals of Europe and the Doctrine of the External Soul*, vol. 1. New York: Macmillan.

———. 1923b. *The Golden Bough.* New York: Macmillan.

———. 1974. *Myths of the Origin of Fire.* New York: Hacker Art Books.

Gabriel, H. W. and F. Tande. 1983. *A Regional Approach to Fire History in Alaska.* BLM-Alaska Technical Report 9. Anchorage: Bureau of Land Management.

Glacken, Clarence J. 1967. *Traces on the Rhodian Shore.* Berkeley: University of California Press.

Goldammer, J. G., ed. 1990. *Fire in the Tropical Biota.* Ecological Studies 84. Berlin: Springer-Verlag.

Goudsblom, Johan. 1992. *Fire and Civilization.* London: Penguin Group.

Grove, R. H. 1994. A historical review of early institutional and conservationist responses to fears of artificially induced global climate change: The deforestation-desiccation discourse 1500–1860. *Chemosphere* 29(5): 1001–1014.

Higgins, K. F. 1984. Lightning fires in grasslands in North Dakota and in pine-savanna lands in nearby South Dakota and Montana. *Journal of Range Management* 37:100–103.

Högbom, A. G. 1934. *Om skogseldar förr och nu och deras roll i skogarnas utvecklingshistoria.* Uppsala.

Hough, Walter. 1926. *Fire as an Agent in Human Culture.* Smithsonian Institution Bulletin 139. Washington, D.C.: Smithsonian Institution.

Houghton, R. A., R. D. Boone, J. R. Fruci, J. E. Hobbie, J. M. Melillo, C. A. Palm, B. J. Peterson, G. R. Shaver, G. M. Woodwell, B. Moore, D. L. Skole, and N. Myers. 1987. The flux of carbon from terrestrial ecosystems to the atmosphere in 1980 due to changes in land use: Geographic distribution of the global flux. *Tellus* 39B:122–139.

Houghton, R. A. and David L. Skole. 1990. Carbon. Pp. 393–408 in B. L. Turner II, W. C. Clark, R. W. Kates, J. F. Richards, J. T. Mathews, and W. B. Meyer (eds.), *The Earth as Transformed by Human Action: Global and Regional Changes in the Biosphere Over the Past Three Hundred Years.* New York: Cambridge University Press with Clark University.

Les incendies de forêts. 1975. *Revue Forestière Française*, special issue.

Iversen, Johannes. 1973. The development of Denmark's nature since the last glacial. *Danmarks Geologiske Undersogelse* 5 rk, vol. 7-C.

Kailidis, D. S. 1992. Forest fires in Greece. Pp. 27–40 in U.N. ECE/FAO/ILO, *Seminar on Forest Fire Prevention.*

Kilgore, Bruce and Dale Taylor. 1979. Fire history of a sequoia mixed conifer forest. *Ecology* 60(1): 129–142.

Komarek, E. V. 1967. Fire—and the ecology of man. *Proceedings, Tall Timbers Fire Ecology Conference* 6:143–170.

Krammes, J. S., tech. coord. 1990. *Effects of Fire Management of Southwestern Natural Resources.* General Technical Report RM-191. U.S. Forest Service.

Kuhnholtz-Lordat, Georges. 1938. *La terre incendiée*. Nîmes: Editions de la Maison Carreo.
———. 1958. *L'écran vert*. Paris: Editions du Muséum.
Lewis, Henry T. 1994. Management fires vs. corrective fires in northern Australia: An analogue for environmental change. *Chemosphere* 29(5): 949–963.
Lotan, James E. et al., tech. coords. 1985. *Proceedings—Symposium and Workshop on Wilderness Fire*. General Technical Report INT-182. U.S. Forest Service.
Marland, G., R. M. Rotty, and N. L. Treat. 1985. CO_2 from fossil fuel burning: Global distribution of emissions. *Tellus* 37B:243–258.
Marsh, George Perkins. 1965. [1864]. *Man and Nature: or, Physical Geography as Modified by Human Action*. Edited by David Lowenthal. Cambridge: Harvard University Press.
Mooney, Harold A. and C. Eugene Conrad, tech. coords. 1977. *Proceedings of the Symposium on the Environmental Consequences of Fire and Fuel Management in Mediterranean Ecosystems*. General Technical Report WO-3. U.S. Forest Service.
Mutch, Robert W. et al. 1993. *Forest Health in the Blue Mountains: A Management Strategy for Fire-Adapted Ecosystems*. General Technical Report PNW-GTR-310. U.S. Forest Service.
Olsson, Roger. 1992. *Levande skog*. Stockholm: Naturskyddsföreningen.
Pallot, Judith and Denis J. B. Shaw. 1990. *Landscape and Settlement in Romanov Russia, 1613–1917*. Oxford: Oxford University Press.
Progulske, Donald R. with Frank J. Shideler. n.d. *Following Custer*. Agricultural Experiment Station, Bulletin 674. Brookings: South Dakota State University.
Pyne, Stephen J. 1982. *Fire in America: A Cultural History of Wildland and Rural Fire*. Princeton: Princeton University Press.
———. 1991. *Burning Bush: A Fire History of Australia*. New York: Henry Holt.
———. 1995. *World Fire*. New York: Henry Holt.
———. 1997. *Vestal Fire: An Environmental History, Told Through Fire, of Europe and Europe's Encounter with the World*. Seattle: University of Washington Press.
Rego, F. Castro. 1992. Fuel management. Pp. 207–221 in U.N. ECE/FAO/ILO, *Seminar on Forest Fire Prevention*.
Sauer, Carl O. 1975. Man's dominance by use of fire. Pp. 1–13 in R. H. Kesel (ed.), *Grasslands Ecology: A Symposium*. Geoscience and Man, vol. 10. Baton Rouge: Museum of Geoscience, Louisiana State University.
Sigaut, François. 1975. *L'agriculture et le feu: Rôle et place du feu dans les techniques de préparation du champ de l'ancienne agriculture européenne*. Paris: Mouton.
Singh, G. 1982. Environmental upheaval. Pp. 90–108 in J. M. B. Smith (ed.), *A History of Australasian Vegetation*. Sydney: McGraw-Hill.
Smith, Kirk R. 1987. *Biofuels, Air Pollution, and Health: A Global Review*. New York: Plenum.
Soininen, A. M. 1974. Vanha maataloutemme. *Historiallisia Tutkimuksia 96*. Helsinki: Scientific Agricultural Society of Finland.
Steensberg, Axel. 1979. *Draved. An Experiment in Stone Age Agriculture: Burning, Sowing, and Harvesting*. Copenhagen: National Museum.
———. 1993. *Fire-Clearance Husbandry: Traditional Techniques Throughout the World*. Herning, Denmark: Paul Kristensen.
Stocks, Brian J. 1991. The extent and impact of forest fires in northen circumpolar countries. Pp. 197–202 in Joel S. Levine (ed.), *Global Biomass Burning*. Cambridge: MIT Press.

Suman, Daniel O. 1991. A five-century sedimentary geochronology of biomass burning in Nicaragua and Central America. Pp. 512–518 in Joel S. Levine (ed.), *Global Biomass Burning*. Cambridge: MIT Press.

Tänd eld på skogen! 1991. *Skog & Forskning* 4, special issue.

Thorarinsson, Sigurdur. 1944. Tefrokronologiska studier på Island, II. Svedjning på Island i forna tider. *Geografiska Annaler* 26:173–215.

U.N. ECE/FAO/ILO. 1990. *Proceedings, International Conference on Forest Fire Research* (Coimbra).

———. 1992. *Seminar on Forest Fire Prevention, Land Use, and People* (Athens).

Weimarck, Gunhild. 1968. *Ulfshult: Investigations Concerning the Use of Soil and Forest in Ulfshult, Parish of Örkened, During the Last 250 Years*. Acta Universitatis Lundensis, Sectio 2, no. 6.

Williams, Michael. 1989. *Americans and Their Forests: A Historical Geography*. Cambridge: Cambridge University Press.

Zackrisson, O. 1977. Vegetation dynamics and land use in the lower reaches of the River Umeälven. *Early Norrland* 9:7–74.

CHAPTER 5

Diachronic Ecotones and Anthropogenic Landscapes in Amazonia: Contesting the Consciousness of Conservation

D ARRELL A. P OSEY

Overspecialization of Western science has helped to obscure human-environmental aspects of coevolutionary relationships between Amerinds and Amazonian ecosystems. Social anthropologists have traditionally been inadequately trained in field aspects of the natural sciences, while biologists seldom consider the cultural components—historical and present—of ecological systems. Biologists and ecologists find it difficult to accept as empirical the knowledge of scientifically untrained folk experts, whose practical experiences have generated well-adapted ecological management systems. On the other hand, romanticists have undermined scientific investigations with simplistic allegations that natives live in harmony with nature (Alvard 1993; Johnson 1989; Redford 1990; Stearman 1994). Thus the day-to-day realities of success and failure in indigenous experiments with strategies of natural resource management have generally been overlooked and understudied.

Much investigation of folk knowledge stops at inventories of native plant and animal names and their uses. Attempts to correlate basic inventories with folk taxonomic systems and related utilitarian patterns of behavior inevitably lead to studies of symbolic and metaphysical concepts that express the logic of other realities (Balée 1994; Descola 1994; Murphy 1960; Reichel-Dolmatoff 1976). Most scientists avoid these subjects because they are nonquantifiable and, consequently, nonscientific. Natural scientists tend to be skeptical of the scientific validity of the social sciences, while social scientists are weary of the cognitive limitations of the analytical paradigms and statistical methodologies of the natural sciences.

A bridge between the natural sciences and the social sciences is, however, hypothesis generation and testing (Posey 1986, 1990). The investigation of folk sciences not only enriches existent fields of Western science, but can also generate more sophisticated hypotheses through the discovery of new categories of knowledge. This requires the elicitation of indigenous cognitive (emic) categories of classification utilizing a generative methodology that seeks native concepts rather than imposing ethnocentric ones (cf. Posey 1983a, 1983c, 1986). The following example will illustrate the importance of this point.

The felling of enormous trees by Amazonian Indians in order to gather honey too high to reach otherwise is thought to be an antiecological practice. Ethnobiologists and cultural historians studying these practices from an *emic* point of view, however, inevitably employ a historical analysis as well. In the case of Kayapó Indians, the result of trees felled for this purpose is a *b`a-krê-ti*. For the Kayapó, a *b`a-krê-ti* is a large forest opening that offers a modified ecological zone into which hundreds of useful medicinal and edible plants can be introduced to form a concentrated "island" of resources near trailsides and campsites (Posey 1985). Game animals are also attracted to these enriched areas, since edible foods are scarce in the lower levels of the surrounding high forest; the Indians even plant some species intentionally to attract birds and mammals. In this example, the felling of one large tree offers an immediate source of honey and provides a new ecological niche for useful plants that develop into a long-term hunting ground. Indigenous concepts of botanical use and conservation in relation to faunal components are integrated into an overall management system that, without the benefit of an emic analysis and a historical perspective, would escape the scientific eye.

A formidable barrier to research is the scientists' lack of credence in folk specialists. This manifests itself in a reluctance to allow the informant to lead the researcher along unfamiliar lines of logic and into areas of research *that the native chooses*. Scientists resist the loss of control of the questioning paradigm and fear leaving the base line of the "reality" that *control* signifies. Concerns about research time also inhibit emic analysis, since restraints on field stays often mean that researchers are reluctant to trade assured results from their project designs for possible "finds" from informants.

Folk categories may be more elaborate than their Western counterparts—or they may have no corresponding Western categories whatsoever, thereby indicating phenomena unknown to our science. The "discovery" of such categories may provide the greatest contribution of ethnobiology to scientific development. The description of the fifty-two folk species of stingless bees named, classified, and utilized by the Kayapó led to the "discovery" of nine species previously unknown to Western science (Posey and Camargo 1985). More than fifty types of diarrhea and dysentery are utilized in Kayapó ethnomedical curing, as compared to less than a dozen utilized by allopathic physicians working in the tropics (Posey and Elisabetsky 1991; Elisabetsky and Posey 1994). The elaboration of taxonomies more complex than scientific ones has also been demonstrated for Amazonian ecological zones (Frechione et al. 1984).

Complex ideas and relationships are often expressed in the highly symbolic codes of myth and ritual and require research over an extended period of time. The

Kayapó recognize, for example, two mythological entities that illustrate how be-liefs can function as ecological concepts. One is Bepkôrôrôti, the spirit of an an-cient shaman unjustly killed by fellow tribesmen while seeking his inherited right to certain parts of a tapir. His spirit now manifests itself in the form of rain, light-ning, and dangerous storms, which can kill people or destroy crops. He becomes angry when people do not share food and resources; fear of his vengeance compels the Kayapó to be generous and sharing. To placate Bepkôrôrôti, Indians leave a portion of honey, pollen, and broodcomb in raided hives. The result of this practice is that some species of stingless bees will return to reestablish their colonies. Thus belief in Bepkôrôrôti helps to preserve and manage bee colonies while insuring continued production.

Another mythological entity is the *mry-ka`ak*, which takes the form of a twenty-or-more-meter-long electric-eel-like animal that lives in deep pools of water. It is the most feared of all creatures, since it can kill with its powerful electric shock from a distance of five hundred or more meters. It is thought to live from minnows, and therefore whenever the Kayapó see schools of spawning fish or minnows, they stay clear of the area for fear of the *mry-ka`ak*. This practice serves to protect the minnows, which are the basic element of the aquatic foodchain of the river.

There is not a single Kayapó who would tell the investigator that Bepkôrôrôti is a bee-management concept, or that the *mry-ka'ak* functions to preserve riverine ecology. I dare say that no Kayapó, not even the wisest and most knowledgeable elder, would even answer "yes" to questions that ask directly about such ecological management functions as I have described them. This does not mean, however, that they do not exist. Researchers cannot avoid interpreting culture in ways unfamiliar to the people they study, since knowledge—and its interpretation—is codified in different ways in different cultures.

Natural Components of Extended Society

The many "components" of nature for indigenous peoples become an extension, not just of the geographical world, but of human society. This is fundamentally dif-ficult for Western society to understand, since the extension of "self" is through "hard technology," not nature (Martins 1993). For indigenous peoples, "natural models" may even serve as templates for social organization, political thought, and modes of subsistence. This also implies radical differences in concepts of time and space. A *Polistes* sp. wasp nest, for example, serves as the natural model for the Kayapó universe: its parallel broodcombs represent the many layers of the upper and lower worlds, while its circular base prescribes the form of an idealized village (Posey 1981).

"Myths" and "folklore" have been analyzed for their structural, metalinguistic, and symbolic components, and have even been shown to regulate ecological as well as social cycles. They have perhaps been less studied as sounding boards for cultural and environmental change.

The Kayapó myth known as "The Journey to Become a Shaman" (Posey 1982) is exemplary of how oral tradition works to explain ecological-social relationships and the changes that occur within them. The myth, dealing with the transformation of the *wayanga* (shaman), goes as follows:

> Listen! Those who become sick from strong fevers lie in death's position; they lie as though they are dead. The truly great ones, the truly strong person who is a *wayanga*, shows the sick how to leave their bodies. They leave through their insides. They pass through their insides and come to be in the form of a stone. Their bodies lie as in death, but beyond they are then transformed into an armadillo. As an armadillo, they assume good, strong health and they pass through the other side, over there [pointing to the east].
>
> Then they become a bat and fly—ko, ko, ko, ko, ko . . . [the noise of flying].
>
> Then they go further beyond in the form of a dove. They fly like a dove—ku, ku, ku, ku . . . [the sound of a dove's flight]. They join the other *wayangas* and all go together.
>
> "Where will we go? What is the way? Go to the east, way over there." Ku, ku, ku, ku. . . .
>
> And way over there is a spider's web. . . . Some go round and round near the spider's web and they just sit permanently. The true and ancient shamans must teach them how to fly through the web. But those who have not been shown how, try to break through the web and the web grabs their wings thusly [the narrator wraps his arms around his shoulders]. They just hang in the web and die. Their bodies are carried by their relatives and are buried without waiting, for the spider's web has entangled them, wrapped up their wings, and they are dead.
>
> Those who have been caused to know themselves, however, go round the spider web. They sit on the mountain seat of the shamans and sing like the dove—tu, tu, tu, tu. . . . They acquire the knowledge of the ancestors. They speak to the spirits of all the animals and of the ancestors. They know [all]. They then return [to their bodies]. They return to their homes. They enter and they breathe.
>
> And the others say: He arrived! He arrived! He arrived! He arrived!
>
> And the women all wail: "*ayayikakraykyerekune*."[1]
>
> [And the shaman says] "Do not bury me, I am still alive. I am a *wayanga*. I am now one who can cure: I am the one who smokes the powerful pipe. I know how to go through my body and under my head. I am a *wayanga*.

The story is centered around the capability of the *wayanga* to leave his or her body (*kà*) and be transformed into other physical forms. Energy (*karon*) can be stored temporarily in rocks, but inevitably gets transformed into armadillos, doves, or bats. The spider's web represents the barrier between the visible and invisible worlds. Armadillos are persistent animals that know to burrow under the web; doves are powerful flyers and can break right through the barrier; while bats are such skillful fliers that they maneuver through the strands.

The sounds of the doves' and bats' flights represent the different frequencies that their vibrations impart. Frequencies have equivalent sounds and colors. Just to analyze the variations in frequencies of bee sounds would require discussing the fifty-two different folk species of stingless bees, each of which has a distinctive sound and curative properties.

The most powerful shamans can transform themselves into not just one of the animals, but all of them. And once on the other side of the spider's web, after they have passed through the endless dark chasm, they enter into the spectral frequencies of different light (or colors). There is a different spectral frequency for each animal (*mry-karon*). The general term for undifferentiated energy is *karon*. Defined energies are given distinctive modifiers (*x-karon*), where *x* might be *mry* for animals (*mry-karon*), *tep* for fish, or *kwen* for birds, and so on.

Some shamans learn the secrets of only a few animals and their energies, while others "know all" (in the words of the myth). They have learned about all of the spectral frequencies and their respective animal energies. Upon return to their bodies, the *wayanga* begin to "work with" (*nhipex*) the animal energies encountered in their transformation. The basis of the "work" is to maintain a balance between animal energies and human energies. Eating the meat of, coming in contact with, or even dreaming about animals can cause an imbalance in these energies—as can, of course, a well-elaborated list of antisocial actions. *Wayanga* use a great variety of techniques for restoring balance (they can also create imbalances—*kané*—that lead to sickness), but plants are the most common "mediators" that manipulate this balance (Posey and Elisabetsky 1991).

Plants themselves have energies (*karon*), but most do not have distinctive energies or spirits (*x-karon*), per se. One exception is some of the *mekrakindjà* ("child-want-thing," plants that aid in conception), which have very powerful spirits and cause the user to dream of a child's conception. Men and women use these dreaming plants, although men are usually the ones who first "conceive"—i.e., first "see" (*kra pumunh*) the child in a dream (Elisabetsky and Posey 1989). Other plants with spirits (i.e., defined energies or *x-karon*) are the *metykdjà* ("the poison plants"), the *mẽudjy* ("witchcraft plants"), and the most deadly and powerful *pitu* (no direct translation): these cause drastic alterations to human beings, such as death, paralyses, blindness, insanity, abortion, etc. Even less-powerful plants have qualities that can either harm or help the balance between human and animal energies (*mẽ-karon* and *mry-karon*)—indeed, it appears that all plants have curative values.

In any case, the Kayapó respect both plants and animals, since their energies are keys to the health of their own society. Permission is asked of the *mry-karon* when taking the life of an animal, and songs of appreciation are offered to the spirits of the dead animals. Likewise, annual rituals extol the importance of plants and instill a great sense of respect for their overall role in the socioecological balance (Posey

1981, 1982b). The Kayapó have no question about the dependence of their existence and future health upon plants and animals and the forces of nature.

Normally, spirits of the dead pass easily into the other world (*mēkaron nhon pyka*) and continue their existence in what is roughly the mirror image of what goes on in this world. "Deceased" (they never really die, for they have already died and just disappear and reappear) *wayanga* live in a special cave in the mountains—hence the reference in the myth to their stone seats. Spirits of dead animals also go to the "other world." Devoted pets are sometimes killed and buried with their "owners" at death, so that the human spirit will not be so lonely (some Kayapó say that dogs are buried with their owners because the dogs can help the human spirit find its way to the "other world").

Those who attempt a shamanistic transformation and do not succccd, however, have a more tragic end: their spirits are lost forever in the spider's web. There is disagreement among the Kayapó as to what this really means, but there is no doubt that it is a terrible fate. It is little wonder that only a small portion of the Kayapó population have ever tried to become a *wayanga*.

Kwyra-ka, one of my shaman/mentors, showed great concern when the first coffin arrived in Gorotire and his nephew was buried in it instead of in the traditional manner. (The Kayapó traditionally bury the body in a crouched position in deep, round pits, covered with logs and soil. Until recently, a secondary burial was practiced four days after the principal burial; this allowed time for the spirit to return to the body in case the "dead" person was only on a shamanistic journey.) He anguished over the possibility that the soul of the child would not be able to escape from the casket to the other world. Likewise, Kwyra-ka and Beptopoop (another shaman) both expressed their serious concern with the plants that were taken during ethnobotanical surveys to be pressed and dried in herbaria. They worried about what would happen to the plants' energies. If the plants were kept in such closed, sterile places, would their spirits be trapped, thereby provoking an imbalance and danger to the Kayapó, as well as to those who "kept" them? Like the small child closed within the casket, would their energy not become imprisoned, thereby impeding the "natural" cycles?

Even deeper concerns are expressed about the massive quantities of plants that would have to be collected to provide the oils, essences, colorings, and the like for the commercialization of plant products. The *wayanga* ask: "Has any one ever consulted the plants?" Would the dreaming that is necessary for the conception of healthy children be jeopardized? Would the plants stop mediating between the human and animal *karons*, thereby leading to the loss of ancient cures and provoking new diseases?

The central concepts of ecological management are deeply embedded and codified in myth from which environmental and social change can also be measured. However, it is important to realize that the forces or energies exemplified in myths are not historical in the Western sense: time may be cyclical, spiral, and/or multidimensional. No matter how hard historical ecologists try, the linearity of time and

space that pervades our categories of interpretation will never capture the nonlin-
earity of some indigenous ecological concepts. This is a serious, inherent limitation
to historical ecology.

Basic Elements of the Kayapó Management Model

The principal elements of Kayapó management have been previously described
in some detail (Posey 1985, 1993). They include the following:

1. Overlapping and interrelated ecological categories
2. An emphasis on ecotone utilization
3. The modification of "natural ecosystems" to create ecotones
4. The extensive utilization of "semidomesticates"
5. The transfer of biogenetic materials between similar ecozones
6. The integration of agricultural with forest management cycles

These principles are permeated by diachronic processes and historical developments that depend upon interactions within and between ecological zones.

Several options are possible for representing indigenous resource management
models. The most inclusive and descriptive representation of the Kayapó system
places savanna or grasslands at one end of a continuum (with *kapôt* as the focal
type, or the ecozone that most typifies the category) and forests at the other (with *bà*
as the focal type). Those *kapôt* types with more forest elements would be represented toward the forest pole, while *bà* types that are more open and with grassy elements would lie toward the savanna end of the continuum. This would put *apêtê* at
the conceptual center, or cognitive interface, since these ecozones introduce forest
elements into the savanna. However, agricultural plots (*puru*) also lie conceptually
at the same place on the continuum, since these bring sun-tolerant vegetation into
forest niches. On this basis, I have also suggested that *apêtê* are conceptual inverses
of *puru*, but function in similar ways since they both serve as "resource islands"
where useful plants can be concentrated in known, managed areas.

Ecological types like high forest (*bà tyk*) or transitional forest (*bà kamrek*) are
not, however, uniform. All forests have edges (*kà*), or openings caused by fallen
trees (*bà krê ti*). These provide zonal variations within the conceptual type and provide transitions between different types. Thus, a plant that likes the margins of a
high forest might also grow well at the margin of a field or an *apêtê*. A plant that
likes light gaps provided by forest openings might also like forest edges or old
fields. Plants from open forest types or forest edges could also be predicted to do
well along edges of trails (*pry*) or thicker zones of *apêtê*. Using the same logic, the
Kayapó would transfer biogenetic materials between matching microzones so that
ecological types are interrelated by their similarities rather than isolated by their
differences. Ecotone recognition and management are, therefore, the uniting elements of the overall system.

In contrast, another interesting dimension to the model emerges when looking diachronically across the system. Agricultural clearings are essentially planted with rapidly growing domesticates, but almost immediately thereafter are managed for secondary forest species. This management depends upon a variety of strategies, which include introducing some varieties (planting and transplanting), removing some elements, allowing others to grow, encouraging some with fertilizer and ash, and preparing or working the soils to favor certain species (Hecht and Posey 1989). The introductions are to provide long-term supplies of medicinals and other useful products, as well as food for humans and animals. The old fields (*puru tum*), sometimes erroneously considered inactive "fallows," are as useful to the Kayapó as are agricultural plots or mature forest. A high percentage of plants in this transition have single or multiple uses (I have estimated 85 percent, although it is only an estimate based upon surveys of sample plots; elsewhere I have argued that fundamentally the Kayapó believe that *each* and *every* plant has a use or potential use).

When the secondary forest grows too high to provide undergrowth as food for animals (and hunting becomes difficult), then the forest is cut again for use as an agricultural plot. It is, therefore, more useful to think of the Kayapó system as focused on the management of transitional forests (*chronological ecotones*). The Kayapó system depends more on NDRs (non-domesticated sources—varieties that have been managed, but not brought into domestication) than on cultivars, although agricultural produce has become increasingly important for the sedentary, postcontact Kayapó.

I have recently hypothesized in Anna Roosevelt's *Amazonian Indians* (1994) that the dependence on "resource islands" is related to hostility and warfare. The more unstable the political system, the greater the dispersal of Kayapó groups and the greater the dependence upon concentrations along forest trails, in forest openings and *apêtê*—an overall pattern that I have called "nomadic agriculture." When peace and stability returns, agriculture begins to dominate and the reliance on non-domesticated resources decreases.

There have been several cycles of warfare-peace-hostility-stability that have caused the shrinkage or expansion of the nomadic or settlement agriculture. Until very recently, the Kayapó were in open warfare against each other and their neighbors; thus, nomadic agriculture was still evident until the early 1980's. Since then, however, settlement agriculture has come to dominate and most trails, *apêtê*, and distant campsites have been abandoned. Nonetheless, the management of old fields (secondary forest vegetation) continues, and useful varieties of NDRs brought from distant villages or other ecozones still play prominent roles in the management cycle (Anderson and Posey 1989).

The flexibility and complexity of the Kayapó system cannot, therefore, be fully appreciated without a profile that shows the fundamental modifications in their historical past. Likewise, the impact of indigenous warfare, travel, and trade would continue to be undervalued without a diachronic analysis of plant distribution and changing subsistence patterns.

Attacks upon the Model

The dynamic of the Kayapó ecological management system is based upon the transfer of useful plants across distances, sometimes great distances, into managed concentrations. In a recent attack on my *apêtê* work, Eugene Parker (1992, 1993) compares the genera and species of managed "forest islands" with "natural" *campo-cerrado* vegetation. Based on these crude comparisons, he concludes that *apêtê* are not vegetatively distinct from nonmanaged areas: the Kayapó do not plant NDRs, nor are they conscious manipulators of nature. Parker's methodology is hopelessly flawed, with no identified control site, no botanical voucher collection available in any herbarium, and confusion over *cf* and *sp* attached to genera as meaning positive field identifications to species. His interviews were with monolingual Kayapó informants, but his limited Portuguese and lack of Kayapó language skills meant that no emic data were obtained.

Just as cognitive categories of illness do not necessarily overlap, neither do botanical classification systems. Species determinations (and even more so, generic classifications) are inadequate to describe folk systems that frequently focus on varieties or subvarieties of species. This is not hard for a gardener who searches for new varieties of tulips to understand. The desired variety may be abundant in the garden next door, but does not yet grace his garden: the frequency of the bulb is not the problem, it is the distribution of it. Ethnopharmacologists also know well how annual cycles and ecosystem differences can significantly modify qualities of even the same variety. And woe to any field-worker who does not recognize the difference between bitter and sweet varieties of *Manihot esculenta*. In other words, distinctions made only at the scientific generic (and sometimes even species) levels may be misleading, if not deadly. This fine-tuning of data analysis makes historical ecology problematic when results from different observers utilizing different methods and levels of analysis are compared.

Parker's attack also fundamentally questions whether indigenous peoples have a conservation mentality and are conscious of the effect of their actions on the environment. This is an old debate in anthropology: how conscious is "common sense"? It is an interesting question only when it evokes detailed investigation into the actual behavior. Indians cast seeds, throw them into termite holes, scatter them in ashes, dig little holes to drop them into while they are defecating, step on them as they walk, and otherwise modify the environment in a myriad of ways. They may not be conscious of these everyday "common sense" acts, but they may become aware of them when they are properly queried about them.

Over many years of research, my informants and I worked out a common understanding of what was meant by the Portuguese verb "to plant" (*plantar*), which does not coincide with any single Kayapó word, phrase, or concept. Someone else asking questions about planting would not necessarily get the same answers because the agreed frame of reference is different. The term "plantable" refers to a plant that is subject to planting—i.e., that the Kayapó in some way propagate or encourage to grow.

Questions about planting and plantable are inherently ambiguous. It is impossible to know, for example, which individual in a patch of daffodils in Central Park was the planted progenitor. We can deduce with certainty, however, that they *were* planted—or their ancestors were planted; in other words, the daffodils are there because of one or more human acts. Likewise, the Kayapó may not be able to consistently identify a botanical specimen as planted, but they know that certain species are plantable and that human beings have molded certain identifiable historic landscapes.

The *apêtê* that Parker attacks are arguably as much human artifacts as Central Park, even if planting (by anyone's definition) does not occur and native species dominate. They are full of trails, paths, human-cut niches, openings, fire pits, charcoal, machete cuts on the trees and shrubs, left-over woven sitting mats, remains of old pots and tools, and discarded food remains: these reflect conscious acts that have a long historical tradition and undeniably create anthropogenic landscapes.

The critical question, then, is not whether indigenous peoples are conscious of their actions, but rather whether scientists consider such acts as conservation—bearing in mind that there is no general agreement within or between scientific specialities as to what conservation is, how it is measured, or whether it is attainable. Archaeologists have similar difficulties in agreeing upon what constitute "sites."

To my knowledge, cultural or anthropogenic landscapes in Amazonia were first detected by Protásio Frickel (1978), who recognized the relationship between historical Indian village sites and enriched forest stands. Nigel Smith (1980), although not the first to recognize the human impact on soils, called attention to the extensiveness and importance of *Terra Preta do Indio* (literally "black soils of Indians," or anthropogenically enriched soils; see Whitehead, chapter 2). Subsequently, I described the Kayapó management principles and processes that affect the Amazonian forest (Posey 1982b), but it was some time later that I realized that indigenous management practices also had significant impacts on scrub savanna and other landscapes (Posey 1983b, 1984).

Once scientists are alerted to the nature of ecological "footprints of the forest" (see Balée 1994), then anthropogenic landscapes become more easily recognized. Dr. James Ratter of the Royal Botanic Garden, Edinburgh, and Professor Peter Furley, Department of Geography, University of Edinburgh, for example, have reported two "*apêtê*-like" mounds of vegetation in the Brasília National Park as having "no other obvious explanation other than being manmade," and they believe that "many more of these islands may exist" (Furley and Ratter, n.d.). Until relatively recently, the *campo-cerrado* around Brasília was certainly inhabited by Gê peoples, including the Kayapó.

William Denevan (1966), Clark Erickson (1995), and William Balée (1995) report thousands of "forest islands" in the Llanos de Mojos of eastern Bolivia. Several thousand of these manmade structures, often including earthworks, have been mapped to date. No floristic inventories have been published, although such surveys are in progress. The distances between these islands in eastern Bolivia and those of the Kayapó give only circumstantial support to *apêtê*, but they show that anthropogenic forest islands are a part of Amazonian aboriginal landscapes.

Political Dimensions of Human Ecology

In Parker's response (1993) to my "Reply to Parker" (Posey 1992),[2] the reader is provided with his attempt to "deconstruct almost every sentence" (p. 721), a careful reading of my corpus (p. 716), and an additional 3,500 or so words that pose as additional data. Alarmingly, his deconstructions have been considered by the former editors of the *American Anthropologist* as having the same scientific validity as field research. This has serious implications for historical ecology and related approaches, since diachronic studies depend upon the highest standards and integrity of historical research and scientific investigation.

Furthermore, the implications of Parker's allegations are extremely dangerous for indigenous peoples, who struggle to maintain their lands and territories against those who see them as detached social entities not linked to their surroundings. The result of such logic is that native peoples have been, and are being, forcefully moved from their homelands without concern for their biocultural environments.

Evidence that indigenous peoples not only interact with, but in fact affect, mold, and can even increase the biological diversity on their lands and territories is, therefore, not a collection of detached, irrelevant data to be merely argued over from opposing ivory towers (see Sponsel 1995). The undermining of such finds would be enthusiastically received by some governments, political and economic decision-makers, banks, and development agencies that would like nothing more than to prove that culture is divorced from the environment, just as ecological conservation would be divorced from human rights, or development from local communities. Developers could then continue to move and remove indigenous peoples from their lands with impunity and with the implicit blessings of science—and historical ecology.

This is not to suggest that scientific data should be manipulated to fit one's political bent, but it is a warning to those who think that their academic pursuits do not have an impact beyond their fellow specialists. The historical links between a people, their society, and the cultural landscapes with which they interact are essential, for example, in determining intellectual property rights over biogenetic resources commercialized from their doorsteps. Even the term *wild*, carelessly bantered about by scientists, has profound implications for indigenous and traditional peoples in relation to the ownership of, access to, and control over resources. "Wild" species and landscapes are presumed to be "natural": local communities can claim no intellectual property rights over biological materials removed from "nature," even if they have used, conserved, and modified them over millennia.

Environmental impact assessments (EIAs) are increasingly important in guiding development where evidence of anthropogenic impact bears heavily upon whether or not projects proceed and local peoples are forcibly resettled. The Convention on

Biological Diversity (CBD) virtually calls on the nationalization of traditional knowledge, innovations, and practices—meaning that the findings of historical ecologists will provide primal evidence for local communities attempting to defend their claim of eminent domain. Finally, as "free trade" (via GATT, TRIPS [Trade-Related Aspects of Intellectual Property], and the new World Trade Organization) further entrenches itself as the basis for the new world order, historical ecologists may find themselves in key roles of defending legacies of biological and cultural diversity against the powerful forces that would reduce the earth to private commodities of raw materials.

Conclusion

Historical ecology has an important role in alerting scientists in many disciplines and subdisciplines to the nature and importance of human influence in the conservation, management, and even stimulation of biological and ecological diversity. Anthropogenic qualities of landscapes are becoming increasingly evident, thereby dismissing the assumption that human societies can only be destructive to nature and that traditional resources are "wild." Such finds also highlight the necessity of including diachronic profiles in all biological and ecological studies.

The Kayapó of Brazil, for example, practice long-term "nomadic agriculture" that includes the management of forest openings, trailsides, and rock outcroppings. They are responsible for the highly anthropogenic *apêtê* ("forest islands") of *campo-cerrado*, and essentially create *chronological ecotones* through their sophisticated management of disturbed ecological zones and secondary forests emerging from agricultural plots. The system is sufficiently flexible to accommodate changing subsistence requirements during periods of peace or hostility. Yet, to overlook the indigenous cognitive (emic) categories and historical processes central to the system is to miss the genius of Kayapó cultural landscapes.

The debate over indigenous peoples' consciousness of conservation draws attention to the difficulties of conducting interdisciplinary and multidisciplinary research, since the consciousness of conservation, like other "common sense" aspects of culture, is difficult to elicit and is frequently encoded in myth and ceremony. Direct questioning alone cannot achieve the "intuitive" nature of research that is based upon another diachronic process affecting historical ecology: that of the long-term relationship between investigators and their cultural consultants.

As important as historical ecology may become in furthering our understanding of environmental processes, it is fundamentally limited by an inextricable tie with Western notions of the linearity of time and space. The metaphysical, cosmic, and invisible forces that permeate indigenous belief will always keep Western scholarship at arm's length from defining the underlying forces of society and nature.

Nonetheless, historical ecology has much to contribute to the human/nature debate through its interdisciplinary and diachronic approaches. Furthermore, the increased global emphasis on locally controlled ecological conservation, countered by a strong resistance from centralized economic and political forces on community empowerment, will thrust historical ecologists, and others who document anthropogenic and cultural landscapes, into significant political debates on the sustained use and management of the components of biological diversity—now, and for many years to come.

Notes

1. This phrase has no direct translation. It roughly means "He has arrived!" For more information on the details of this myth, see Posey 1982a.

2. My reply was prompted by his initial attack on my Kayapó research (Parker 1992).

References

Alvard, M. S. 1993. Testing the "ecologically noble savage" hypothesis. *Human Ecology* 21(4): 355–387.

Balée, William. 1985. Ka'apor ritual hunting. *Human Ecology* 13(4): 485–510.

———. 1994. *Footprints of the Forest: Ka'apor Ethnobotany—The Historical Ecology of Plant Utilization by an Amazonian People*. New York: Columbia University Press.

———. 1995. Historical ecology of Amazonia. Pp. 97–110 in Sponsel (ed.), *Indigenous Peoples and the Future of Amazonia*.

Denevan, W. 1966. *The Aboriginal Cultural Geography of the Llanos de Mojos of Bolivia*. Ibero-Americano 48. Berkeley: University of California Press.

Descola, P. 1994. *In the Society of Nature: A Native Ecology in Amazonia*. Cambridge: Cambridge University Press.

Elisabetsky, E. and D. A. Posey. 1989. Use of contraceptive and related plants by the Kayapó Indians (Brazil). *Journal of Ethnopharmacology* 26:299–316.

———. 1994. Ethnopharmacological search for anti-viral compounds: Treatment of gastrointestinal disorders by Kayapó medical specialists. Pp. 90–101 in *Ethnobotany and the Search for New Drugs*, CIBA Foundation Symposium 185. London: CIBA Foundation.

Erickson, C. 1995. Archaeological methods for the study of ancient landscapes of the Llanos de Mojos in the Bolivian Amazon. Pp. 66–95 in P. W. Stahl (ed.), *Archaeology in the Lowland American Tropics*. Cambridge: Cambridge University Press.

Frechione, J., J. Eddins, L. F. da Silva, and D. A. Posey. 1984. Ethnoecology as applied anthropology in Amazonian development. *Human Organization* 43(2): 95–107.

Frickel, P. 1978. Areas de aboricultura pré-agrícola na Amazônia: Notas preliminares. *Revista Antropológica* 21(1): 45–52.

Furley, P. and J. Ratter. n.d. Unpublished report of an ecological survey from central Brazil. Royal Botanical Garden, Edinburgh.

Hecht, S. and D. A. Posey. 1989. Preliminary findings on soil management of the Kayapó Indians. Pp. 174–188 in Posey and Balée (eds.), *Resource Management in Amazonia*.

Johnson, A. 1989. How the Machiguenga manage resources: Conservation or exploitation of nature? Pp. 213–222 in Posey and Balée (eds.), *Resource Management in Amazonia*.

Martins, H. 1993. Hegel, Texas: Issues in the philosophy and sociology of technology. Pp. 226–249 in H. Martins (ed.), *Knowledge and Passion: Essays in Honour of John Rex*. London: Tauris.

Murphy, R. F. 1960. *Headhunter's Heritage: Social and Economic Change among the Mundurucu Indians*. Berkeley: University of California Press.

Parker, E. 1992. Forest islands and Kayapó resource management in Amazonia: A reappraisal of the *apêtê*. *American Anthropologist* 94:406–428.

———. 1993. Fact and fiction in Amazonia: The case of the *apêtê*. *American Anthropologist* 95:715–723.

Posey, D. A. 1981. Ethnoentomology of the Kayapó Indians of central Brazil: Wasps, warriors, and fearless men. *Journal of Ethnobiology* 1(1): 165–174.

———. 1982a. The journey of a Kayapó shaman. *Journal of Latin American Indian Literatures* 6(3): 13–19.

———. 1982b. Keepers of the forest. *New York Botanical Garden Magazine* 6(1): 18–24.

———. 1983a. Ethnomethodology as an emic guide to cultural systems: The case of the insects and the Kayapó Indians of Amazônia. *Revista Brasileira de Zoologia* 1(3): 135–144.

———. 1983b. Indigenous ecological knowledge and development in the Amazon. Pp. 225–257 in E. Moran (ed.), *The Dilemma of Amazonian Development*. Boulder, Colo.: Westview Press.

———. 1983c. Indigenous knowledge and development: An ideological bridge to the future? *Ciência e Cultura* 35(7): 877–894.

———. 1984. Keepers of the campo. *Garden* 8(6): 8–12, 32.

———. 1985. Indigenous management of tropical forest ecosystems: The case of the Kayapó Indians of the Brazilian Amazon. *Agroforestry Systems* 3:139–158.

———. 1986. Topics and issues in ethnoentomology, with some suggestions for the development of hypothesis generation and testing in ethnobiology. *Journal of Ethnobiology* (special vol.: *New Directions in Ethnobiology*) 6(1): 99–120.

———. 1990. Introduction to ethnobiology: Its implications and applications. Pp. 1–8 in D. A. Posey and W. L. Overal (eds.), *Proceedings of the First International Congress of Ethnobiology (Belém, Pará)*. Belém: Museu Paraense Emílio Goeldi/CNPq (Brazil).

———. 1992. Reply to Parker. *American Anthropologist* 94(2): 441–443.

———. 1993. The importance of semi-domesticated species in post-contact Amazonia: Effects of the Kayapó Indians on the dispersal of flora and fauna. Pp. 63–72 in C. Hladik, H. Pagezy, O. Linares, A. Hladik, and H. Hadley (eds.), *Tropical Forests, People, and Food*. Man and the Biosphere Series, vol. 13. Paris: UNESCO and Parthenon.

———. 1994. Environmental and social implications of pre- and postcontact situations on Brazilian Indians: The Kayapó and a new Amazonian synthesis. Pp. 271–286 in Roosevelt (ed.), *Amazonian Indians from Prehistory to the Present*.

Posey, D. A. and W. Balée, eds. 1989. *Resource Management in Amazonia: Indigenous and Folk Strategies*. Advances in Economic Botany, vol. 7. Bronx: New York Botanical Garden.

Posey, D. A. and J. M. F. Camargo. 1985. Additional information regarding the keeping of stingless bees (*Meliponinae*) by the Kayapó Indians of Brazil. *Annals of the Carnegie Museum* 54(8): 247–274.

Posey, D. A. and E. Elisabetsky. 1991. Conceitos de animais e seus espíritos em relação a doenças e curas entre os indios Kayapó da Aldeia Gorotire, Pará. *Boletim do Museu Paraense Emílio Goeldi* 7(1): 21–36.

Redford, K. 1990. The ecologically noble savage. *Orion Nature Quarterly* 9(3): 24–29.

Reichel-Dolmatoff, G. 1976. Cosmology as ecological analysis: A view from the rain forest. *Man* 11:307–318.

Roosevelt, A., ed. 1994. *Amazonian Indians from Prehistory to the Present: Anthropological Perspectives*. Tucson: University of Arizona Press.

Smith, N. J. H. 1980. Anthrosols and human carrying capacity in Amazonia. *Annals of the Association of American Geographers* 70(4): 553–566.

Sponsel, L. E., ed. 1995. *Indigenous Peoples and the Future of Amazonia: An Ecological Anthropology of an Endangered World*. Tucson: University of Arizona Press.

Stearman, A. M. 1994. Revisiting the myth of the ecologically noble savage in Amazonia: Implications for indigenous land rights. *Newsletter, American Anthropological Association*, pp. 2–6.

CHAPTER 6

Metaphor and Metamorphism: Some Thoughts on Environmental Metahistory

Elizabeth Graham

In this chapter I shall do three things. First, I explain the chapter title, because it reflects the antonymous routes I am following in the design and implementation of research concerning the long-term impact of humans on the environment. Next I describe how I envision the concept of "historical ecology." Although my perspective may turn out to be idiosyncratic, it is nonetheless relevant to describe it here in order to place the discussion of my research in an intellectual context. In any event, I keep this section brief, because both William Balée's and Neil Whitehead's contributions to this volume will serve to integrate more fully the variety of ideas and approaches concerning historical ecology.

In the remainder of the chapter, I focus on ideas that have grown out of my archaeological and environmental research in Belize, in Central America. Since I began excavating Maya archaeological sites in Belize in 1973 (Heighway et al. 1985) my aims have included, and they continue to include, the recovery of information on the culture history of ancient Maya civilization. More recently I have been involved in the excavation of sites dating to the Spanish colonial period (Graham, Pendergast, and Jones 1989; Graham 1991), and I have been irrevocably drawn to and influenced by the work of my colleague and coinvestigator Grant Jones (1989) concerning the range of community responses to Spanish rule in the frontier regions of Yucatan. Underlying the archaeologically oriented work, however, and coming much more strongly to the fore in my recent research on Ambergris Caye (Graham and Pendergast 1989), has been my interest in the immense transformative processes of human occupation.

I credit William Balée with helping me to crystallize my thinking on just how extensive and intensive the human impact on the environment has been in my area of research. We discovered the closeness of our ideas on the complexities of human/environmental interaction when we met at a Forest History Conference in Costa Rica in 1991 (Steen and Tucker 1992). The result of this meeting was the extension of my interest to the Amazon region, a visit to Brazil and to the Amazon in 1993, and the firm conviction that approaches to assessing long-term human impact would be far more fruitful if scholars in Mesoamerica and Brazil cooperated on human-impact studies.

Therefore, what I will discuss in the third section is an attempt to lay the theoretical groundwork for cooperative research. The specific topic of this section falls largely within the purview of a natural science paradigm, but it is designed to contribute to approaches at other levels that are guided by humanistic concerns. In addition, I hope that this contribution will be seen to strengthen the value of a historical-ecological framework.

Metahistory, Metamorphism, Metaphor

In composing a title, I wanted to include something of each of the themes I hoped to explore in this chapter. I used *meta*history because the project that I have started in Belize and hope to extend to Brazil grows out of my interest in long-term environmental change, the kind of change that cannot be envisaged without incorporating sweeping expanses of time. The sweep is not quite geological time, though, and its range is firmly the human experience—which is why I have used the term *history*, as in the history of human/land relationships, rather than, say, the *evolution* of human/land relationships. History puts the spotlight on human decision-making, and all outcomes are important. Evolution, although a potentially more powerful and encompassing explanatory paradigm, is also (by nature) more selective, and some outcomes and processes, from the perspective of time, are inevitably privileged over others. In keeping with the theme of this volume, my emphasis lies with human decision-making and not with retrospective judgments about the adaptability or fitness of outcomes.

Even so, I have used *history* in its loose sense of "*all* recorded events of the past" (*Webster's New World Dictionary*, 2d college ed.) rather than in the strict sense of "written records" (also see Crumley, foreword). This should not be surprising: as an archaeologist I accord the stratigraphic record as much importance as the epigraphic—so that *how* history is recorded, which is the basis for the distinction between history and prehistory, is of less consequence than the fact that there is something there to be read or interpreted. If I had to pinpoint the period in which the process of human (or hominid) alteration of the face of the earth was set in motion, I would begin with the Pliocene (White, Suwa, and Asfaw 1994; Kidder, chapter 7, this volume).

My title also includes the terms *metaphor* and *metamorphism*. This is obviously a play on words—but both the play and the words have meaning. In recent years I have been trying to develop a working model that would enable me to quantify particular sorts of changes wrought by human experience in the Maya area. One way to do this is to reduce the changes, no matter how tied they might be to cultural context, to a common denominator that will facilitate quantitative analysis. In other words, I am proposing to reduce human experience to its physical properties—specifically, by arguing for the inclusion of the decay products of the world's civilizations in soil formation processes. At first, this may seem to imply that the natural sciences can account for all that is human. Ironically, however, by proposing to expand a *scientific* concept, in particular the concept of "metamorphism" in geology, to include within its *natural* purview what scientists have historically rejected as *cultural*—the built environment—I hope to engender a metaphorical change in the image of nature (Merchant 1990:683). By using the argument that scientists have not been sufficiently rigorous, and indeed are guilty of excluding particular avenues of scientific enquiry on the basis of their *cultural* perception, I hope to make a point about "nature" as a construct of culture and education. And of course as a construct, "nature" must continually be reinvented and defended. This is by no means an original thought (see Bird 1987; and especially Haraway 1991 and Dwyer 1996), but my approach is perhaps an original example.

Historical Ecology

Before discussing soil formation processes, I would like to consider what kind of anthropology we mean when we refer to "historical ecology," particularly since neither "history" nor "ecology" is related to anthropology in either disciplinary goals or methods. The kind of ecology that is best complemented by the descriptor *historical* is the original and base-line concept of ecology, which focuses on the diverse roles and relationships of organisms. Although Ernst Haeckel was the first to use the word (*Ökologie*) (1911:793–794, as cited in Ellen 1989:66), Gilbert White's *Natural History of Selborne*, published in 1789, is said to be the first book on ecology because it set the stage for viewing organisms not as isolated individuals but as parts of a community, interacting with other organisms and with the environment (May and Seger 1986:256). This broad view of ecology clearly had a place for the study of historical influences on community and species relationships, but the consideration of history in ecology apparently steadily diminished from the time of the neo-Darwinian synthesis of the 1940s until only very recently (Losos 1991:1002; Brooks and McLennan 1991). We could say, then, that the 1990s may mark a return to a consideration of history in both ecology and anthropology, and therefore our use of the term *historical* in tandem with *ecology* is not as antithetical as it might have appeared ten years ago.

   There are more programmatic approaches extant within ecology that have been adopted in the social sciences to varying degrees. One approach focuses on populations that are thought of as "mechanistic dynamical systems with characteristic ways of reacting to disturbance" (May and Seger 1986:256); another is structured by the way in which natural selection is seen to influence social behavior (264). Neither of these approaches directly concerns us here, although evolutionary ecology, practiced by a number of anthropologists, seems tied more closely to an interest in populations as dynamical systems.

Balée (1992b:35, 52) has criticized evolutionary ecology as being ahistorical, and indeed it is; but this feature is not a reflection of misuse but rather of design (Winterhalder and Smith 1992:3–11). In evolutionary ecology, and in the modeling of dynamical systems, the *origin* of a particular type of disturbance or the *source* of a change in resource supply is simply not of primary interest to the investigator. Research concentrates on a population and on the nature of its reaction to disturbance, and in anthropology the "disturbance" most often takes the form of resource fluctuation. The best of these studies in anthropology focus on populations that are marginal to industrial societies and/or operate in contexts in which there is considerable dependence on hunting or foraging (e.g., E. A. Smith 1981, 1991; Stewart 1993). The intent seems to be to select situations in which "signals"—and by this I mean some observable behavior, often related to risk-taking or to survival strategies—are not masked by "noise" in the form of unwanted signals that interfere with the observation but do not bear on the behavior (Gauch 1993). Evolutionary ecology is designed to deal with prediction in the strict scientific sense of model-building, so that it seeks to isolate patterns in order to predict the future of behavior (e.g., Winterhalder and Goland 1993); the recovery of "noise" is thereby penalized in an effort to amplify patterns (Gauch 1993:468, 477).

One could say that "noise" is the stuff of historical ecology. As Karl Butzer (1990:698) has suggested with reference to his long-term research on Aín, a mountain village in eastern Spain: "The purpose of historical studies is not to extrapolate from the past to predict future scenarios. The goal is to understand how communities and societies cope with crises or respond to change." The second part of this statement is actually rather ambiguous, but Butzer makes the point that history does not turn on prediction. Whether it ought to is a matter that has been vigorously debated (Dray 1966; Gardiner 1982), but in the intellectual climate of the 1990s it is safe to assume that history need not center on prediction or explanation in the positivist sense. Despite the absence of predictive potential, we nevertheless learn from history. We let insight gained from the historical experience affect our decision-making, and we use information from history in the hope of influencing others; but history is not about building models to predict decisions that people will make.

At most, the methods of history, as well as the methods of anthropology, enable us to distinguish patterns in human behavior, but these patterns are usually complex, "noisy," and sometimes apparent only from the perspective of the passage of time. Balée's (1992b:37–41) finding concerning what he terms the regression of some Tupi-Guarani groups from agricultural to exclusively foraging societies is an excellent example of just such a pattern, as is his observation of the dependence of many modern foragers on old fallows and anthropogenic forests (1992a; 1992b: 41–48; 1989:6–16)—areas that were once part of cultivation or land-use cycles but have since been abandoned, perhaps as long ago as pre- or protohistoric times, to long-term forest regrowth.

Evolutionary ecology, with its treatment of the environment in terms of the range of resources it provides, is not designed to generate questions about the historical transformation of resources or resource practices, so that the role of old fallows in the subsistence practices of Amazonian foragers or the changing of Tupi-Guarani groups from agricultural to exclusively foraging societies are patterns that become apparent only through methods of historical inquiry. On the other hand, evolutionary ecology provides one paradigmatic means by which resource use and the nature of risk-taking at different periods of time can be compared and even evaluated. Thus evolutionary ecology can be used productively to interpret particular historical episodes. For example, Balée's (1992b:52) proposal that agriculture becomes increasingly implausible for a group approaching band size seems closer to evolutionary ecology, and is a good example of the kind of knowledge that can result from an evolutionary perspective. (See also Dyson-Hudson and Smith 1978; Winterhalder 1981.)

From the 1950s to the 1970s, evolution was emphasized in both anthropology and archaeology at the expense of history. The negative impact this has had on Amazonian studies is now widely recognized (see Balée 1989:2–6), and a different picture of Amazonia is resulting from new fieldwork and the restudy of older work (Roosevelt 1993:255). Our mistake as anthropologists now would be to emphasize history at the expense of evolution. History and evolutionary ecology aim to develop different kinds of knowledge, and each has advantages and pitfalls. At the present time, in the 1990s, we are in the process of reacting against the errors of past evolutionary approaches (e.g., Meggers 1954)—not because all conclusions drawn about environments and resource use were wrong (comparing tropical to temperate climate agriculture remains a relevant if problematic issue [Janzen 1973]), but because it was assumed, in effect, that environments and resource use had no history (Wolf 1982:4).

Historical Ecology and an Alternative to Ecosystems

Where will historical ecology then lead us? One of the things I hope to see is a consideration of models that are not, strictly speaking, systems models, which are common in both cultural and evolutionary ecology (Burnham and Ellen 1979). This is not because I fail to recognize that at least one systems concept, the "ecosystem" (Tansley 1946), has been instrumental in beneficially altering the character of ecological work in anthropology (Ellen 1989:74). The problem is that concepts from different systems models are sometimes transferred without sufficient consideration of their utility (a point made by Ellen 1989:200–201; Golley 1984; E. A. Smith 1984), and the term "system" itself is often used rather loosely—which is perfectly acceptable, as long as equilibrium, and especially homeostasis, is not assumed under these approximate or analogical conditions. But even when the term is used in the looser sense of an open system, because the "system" constitutes the problem or area of interest, then we should define how we envisage its organization, its operation, or its hierarchy (E. P. Odum 1989:177).

The working concept of a balance or equilibrium in natural ecoystems is a plausible one, but its existence is something that should be argued for and not taken for granted. Or, at least, some effort should be made to explain why an underlying assumption of a balance or equilibrium is useful (see Moran 1984:14–15 and Netting 1986:102 for rethinking the ecosystem concept along these lines). Simply assuming that all parts of a system work to maintain a balance or homeostasis via adaptation is what led to the functionalism that characterized much of cultural ecology (Ellen 1989:177–203).

To bring home this point, I am going to turn to the natural sciences for a claim to authority. Drawing inspiration from biology is hardly new in anthropology, and it does not come without problems (Ellen 1989:66–94), but ecology as a discipline was developed by botanists and zoologists (66), and it seems fair to take heed of their advice where ecology is concerned. To quote an observation: "If a balance of nature exists, it has proved exceedingly difficult to demonstrate" (Connell and Sousa 1983:808). It should be as important for anthropology that stability not be assumed a priori; yet in large part we make this assumption when we talk about humans or their behavior as a system, because such talk implies (1) that stability is at least a hypothetical state that humans attempt to achieve, and (2) that we all agree on what is meant by "stability."

In the paper from which I just quoted, the ecologists present their data on the evidence that is needed in order to judge ecological stability, and they offer concrete suggestions as to how a variation on the concept of stability, which they term "persistence," might be more applicable to real ecological systems (see also Holling 1973 and Ellen 1989:187–188). One of the main points for us is not the details of their conclusions, but the fact that they started out by questioning assumptions of stability in living systems and came up with an alternative dynamic of systems interaction. They also draw a contrast with the physicist's classic ideas of stability, which they feel are not suitable for real ecological systems. It is interesting, though, that it is the physicist's concept of stability, with its attendant concepts of homeostasis and maintenance of equilibrium, that has loomed large in anthropological systems (Ellen 1989:177–203, esp. 186–189).

Where ecosystems are concerned, it is my hope that historical ecology will widen the scope of anthropological ecology to include alternatives to the concept of stability. I also see historical ecology as opening up anthropological ecology to the idea that the models we construct may work only at particular analytical levels, and that level-shifting ought to involve new models, a point made by Emilio Moran a decade ago (Moran 1984). What I like best about the term *historical ecology* itself, in fact, is that it embodies contradiction. I see the contingency, directionality, and agency inherent in history as a stark contrast to ecology's emphasis on regularities, cyclical processes, and general behaviors. There is no denying that this terminological mixing bowl holds concepts that may inevitably separate, like oil and water, each according to its own specific gravity; but in certain circumstances the force of a particular question or inquiry, like the rapid whisking of an oil-and-vinegar mixture, can bring about an insight or result that would not have been possible with either conceptual paradigm alone.

Another way to view historical ecology emphasizes its flexibility. In a discussion of human/nature theory, Robert Sack (1990:660–662) divides our intellectual approaches to human/environmental relationships into the realms of society, nature, and meaning—bedfellows as odd as "history" and "ecology." These realms coincide roughly with the academic domains of the social sciences, natural sciences, and the humanities, respectively. Cultural ecology has attempted mainly to integrate the realms of natural and social relations. Historical ecology, although not the only alternative (see Ellen and Fukui 1996), has the potential to integrate all three. By this I do not mean that historical ecology's mandate is to emerge with an overarching theory of nature, society, and mind. However, by embodying contradiction, historical ecology may allow us to move more freely among the three realms because it does not require a macrotheory to integrate them.

Having said this, I will now to turn to my own approach, which is best described as ecosystemic. My method is not based on a classic systems approach because in addition to the problems, noted above, that can arise with such an approach in anthropology, classic systems theory works best when one common item can be seen to be exchanged or regulated throughout the system, and the ideal item in my view is energy (H. T. Odum 1988). Therefore I do not seek to connect humans and nature by focusing on characteristics that both systems possess, as in systems theory, but by reducing human actions to physical ones (Sack 1990:665). This seems to be the ultimate in reductionism—but it is a means, not an end; my hope, as I have already explained, is that the metaphor will ultimately be turned inside out to reveal the world through a most human perspective.

Archaeology and Anthropogenic Soils

The model I shall propose has as its focus the study and analysis of anthropogenic soils—soils heavily influenced by human action. Why, as an archaeologist, have I attached so much importance to soils rather than to some other aspect of human prehistory with which I am familiar?

First of all, I believe that a rigorous approach to the study of anthrosols can contribute substantially to assessments of the long-term impact of humans on the environment, and thereby help fine-tune larger-scale models of the earth as transformed by human action (Turner et al. 1990).

Second, I believe that the way in which human occupation has contributed to processes of soil formation and transformation in the humid tropics is widely underestimated and misunderstood. This is partly because our models of urban development and agricultural health are strongly rooted in the temperate-climate/Near Eastern experience (e.g., Adams 1981). In spite of the attention devoted to tropical agriculture and tropical soils, we see the human/land relationship in terms of this experience, and we associate humans almost universally with degradation. This point of view affects both agricultural development and urban planning in the tropics. Soil *formation*, not just soil degradation, is an important by-product of the human experience. Although this is occasionally mentioned by soils scientists, particularly in relation to organic input into soils (Rozanov, Targulian, and Orlov 1990:204), it pales in comparison to the attention given to degradation. A rigorous assessment of anthropogenic soils ought to contribute to the development of quantifiable assessments of human-influenced soil formation, and this in turn should become a factor to consider in land-use studies.

Third, anthropogenic soil studies will invariably yield new insights into the chemistry of the decomposition of the built environment, and thereby have implications for modern waste-management studies and long-term land management (Egziabher et al. 1994). So far this is the least developed of my avenues of research, but it is one that will widen as time progresses. There is no generally agreed-upon definition of the term "built environment." Here I refer specifically to the processes by which humans accumulate materials for the express purpose of physically and/or chemically altering the shape and structure of the world around them. Thus skyscrapers, sewage systems, land fills, reservoirs, and agricultural terraces are all part of the built environment.

Therefore anthrosol research is not just about soil chemistry, but about larger-scale processes in which soil chemistry is one clue among many that will lead to a greater understanding of global transformative processes. Such processes are perhaps best understood in terms of specifics, and I will now turn to the landscape of the Maya lowlands: how it has been altered by human activity, and how I see the anthrosol research within the context of my archaeological work in Belize.

Landscape in Mesoamerica Altered by Human Activity

The research that I have carried out at a number of archaeological sites in Belize over the past twenty years has revealed a contemporary landscape extensively molded by pre-Columbian activity (e.g., Graham 1989; Graham and Pendergast 1992; Graham 1994). Forests, savannas, swamps, estuaries, and the coral reef, which are relatively lightly inhabited today, were zones of extensive habitation and/or utilization in the past.

The extent and scale of the ancient contribution to the modern Mesoamerican landscape is becoming a common feature of environmental reassessments (Denevan 1992; Whitmore and Turner 1992). Generally speaking, ancient impact is assessed in terms of the extent of land under cultivation, but also in terms of human impact on forests and savannas through the use of fires, the impact on animal populations, and estimates of the general extent of the built environment. Anthropogenic forests are described, and are seen as a reflection of a combination of factors: ancient clearing, burning, culling, and possibly other, more inadvertent human activities (Denevan 1992:373–375).

On the whole, however, human influence is seen as more destructive than constructive in the long term—if not in the Maya lowlands in particular, then globally (Turner et al. 1990). Where actual measurements of human impact are hazarded, they overwhelmingly relate to destructive processes (e.g., Rozanov, Targulian, and Orlov 1990). Whether human influence is seen as constructive or destructive, any dynamic assessment has yet to work its way into natural scientists' methods of studying long-term vegetational succession or soils (e.g., Whitmore 1990; Mabberley 1992). For the most part, whatever humans do, it is considered a distinct, qualitatively different disturbance that cannot be included in what is assumed to be the "natural" state of forest or soil dynamics (Whitmore 1990:99–132; Mabberley 1992:31–51).

One important exception exists in the foundational work of the ecologist Joseph Connell (1978). Connell has set forth a hypothesis concerning the maintenance of diversity in tropical rain forests and coral reefs. He suggests that the high diversity of trees and corals is maintained, not in an equilibrium or climax state, but in a non-equilibrium state as a consequence of disturbance. He suggests that diversity is higher when disturbances are intermediate, and lower when equilibrium is reached or when severe perturbation persists. In this model, not only do humans have a place along with hurricanes as initiators of disturbances that can be measured and quantified, but the focus on the intensity of the disturbance, rather than on the species initiating the disturbance, leaves open the possibility that global biodiversity can be as much the *result* of human activity as it is thought to result from the *absence* of human activity. In my analysis, Connell's work will help significantly in structuring hypotheses concerning the range of inferences we can draw from the studies of vegetation communities associated with anthrosols.

Anthrosols, Metaphor, and Metamorphism

By the term *anthrosol* I refer to soils whose genesis is influenced by human activity (see, e.g., N. J. H. Smith 1980). I have elected to use the term *influence* because soils are generally considered to be the product of interactions between climate and the geologic formations of the earth's crust (Van Wambeke 1992:31). Russian scientists define soils more elegantly as the "solid-phase products of the continuous functioning of the biosphere" (Rozanov, Targulian, and Orlov 1990:203)—which functioning includes, at least nominally, both natural and anthropogenic processes. These scientists subsume both processes in the metaphor of soil as a membrane: "[Soil] is not only a friable layer and a stock of plant nutrients at the land surface, but also a specific membrane, regularly differentiated in space and depth, that regulates biosphere-geosphere interactions" (203).

It is interesting that Russian soil scientists generally, based on the literature with which I have come in contact, seem to have less aversion to seeing plants and animals as having an interactive role with geology and climate in creating a specific soil (Golley 1984:35; Rozanov, Targulian, and Orlov 1990). In Western science, biological factors such as vegetation and fauna are not seen as essential contributors to soil formation, but only as affecting soils near the surface through chemical reactions and physical processes. Humans, however, even as "biological factors," are not noted in standard soils texts as having affected the formation of soil horizons.

The premier organism recognized as significantly affecting soil formation is the termite (Lee and Wood 1971). But even termite activity is assessed, not as "contributing to" soil formation, but as "interfering with" what are otherwise natural processes (Van Wambeke 1992:115). As an archaeologist who has excavated deep cultural deposits and substantial architecture in many areas of Belize that are now forested or covered by alluvial deposits (Graham 1994), I find it difficult to understand why cultural material and human activity should be upstaged by the termite.

Why are cultural phenomena in the form of the debris of house construction, garbage and waste deposition, body decomposition, cooking, processing, craft and other production activities, farming, metalworking, landscape manipulation, and road-building ignored as factors in soils genesis? Why is it that, if human activity is seen at all to figure in soils processes, it is when humans are held responsible for conditions of degradation, erosion, and soil depletion?

There are several reasons why "culture" is interpreted in this way by natural scientists, and why the culture/nature dichotomy is upheld by both natural scientists and anthropologists. The human factor in soil formation processes has been excluded because scientists argue that in geologic time, what humans do is irrelevant. Granted, humans had nothing to do with the formation of the Cambrian shield or the uplift of Cretaceous limestones. However, many soils studies are tied specifically to assessments of land-use for agricultural purposes. At this timescale, surely the effects that thousands of years of human occupation have had on soils should have found their way into the literature on soils genesis. But as far as I know, except for Russian macrotheory, human occupation is excluded as a factor.

This exclusion is not based on a scientific reality, but on a cultural or historical one. With regard to history, tropical forests have been studied as pristine or natural—not because their history was investigated, but because no questions were asked (see reference to Paul Richards in Denevan 1992:373). Not only are there people without history (Wolf 1982), there are landscapes without history as well! The fact that no questions were asked is a reflection of the way that natural science has evolved in the West, as a field in which cultural forces are seen to be operationally separate from natural ones.

With regard to culture, it may simply be that twentieth-century individuals, natural scientists and anthropologists alike, are not accustomed to thinking about their home towns or cities or skyscrapers as temporary or ephemeral. Archaeologists may think about what it might be like in the year 3000 to excavate a shopping center, or to have the bad luck to put your first test pit in a ruined city over the airport runway, but not many people see their surroundings as so transitory. Earth scientists can conceive of the earth's plates shifting, and geomorphologists envision mountains folding, but the artifacts of everyday life—parking lots, towers of concrete and glass, amusement parks, shopping malls, city dumps—are accorded a reality and a stability that are at odds with what we know to be the long-term processes of urbanization and decay, and the growth and decline of civlizations.

To bridge this culture/nature divide, the built environment must be incorporated into theories of soil formation processes. Organic remains, as biological factors, already are included to some extent, although their overall importance is minimized. But as far as I know, the physical aspects of the built environment have been excluded entirely. One way to include the physical aspects of the built environment would be to expand the concept of metamorphism in geology to include changes in materials brought about by human intervention. As it is now, the concept of metamorphism applies only to the earth's transformative processes—when basic rocks are subjected to heat and pressure to such an extent that many of their properties are chemically altered.

All the materials that humans use for building and production are mined in one way or another from the earth. Sometimes they are subjected to alteration through heat and pressure and manipulation to form other substances, such as cements or plastics; from Mesoamerica's past, pottery, metals, and plaster leap to mind as examples. At other times, the earth's materials are merely physically altered, as in the quarrying and cutting of limestone blocks for building construction by the Maya. Whatever the case, ultimately these materials formed part of a built environment that in many places was abandoned and allowed to decay; vegetation regrew, and forests replaced what had once been cities.

The Impact of the Built Environment

I first became aware of the extensive impact of the ancient built environment on the modern landscape when I was carrying out dissertation research in the Stann Creek District of southern Belize, from 1975 to 1977 (Graham 1994). The area in which I worked, unlike most of the Maya lowlands, has virtually no limestone, so that the conditions for archaeological preservation are very different from those in limestone-influenced areas. The acidic soils (see Graham 1994: app. 2) derived largely from granitic parent materials; they preserve pottery extremely badly, and indeed it is rare to find even the most rudimentary sherd on the surface. The dearth of limestone also means that chert sources are lacking, and material for the manufacture of cutting tools had to be imported from outside the region. Under these conditions, the ancient Maya were extremely conservative in the manufacture of tools; items were usually recycled, so that chert chipping debris and broken tools are rare artifacts to find on the surface. Most of the artifacts that are found were made of the locally abundant ground stone or slate—but compared to areas of Maya sites elsewhere, the Stann Creek District, literally on the surface of it, reveals little evidence of Maya occupation, at least in terms of artifact presence. Telltale "mounds," the hallmarks of Maya sites, are not even easy to spot, and bulldozing compounds the problem in the river valleys. The rolling hills of the district mask artificial construction; in some areas natural features were used or only slightly modified by the Maya, so that habitation is not easy to detect. In riverine zones, our excavations revealed that huge expanses of structures were buried by alluvial deposits.

Working in the district under these conditions forced me to become sensitized to features of the landscape and vegetation that might yield evidence of human presence, but were not as obvious as the artifacts and architectural clues we archaeologists are used to. Odd configurations of hillsides, stones foreign to local outcrops, and stands of vegetation replaced artifacts and mounds as indicators of settlement. Perhaps the best example (see Graham 1989:139–143, figure 10.4) is along the Sittee River, where geographers characterize a vegetation zone in terms of the presence of moderately lime-loving species (Wright et al. 1959: natural vegetation map, sheet 2). This zone turns out to coincide with the ruins of the site of Kendal, where calcareous clays, probably mined a good distance upriver, were used as material for floors, plastering, and mortar in monumental and residential construction. The disintegration of Maya buildings clearly affected the soil chemistry enough to permit colonization by lime-loving species that otherwise would not thrive. Our excavations at Kendal, and at the site of Pomona in the North Stann Creek Valley, revealed the extent to which alluvial soils have buried cultural deposits, thus masking the volume of pre-Columbian construction so that, to all appearances, the environment is a "natural" one.

It was in the process of carrying out the Stann Creek investigations that I first became interested in the anomalous stands of vegetation and the presence of black soils along Belize's coast and on the cayes. Broadleaf forest, for example, can be found amidst mangrove swamp, or as stands in areas otherwise characterized by coral sands over limestone. At the present time, David Pendergast and I are excavating sites on Ambergris Caye in northern Belize, in part because of my interest in anthrosols (Graham 1989; Graham and Pendergast 1989). In many of the areas in which I suspect that anthropogenic soil processes have been at work, soils are cultivable and are planted today in orchard and root crops. Our excavations so far have revealed occupation that is more or less continuous from about 300 B.C., and possibly earlier, through the Spanish and British colonial periods, to the present day. Cultural deposits can be as thin as about 50 cm to as thick as 4 m, and they are even thicker where architecture is present.

What interests me most is the idea that anthropogenic processes on the cayes may have enabled the Maya to bring soils under cultivation that, under normal conditions, would support only beach grasses or sedges. Anthropogenic processes such as these may have affected the landscape as early as the Classic period (A.D. 250–900); less-equivocal stratigraphic evidence suggests that cultivation of these soils was in full swing by the Postclassic period (after A.D. 900). My concomitant hypothesis is that the presence of anthropogenic, *terra preta*–type soils enabled the expansion of settlements on the cayes that were formerly dependent on marine resources and imported goods.

Although studies of anthropogenic soils, and especially of the *Terra Preta do Indio*, are most closely associated with Amazonian investigations (see Whitehead, chapter 2; Posey, chapter 5), the Maya lowland zone is also an area where research on anthropogenic soils and human alteration of the landscape should be a prime directive. In fact, parallel studies that are part of a well-integrated, theoretically consistent research inquiry would be ideal. The reasons why research in both areas is important are as follows:

1. The Maya lowlands were an area of high population density and the landscape was intensively manipulated. Therefore the lowlands provide an excellent test case for the long-term effects of urban populations on the wet tropical environment. This complements Amazonian research, where populations were also dense and chiefdoms were established (Roosevelt 1992, 1993), but where the landscape was probably not as urbanized.

2. Maya archaeologists have already accumulated a considerable body of knowledge about the nature of the built environment through time, and about the breadth of cultural processes that gave rise to the built environment. Thus Maya research provides a basis for gauging in specific terms how landscapes were modified and how this changed through time. The archaeology in much of the Amazon area is not as well known, so that Maya research can provide guidelines and models for Amazonian projects.

3. In both the Maya and the Amazonian regions, much of the previously intensively occupied zones now lies under bush, and both regions are under considerable deforestation and development pressure. No time should be lost in undertaking projects in these regions. Natural scientists as well ought to be interested in the nature of tropical forest recovery after heavy urban perturbation. Such opportunities are rare in world history, and their significance is not widely recognized—although closer inspection would undoubtedly reveal parallel circumstances elsewhere, perhaps in the zones of Yoruba urbanism in Nigeria or in the Khmer region of Cambodia.

The Maya Region and the Amazon

Although the Dutch have carried out a considerable amount of research into modern anthropogenic soils—that is, the creation of soils for modern agriculture in Holland—it is in South America, and particularly in the Amazon Basin, that quantitative studies of extant anthropogenic soils have been concentrated (Eden et al. 1984; Kern 1988; Kern and Kampf 1989; Pabst 1991; N. J. H. Smith 1980). Patterns in the geochemistry of anthropogenic soils are emerging; but the significance of these patterns has not been fully grasped, and the results float in a kind of limbo because the soils studies are not conceptually integrated into larger-scale ecological questions concerning the organic and the physical processes of landscape alteration. In many cases in the Amazon, the archaeology and prehistory of an area are not well enough known, so that the relationship of cultural to natural stratigraphy is not understood. In short, anthropogenic soils studies are in danger of persisting in an intellectual vacuum in which the full implications of the research are not realized.

On the whole, anthrosol research has fallen within the purview of geography and the natural sciences rather than archaeology. The advantage has been the recognition of anthropogenic soils as a legitimate subject of study, and the discovery of patterns in anthrosol geochemistry. There is a disadvantage, however, and Nigel Smith's (1980) key article on anthrosols is a case in point. As is sometimes characteristic of the discipline (see Denevan 1992), the range of ancient activity that geographers are prepared to envision seems to be restricted to humans setting fires. Some of these perceptions are changing with archaeologists-turned-geochemists, as in Dirse Kern's work (Kern 1988; Kern and Kampf 1989), but even in this case the research is hampered to some extent because the archaeological stratigraphy of the areas studied is not well known. It is clear that only with the cooperation of scholars from several disciplines and with a focus on more than one region will the significance of anthrosol and related vegetational studies take its rightful place in global environmental research.

Preliminary Research on Ambergris Caye

Large-scale archaeological investigations, based on earlier testing (Graham and Pendergast 1989), began on Ambergris Caye at the site of Marco Gonzalez in 1990. In 1991 we completed an initial vegetation survey in which some tentative ideas were put forward on how a more extensive survey might proceed, and how vegetation assemblages associated with archaeological sites might provide us with indicators of plant associations to expect over soils that lacked artifacts but that nonetheless might be anthropogenic.

In 1992 we carried out a test case of geochemical analysis to see if we could detect mineralogic and crystallographic evidence of lime-processing (Mazzullo, Teal, and Graham 1994). Detecting the presence of lime-processing is important from an entirely archaeological point of view, but because layers of calcium carbonate can occur naturally in soils, the ability to detect what might be called the anthropogenic origin of calcium carbonate layers, especially in the absence of artifacts, would provide an important indicator of human-influenced soils that would otherwise go undetected.

In most cases, circumstantial evidence such as the occurrence of calcium carbonate within levels that contain artifacts forms the basis for the assumption that the calcium carbonate is the result of rehydrated lime. Of more interest to us was geochemical or mineralogic evidence that the calcium carbonate layer was unequivocally the product of rehydrated lime. Salvatore Mazzullo and Chellie Teal therefore set out to determine the mineralogy of the calcium carbonate layers revealed through excavation by X-ray diffraction, and the crystallography by scanning electron microscopy. Ultimately, they were able to present criteria by which recarbonated lime can be distinguished from naturally occurring calcium carbonate deposits.

The specific methods are not of interest here, but it is important to note that the results we achieved on Ambergris Caye were dependent on Mazzullo's many years of familiarity with the geology of the island (e.g., Dunn and Mazzullo 1993; Mazzullo, Reid, and Gregg 1987; Mazzullo and Dunn 1990; Mazzullo and Bischoff 1992): he was familiar with the by-products of the burning of conch for lime, and he knew the specific mineralogy of the Pleistocene rocks and dolomitic rocks in the area. My expertise provided the archaeological context, the interpretation of the stratigraphy, and the dating. We can now apply the method devised by Mazzullo and Teal to studies of soils elsewhere on the caye where vegetational associations suggest pre-Columbian presence, where no artifacts are present, but where calcium carbonate levels occur. My dating of the original levels and reconstruction of past landscapes will not bear on calcium carbonate levels elsewhere, but in the zone of the original test pit, it will affect the conclusions we ultimately draw about the time factors involved in the development of the modern soils profile, and the time factors involved in vegetational succession.

Conclusion

The calcium carbonate research is small-scale and locally focused, but it has larger-scale implications. It provides an example of how actual fieldwork on long-term environmental change must proceed. Work on anthrosols requires archaeologists who know the local and regional archaeology, geologists who know the local and regional geology, geographers who are familiar with the region's biogeography and landscape history, and soils scientists and botanists who do not mind considering that humans may have left enough debris around to affect their subjects of study, soils and vegetation, over the long term. Under these conditions, research into the positive as well as the negative long-term impact of pre-Columbian populations on modern soils and vegetation will take its rightful place in ecological studies.

References

Adams, Robert McC. 1981. *The Heartland of Cities*. Chicago: University of Chicago Press.

Balée, William. 1989. The culture of Amazonian forests. Pp. 1–21 in Darrell A. Posey and William Balée (eds.), *Resource Management in Amazonia: Indigenous and Folk Strategies*. Advances in Economic Botany, vol. 7. Bronx: New York Botanical Garden.

———. 1992a. Indigenous history and Amazonian biodiversity. Pp. 185–197 in Steen and Tucker (eds.), *Changing Tropical Forests*.

———. 1992b. People of the fallow: A historical ecology of foraging in lowland South America. Pp. 34–57 in Kent H. Redford and Christine Padoch (eds.), *Conservation of Neotropical Forests: Working from Traditional Resource Use*. New York: Columbia University Press.

Bird, Elizabeth A. R. 1987. The social construction of nature: Theoretical approaches to the history of environmental problems. *Environmental Review* 11(4): 255–264.

Brooks, Daniel R. and Deborah A. McLennan. 1991. *Phylogeny, Ecology, and Behavior*. Chicago: University of Chicago Press.

Burnham, P. C. and R. F. Ellen. 1979. *Social and Ecological Systems*. London: Academic Press.

Butzer, Karl W. 1990. The realm of cultural-human ecology: Adaptation and change in historical perspective. Pp. 685–701 in Turner et al. (eds.), *The Earth as Transformed by Human Action*.

Connell, Joseph H. 1978. Diversity in tropical rain forests and coral reefs. *Science* 199:1302–1309.

Connell, Joseph H. and Wayne P. Sousa. 1983. On the evidence needed to judge ecological stability or persistence. *The American Naturalist* 121(6): 789–824.

Denevan, William M. 1992. The pristine myth: The landscape of the Americas in 1492. *Annals of the Association of American Geographers* 82(3): 369–385.

Dray, William H., ed. 1966. *Philosophical Analysis and History*. New York: Harper and Row.

Dunn, Richard K. and S. J. Mazzullo. 1993. Holocene paleocoastal reconstruction and its relationship to Marco Gonzalez, Ambergris Caye, Belize. *Journal of Field Archaeology* 20:121–131.

Dwyer, Peter D. 1996. The invention of nature. Pp. 157–186 in Ellen and Fukui (eds.), *Redefining Nature*.

Dyson-Hudson, Rada and Eric Alden Smith. 1978. Human territoriality: An ecological reassessment. *American Anthropologist* 80:21–41.

Eden, M. J., W. Bray, L. Herrera, and C. McEwan. 1984. *Terra preta* soils and their archaeological context in the Caqueta Basin of southeast Colombia. *American Antiquity* 49:125–140.

Egziabher, Axumite G., Diana Lee-Smith, Daniel G. Maxwell, Pyar Ali Memon, Luc J. A. Mougeot, and Camillus J. Sawio. 1994. *Cities Feeding People*. Ottawa: International Development Research Centre.

Ellen, Roy. 1989. *Environment, Subsistence, and System: The Ecology of Small-Scale Social Formations*. Cambridge: Cambridge University Press.

Ellen, Roy and Katsuyoshi Fukui. 1996. *Redefining Nature*. Oxford: Berg.

Gardiner, Patrick, ed. 1982. *The Philosophy of History*. Oxford: Oxford University Press.

Gauch, Hugh G., Jr. 1993. Prediction, parsimony, and noise. *American Scientist* 81(5): 468–478.

Golley, Frank B. 1984. Historical origins of the ecosystem concept in biology. Pp. 33–49 in Moran (ed.), *The Ecosystem Concept in Anthropology*.

Graham, Elizabeth. 1989. Brief synthesis of coastal site data from Colson Point, Placencia, and Marco Gonzalez, Belize. Pp. 135–154 in Heather McKillop and Paul F. Healy (eds.), *Coastal Maya Trade*. Occasional Papers in Anthropology 8. Peterborough, Ont.: Trent University.

———. 1991. Archaeological insights into colonial period Maya life at Tipu, Belize. Pp. 319–335 in David Hurst Thomas (ed.), *Columbian Consequences*, vol. 3: *The Spanish Borderlands in Pan-American Perspective*. Washington, D.C.: Smithsonian Institution Press.

———. 1994. *The Highlands of the Lowlands: Environment and Archaeology in the Stann Creek District, Belize, Central America*. Monographs in World Archaeology no. 19. Madison, Wis.: Prehistory Press and the Royal Ontario Museum.

Graham, Elizabeth and David M. Pendergast. 1989. Excavations at the Marco Gonzalez site, Ambergris Cay, Belize, 1986. *Journal of Field Archaeology* 16:1–16.

————. 1992. Maya urbanism and ecological change. Pp. 102–109 in Steen and Tucker (eds.), *Changing Tropical Forests.*

Graham, Elizabeth, David M. Pendergast, and Grant D. Jones. 1989. On the fringes of conquest: Maya-Spanish contact in early Belize. *Science* 246:1254–1259.

Haeckel, E. 1911. [1868]. *Natürliche Schöpfungsgeschichte.* Berlin: Georg Reimer.

Haraway, Donna J. 1991. *Simians, Cyborgs, and Women: The Reinvention of Nature.* New York: Routledge.

Heighway, Carolyn, Iris Barry, Elizabeth Graham, and Norman Hammond. 1985. Excavations in the Platform 137 group, 1973–1974. Pp. 235–385 in Norman Hammond (ed.), *Nohmul: A Prehistoric Maya Community in Belize.* BAR International Series 250(i). Oxford: B.A.R.

Holling, C. S. 1973. Resilience and stability of ecological systems. *Annual Review of Ecology and Systematics* 4:1–23.

Janzen, Daniel H. 1973. Tropical agroecosystems. *Science* 182:1212–1219.

Jones, Grant D. 1989. *Maya Resistance to Spanish Rule.* Albuquerque: University of New Mexico Press.

Kern, Dirse C. 1988. Caracterição pedológica de solos com terra preta arqueológica na região de Oriximiná, Pará. Master's thesis, Universidade Federal do Rio Grande do Sul.

Kern, Dirse C. and N. Kampf. 1989. Antigos assentamentos indígenas na formação de solos com terra preta arqueológica na região de Oriximiná, Pará. *Revista Brasileira de Ciências do Solo* 13:219–225.

Lee, K. E. and T. G. Wood. 1971. *Termites and Soils.* London: Academic Press.

Losos, Jonathan B. 1991. The phylogenetic perspective. *Science* 252:1002–1003.

Mabberley, D. J. 1992. *Tropical Rain Forest Ecology.* 2d ed. London: Blackie.

May, Robert M. and Jon Seger. 1986. Ideas in ecology. *American Scientist* 74(3): 256–267.

Mazzullo, S. J. and W. D. Bischoff. 1992. Meteoric calcitization and incipient lithification of recent high-magnesium calcite muds, Belize. *Journal of Sedimentary Petrology* 62:196–207.

Mazzullo, S. J. and Richard K. Dunn. 1990. Holocene evolution of a carbonate barrier island, Ambergris Cay, northern shelf of Belize. *Geological Society of America South-Central Meeting, Abstracts with Program* 22:27.

Mazzullo, S. J., A. M. Reid, and J. M. Gregg. 1987. Dolomitization of Holocene Mg-calcite supratidal deposits, Ambergris Cay, Belize. *Geological Society of America Bulletin* 98:224–231.

Mazzullo, S. J., C. S. Teal, and Elizabeth Graham. 1994. Mineralogic and crystallographic evidence of lime processing, Santa Cruz Maya site (Classic to Postclassic), Ambergris Caye, Belize. *Journal of Archaeological Science* 21:785–795.

Meggers, Betty J. 1954. Environmental limitations on the development of culture. *American Anthropologist* 56:801–824.

Merchant, Carolyn. 1990. The realm of social relations: Production, reproduction, and gender in environmental transformations. Pp. 673–684 in Turner et al. (eds.), *The Earth as Transformed by Human Action.*

Moran, Emilio F. 1984. The problem of analytical level shifting in Amazonian ecosystem research. Pp. 265–288 in Emilio F. Moran (ed.), *The Ecosystem Concept in Anthropology.* Boulder, Colo.: Westview Press.

Netting, Robert M. 1986. *Cultural Ecology.* 2d ed. Prospect Heights, Ill.: Waveland Press.

Odum, Eugene P. 1989. Input management of production systems. *Science* 243:177–182.

Odum, Howard T. 1988. Self-organization, transformity, and information. *Science* 242: 1132–1139.

Pabst, Erich. 1991. Critérios de distinção entre terra preta e latossolo na região de Belterra e os seus significados para a discussão pedogenética. *Boletim do Museum Paraense Emílio Goeldi*, Série Antropologia, 7(1): 5–19.

Roosevelt, Anna C. 1992. Secrets of the forest. *The Sciences* 32(6): 22–28.

———. 1993. The rise and fall of the Amazon chiefdoms. *L'Homme* 33(2–4): 255–283.

Rozanov, Boris G., Viktor Targulian, and D. S. Orlov. 1990. Soils. Pp. 203–214 in Turner et al. (eds.), *The Earth as Transformed by Human Action.*

Sack, Robert D. 1990. The realm of meaning: The inadequacy of human-nature theory and the view of mass consumption. Pp. 659–671 in Turner et al. (eds.), *The Earth as Transformed by Human Action.*

Smith, Eric Alden. 1981. The application of optimal foraging theory to the analysis of hunter-gatherer group size. Pp. 35–65 in Bruce Winterhalder and Eric Alden Smith (eds.), *Hunter-Gatherer Foraging Strategies.* Chicago: University of Chicago Press.

———. 1984. Anthropology, evolutionary ecology, and the explanatory limitations of the ecosystem concept. Pp. 51–85 in Moran (ed.), *The Ecosystem Concept in Anthropology.*

———. 1991. *Inujjuamiut Foraging Strategies: Evolutionary Ecology of an Arctic Hunting Economy.* New York: Aldine de Gruyter.

Smith, Nigel J. H. 1980. Anthrosols and human carrying capacity in Amazonia. *Annals of the Association of American Geographers* 70:553–566.

Steen, Harold K. and Richard P. Tucker, eds. 1992. *Changing Tropical Forests: Historical Perspectives on Today's Challenges in Central and South America.* Proceedings of a conference sponsored by the Forest History Society and IUFRO Forest History Group. Durham, N.C.: Forest History Society.

Stewart, Andrew McLean. 1993. Caribou Inuit settlement response to changing resource availability on the Kazan River, Northwest Territories, Canada. Ph.D. diss., University of California, Santa Barbara.

Tansley, A. G. 1946. *Introduction to Plant Ecology.* London: Allen and Unwin.

Turner, B. L. II, William C. Clark, Robert W. Kates, John F. Richards, Jessica T. Mathews, and William B. Meyer, eds. 1990. *The Earth as Transformed by Human Action: Global and Regional Changes in the Biosphere over the Past Three Hundred Years.* New York: Cambridge University Press with Clark University.

Van Wambeke, Armand. 1992. *Soils of the Tropics.* Toronto: McGraw-Hill.

White, Tim D., Gen Suwa, and Berhane Asfaw. 1994. *Australopithecus ramidus*, a new species of early hominid from Aramis, Ethiopia. *Nature* 371:306–312.

Whitmore, Thomas M. and B. L. Turner II. 1992. Landscapes of cultivation in Mesoamerica on the eve of the conquest. *Annals of the Association of American Geographers* 82(3): 402–425.

Whitmore, T. C. 1990. *An Introduction to Tropical Rain Forests.* Oxford: Clarendon Press.

Winterhalder, Bruce. 1981. Foraging strategies in the boreal environment: An analysis of Cree hunting and gathering. Pp. 66–98 in B. Winterhalder and E. A. Smith (eds.), *Hunter-Gatherer Foraging Strategies.* Chicago: University of Chicago Press.

Winterhalder, Bruce and Carol Goland. 1993. On population, foraging efficiency, and plant domestication. *Current Anthropology* 34(5): 710–715.

Winterhalder, Bruce and Eric Alden Smith. 1992. Evolutionary ecology and the social sciences. Pp. 3–23 in Eric Alden Smith and Bruce Winterhalder (eds.), *Evolutionary Ecology and Human Behavior*. New York: Aldine de Gruyter.

Wolf, Eric R. 1982. *Europe and the People Without History*. Berkeley: University of California Press.

Wright, A. C. S., D. H. Romney, R. H. Arbuckle, and V. E. Vial. 1959. *Land in British Honduras*. London: Her Majesty's Stationery Office.

PART TWO

Regional Research and Landscape
Analyses in Historical Ecology

CHAPTER 7

The Rat That Ate Louisiana: Aspects of Historical Ecology in the Mississippi River Delta

TRISTRAM R. KIDDER

> I wish to speak a word for Nature, for absolute freedom and wildness, as contrasted with a freedom and culture purely civil, to regard man as an inhabitant, or a part and parcel of Nature, rather than a member of society.
>
> HENRY DAVID THOREAU, *WALKING*

Introduction: The Rat That Ate Louisiana

According to popular legend, a 1941 hurricane caused an ecological disaster in Louisiana's coastal marshes. Surprisingly, it was not the storm itself that did the real damage—for worse than the winds was the fact that 150 nutria (*Myocastor coypus*) escaped from their pens at Avery Island in the central coastal zone (Conniff 1989; Jackson 1990). These South American rodents rapidly expanded in number, achieving populations estimated to be as high as 20 million by 1959. Nutria have been implicated in the alteration of the ecology of the marsh area in a number of ways. First, although their precise role is still debated, they may have led to the rapid decline in the indigenous muskrat (*Ondatra zibethicus*) populations in coastal Louisiana by outcompeting the native species in essentially the same niche. In 1945–1946 eight million muskrat were taken by trappers in the state, but six years later the catch was down to only one million. By 1962–1963 more nutria were being taken than muskrat, despite the fact that muskrat pelts were desired by the fur industry because of their better quality and commercial appeal. Second, nutria, which are voracious eaters, will literally eat-out and denude the vegetation in certain areas where populations grow to exceed the carrying capacity of the marsh. The eat-outs lead to the long-term destruction of large parts of the marsh, including the root structure (Visser 1994). Both nutria and muskrat eat-outs can lead to local-ized erosion and habitat loss. In some cases these localized conditions are essen-tially permanent, resulting in the formation of open-water lakes and ponds.

Beyond the fact that it was people who introduced nutria into the coastal zone of Louisiana, additional human influences have affected the explosion of nutria popu-lations in the area. One such influence was the severe overhunting of the American alligator, which resulted in a diminished predatory threat to the nutria. As one wildlife biologist noted, young nutria are like popcorn for alligators (Jackson 1990:89); the absence of significant predatory pressure on the nutria was certainly to this animal's advantage. Further, trends in fur use and style (largely dictated in Europe, but also in Asia) emphasized the use of muskrats as opposed to nutria until the mid-to-late 1960s. Thus, not only were natural predators removed, but human predators too shunned the nutria. In this unnatural setting, quite similar to their na-tive habitat in temperate southern South America (Lowery 1974:294–295, map 46), the nutria flourished, and today they are considered an urban nuisance and a con-tributor to accelerated land loss in the coastal zone (Visser 1994).

What do nutria have to do with historical ecology? The present deplorable situa-tion in the Louisiana coastal zone, where land loss is estimated to be roughly 100 km^2 per year, is partly (albeit perhaps minutely) due to the presence of the nutria. Their existence here, however, is not a historically predictable, expected, or natural phenomenon; rather, it is due to human intervention. The results were wholly acci-dental and unexpected, but the consequence is still the same. Furthermore, the other complications caused by humans, such as overhunting alligators, additionally im-plicate human agency in the shaping of modern coastal ecology in Louisiana. If fur-ther proof of this historical contingency is needed, I would point out that while the so-called great nutria escape of 1941 may have been the proximate cause of the nu-tria population boom in the coastal area, these animals were imported into the Louisiana coastal zone as a means of controlling the spread of the water hyacinth—which was in turn introduced into the area from South America during the New Or-leans Cotton Exposition in 1884 (Jackson 1990).

Of course, one could protest that the use of the nutria as an archetypical vehicle of historical ecology is misguided—noting, for example, that their presence and spread is really the result of the modern capitalist economies of the industrial era. The fluctuations of nutria and muskrat populations and their relationship to land loss in Louisiana are tied intricately to the needs of modern fur producers who are, in turn, driven by styles set in distant locations. It seems to me, however, that the nutria, lowly as they may be, stand as an example of the theory of historical ecology in action. Historical ecology has as a fundamental underpinning the notion that his-torical processes are active determinant agents in the expression of an ecosystem at any point in time. My use of the word *determinant* is not meant to indicate that I be-lieve these processes to be inevitable, or to suggest that they are necessarily pur-poseful. The agent and agency may vary, they may or may not be conscious of their actions, and they may be active or passive in their role—but they are always there.

The accident of the nutria is such a nice example precisely because it was unplanned and its end result today could not have been foreseen.

Another protest that the nutria are undeserving of our attention is that the present situation in the Louisiana coast—one of hastened coastal erosion, threats to the native wildlife, pollution, habitat loss or reduction, and a myriad of other other problems—is due only to the modern human ability to vastly alter the ecosystem. While it is fundamentally true that people today have a greater technological (and, one could argue, ethical and philosophical) capacity to alter our world, to suggest that this capacity was beyond the scope of premodern peoples belies the evidence that humans through time, regardless of their technologies, have affected their so-called natural environment. In fact, I would argue that we need to take it as axiomatic that the very nature of human existence has an effect on the environment.

From the earliest members of the genus *Homo*, over two million years ago, to today, humans have been and continue to be, if nothing else, untidy. Archaeologists and paleoanthropologists recognize a fundamental revolution in the course of becoming human when they can identify archaeological sites. These features on the landscape are, after all, only the garbage left behind by our earliest ancestors. Throughout human history we have left our garbage behind, and in doing so we have both subtly and not so subtly altered the ecology of every inhabited continent. Of course, even premodern peoples intervened in their ecosystems to greater extents than just leaving behind their garbage (Butzer 1982, 1990; Hughes 1975). My point, though, is that what has changed during the evolution from *Homo habilis* to contemporary humans is less the quality of our garbage than the quantity of it: our mark on the landscape is more obvious, but not necessarily more consequential.

There is also a philosophical issue embedded in these topics. The notion that premodern peoples did not (or could not) effectively or significantly alter their environment is found in two separate strands of popular myth-history (Heehs 1994; McNeill 1986). In North America today the Native American inhabitants are acclaimed as the first ecologists and environmentalists (Crosby 1986; Sale 1990). A heroic myth has emerged in which the Native Americans lived in harmony with the land, neither taking more than they needed, nor despoiling or altering the land. The prevalence of this picture in the popular media and in Native American myth-history is significant and shapes the present-day discourse over the fate of America's remaining wilderness. Similarly, however, the myth-history of the European settlers of North America worked also to minimize the role that premodern Native Americans had in affecting their environment. This notion dates back to the first colonists and is perpetuated even today. As M. J. Bowden (1992:20) observes:

> The grand invented tradition of American nature as a whole is the pristine wilderness, a succession of imagined environments which have been conceived as far more difficult for settlers to conquer than they were in reality. . . . The ignoble savage . . . was invented to justify dispossession . . . and to prove that the Indian had no part in transforming America from Wilderness to Garden.

In a curious irony of history, the myth-histories of groups at vastly different ends of the philosophical spectrum coincided to perpetuate two myths: at one end was the notion of the precontact Forest Primeval, untouched and unaltered by the natives and heroically conquered by the European colonists; at the other end was the myth of the passive, ecologically invisible Indian (MacCleery 1994:7), the prototype for today's environmentalist. The combined myths resulted in a popular image of

> Indians who lived . . . in harmony with nature, making no irremediable changes in the environment, and handing over to the Europeans a virgin land. Whether denigrated as ignoble savages or idealized as native [*sic*] Americans living in perfect equilibrium and harmony with the environment, the Indians are given no credit for opening up the Eastern Woodlands, for creating much of America's grassland, and for transforming hardwoods to piney woods. (Bowden 1992:20)

In the case of both these myths—the Forest Primeval and the Invisible Indian—historical evidence and modern scholarship demonstrate their patent falsity (Butzer 1990; Denevan 1992). Principally, these claims ignore the basic fact that all humans are a part of their environment, not apart from it. But how these "invented traditions" (Hobsbawm 1983) are created and perpetuated should be an important concern for historical ecologists too. The representation of nature often defines what nature becomes. This is one of the facts of human existence as a species: we impose our vision of what is and should be on the land, and thus create the landscape to suit our needs (Crosby 1994). History, loosely conceived, allows cultures to generate and regenerate the past in the present, and in doing so history serves the ends of those representing the past, not the past itself (Crombie 1988; Hacking 1994:35). In this regard practitioners of ecological history and historical ecology run the risk of creating environments as we think they were. Ian Hacking (1994:47) makes the penetrating observation that "he who describes a certain vision of ourselves and our ecology has that vision himself . . . philosopher and historian alike are part of the ecosystem that has been transformed by bearers of that vision in their interactions with nature as they saw it."

The paradigm of historical ecology is powerful, I think, because it allows us to situate humans in the context of their environment and to explore the ramifications of their participation in the natural world at a number of levels. In the rest of this chapter I investigate how precontact Native Americans subtly (and perhaps not so subtly) altered the ecology of the Mississippi River Delta. What I hope to show is that the effects of human behaviors begun hundreds (and even thousands) of years ago are still with us today, both as marks on the landscape and as influences on the modern environment. The results of these influences on the environment are important, in part, because they survive even in the face of vast and dramatic changes brought on by recent quantitatively dramatic changes in the ecology of this region.

The Great Sewer

The Mississippi River deltaic plain comprises an area of some 28,000 km^2 and serves as home for some four million people. Today this coastal lowland is the basis for a vast industrial and agricultural complex, and accounts for nearly one-quarter of the commercial landings of U.S. fisheries (Van Beek and Gagliano 1984). The land itself, however, is only some 4,000–6,000 years old at the oldest, and parts of the deltaic plain are being formed as this is being written. Of course, at the same time that land is being created, it is being destroyed, too. On the one hand this cycle is a completely natural process, but on the other hand, land loss is rapidly accelerated due to modern human intervention in the ecosystem.

The deltaic plain of the Mississippi is a vast lowland comprised of swamps, marshes, lakes, and bays, interlaced by natural levee ridges of relict and active distributaries of the river. The morphology and ecological structure of the deltaic plain is the result of a long-term process of delta growth and decline that has occurred in an overlapping cycle over the last 6,000 years. The modern coastline of Louisiana is comprised of a series of geologically discrete delta formations, termed *lobes*. Each lobe has its own characteristics and morphology, yet all are created in a similar manner. This pattern of delta lobe evolution is well known and can be predicted with relatively great precision (Coleman 1988, 1981; Gagliano 1984). The four-stage cycle of development and decline of the delta is driven by the introduction of water and sediment via the river and its interaction with geophysical forces, such as subsidence and compaction, combined with the actions of waves and currents in the Gulf of Mexico. Models allow us to comprehend how, at each stage in the overall cycle, the ecological structure and diversity and the biological productivity of the region are manifest (Gagliano 1984; Neill and Deegan 1986; Van Beek and Gagliano 1984).

The first stage in the formation of the delta occurs when a gradient advantage between an upstream portion of the river and the Gulf of Mexico results in a major diversion of the river's flow and a subsequent change in its course to the sea. With this course change comes a relocation of sedimentary deposition and the onset of delta formation. Initially the growth of the delta is subaqueous, but soon natural levee ridges emerge as energy flow diminishes due to the progradation of the river mouth. The natural levees are initially flanked by narrow bands of salt-tolerant wetlands. As distributaries extend seaward, the second phase is initiated. Natural levees increase in elevation, and interdistributary basins become defined. Overbank flooding provides sediments in these basins and encourages the formation of freshwater marshes and swamps. The third phase begins when another upstream diversion occurs, gradually reducing water flow and rates of sedimentation in the system. The absence of significant overbank sedimentation in the interdistributary basins leads to the development of greater salinity gradients and promotes a marked wetland zonation of plants, from saline marsh to freshwater swamp as one moves farther into the interior. The reduction of sediments at the distributary mouth and lateral edges results in increased reworking and net sediment removal by wave erosion, tidal currents, compaction, and subsidence. At this time we see the enlargement of brackish and saline bays and the formation of transgressive barrier islands.

The final stage of the delta cycle is found when the complex is completely, or essentially completely, cut off from river water and sediments. Net sediment loss increases dramatically as the entire delta is slowly eroded by wind, waves, and currents. Barrier islands erode away with the loss of sediments, and bays become progressively larger due to coastal erosion. As these processes are amplified through time, wetlands erode away and natural levee ridges sink beneath the marshes. Subsidence and compaction due to the consolidation of sediments, faulting, and the weight of overlying deposits combine with these natural elements to accelerate the process of destruction. Eventually the entire delta complex disappears and becomes an open bay, and the cycle of delta formation is complete.

The complete cycle of delta development is illustrated in coastal Louisiana today. There are five recognized delta complexes (lobes) superficially extant in the Louisiana coastal zone. From oldest to youngest they are the Teche, St. Bernard, Lafourche, Plaquemines (also known as the Plaquemines-modern), and Atchafalaya (Autin et al. 1991; Coleman 1988). The first and second stages of delta formation are evident in the Plaquemines lobe, which constitutes the present-day mouth of the Mississippi, and especially in the newly developed Atchafalaya River delta, which began forming only in the last fifty years (Autin et al. 1991; Shlemon 1975). Most of southeast coastal Louisiana exhibits characteristics of the third stage of delta formation. The Lafourche and southern segments of the St. Bernard delta lobes are good representatives of what can be termed the mature stage of the cycle. The eastern distal ends of the St. Bernard delta lobe and most of the coastal portion of the Teche lobe demonstrate the final stage of the delta sequence: here the lack of sedimentation results in a net land loss through time as the delta fronts are reworked and eroded by coastal processes, and as subsidence and compaction act to lower the land surface, further hastening the destruction of these coastal marshes.

A model of the relationship between delta development and biological productivity has been developed and tested over the last several decades (Gagliano 1984; Gagliano and Van Beek 1975; Van Beek and Gagliano 1984). A predictable sequence of biological events characterizes each stage of delta development. Best described graphically (Gagliano 1984: figure 1.15; Neill and Deegan 1986: figures 2, 4), this model suggests that biological productivity, measured by the number of habitats and the species diversity of those habitats (Neill and Deegan 1986), will increase as the delta cycle matures, and will reach its maximum in the third stage of delta development. These measures of biological productivity will decline as the fourth stage advances and the delta lobe deteriorates (Gagliano 1984; Gagliano and Van Beek 1975; Neill and Deegan 1986). The delta development sequence also carries important implications for human settlement above and beyond habitat diversity.

Newly emerging and recently emerged delta lobes (stages 1 and 2) are extremely dynamic landforms. Distributary channel diversion, overbank flooding, and rapid habitat changes consequent to landform alterations make these environments unsuitable for permanent human occupation. Only with the advent of stage three in the delta-building cycle will a delta lobe become conducive to extensive and/or intensive human occupation. As the lobe matures, the environment becomes more stable and thus more predictable. As noted above, biological productivity increases, further enhancing the opportunities for human settlement. With the onset of the fourth stage of the cycle the delta lobe will become less hospitable to humans for long-term inhabitation. The cessation of permanent flowing freshwater, the subsidence of distributary systems, the encroachment of salt marshes, coastal erosion, and diminished habitat diversity generally reduce the opportunity for permanent occupation in these now terminal, eroding delta lobes. These final-stage deltas are still usable by humans, however: they present abundant opportunities for exploitation by fisher-forager groups, and in some cases by horticulturalists as well.

Vegetation types in south Louisiana occur in bands roughly parallel to the shoreline, with their abundance and distribution being regulated by local conditions. Distributary channels cut through these vegetation zones as they flow roughly perpendicular to the coast. These distributaries are flanked by elevated natural levees, which slope downward from the stream bank. Elevation above water level and tolerance to salinity have been identified as the principal factors regulating plant distributions in the deltaic plain (Penfound and Hathaway 1938; White, Darwin, and Thien 1983). The diversity of plant species present in a given habitat zone is directly correlated with elevation and salinity, such that 146 plant species are found on the relatively well drained natural levees, but the number declines to only 25 species in the salt marsh (Conner, Sasser, and Barker 1986: table 2).

Interior areas with strictly fresh water and segments of well-elevated levees support hardwood forests in the highest elevations, and are bordered at lower elevations by seasonally inundated swamps. The oak forests are usually comprised of American elm (*Ulmus americana*), water oak (*Quercus nigra*), sweetgum (*Liquidambar styraciflua*), hackberry (*Celtis laevigata*), swamp red maple (*Acer rubrum* var. *drummondii*), and rarely, live oak (*Quercus virginiana*). Understory growth includes a number of herbaceous plants and vines, especially in areas where openings in the canopy allow light to penetrate to the forest floor (Conner, Sasser, and Barker 1986:114). These plants include palmetto (*Sabal minor*), poison ivy (*Rhus radicans*), briar (*Smilax* spp.), pepper-vine (*Ampelopsis arborea*), Virginia creeper (*Parthenocissus quinquefolia*), and green haw (*Crataegus viridis*) (Conner, Sasser, and Barker 1986; Penfound and Hathaway 1938; White, Darwin, and Thien 1983:103–104). Depending on the size of the natural levee and its elevation and slope, the hardwood bottomland forest can be fringed with black willow (*Salix nigra*) and cut-grass (*Leersia lenticularis*). In interior sections of the delta the levees are frequently flanked by cypress-tupelo swamps (Conner, Sasser, and Barker 1986; Swanson 1991), which may support virtually pure stands of cypress (*Taxodium distichum*) or water tupelo (*Nyssa aquatica*), or a mixture of the two. Other tree species found in this habitat include water ash (*Fraxinus caroliniana*), green ash (*F. pennsylvanica*), and swamp red maple. Farther toward the coast the freshwater marsh extends from the fringe of the levee and is comprised of bulltongue

(*Sagittaria lancifolia*), swamp-potato (*S. platyphylla*), giant bulrush (*Scirpus californicus*), and common cattail (*Typha latifolia*) (Conner, Sasser, and Barker 1986; Penfound and Hathaway 1938:51–52; White, Darwin, and Thien 1983:106–107). Even closer to the coast, elevations decrease and salinity increases. These slightly brackish water habitats include hardwood forests only on the highest elevations, with cypress swamp at the lower elevations. The swamp directly abuts the intermediate (or brackish) marsh where wire grass (*Spartina patens*) and salt grass (*Distichlis spicata*) are the most abundant species.

Considerable parts of the inundated interior section of the coastal zone consist of floating mats of either fresh or salt marsh vegetation, called flotant, overlying saturated organic muck and peat (Kolb and Van Lopik 1958:51, figure 14). In some contexts, usually in more moderately brackish settings, the oak forest at the highest elevation is fringed initially by a shrub belt characterized by a predominance of marsh elder (*Iva frutescens*), buckbrush (*Baccharis halimifolia*), and palmetto. At slightly lower elevations there is a cane zone distinguished by the presence of quill cane (*Spartina cynosuroides*) as the dominant species. Beyond the cane zone lies the brackish marsh (Bahr and Hebrard 1976; Penfound and Hathaway 1938). In strongly brackish habitats, cane gives way to salt grass and then brackish-water marsh communities as elevation decreases. Wire grass is the most common species in this zone, with roseau (*Phragmites australis*) and switch grass (*Panicum virgatum*) as minor constituents (Bahr and Hebrard 1976: table 4; Conner, Sasser, and Barker 1986). The salt grass belt is made up almost exclusively of salt grass, while oyster grass (*Spartina alterniflora*), salt grass, and needle rush (*Juncus roemerianus*) are the predominant species in the salt marsh segment (Conner, Sasser, and Barker 1986).

Fire on the Bayou

Recognition that these delta cycles result in relatively predictable (at least on the gross level) ecological responses prompted the formation of the so-called Man-Land Relationship school of cultural geography and archaeology, best represented in the Mississippi Delta region by the works of Sherwood Gagliano and his colleagues (Gagliano 1984; Gagliano and Van Beek 1975; Gagliano, Weinstein, and Burden 1975; Gagliano, Weinstein, et al. 1979; Weinstein 1981; Weinstein et al. 1978, 1979; Wiseman, Weinstein, and McCloskey 1979). In these works human behavior is represented as a part of a larger system of relationships between variables. On the highest level of the system are the global parameters, such as sea-level variation, continental glaciation, tectonics, and the like. These long-term trends control lower-level subsystems, which tend to be cyclical in response to system state shifts in the higher-order systems. Delta lobe formation, maturation, and destruction exemplify these lower-order systems. Human behavior within these systems is perceived to be constrained by the natural environment. In a summary statement concerning man-land relations, Gagliano (1984:40, emphasis added) suggests that,

> while culture is not dictated by environment, the environment does provide the physical framework and resource base on which culture operates and, therefore, is a key *parameter* in archaeological interpretation. . . . The distribution of archaeological sites in the coastal zone of the northern Gulf of Mexico is intimately related to the characteristics of the natural systems. . . . Site locations are related to the character of landforms and neighboring habitats at the time of occupation. Activities at a given site may be related to changes in culture and technology as well as to changes in processes and habitat conditions.

Typical of the viewpoint that culture functions as an adaptation to the environment (see Trigger 1989: ch. 8), Gagliano's systems-theory approach to man-land relationships presents a static view of human behavior within the ecosystem. Although he quite articulately notes the role that the environment plays in shaping human behavior, he and others in this school of thinking are unwilling to concede the opposite dimension, that humans may have played a role in shaping their environment. The weight of the evidence, however, demonstrates that through time humans in the southeastern United States, and in southern Louisiana specifically, have influenced their environment in a number of ways. Perhaps the best-known manner in which Native Americans are thought to have modified the so-called natural ecosystem is through the use of fire and selective burning (MacCleery 1994; Pyne 1982; Sauer 1975).

According to Stephen Pyne (1982:81), "the evidence for aboriginal burning in nearly every landscape of North America is so conclusive . . . that it seems fantastic that a debate about whether Indians used broadcast fire or not should ever have taken place" (see also Pyne, chapter 4, this volume). Many researchers believe that aboriginal and early historic practices of deliberately firing forest undergrowth have considerably altered the succession of species and, in the Southeast, have resulted in the expansion of pine forests in the Gulf Coastal plain at the expense of mixed deciduous species (Williams 1989:47–48; see also Whitehead and Sheehan 1985). Indeed, Carl Sauer (1975) noted that the evidence showed that Native Americans set fires in every locality except where it was too wet (and therefore not especially flammable) or too arid (and thus lacking in fuel) (see also Pyne 1982:71–83). While these generalizations allow for considerable speculation on the role of fire in structuring the southeastern landscape before the arrival of Europeans, they do us little good when examining the specific context of south Louisiana. The marshes and forests of this area are often inundated, or partly so, and thus we need to examine what role, if any, fire might have played in shaping the landscape.

Despite the fact that the marshes of the coast are frequently inundated, fire is and has been an effective means of modifying these environments. Beginning in the nineteenth century, trappers recognized the value of controlled burning in developing ideal habitats for muskrat and other fur-bearing game (O'Neil 1949; Penfound and Hathaway 1938:46–47). The marshes are usually fired in fall or early winter, most commonly from October to January. Historically, at least, burning was done, as one writer says, "just for 'the hell of it'"; to hunt alligators, raccoon, mink, and rabbits; to make traveling easier; and to satisfy a desire to burn billions of mosquitoes, flies and gnats" (O'Neil 1949:27). Although many writers attribute the use of controlled fire to mid- or even late-nineteenth-century trappers and marsh dwellers (O'Neil 1949), evidence from early historic accounts exists to suggest that Native Americans also burned the marshes. In 1684 Minet, for example, observed fire "that had burned over all these places" along the eastern Gulf Coast of Texas (Bell 1987:105). This vague reference to the burning of the marshes or prairies is further substantiated by Pierre LeMoyne d'Iberville, who in January 1699 observed fires burning inland in the approximate location of modern-day Choctawahatchee Bay, Florida. Iberville specifically noted that "it is prairies that the Indians are burning off at this season for the buffalo hunt" (McWilliams 1981:29). In February of that same year he observed the burning of the forest undergrowth along the Mississippi Gulf Coast (McWilliams 1981:49), and on 5 March, during his ascent of the lower part of the Mississippi River, he "discovered a fire on the left [west?] side of the river, in prairies, and . . . noticed that an Indian had passed along" (McWilliams 1981:54). Iberville also indicated that he had seen fire used to clear the countryside when he observed, in relation to the cane lining the riverbank: "It is impenetrable country, which would be easy to clear. Most of the canes are dry; when set on fire they burn readily" (McWilliams 1981:56).

Burning the marsh has a number of effects. It serves to remove dead vegetation, it promotes vigorous new growth, and it facilitates travel and the location of faunal resources within the marsh. Depending on the time of year of the burn, the results can favor the development of particular floral successions. For example, bayonet rush (*Scirpus olneyi*), a significant plant food for both muskrats and blue geese (O'Neil 1949:19), grows in the winter and is thus favored by fall burns. Fall burning of the flotant leads to the succession of bayonet rush over the so-called climax species, cord grass (*Spartina patens*), because it is only during the fall or early winter that the root system of *S. patens* is exposed and is thus affected by the fire (at other times of the year its roots are submerged beneath the water table). Ted O'Neil (1949:29–30) suggests that these plant successions in the brackish-water marsh occur over a twenty-to-thirty-year cycle and are largely dependent on fire as a means of creating species turnover. Burning the vegetation has the added effects of promoting specific plant successions, and of helping to expose muskrat dens, thus aiding in the hunting of these animals.

Muskrats were important to precontact Native Americans in the Mississippi River delta; they comprise up to 20 percent of the total estimated animal food biomass at some sites in the region (Jones, Franks, and Kidder 1994; see also Byrd 1974). Since bayonet rush makes up 90 percent of the Gulf Coast muskrat's food supply (O'Neil 1949:19), we can infer that encouraging the growth of this plant would have been in the best interest of the Native American inhabitants of the marsh. Fur trappers in the early twentieth century utilized fire to encourage specific environmental conditions, and one might reasonably assume that the Native Americans had the same capacity. Of course, the permanence of the ecosystem changes brought on by these marsh burns is uncertain, but evidence suggests that periodic burning of the marsh may have important consequences for the composition of floral communities (Penfound and Hathaway 1938:46–47).

Another impact of Indians on the environment comes from their clearing of forests and landscapes for agricultural fields and for fuel for domestic fires. Numerous historical records indicate that Native American communities in the Southeast were relatively open, with abundant cleared fields and controlled understory growth (Hammett 1992; MacCleery 1994; Williams 1989; Yarnell 1982:5–6). Long-term evidence from pollen records and geomorphic and paleoethnobotanical research in the interior of the Southeast implicates land clearing as a major factor in floral composition (Chapman et al. 1982; Delcourt and Delcourt 1985). Analyses of pollen samples and archaeologically derived plant remains from the Little Tennessee River Valley in eastern Tennessee demonstrate a long-term pattern of increasing anthropogenic disturbance through time (Chapman et al. 1982). High percentages of ragweed (*Ambrosia*-type) pollen, along with pollen from herbs indicative of disturbed, open ground, become increasingly prevalent in the river terraces after ca. 1500 B.P. At Tuskegee Pond, in the Little Tennessee Valley area, influxes of charcoal particles in the sediments increase greatly during the period ca. 1000–600 B.P. This increase in charcoal particles is linked to intensification of the use of fire by Indians for cooking food and clearing land (Delcourt and Delcourt 1985:22). After contact, the influx of charcoal particles and mineral sediments increases by an order of magnitude, reflecting the conflicts between British soldiers and the Cherokee, and subsequent deforestation associated with land clearance by European settlers (Delcourt and Delcourt 1985:21–22, figure 13).

Similarly, Donald Whitehead and Mark Sheehan (1985) demonstrated an increase in pollen from weedy taxa associated with disturbed, open ground in the Upper Tombigbee River drainage in western Alabama and eastern Mississippi. Pollen from deposits dated ca. 2400–500 years B.P. demonstrate a distinct maximum of herbs, sharply increasing pine percentages, and declining frequencies of *Nyssa* and oak (Whitehead and Sheehan 1985:129, figure 5). The authors conclude that increasing clearance associated with the development of Native American agriculture and an increased frequency of fires were responsible for the expansion of pine (Whitehead and Sheehan 1985:134). Pollen cores from the Powers Fort Swale in southeast Missouri document a similar pattern of increasing frequency of ragweed and other herbs indicative of disturbed, anthropogenic environments, beginning around 700–600 years B.P. Corn pollen appears in the sample at roughly the same time, and archaeological data from an adjacent prehistoric site indicate that it was in this same period that occupation achieved its greatest extent (Royall, Delcourt, and Delcourt 1991:169, figures 7–8).

These data indicate that land clearance becomes most significant after the advent of agricultural economies, starting ca. A.D. 800. Closer to home, so to speak, Iberville again provides a good description of the result of clearing, noting, for example, that the Bayougoula village near New Orleans was situated in country that was "very fine, with tall trees, all kinds mixed except pines" (McWilliams 1981:64). Near Mobile, Alabama, he notes that there were a number of abandoned

Indian settlements "where one has only to settle farmers, who will have no more to do than cut canes or reeds or brambles before they sow" (McWilliams 1981:169). The significance of land clearing for agriculture in the delta zone may be questioned, however, given the scant evidence that populations living in this region practiced horticulture (Fritz and Kidder 1993; Wetterstrom 1987). Recent archaeological investigations of a small (ca. 500 m^2) site on Bayou Des Familles, south of New Orleans, indicate that maize agriculture was being practiced by about A.D. 1400 (Kidder 1995). Ethnohistoric data suggest that Chitimacha Indians living in south Louisiana practiced small-scale horticulture by clearing patches of land along high, dry levees. The total area exposed by these agricultural practices may have been minimal, but the cumulative amount of land clearance may have been relatively significant over time as new plots were added and others were allowed to fallow. On a narrow levee segment such as is found along Bayou Des Familles, where total arable land is restricted, agricultural practices may have had a significant impact on the landscape. Certainly this was the case in the early historic period in the same area (Swanson 1991). Land clearing for the acquisition of fuel for domestic fires was also almost certainly a factor in the composition of local floral communities, but the impact of this activity cannot be adequately gauged with the available data.

One of the least appreciated aspects of Native American impact on the ecology of the Mississippi River delta lies in the composition of the archaeological sites themselves. In this part of the world most archaeological sites are comprised of a combination of midden refuse mixed with the shells of the brackish-water clam, *Rangia cuneata*. These shell middens not only mark the most obvious aspect of prehistoric human behavior, they also represent evidence for significant and long-term human impact on the environment. Studies of shell-midden flora demonstrate that these localities represent unique microhabitats supporting highly diverse plant communities that are not evidently represented in any other habitat in the region. Unlike macro processes such as fire or land clearing, the results of which are hard to quantify, shell-midden sites demonstrate tangible remains of Native American behavior. Separate studies of shell-midden flora have been undertaken, and collectively they provide a case study for the impact that precontact natives had on their environment.

The earliest analysis of flora associated with Indian sites was conducted by Claire A. Brown (1936) and was entitled "The Vegetation of the Indian Mounds, Middens, and Marshes in Plaquemines and St. Bernard Parishes." This work was a groundbreaking study of the microhabitats associated with the built environment, in the marshes of southeast Louisiana. Brown began his study by noting that "the vegetation of the Indian mounds and middens is conspicuous because it presents a striking contrast to the vegetation of the surrounding marshes" (1936:423). He noted the general composition of the marsh plant communities, observing that plant distributions were determined by elevation and salinity, but that the overall diversity of the marsh was relatively low (1936:423–427). The study sample of Indian sites was comprised of sixteen mounds and middens, visited during two trips made in April and July. These archaeological sites have been subsequently dated to the

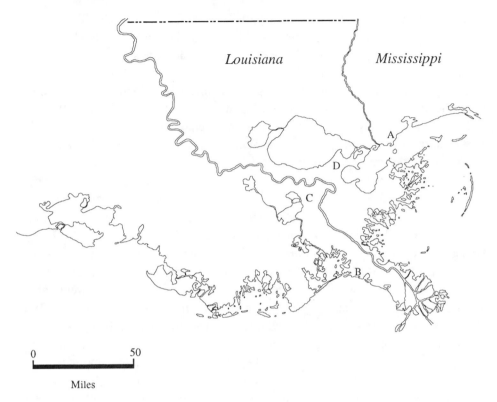

Fig. 7.1 *The Lower Mississippi River Delta, Louisiana and Mississippi.* Capital letters mark the locations of sites or regions noted in the text. *A*: The Hancock County, Mississippi, area (including the Cedar Island site [Eleuterius and Otvos 1979]). *B*: The Lower Bayou Robinson, or Bayou Ronquille, site (Brown 1936). Brown also sampled a number of other sites throughout the delta. *C*: The Coquilles site (Dunn 1983). *D*: The Big Oak Island and Little Oak Island sites (Shenkel 1981).

period ca. A.D. 400 to contact (Gagliano, Weinstein, et al. 1979; Kidder 1994; Kniffen 1936; Neuman 1977; Wiseman, Weinstein, and McCloskey 1979). Brown warned that his study was limited by the short temporal span of his visits and their restriction to the spring and summer months. Still, he observed that "while the vegetation on the different sites is quite similar, most of the mounds have some individuality" (1936:427).

The region analyzed by Brown consisted of both older marshes of the St. Bernard delta lobe, and more recent marshes in the modern or Plaquemines lobe. Native vegetation in these areas is comprised of brackish or intermediate marsh associations in the St. Bernard delta lobe sites, and principally saline marsh in the modern delta lobe. In the brackish marsh, plant diversity is relatively low in comparison to higher areas (Conner, Sasser, and Barker 1986: table 2; Lemaire 1961; Penfound and Hathaway 1938). Inventories of non-Indian brackish marsh sites revealed a maximum of 82 plant species present in this zone, but the floral composition is dominated by 4 species: *Distichlis spicata, Juncus roemerianus, Scirpus olneyi,* and *Spartina patens.* No trees, and few shrubs or woody vines, were reported

(Conner, Sasser, and Barker 1986; Lemaire 1961; Penfound and Hathaway 1938: table 8; White, Darwin, and Thien 1983:106–107). In the saline marsh, plant diversity is very low, being comprised largely of monotypic stands of grasses; only 35 plant species have been recorded in this zone, and only 4 species are dominant: *Distichlis spicata*, *Juncus roemerianus*, *Spartina alterniflora*, and *Spartina patens* (Lemaire 1961; O'Neil 1949:17–26; Penfound and Hathaway 1938: table 9).

Brown (1936:428–433) identified a total of 99 plant species associated with the sixteen archaeological sites. Of these 99 species, 96 were native to North America; the introduced plants included one chinaberry tree (*Melia azedarach*) and one orange tree (*Citrus trifoliata*) that were reported from two separate mounds, while sow thistle (*Sonchus asper*) was identified at five sites. Not all sites supported similarly diverse floras, however. The largest number of endemic plant species recorded at a single site was 39 from the Lower Bayou Robinson (also called Bayou Ronquille) site, located in the saline marsh zone, which dates to the late prehistoric period (Kniffen 1936; McIntire 1958). The mean number of species per site was 18.44, with a range between 39 and 5. Of the six sites with more than twenty plant species, four were associated with saline marsh environments. At least four of the six sites were either shell mounds or earthen mounds built on shell deposits (Brown 1936; Kniffen 1936). A subsequent study undertaken in St. Bernard Parish in the St. Bernard delta lobe revealed that two of the larger and higher Indian mounds in the marsh supported a total of 74 different species (Lemaire 1961:58). Regrettably, in this study the counts of species were not broken down by site, but both were shell middens or mounds and were located in the intermediate marsh zone.

Farther east, in the salt marshes of coastal Mississippi, Lionel Eleuterius and Ervin Otvos identified 62 species of flowering plants on six shell middens (Eleuterius and Otvos 1979:103–105). This area of marsh was initially formed by marine sands of the developing Mississippi barrier island-shoal chain combined with alluvial deposits from the Pearl River. An east-to-west series of barrier islands developed along the Mississippi Gulf Coast and extended as far west as present-day eastern New Orleans (Gagliano, Fulgham, and Rader 1979:2.17–2.22, figures 2.7 and 2.8). This series of barrier islands and shoals is referred to as either the Bayou Sauvage–Hancock County, or the Pine Island trend (Eleuterius and Otvos 1979: 103–105). Radiocarbon dates obtained from surface-collected *Rangia cuneata* shells from these middens indicate that some formed as early as 2900 ± 70 B.P., and are associated with early Woodland Tchefuncte pottery. More recent radiocarbon dates and ceramics indicate that occupation continued into the later Woodland period (Eleuterius and Otvos 1979: table 1). Subsidence due to the compaction of underlying Holocene sediments has reduced the surface exposure of many of these middens. The largest midden studied by Eleuterius and Otvos, the Cedar Island site, has a surface height of approximately 1 m today; borings indicate that archaeological deposits continue to roughly 3.5 m below the present-day marsh surface (Eleuterius and Otvos 1979:105).

According to Eleuterius and Otvos (1979:107), "the vegetation of the Indian middens contrasts conspicuously with the low profile, homogeneous, and rather monotonous vegetation of the surrounding salt marshes." Like the saline marshes to the west, the Mississippi coast marsh vegetation is relatively limited in diversity and principally is comprised of *Distichlis spicata*, *Juncus roemerianus*, and *Spartina alterniflora* (Eleuterius 1972; Eleuterius and McDaniel 1978; Eleuterius and Otvos 1979). Of the six sites surveyed, the largest, Cedar Island, had the highest topographic relief and the most diverse flora. Forty-seven species of flowering plants were found at this site, of which twelve plants are not found in the adjacent marsh (Eleuterius and Otvos 1979: table 2). The authors of this study believe that these twelve species are good indicators of the shell deposits at the middens. Although these species occur in other areas or habitats, "several of these indicator species are more frequently found on Indian occupation sites. . . . They are also found on . . . sites . . . on the adjacent terrestrial mainland and may serve as indicator species for both marsh and terrestrial occupation sites" (Eleuterius and Otvos 1979:111). Eleuterius and Otvos also report that one plant, *Sagittaria minutiflora*, which has not been identified in previous floristic inventories of the area, was found on the shell middens. The only other occurrences of this plant in the northern Gulf Coast have been on shell middens in Alabama, suggesting that these sites provide a unique habitat for at least one previously unreported plant (Eleuterius and Otvos 1979:109–110; Lucas 1995:9).

Farther west, in eastern New Orleans, there is a series of small natural elevations situated in the midst of brackish-water marshes marking the western end of the relict barrier island trend. These are the so-called Pine Islands. Archaeological sites, principally comprised of midden debris mixed with shells of the clam *Rangia cuneata*, are found on the natural elevations closest to modern-day Lake Pontchartrain. All of these sites were first occupied during the early Woodland period, beginning ca. 500 B.C. (Gagliano, Fulgham, and Rader 1979:2.19–2.22; Shenkel 1974, 1981, 1984; Shenkel and Gibson 1974). These archaeological sites manifest a very different floral composition today than those elevations which do not support archaeological sites. The Big Oak, Little Oak, and Little Woods sites (Czajkowski 1934; Gagliano, Fulgham, and Rader 1979:2.19–2.22; Shenkel 1981, 1984) support stands of hardwood trees—most notably live oak, but also hackberry, cypress, willow, and chinaberry. These islands are fringed by marsh elder, palmetto, and buckbrush. At the Little Oak Island site the influence of prehistoric human habitat alteration is most notable. Little Oak Island is one of three semiconnected sand islands (Shenkel 1981:8). Its archaeological site is a horseshoe-shaped midden, largely comprised of earth, sand, and midden debris; *Rangia* shells are sparse in comparison to the nearby Big Oak Island site (Shenkel 1981:7–9). Immediately to the south of Little Oak Island is Little Pine Island, with Pine Island to its south (Shenkel 1981: plate 2). Neither of these latter islands was occupied during the prehistoric period. In contrast to the adjacent and physically connected Little Oak Island, the two Pine Island elevations support, as their name suggests, a very different flora, especially notable for the presence of stands of pine (*Pinus* spp.) and juniper (*Juniperus* spp.) (Gagliano, Fulgham, and Rader 1979: table 2.2). Although the presence of chinaberry at Big Oak and Little Oak islands means that some of the vegetation has been changed in the historic period, there seems to be no reason to suggest that the flora of these barrier islands is not the result of long-term alteration of the environment, beginning several centuries before the time of Christ. These sites ceased to be major occupations after the early Woodland period, but archaeological data indicate that they were periodically reoccupied into late prehistoric times (Shenkel 1981).

All of the sites discussed so far are located in what is now brackish or saline marsh. In most instances the archaeological and geological data indicate that these sites were once situated on topographic high spots, mostly natural levee crests. Subsidence due to the compaction of Holocene deposits has drowned most of these higher elevations, leaving the Indian sites as the last remnants above the marsh surface. As noted by numerous researchers, there is a direct and significant correlation between elevation and floral diversity (Eleuterius and Otvos 1979; Lemaire 1961; Penfound and Hathaway 1938; White, Darwin, and Thien 1983). The principal relationship in this correlation is between elevation and salinity (Penfound and Hathaway 1938:34, 36–40), although soil conditions and tolerance of standing water are also important factors (Penfound and Hathaway 1938; White, Darwin, and Thien 1983). Indian sites elevated above the marshes might be logical localities, then, for a relatively diverse flora simply as a result of edaphic factors associated with their topographic situation.

Analysis of the Coquilles site in Jefferson Parish, Louisiana, however, demonstrates that Indian sites located in well-elevated hardwood levee segments beyond the fringes of the marsh also support an abundantly diverse and in some ways unique floral inventory (Dunn 1983). The Coquilles site consists of a series of shell middens, midden deposits, and human-constructed mounds at the junction of Bayou Des Familles and Bayou Coquilles. This site supported a significant occupation during the Middle Woodland period, ca. A.D. 1 to A.D. 400, and was evidently occupied, at least intermittently, through the late prehistoric period (Beavers 1982a,b; Giardino 1984, n.d.; Speaker et al. 1986). It is is situated on a well-elevated section of the former course of the Mississippi River now occupied by Bayou Des Familles. A floristic study of the site area, measuring roughly 100 by 300 meters, was undertaken by Mary E. Dunn over four months in the fall and spring/early summer of 1977, and again in 1981 (Dunn 1983:349–351). Dunn (1983:351, table 1) recorded 68 plants in her study area, of which 3 were historically introduced into the region. Although she indicated that 32 plants in her sample had not been reported for the region in previous floristic inventories, comparison to a more recent study of the same area indicates that only 9 plants found on the archaeological site are not recorded elsewhere in the Bayou Des Familles/Bayou Coquilles area (Dunn 1983; White, Darwin, and Thien 1983). The Coquilles site shares 24 species with Indian mounds and middens reported by Brown (1936), 15 species associated with two shell middens in St. Bernard Parish reported by Lemaire (1961), and 15 of the plant species reported by Eleuterius and Otvos (1979) for the Mississippi Gulf Coast middens.

Dunn also studied the potential economic value of the 65 plants endemic to the area, and found that there were only 9 for which there was no documented economic use (Dunn 1983:351, tables 2–3); furthermore, at least 2 of these 9 plants are known to be edible, and may have been used by the Indians in the past (Dunn 1983:351). Excluding the use of plants or plant parts for such purposes as shelter, tool making, and clothing, Dunn noted that 26 species of plants could be used for making beverages or food, and 35 had documented use in curative or medicinal contexts (Dunn 1983: tables 2–3). Of the 9 endemic plants unique to the Coquilles site, 7 have documented uses for food and beverage, or as curative or medicinal plants. The remaining 2 species, white ash (*Fraxinus americana*) and wood grass (*Oplismenus setaris*), have no documented economic value, but white ash could have been used for building shelters and for firewood (Dunn 1983: table 4). It is likely, therefore, that 8 out of 9 plants unique to this site had some economic value to the Indians, and that the bulk of the plant species found at the site would have been useful to the Native American inhabitants. It is worth noting, however, that none of the plants that appear to be unique to shell mounds are domesticates or even semidomesticates (Brown 1936; Dunn 1983; Eleuterius and Otvos 1979; Lemaire 1961).

Discussion

How do we account for the vegetation associated with Indian sites in the Mississippi River Deltaic Plain? The earliest explanation was offered by Claire Brown, who observed (1936:436) that

> the vegetation of the mounds is very similar to what one may find on the higher alluvial soils anywhere in lower Louisiana. Inasmuch as the marshes are of alluvial origin, it is natural to expect plants which inhabit alluvial soils. The vegetation on the mounds therefore possibly represents a relict vegetation which once had a much wider distribution, but because of the subsidence of the land, remains only on the high points above salt water.

Brown noted that the mounds are merely favorable habitats in a region that is generally unsuitable for normal growth (1936:437). In this sentiment he was echoing a popular notion that the Indians could not have been an agent in the transformation of the present-day natural habitat. But what evidence is there that these mounds and shell middens are not localities of relict flora? Brown is certainly correct in his assertion that these sites were once located on the levees of distributary channels, and that these distributaries must have supported a more diverse flora than exists in the marshes today. Furthermore, many of the mounds and middens are topographic highs in an otherwise permanently flooded salt marsh.

There are two possible ways to explore the consequences of Indian activities as a contribution to the flora of the shell mounds. One is to examine the flora found on shell deposits that are known not to have been constructed by Indians. Regrettably, although such shell deposits are known (Gagliano, Pearson, et al. 1982:60–61), no floristic studies have been conducted that would allow for a comparison between Indian and non-Indian shell deposits. Another possibility is to compare the flora associated with the mounds in the salt marsh to that found on recently formed natural levees and modern spoil deposits from dredging and canal building in the same area: such studies suggest that the flora of the mounds and middens is at times more diverse and, more importantly, clearly differs in composition from that of the spoil banks and natural levees (Brown 1936; Dell and Chabreck 1986).

According to David Dell and Robert Chabreck, an inventory of the flora on recently formed levees and modern spoil deposits located in the salt marsh yields a total of 36 species. Similarly, Brown's inventory of the plants of the mound on Lower Robinson's Bayou (Bayou Ronquille) yielded 36 species. The two localities, both situated in the salt marsh, shared only 10 species, however. Excluding economic uses for such items as shelter, tools, or firewood, the natural levee/spoil deposits contained 9 plants with documented use for food, beverage, or medicinal purposes. The Indian site contained 15 plants with documented aboriginal use for these functions. If this vegetation is relict, it was evidently derived from different contexts than are present on nearby surviving levee segments located in similar ecological settings.

Further militating against Brown's hypothesis that these sites are localities for relict levee vegetation is the fact that many of the plants found on these sites are calciphiles—that is, plants that favor calcium-rich soils or substrate. This fact is especially true in the middens of the Mississippi Gulf Coast, where Eleuterius and Otvos note (1979:111) that "the investigated Hancock [County] marsh mound vegetation cannot be relict, since most of the plant species, especially those considered indicators, are calciphiles. Their presence is apparently favored and determined by the large amount of calcium contained in the clam shells." Thus we introduce another possible intervening variable associated with the unique plant distributions on the mounds. The soils of these sites are invariably anthropogenically enriched, and in most cases they contain abundant calcium as a result of the many clam shells incorporated in the midden matrix. The possibility exists, then, that these habitats are unusual largely because the edaphic factors underlying the locations are peculiar and unique to archaeological sites. Arguments of this sort are not new to the study of the relationship between plant distributions and human environments (Yarnell 1964, 1982) and carry significant weight, especially in a region where differences in elevation of 20–30 cm can result in very different plant associations (Eleuterius and Otvos 1979; Penfound and Hathaway 1938). Other factors, such as the dispersal of seeds by means of wind, water, or animal agency, may also play a role in the composition of flora associated with Indian sites (Dunn 1983). The presence of European introduced species, such as orange and chinaberry, on Indian midden sites in the salt marsh zone demonstrates further the importance of human agency in creating these landscapes, and also points to the continuation of a trend begun hundreds, if not thousands, of years ago.

Although it is one of the most obvious potential agencies influencing the nature and composition of flora at Indian sites, human action is surprisingly understated in explanations of how these plant communities evolved (Dunn 1983; Yarnell 1982). Archaeologists working in the delta region of the Lower Mississippi Valley have invested a great deal of time and energy exploring the relationship between humans and their environment, but rarely have they considered the role played by our species as part of the environment. Shell middens are a classic example. Numerous authors have noted that the brackish-water clam, *Rangia cuneata*, is a poor subsistence staple. One analysis indicated that to obtain 50 lb of clam meat—roughly the equivalent of the edible biomass of one deer—25,310 clams are required (Byrd 1976: table 1). Not only are large quantities necessary in order to obtain significant amounts of food, these clams are very low in caloric and nutritional values, supplying only 12.6 g of protein, 1.6 g of fat, 2 g of carbohydrates, and only 76 calories per 100 grams of flesh (Byrd 1976: table 2). Yet this clam is nearly ubiquitous at archaeological sites in the region, and has been since the Archaic period. Traditional models suggest that these clams were integrated into the subsistence system as part of seasonal scheduling, or as a famine food in times of stress (Byrd 1976). Recently, however, such models have come into question in the context of shell-midden archaeology around the world. Evaluation of the dietary contribution of molluscs is an important research topic and requires considerably more study before these food resources are written off as solely a famine food (Claassen 1991; Waselkov 1987).

Another possibility needs to be explored, however. Perhaps these clams were used, at least in part, as a means of creating an artificial environment (Claassen 1991; Upchurch, Jewel, and DeHaven 1992; Waselkov 1987). In the first place, they would have been useful in creating and maintaining a living surface in an environment well known for frequent inundation and flooding. Ian Brown (1984, 1988) has noted the presence of numerous small shell middens in the central Louisiana coast area and has identified some as containing circular depressions that he believes represent houses or structures. These so-called Deadman's Island features represent one example of the use of shell to support an occupation in the marsh. There are numerous other examples of shell mounds and earthen mounds built on shell ridges or shell bases in south Louisiana; in fact, it is reasonable to suggest that virtually every archaeological site south of Lake Pontchartrain is built of or on a shell layer. Early historical maps of the New Orleans region indicate the presence of numerous shell middens and mounds (Swanson 1991:60; Trudeau 1803). The Trudeau map of New Orleans and its environs (1803) shows a series of oak stands in the marshes and along the margins of the lakes and bayous south of the city. These oak stands, noted for their economic potential to the colonists, also precisely map the distribution of the largest shell middens known in the area.

Did the Indians recognize the potential of these deposits to support a unique and highly diverse flora? We can only infer what the intentions of these people may have been. Analysis of Native American behavior in the use of fire throughout the continent suggests that the Indians were fully aware of their impact on the environment, and that they recognized their role in shaping the landscape (Hammett 1992). Most shell middens and mound sites in the delta were occupied for a considerable period of time. Initial occupation was minimal, and usually occurred when distributary

channels were still forming. The most intensive occupations occurred during a relatively short period of time after the distributary channels had ceased being active,
but before they began to sink beneath the marsh or to be reclaimed by the Gulf. In
most cases, though, human use of these shell middens continued well after the sites
had been abandoned as full-time occupations. Archaeological data indicate that
these localities were revisited over and over, and that the use shifted from habitation site to a short-duration camp (Kidder 1995). Part of this cycle of reuse was dictated by strategic factors: these sites were usually well situated with respect to local
water courses, and they were topographically high, relatively dry places to camp
(Gagliano 1984). I suspect also that they served as good localities to return to because they were important and unique habitats. They may have served as localities
where terrestrial fauna could be easily obtained (Dell and Chabreck 1986), and they
were near or adjacent to good fishing grounds; they also contained, as we have
seen, an important and diverse flora.

In this sense I suspect that Claire Brown (1936) was correct in identifying the
flora of these sites as relict, but he did so for the wrong reason. The flora was the
relict of the human and natural processes associated with the creation and use of
these shell middens. As an environment, the middens are uniquely human. The
processes by which plants colonized, occupied, and perpetuated themselves in
these habitats were multiple and cannot be isolated to one specific causal agent.
That these sites are frequently the location of historic and modern camps is also notable, as is the fact that the floral and faunal composition is still being modified by
human agency today.

In south Louisiana, shell middens are rapidly being destroyed for use as construction fill; they are also being dredged for oil field canals. Natural and human
agencies seem to conspire to reduce these unique habitats and remove them from
the face of the marsh. Although they are but a tiny fraction of the total ecology of
the region today, this was not necessarily always the case. In the past, shell middens
and their associated flora and fauna were considerably more common and widespread (Trudeau 1803). John W. Foster (1874:157–158) described the situation in
southeast Louisiana in 1874:

> These [shell] ridges and occasional mounds are very numerous near the city of
> New Orleans and along Lakes Pontchartrain and Maurepas, and on the small
> bayous that pass from one into the other. . . . Shell-mounds and shell accumu-
> lations abound along the Metairie, the Gentilly, and the lake-shores, but none
> along the Mississippi. . . . Along the banks of [Bayou Barataria] are vast shell
> accumulations, which for years, like the others I have named, have been used
> for street grading and garden-walks in New Orleans . . . this trade is fast ex-
> hausting these supplies.

This quotation suggests that at one time these anthropogenically altered environments were far more common than they are now and would have had a more significant impact on the distribution of plants and plant communities in the region. It is a

cruel irony of history, then, that has us today writing of these topics while walking on sidewalks and driving on streets built from the remains of the Indian middens.

Conclusion

In archaeology, the paradigm of positivism has, over the last thirty years, developed a decidedly antihistorical bias (Trigger 1989). Fundamental to the conceptualization of processual archaeology is the notion that the science of archaeology requires the study of processes and leads to large-scale generalization (Binford 1962, 1964, 1968; Watson, LeBlanc, and Redman 1971, 1984). These ideas as initially formulated were contrasted with history, which was seen to be narrowly focused and particularistic (Smith 1992; Trigger 1989). This dichotomization of science and history has been argued to represent a profound misunderstanding of the nature of both history and science (Smith 1992:24). Archaeologists are increasingly aware of the role that historical theory can play in our understanding of past and present (Bintliff 1976, 1988, 1991; Fletcher 1992; Gardin and Peebles 1992; Knapp 1992). Especially relevant in this context of shifting theories and paradigms is the conceptualization that historical principles operate at different temporal rhythms (Braudel 1949, 1980; Knapp 1992; Smith 1992). Central to the notion of the variation of temporal patterns are what Fernand Braudel referred to as the "Event," which characterizes individual and group behavior over a short period of time, and the "Longue Durée," which encompasses human behavior "in relationship to the environment, a history in which all change is slow, a history of constant repetition, ever-recurring cycles" (Braudel 1949:20).

The concept of duration is useful in our understanding of the historical ecology of the Mississippi River delta, for it presents us with a theoretical tool with which to frame the interaction between humans and the environment. Seen in the long run, the "ever-recurring cycles of time" are represented by the shifting patterns of delta lobe formation against which all human action is measured. But within this pattern of recurrent natural cycles we must contrast human action and its consequences (Smith 1992:25–26). As a species, we have altered the long-term cycles, changing natural habitats into ones that better suit our immediate needs and reflect our actions. Karl Butzer (1982) has modified Braudel's scheme of duration by extending its temporal depth to match the needs of archaeological time, and by placing greater emphasis on the dynamic nature of human interaction in the landscape. The "adaptive modifications" and "adaptive transformations" of Butzer remain trapped, however, in the framework of perceiving the environment as the limiting factor in human cultures (Trigger 1989:305).

Here the problem is that in the discourse of Braudel and others the long- and short-term never meet. They are portrayed as linear historical phenomena, acting as if they were waves on an oscilloscope. We may measure amplitude and frequency, but the historical patterns, like waves, march off the screen, returning only as we can capture them on paper. History, though, is not a simple linear process: it is better described in terms of nonlinear phenomena, much as physicists describe chaos (Dyke 1990; Reisch 1991; Shermer 1993). The duration of Braudel is only possible because of the event, yet the event can only be framed in the context of the larger pattern. Each event alters the pattern and in turn alters the next event; the rhythms of historical duration are intertwined and intersecting. Historical processes are exquisitely sensitive to their initial conditions (Reisch 1991); as a result, subsequent events are dependent on previous ones yet are not predictable, short of knowing the total range of parameters affecting the initiating historical event. For archaeology the event is the domain of the individual and his or her actions and interactions. This domain is not a linear system, as is seen in the idea that culture is a system for adapting to external environments (Trigger 1989:303–319). Change in linear systems always requires energy input to move to the next system state. Yet if this input is measured solely as energy, we must revert to relying on external sources (environmental change being a favored one) to cause change. If, however, we view change as partly a result of the intersection between the event and the long-term, and the consequences of this intersection, we need not require wholly external causality in our explanation of historic behavior.

The historical ecology of shell middens in southeast Louisiana is a minor and perhaps trivial mirror, but one worth reflecting on (or in). There is no doubt that these middens represent an indigenous contribution to regional biodiversity. Similar results have been documented elsewhere in the southeastern United States (Lucas 1995), notably in coastal southwest Florida (Marquardt 1992). The shell middens were imposed on the landscape and, in fact, became part of the landscape. The notion of consciousness of action is often debated but can never be fully resolved with archaeological data. Ethnohistoric data support the notion that some degree of conscious choice was exercised in selecting the timing, location, and extent of field clearing and forest burning (Hammett 1992). What we can note, however, is that as the natives altered their environment they too were changed by their experiences and actions. The consequences may have been minimal: they did not lead to agricultural revolutions or dramatic changes in social organization; yet they had, and continue to have, an important effect. Native Americans were no less ecological imperialists (Crosby 1986) than were their European successors—they just were not as dramatic about it as were the newcomers. We cannot change these Indian-created habitats today without changing ourselves and the world we live in. In historical ecology we see the tangible meeting of the event and the *longue durée*. Rather than moving inexorably off the stage, this history shows us that the past and the present are linked in ways we are only beginning to perceive.

Historical ecology has as its most persuasive argument the notion that humans are part of the dynamic environment, and thus not necessarily limited because of the natural world. Change in human behavior through time is in part the reflection of humans as they live and adapt within their natural world, but critically too it is the result of humans as they transform and reach beyond their constraints, natural or otherwise. Archaeologists are moving beyond antihistorical processualist biases and beginning to incorporate the notion of temporal scaling and dynamic participation in shaping the natural world. In the Mississippi Delta this theoretical shift means moving beyond static man-land relationships to concepts that better approximate the role played by prehistoric peoples in shaping and transforming their natural world. The myth-history that Indians did not alter their landscape we now know to be untrue (Silver 1990). At the same time, acknowledging this fact reminds us that the natural world of the Indian may not have been so natural after all.

References

Autin, Whitney J., Scott F. Burns, Bobby J. Miller, Roger T. Saucier, and John I. Snead. 1991. Quaternary geology of the Lower Mississippi Valley. Pp. 547–582 in R. B. Morrison (ed.), *The Geology of North America*, vol. K-2, *Quaternary Nonglacial Geology: Conterminous United States*. Boulder: Geological Society of America.

Bahr, L. M. and J. J. Hebrard. 1976. *Barataria Basin: Biological Characterization*. Sea Grant Publication, LSU-T-76–005. Baton Rouge: Center for Wetland Resources, Louisiana State University.

Beavers, Richard C. 1982a. *Archaeological Site Inventory: Barataria Basin Marsh Unit - Core Area, Jean Lafitte National Historical Park, Jefferson Parish, Louisiana*. New Orleans: Jean Lafitte National Historical Park.

———. 1982b. *Data Recovery for Area of Adverse Impact by Proposed Public Access Facilities: The Barataria Basin Marsh Unit - Core Area*. New Orleans: Jean Lafitte National Historical Park.

Bell, Ann L. 1987. Journal of our voyage to the Gulf of Mexico. Pp. 83–126 in R. S. Weddle (ed.), *La Salle, the Mississippi, and the Gulf: Three Primary Documents*. College Station: Texas A&M University Press.

Binford, Lewis R. 1962. Archaeology as anthropology. *American Antiquity* 28:217–225.

———. 1964. A consideration of archaeological research design. *American Antiquity* 29:425–441.

———. 1968. Some comments on historical versus processual archaeology. *Southwestern Journal of Anthropology* 24:267–275.

Bintliff, John L. 1976. Archaeology at the interface: An historical perspective. Pp. 4–31 in J. L. Bintliff and C. F. Gaffney (eds.), *Archaeology at the Interface: Studies in Archaeology's Relationships with History, Geography, Biology, and Physical Science*. B.A.R. International Series, vol. 300. Oxford: Oxford University Press.

———. 1988. A review of contemporary perspectives on the meaning of the past. Pp. 3–36 in J. L. Bintliff (ed.), *Extracting Meaning from the Past*. Oxford: Oxbow Books.

———. 1991. The contribution of an annaliste/structural history approach to archaeology. Pp. 1–33 in J. Bintliff (ed.), *The Annales School and Archaeology*. Leicester: Leicester University Press.

Bowden, Martyn J. 1992. The invention of American tradition. *Journal of Historical Geography* 18:3–26.

Braudel, Fernand. 1949. *La Méditerranée et le monde méditerranéen à l'époque de Phillipe II*. Paris: Librairie A. Colin.

———. 1980. *On History*. Chicago: University of Chicago Press.

Brown, Claire A. 1936. The vegetation of the Indian mounds, middens, and marshes in Plaquemine and St. Bernard Parishes. Pp. 423–440 in Russell et al. (eds.), *Lower Mississippi River Delta*.

Brown, Ian W. 1984. Late prehistory in coastal Louisiana: The Coles Creek period. Pp. 94–124 in Davis (ed.), *Perspectives on Gulf Coast Prehistory.*

———. 1988. Coles Creek on the western Louisiana Coast. Paper presented at the 45th Southeastern Archaeological Conference, 20 October 1988, New Orleans.

Butzer, Karl W. 1982. *Archaeology as Human Ecology.* Cambridge: Cambridge University Press.

———. 1990. The Indian legacy in the American landscape. Pp. 27–50 in M. P. Cozen (ed.), *The Making of the American Landscape.* London: Unwin Hyman.

Byrd, Kathleen M. 1974. Tchefuncte subsistence patterns: Morton Shell Mound, Iberia Parish, Louisiana. M.A. thesis, Louisiana State University.

———. 1976. The brackish water clam (*Rangia cuneata*): A prehistoric staff of life or a minor food resource. *Louisiana Archaeology* 3:23–31.

Chapman, Jefferson, Paul A. Delcourt, Patricia A. Cridlebaugh, Andrea B. Shea, and Hazel R. Delcourt. 1982. Man-land interaction: 10,000 years of American Indian impact on native ecosystems in the Lower Little Tennessee River Valley, eastern Tennessee. *Southeastern Archaeology* 1:115–121.

Claassen, Cheryl. 1991. Normative thinking and shell-bearing sites. Pp. 249–298 in M. Schiffer (ed.), *Archaeological Method and Theory*, vol. 3. Tucson: University of Arizona Press.

Coleman, James N. 1981. *Deltas: Processes of Deposition and Models for Exploration.* Minneapolis: Burgess.

———. 1988. Dynamic changes and processes in the Mississippi River Delta. *Geological Society of America Bulletin* 100:999–1015.

Conner, William H., Charles Sasser, and Nancy Barker. 1986. Floristics of the Barataria Basin wetlands, Louisiana. *Castanea* 51:111–128.

Conniff, Richard. 1989. Keeping an immigrant in check. *National Wildlife* 27(1): 43–44.

Crombie, Alistair C. 1988. Designed in the mind: Western visions of science, nature, and humankind. *History of Science* 26:1–12.

Crosby, Alfred W. 1986. *Ecological Imperialism: The Biological Expansion of Europe, 900–1900.* Cambridge: Cambridge University Press.

———. 1994. *Germs, Seeds and Animals: Studies in Ecological History.* Armonk, N.Y.: M. E. Sharpe.

Czajkowski, J. Richard. 1934. Preliminary report of archeological investigations in Orleans Parish. *Louisiana Conservation Review* 4:12–18.

Davis, D. D., ed. 1984. *Perspectives on Gulf Coast Prehistory.* Gainesville: University Presses of Florida.

Delcourt, Hazel R. and Paul A. Delcourt. 1985. Quaternary palynology and vegetational history of the southeastern United States. Pp. 1–37 in V. M. Bryant and R. G. Holloway (eds.), *Pollen Records of Late Quaternary North American Sediments.* Dallas: American Association of Stratigraphic Palynologists Foundation.

Dell, David A. and Robert H. Chabreck. 1986. *Levees and Spoil Deposits as Habitat for Wild Animals in the Louisiana Coastal Marshes.* Research Report, 7. Baton Rouge: Louisiana Agricultural Experiment Station.

Denevan, William M. 1992. The pristine myth: The landscape of the Americas in 1492. *Annals of the Association of American Geographers* 82(3): 369–385.

Dunn, Mary E. 1983. Coquille flora (Louisiana): An ethnobotanical reconstruction. *Economic Botany* 37:349–359.

Dyke, Charles. 1990. Strange attraction, curious liaison: Clio meets chaos. *Philosophical Forum* 21:369–392.

Eleuterius, Lionel N. 1972. The marshes of Mississippi. *Castanea* 37:153–168.

Eleuterius, Lionel N. and Sidney McDaniel. 1978. The salt marsh flora of Mississippi. *Castanea* 43:86–95.

Eleuterius, Lionel N. and Ervin G. Otvos. 1979. Floristic and geologic aspects of Indian middens in salt marshes of Hancock County, Mississippi. *Sida* 8:102–112.

Fletcher, Roland. 1992. Time perspectivism, annales, and the potential of archaeology. Pp. 35–49 in Knapp (ed.), *Archaeology, Annales, and Ethnohistory.*

Foster, John W. 1874. *Pre-Historic Races of the United States of America.* Chicago: Griggs.

Fritz, Gayle J. and Tristram R. Kidder. 1993. Recent investigations into prehistoric agriculture in the Lower Mississippi Valley. *Southeastern Archaeology* 12:1–14.

Gagliano, Sherwood M. 1984. Geoarchaeology of the northern Gulf Shore. Pp. 1–40 in Davis (ed.), *Perspectives on Gulf Coast Prehistory.*

Gagliano, Sherwood M., Susan Fulgham, and Bert Rader. 1979. *Cultural Resources Studies in the Pearl River Mouth Area, Louisiana-Mississippi.* Baton Rouge: Coastal Environments.

Gagliano, Sherwood M., Charles E. Pearson, Richard A. Weinstein, Diane E. Wiseman, and Christopher M. McClendon. 1982. *Sedimentary Studies of Prehistoric Archaeological Sites: Criteria for the Identification of Submerged Archaeological Sites of the Northern Gulf of Mexico Continental Shelf.* Baton Rouge: Coastal Environments.

Gagliano, Sherwood M. and Johannes L. Van Beek. 1975. An approach to multiuse management in the Mississippi Delta system. Pp. 223–238 in M. L. Broussard (ed.), *Deltas: Models for Exploration.* Houston: Houston Geological Society.

Gagliano, Sherwood M., Richard A. Weinstein, and Eileen K. Burden. 1975. *Archaeological Investigations Along the Gulf Intracoastal Waterway: Coastal Louisiana Area.* Baton Rouge: Coastal Environments.

Gagliano, Sherwood M., Richard A. Weinstein, Eileen K. Burden, Katherine L. Brooks, and Wayne P. Glander. 1979. *Cultural Resources Survey of the Barataria, Segnette, and Rigaud Waterways, Jefferson Parish, Louisiana.* Baton Rouge: Coastal Environments.

Gardin, Jean-Claude and and Christopher S. Peebles, eds. 1992. *Representations in Archaeology.* Bloomington: Indiana University Press.

Giardino, Marco J. 1984. *Report on the Ceramic Materials from the Coquille Site (16JE37), Barataria Unit, Jean Lafitte National Historical Park.* New Orleans: National Park Service, Jean Lafitte National Historical Park.

———. n.d. *Overview of the Archaeology of the Coquilles Site, Barataria Unit, Jean Lafitte National Historical Park, Louisiana.* New Orleans: National Park Service, Jean Lafitte National Historical Park.

Hacking, Ian. 1994. Styles of scientific thinking or reasoning: A new analytical tool for historians and philosophers of the sciences. Pp. 31–48 in K. Gavroglu, J. Christianidis, and E. Nicolaidis (eds.), *Trends in the Historiography of Science.* Boston Studies in the Philosophy of Science, vol. 151. Dordrecht: Kluwer.

Hammett, Julia E. 1992. Ethnohistory of aboriginal landscapes in the southeastern United States. *Southern Indian Studies* 41:1–50.

Heehs, Peter. 1994. Myth, history, and theory. *History and Theory* 33:1–19.

Hobsbawm, Eric. 1983. Introduction: Inventing traditions. Pp. 1–14 in E. Hobsbawm and T. Ranger (eds.), *The Invention of Tradition.* Cambridge: Cambridge University Press.

Hughes, J. Donald. 1975. *Ecology in Ancient Civilizations.* Albuquerque: University of New Mexico Press.

Jackson, Donald D. 1990. Orangetooth is here to stay. *Audubon* 92 (July): 89–94.

Jones, Kenneth R., Herschel A. Franks, and Tristram R. Kidder, eds. 1994. *Cultural Resources Survey and Testing for Davis Pond Freshwater Diversion, St. Charles Parish, Louisiana.* 2 vols. Cultural Resources Series, Report No. COELMN/PD-93/01. New Orleans: Earth Search.

Kidder, Tristram R. 1994. Ceramic and cultural chronology. Pp. 397–437 in Jones, Franks, and Kidder (eds.), *Cultural Resources Survey and Testing,* vol. 2.

———. 1995. *Archaeological Data Recovery at 16JE218, Jefferson Parish, Louisiana.* Cultural Resources Series, Report No. COELMN/PD-95/03. New Orleans: Earth Search.

Knapp, A. Bernard. 1992. Archaeology and annales: Time, space, and change. Pp. 1–21 in A. B. Knapp (ed.), *Archaeology, Annales, and Ethnohistory.* Cambridge: Cambridge University Press.

Kniffen, Fred B. 1936. A preliminary report on the Indian mounds and middens of Plaquemines and St. Bernard Parishes. Pp. 407–422 in Russell et al. (eds.), *Lower Mississippi River Delta.*

Kolb, Charles R. and J. R. Van Lopik. 1958. *Geology of the Mississippi River Deltaic Plain of Southeastern Louisiana.* Vicksburg: U.S. Army Corps of Engineers Waterways Experiment Station.

Lemaire, Robert J. 1961. A preliminary annotated checklist of the vascular plants of the marshes and included higher lands of St. Bernard Parish, Louisiana. *Proceedings of the Louisiana Academy of Sciences* 24:56–70.

Lowery, George H., Jr. 1974. *The Mammals of Louisiana and Its Adjacent Waters.* Baton Rouge: Louisiana State University Press.

Lucas, Greg. 1995. South Carolina Heritage Trust protects late Archaic period shell ring. *Southeastern Archaeological Conference Newsletter* 37(1): 8–9.

MacCleery, Doug. 1994. Understanding the role the human dimension has played in shaping America's forest and grassland landscapes. *Eco-Watch,* 10 February, pp. 1–12.

Marquardt, William H. 1992. *Culture and Environment in the Domain of the Calusa.* Institute of Archaeology and Paleoenvironmental Studies, Monograph 1. Gainesville: University of Florida.

McIntire, William G. 1958. *Prehistoric Indian Settlements of the Changing Mississippi River Delta.* Coastal Studies Series, 1. Baton Rouge: Louisiana State University Press.

McNeill, William H. 1986. Mythhistory, or truth, myth, history and historians. *American Historical Review* 91:1–10.

McWilliams, Richebourg G., trans. 1981. *Iberville's Gulf Journal.* University, Ala.: University of Alabama Press.

Neill, Christopher and Linda A. Deegan. 1986. The effects of Mississippi River lobe development on the habitat composition and diversity of Louisiana coastal wetlands. *American Midland Naturalist* 116:296–303.

Neuman, Robert W. 1977. An archaeological assessment of coastal Louisiana. *Mélanges* 11:1–43.

O'Neil, Ted. 1949. *The Muskrat in the Louisiana Coastal Marshes.* New Orleans: Louisiana Department of Wildlife and Fisheries.

Penfound, William T. and Edward S. Hathaway. 1938. Plant communities in the marshlands of southeastern Louisiana. *Ecological Monographs* 8:1–56.

Pyne, Stephen J. 1982. *Fire in America: A Cultural History of Wildland and Rural Fire*. Princeton: Princeton University Press.

Reisch, George. 1991. Chaos, history, and narrative. *History and Theory* 30(1): 1–20.

Royall, P. Daniel, Paul A. Delcourt, and Hazel R. Delcourt. 1991. Late Quaternary paleoecology and paleoenvironments of the central Mississippi alluvial valley. *Geological Society of America Bulletin* 103:157–170.

Russell, R. J., H. V. Howe, J. H. McGuirt, C. F. Dohm, W. Hadley, Jr., F. B. Kniffen, and C. A. Brown, eds. 1936. *Lower Mississippi River Delta: Reports on the Geology of Plaquemines and St. Bernard Parishes*. Geological Bulletin, vol. 8. New Orleans: Department of Conservation, Louisiana Geological Survey.

Sale, Kirkpatrick. 1990. *The Conquest of Paradise*. New York: Knopf.

Sauer, Carl O. 1975. Man's dominance by use of fire. Pp. 1–13 in R. H. Kesel (ed.), *Grasslands Ecology: A Symposium*. Geoscience and Man, vol. 10. Baton Rouge: Museum of Geoscience, Louisiana State University.

Shenkel, J. Richard. 1974. Big Oak and Little Oak Islands: Excavations and interpretations. *Louisiana Archaeology* 1:37–65.

———. 1981. *Oak Island Archaeology: Prehistoric Estuarine Adaptations in the Mississippi River Delta*. New Orleans: National Park Service, Jean Lafitte National Historical Park.

———. 1984. Early Woodland in coastal Louisiana. Pp. 41–71 in Davis (ed.), *Perspectives on Gulf Coast Prehistory*.

Shenkel, J. Richard and Jon L. Gibson. 1974. Big Oak Island: An historical perspective of changing site function. *Louisiana Studies* 8:173–186.

Shermer, Michael. 1993. The chaos of history. *Nonlinear Science Today* 2(4): 1–13.

Shlemon, Roy J. 1975. Subaqueous delta formation—Atchafalaya Bay, Louisiana. Pp. 209–221 in M. L. Broussard (ed.), *Deltas: Models for Exploration*. Houston: Houston Geological Society.

Silver, Timothy. 1990. *A New Face on the Countryside: Indians, Colonists, and Slaves in South Atlantic Forests, 1500–1800*. Cambridge: Cambridge University Press.

Smith, Michael E. 1992. Braudel's temporal rhythms and chronology theory in archaeology. Pp. 23–34 in Knapp (ed.), *Archaeology, Annales, and Ethnohistory*.

Speaker, John Stuart, Joanna Chase, Carol Poplin, Herschel A. Franks, and R. Christopher Goodwin. 1986. *Archeological Assessment of the Barataria Unit, Jean Lafitte National Historical Park*. Professional Paper 10. Santa Fe: Southwest Cultural Resources Center, National Park Service, Southwest Region.

Swanson, Betsy. 1991. *Terre Haute De Barataria*. Monograph 11. Harahan, La.: Jefferson Parish Historical Commission.

Trigger, Bruce G. 1989. *A History of Archaeological Thought*. Cambridge: Cambridge University Press.

Trudeau, Carlos. 1803. *Plan del local de las tierras que rodean la ciudad de Nueva Orleans*. New Orleans: Historic New Orleans Collection.

Upchurch, Sam B., Pliny Jewel IV, and Eric DeHaven. 1992. Stratigraphy of Indian mounds in the Charlotte Harbor area, Florida: Sea-level rise and paleoenvironments. Pp. 59–103 in Marquardt (ed.), *Culture and Environment in the Domain of the Calusa*.

Van Beek, Johannes L. and Sherwood M. Gagliano. 1984. Renewal and use of the Mississippi River deltaic plain. *Water Science and Technology* 16:699–705.

Visser, Jenneke M. 1994. Erosion increased by nutria grazing. *Louisiana Environmentalist* 2(4): 18–21.

Waselkov, Gregory A. 1987. Shellfish gathering and shell midden archaeology. Pp. 93–210 in M. B. Schiffer (ed.), *Advances in Archaeological Method and Theory*, vol. 11. Orlando: Academic Press.

Watson, Patty Jo, Steven A LeBlanc, and Charles L. Redman. 1971. *Explanation in Archeology: An Explicitly Scientific Approach*. New York: Columbia University Press.

———. 1984. *Archaeological Explanation: The Scientific Method in Archaeology*. New York: Columbia University Press.

Weinstein, Richard A. 1981. Meandering rivers and shifting villages: A prehistoric settlement model in the Upper Steele Bayou Basin, Mississippi. *Southeastern Archaeological Conference Bulletin* 24:37–41.

Weinstein, Richard A., Eileen K. Burden, Katherine L. Brooks, and Sherwood M. Gagliano. 1978. *Cultural Resource Survey of the Proposed Relocation Route of U.S. 90 (LA 3052), Ascension, St. Mary, and Terrebonne Parishes, Louisiana*. Baton Rouge: Coastal Environments.

———. 1979. *Cultural Resources Survey of the Upper Steele Bayou Basin, West-Central Mississippi*. Baton Rouge: Coastal Environments.

Wetterstrom, Wilma. 1987. Botanical remains from the Morgan site (16VM9) recovered during the 1986 excavations. Pp. 369–412 in R. S. Fuller and D. S. Fuller (eds.), *Excavations at Morgan: A Coles Creek Mound Complex in Coastal Louisiana*. Lower Mississippi Survey, bulletin 11. Cambridge: Peabody Museum, Harvard University.

White, David A., Steven P Darwin, and Leonard B. Thien. 1983. Plants and plant communities of Jean Lafitte National Historical Park, Louisiana. *Tulane Studies in Zoology and Botany* 24:100–129.

Whitehead, Donald R. and Mark C. Sheehan. 1985. Holocene vegetational changes in the Tombigbee River Valley, eastern Mississippi. *American Midland Naturalist* 113:122–137.

Williams, Michael. 1989. *Americans and Their Forests: A Historical Geography*. Cambridge: Cambridge University Press.

Wiseman, Diane E., Richard A. Weinstein, and Kathleen G. McCloskey. 1979. *Cultural Resources Survey of the Mississippi River-Gulf Outlet, Orleans and St. Bernard Parishes, Louisiana*. Baton Rouge: Coastal Environments.

Yarnell, Richard A. 1964. *Aboriginal Relationships Between Culture and Plant Life in the Upper Great Lakes Region*. Anthropological Papers, 23. Ann Arbor: Museum of Anthropology, University of Michigan.

———. 1982. Problems of interpretation of archaeological plant remains of the eastern Woodlands. *Southeastern Archaeology* 1:1–7.

CHAPTER 8

Cultural, Human, and Historical Ecology in the Great Basin: Fifty Years of Ideas About Ten Thousand Years of Prehistory

ROBERT L. BETTINGER

Environment is a central and obvious feature of Great Basin aboriginal life and of anthropological interpretations of its meaning. So vivid is the ethnography of this region, and so vivid the force of environment in it, that one quite naturally wants to ask whether what was observed had to be so, or whether it was merely so because of chance or history. The case that shares material circumstances most in common with Great Basin ethnography—and hence the one most likely to shed light on this question as a control—is Great Basin prehistory. It is not surprising therefore, that Great Basin archaeologists have devoted most of their efforts over the last half century to just this endeavor. They have asked, in effect, "How well does Great Basin ethnography account for Great Basin prehistory, and why?" That the environment is much of the "why?" has, of course, always been assumed. In pursuing this argument the path of Great Basin archaeological research parallels that of ecological anthropology generally, moving from early (and from the current vantage, naive) interpretations in which environment is essentially all that matters, to a more complex and truly ecological view in which culture and environment interact in complex ways that benefit from an evolutionary analysis in which history is important. The path begins outside archaeology with the cultural-ecological approach of Julian Steward.

Cultural Ecology: The Steward Model

Julian Steward (1936, 1937, 1938, 1955) offered *cultural ecology* as a strategy for understanding why the ethnographic occupants of the intermontane region between the Sierra Nevada and Rocky Mountains behaved as they did, and how environment figured in that behavior. His *Basin-Plateau Aboriginal Sociopolitical Groups* (Steward 1938) was a landmark work that was profoundly influential within both general anthropology and its major subdiscipline, social-cultural anthropology. It was equally influential, and arguably in a more lasting way, in archaeology, because it dealt with hunter-gatherers and material conditions—subjects that have always attracted archaeologists more than ethnologists (Bettinger 1991).

In western North America, and more specifically within the Great Basin itself, Steward's monograph was doubly influential because it could be seen as serving an established research strategy that sought to connect ethnography and prehistory through what is known as the direct historical approach. The methodology was perfected by scholars trained at Berkeley (Steward, Philip Drucker, Isabel Kelly, William Strong, Waldo Wedel, R. F. Heizer) under Alfred Kroeber, and it directly reflected Kroeber's conviction that ethnography was always the last chapter of culture history and therefore the logical place to begin culture historical inquiry. In the years following World War II, however, Great Basin archaeologists would use the direct historical approach not as the point of departure for culture history, as Kroeber and his students had intended, but rather as a basis for inquiring about prehistoric culture process.

Cultural ecology presented a processual account of Great Basin ethnographic behavior. If that account could be extended into the past, as the direct historical approach suggested, it followed that it could also provide a processual account of prehistory. Whether in doing this one would also be doing culture history was not clear, nor was it clear whether (or how) processual continuity could be assumed in the same sense that historical continuity was assumed in the direct historical approach. Great Basin archaeology continues to grapple with the relationship between ethnographic and prehistoric Great Basin peoples. This is a theoretical endeavor in the sense that continuity (or degree of relatedness) is always measured by the variables that matter most in theory; theory thus justifies the terms for equating prehistory with ethnography. Ethnography, however, enjoys a privileged position in the relationship because it is regarded, in the main, as more readily accessible and therefore the source to which one turns first in a historical analysis. It follows that what one can say about the relationship between ethnography and prehistory is in the first instance limited by the level at which one is willing to generalize from ethnography.

It suited Steward's purpose that the cultural ecology of the Great Basin was simple and relatively uniform, in comparison to the other places where knowledge of hunter-gatherers was fairly complete—California and the Northwest Coast, for example. The Great Basin is relatively homogeneous whether taken as a biotic, climatic, physiographic, or hydrographic province (Cronquist et. al 1972:1–18; Harper 1986). It is an internally draining, semiarid, cold winter–dry summer, basin-and-range steppe bounded on the west by the crest of the Sierra Nevada, on the east by the Wasatch Front (western margin of the Rocky Mountains), on the north by the Snake River Plain, and on the south by the Colorado River (figure 8.1). It centers on

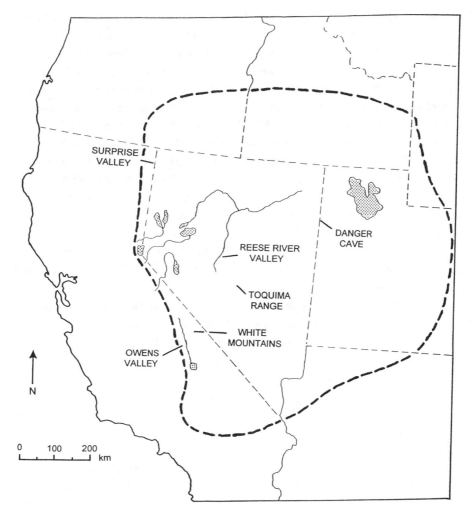

Fig. 8.1 *Geographical Limits of the Great Basin, and Localities Mentioned in the Text.*

a vast area in which lower elevations are dominated by sagebrush (*Artemisia tridentata*) and various species of saltbush (*Atriplex*) (Cronquist et. al 1972:2). Throughout this area, the plants and animals most ethnographically important as native food or raw material were widely distributed as species (e.g., *Ovis*) or ecotypes (*Chenopodium* spp.).

At the time of Euro-American contact nearly all of the area was in the hands of peoples speaking closely related languages of Numic—a recent and minimally differentiated branch of Uto-Aztecan, of which there are three groups (Western, Central, and Southern) each consisting of two languages (Western: Mono, Northern Paiute; Central: Panamint, Shoshoni; Southern: Kawaiisu, Ute-Southern Paiute [Miller 1983]). Numic technology was uniform and comparatively simple (Steward 1940; Zingg 1939; Delacorte 1990). The same kinds of wood, fiber, and stone tools, implements, and facilities were used everywhere for plant and animal procurement

(Driver and Massey 1957; Kelly 1932, 1964; Jorgensen 1980; Steward 1941, 1943; Stewart 1941, 1942). Sociopolitical organization was equally simple and uniform (Steward 1938:239–258). Almost without exception, family bands, or kin-cliques, constituted autonomous and independent social, economic, and political units (D. Fowler 1966). Prior to Euro-American contact, political leaders could persuade but not coerce, and gatherings larger than the family were brief and occasioned by the need for cooperation in some short-term, mutually beneficial endeavor.

There was certainly environmental and cultural diversity within the Great Basin. Steward (1938:19, 104) pointed out, for instance, that such key resources in the central and southern Great Basin as ricegrass (*Oryzopsis*) and pinyon pine (*Pinus monophylla*) were replaced by roots (e.g., camas, *Camassia*) and berries (e.g., *Shepherdia*) in the colder and more mesic northern Great Basin. Overlying these natural patterns were such cultural ones as environmental manipulation (Downs 1966), which included the burning and sowing of seed patches in some places (Steward 1938:104, 115, 119) and the irrigation of root and seed plots in others (Lawton et al. 1976).

Steward, however, perceptively identified, and chose to work with, a block of cultural and natural features that remained quite uniform across the Great Basin. These he crafted into a simple argument in which environment and technology are independent "givens." This distinguishes cultural ecology from *environmental determinism* (Huntington 1945, 1963; Semple 1911), which makes technology dependent on environment. This, of course, reduces technology—and so everything cultural supported by technology—to environment. In cultural ecology, by contrast, because technology is independent from environment (i.e., not simply a function of environment), culture does not interact with the environment at large but with the part that is accessible given the available technology—the "effective environment," so to speak. It is through technology and the process of work that humans interact with this "effective environment." Social and economic organization and other elements of this "culture core" are shaped in accord with the demands of the work process.

Culture ecology, then, stipulates the importance of cultural and historical forces, which are the source of technology. By this view—and in contrast to environmental determinism and possiblism (Kroeber 1939), in which culture remains passive with respect to environment—Steward envisioned a connection between culture and environment that is active in the sense that, via technology (which determines effective environment), culture chooses its environment. This is why Steward's analysis of Great Basin cultural ecology revealed patterns broader and more uniform than would be expected from environment alone. The uniformity of Great Basin technology is why, despite the aforementioned variations in natural environment, one finds everywhere essentially the same important combination of seeds, roots, and small game.

Steward (1938:1) argued that, given the limited technology at their disposal, the aboriginal inhabitants of the Great Basin were especially subject to the vicissitudes of their marginal environment. Resources were seldom plentiful, and never reliable.

Dependence was mainly on plants that relied on rainfall, which varied substantially in amount from year to year in a given locality. This resulted in much annual and seasonal movement, the speed and amount of which were limited by the absence of any means of transportation other than foot. Since groups could neither predict where resources would next be abundant, nor move to those places at will, population was necessarily small and thinly spread. Durable social units had to be large enough to carry out routine extraction and production, which required minimal cooperation, but no larger than could be sustained by the minimum quantity of resources likely to be available within the area that could be covered efficiently on foot. For the most part, these requirements were satisfied by the *family band*, which typically restricted its movement to a home range of its liking without marking its own fixed territory from which other groups were excluded. Family bands (Fowler 1966) were built around a nuclear family core (mother-father-children) to which might be briefly attached a small number of related hangers-on: generally the elderly, infirm, unmarried or widowed relatives of either parent. It was not uncommon for related family bands—of two brothers, or of a father and son—to throw in together for short periods, but the alliance seldom lasted more than a winter or two. Similarly, family bands periodically cooperated in the conduct of communal antelope or, more commonly, rabbit drives. These affairs were organized under the aegis of a single, widely respected individual (the antelope or rabbit "boss") who directed all phases of the activity and to whom all participants were required to defer. Individuals were free to leave the group at any time, however.

What is conspicuously lacking in Steward's account is any sense of "culture" or "ethnicity"—"Numic-ness," so to speak. Steward was clearly aware that such features existed and were important. They were not, however, a part of the "culture core," and thus they were beyond the reach of cultural-ecological explanation and were largely ignored in *Basin-Plateau Aboriginal Sociopolitical Groups*. Steward's characterization of Numic culture as mainly "gastric" makes the bias clear. His treatment of the Numic culture core itself glosses specifics with similar effect. Environment is cast not in terms of individual resources (although these are sometimes discussed as examples), but in terms of key attributes (e.g., reliability) as these pertained to equally general observations about work and organization. What emerges is an argument that only incidentally concerns Numic peoples, who are seen as doing what any group of similarly equipped hunter-gathers would be expected to do under the same or similar circumstances. This is clearly evident in the nonlocalized account of Shoshonean lifeways that appears in *Theory of Culture Change* (Steward 1955). As David H. Thomas (1983b) notes, these tendencies grow stronger in Steward's later writings, which are concerned less with adaptive variation within the Great Basin and more with the characterization of the Great Basin as single instance showing how hunter-gatherers solved problems in marginal environments, and how this determined sociopolitical organization. A large part of latter-day criticism directed toward Steward deals with that issue: the degree to which he is faithful to the variation in Great Basin ethnology in light of his tendency toward generalization (C. S. Fowler 1982, 1990).

The Desert Culture Hypothesis

The first formal attempt to apply Steward's ethnographic account of the Great Basin to prehistory was essayed by Jesse Jennings (1957; Jennings and Norbeck 1955) in reference to the results of excavations conducted over a lengthy period at Danger Cave, a stratified, dry cave site near the modern town of Wendover on the Nevada-Utah border. The nondescript assemblage of chipped- and ground-stone tools, perishables, and other debris recovered there, which showed little variation from the top of the deposit to the bottom, suggested a generalized pattern of hunting and gathering similar to that which could be inferred from sites throughout the Desert West, and quite distinct from the pattern of specialized big-game hunting said at the time to characterize the earliest occupants of the New World (Jennings 1957:266–271). The received wisdom was that the big-game pattern had given way to generalized hunting and gathering following the extinction of the Pleistocene megafauna upon which the hunting pattern depended. Danger Cave made that idea untenable, however, because it was occupied by at least 10,000 B.P., prior to the wholesale extinction of megafauna, and it exhibited essentially the same pattern of generalized hunting and gathering from that time until the site was finally abandoned shortly after Euro-American contact. Since the specialized hunting pattern could no longer be viewed as antecedent to, and the progenitor of, the generalized Danger Cave–type pattern, Jennings argued that the two were equally ancient—having evidently spread at essentially the same time from the Old World to the New, where the Danger Cave pattern persisted into historic times in those marginal circumstances to which it was especially well suited (Jennings and Norbeck 1955:7; Wauchope 1956:71).

Jennings was quite prepared to move beyond what could be inferred directly from the Danger Cave deposits to more-encompassing speculation about adaptation and lifeway (Jennings 1957:276). Accepting Steward's thesis about the relationship between environment, technology, subsistence, settlement, and sociopolitical organization in the Great Basin ethnographic present, Jennings reasoned that if, during the time when Danger Cave was occupied, environment and technology had remained stable and within the range reported by Steward, it would follow that the social, political, and economic behaviors Steward had explained by means of environment and technology could be inferred on the strength of the ecological argument alone (Jennings 1957:7–8, 278). Jennings mounted arguments on both counts: with regard to technology, "most artifact types persist throughout the deposit; no dramatic changes in content occur" (Jennings 1957:287); and as for environment, "the evidence is that 10,000 years ago the desert was much as it exists today. Human existence was in precarious balance" (Jennings 1957:6). Either major Holocene climatic change was denied, or its importance was discounted.

Since, as we have seen, the Danger Cave assemblage had parallels across the Desert West, which was said to be relatively uniform environmentally, Jennings was prepared to extend this argument beyond Danger Cave and to argue that Steward's ecological algorithm applied to the Great Basin generally (Jennings 1957: 279–280). Indeed, given this level of generality, it should apply wherever simple, pedestrian hunter-gatherers faced relatively severe environmental hardships—which Jennings was also prepared to argue (1957:284).

Responses to the Desert Culture were predictably strident. The initial reaction came from R. F. Heizer (1956), a student of Kroeber's who reflected his mentor's abiding interest in historical trajectories, especially those connecting ethnography and prehistory in western North America (e.g., Kroeber 1923). Like so many other Kroeber students, Heizer was an advocate of the culture historical methodology termed the *direct historical approach,* in which one traces the histories of specific ethnographically recognized ethnic groups or cultures, starting from historical accounts and proceeding backward in time into the archaeological record (Heizer 1941; see also Wedel 1938; Steward 1942). Given this bias, Heizer could not get past Jennings's use of the term *culture,* and he observed that there were at the time of Euro-American contact many discrete social entities in the Great Basin that anthropologists had legitimately recognized as *cultures.* This, of course, misread what Jennings meant by "culture," which conformed to Steward's cultural-ecological usage (*culture core*) and had nothing to do with culture history. Heizer and other Desert Culture critics, however, voiced two additional, and more germane, objections: (1) that Great Basin climates had changed enough over time to affect environment, and hence human adaptation (e.g., Antevs 1948; Baumhoff and Heizer 1965); and (2) that, holding climate constant, Great Basin environments differed enough over space to produce structurally different adaptations.

These and other objections led Jennings gradually to adopt more-heuristic versions of the Desert Culture hypothesis (Jennings 1964; and see Aikens 1970). In the final version, the Desert Culture (Jennings 1968) became the Western expression of a continent-wide Archaic stage—i.e., Desert or Western Archaic, which united locally varied hunter-gatherer adaptations under the explanatory umbrella of "efficiency." Increasing adaptive efficiency during the Archaic produced identifiable patterns of change in the form of emerging local resource specializations, population growth, and movement into previously uninhabited environments.

The Regional Survey Version

The level of generality at which Steward had articulated his model, and the still greater level of generality at which Jennings sought to extend it into the past, presented problems for those who attempted to use it for their own purposes. Jennings argued, for instance, that "no fundamental change" had occurred in regional human ecology—yet in a strict sense that was surely not so. Among other things, the bow and arrow had replaced the atlatl and dart (Jennings 1957:278), and pottery had been added to the material culture inventory (Jennings 1957:180–181). Since these had been traditionally regarded as major technical advances in most schemes of New and Old World prehistory (e.g., Kidder 1927; Clark 1953), it was unclear just how much technological change would be needed to alter the technoenvironmental algorithm.

Part of the problem here was that Jennings's warrant to speculate about the specifics of prehistoric lifeways was severely constrained by the materials that Danger Cave actually contained. His options were limited. He could describe the Danger Cave assemblage in detail and wait for further information, as others had done. Alternatively, he could tie his limited data to a theoretical framework, provided it was general enough. Steward's cultural ecology filled the bill. Because Jennings was required to argue that Danger Cave was fully representative of the pattern he sought to construct, however, that pattern could be no more specific than Steward's generalizations about the Numic culture core. In short, Jennings was arguing about a very general pattern from very specific data. This, of course, did not go unnoticed by critics, who doubted whether Danger Cave provided convincing support for a Desert Culture lifeway too mobile, flexible, and seasonally varied to be represented in full at any single locality (Bettinger 1977; Heizer and Napton 1970:43; O'Connell 1975:12; Thomas 1973:174; Weide 1974:77–78). Given that the role of Danger Cave in the seasonal round was unknown, it was quite thinkable that its use might change very little despite profound adaptive restructuring in other aspects of the subsistence-settlement system not monitored by that site.

Clearly, assessing the merit of the Desert Culture hypothesis required more-specific predictions about prehistoric lifeways that could be evaluated against specific archaeological data representing the full range of activities conducted by prehistoric groups in their home ranges. Steward, of course, had argued that his technoenvironmental algorithm applied to the aboriginal inhabitants of the Great Basin generally (Steward 1938:256–258; 1955). It followed that the behavior of specific Numic groups could be regarded as local variants of the more general pattern (e.g., Thomas 1973:156–157). As such, local ethnography could serve as a basis of inferences about the prehistoric behaviors to be expected in the same places if the Desert Culture model were valid. Simply put, the Desert Culture hypothesis could be read as implying that local prehistory should recapitulate local ethnography with respect to material conditions and adaptive behavior.

In the late 1960s and 1970s, workers in several localities sought to put this approach into operation through programs of surface survey and excavation designed to shed light on prehistoric regional subsistence-settlement systems. The best known of these was conducted by David Hurst Thomas (1971, 1973) in the Reese River Valley of central Nevada. Thomas compared the data obtained by probabilistic surface survey against quantitative predictions derived from a thorough reanalysis and simulation of Reese River Shoshone subsistence and settlement patterns as they had been reported by Steward. Other research programs, mostly less formal than Thomas's, were conducted in Surprise Valley in northeastern California (O'Connell 1971), Owens Valley in central eastern California (Bettinger 1975), and Warner Valley in southern Oregon (Weide 1968).

Despite the great care invested in their design and execution, these studies produced equivocal results in relation to the Desert Culture hypothesis. This was, in itself, telling. In the Reese River case (e.g., Thomas 1971, 1973, 1974), for example, the Desert Culture hypothesis was supported to the extent that prehistoric surface assemblages conformed to ethnographic expectation during the 4,500 years of detectable human presence. At the same time, the hypothesis was refuted to the extent that Reese River, as a relatively rich environment, should have been, but was not, intensively occupied prior to that—perhaps owing to intervals of catastrophic drought (Thomas 1974:15).

The prehistoric subsistence and settlement patterns observed in Surprise Valley, northeastern California (O'Connell 1971, 1975; O'Connell and Hayward 1972), more directly contradicted the hypothesis because they changed profoundly over time. Excavated houses and faunal assemblages from a series of Surprise Valley sites documented a shift in subsistence from large to small game, and in residence type from large, multifamily pit houses to rudimentary, family-sized brush wikiups, between about 5000 and 4000 B.P. James O'Connell linked these changes in diet and residential-group size to climatically induced changes of environment, specifically increasing aridity. Still, since those archaeological changes could be considered "rational" responses to changing environmental circumstances, they could be construed as supportive of the Desert Archaic hypothesis, which stressed adaptive efficiency as its unifying theme (O'Connell 1975:52). Part of the problem, clearly, was that read narrowly (i.e., as arguing that absolutely no change had occurred), the Desert Archaic hypothesis could not possibly be right; which suggested that it was intended to be read broadly—in other words, as arguing only that no *fundamental* change had occurred. This could, in turn, be construed as contending, as in Jennings's Archaic stage concept, that humans will tend to adapt efficiently—in which case it could not possibly be wrong (for functionalists, at least).

Even in this general form, however, the Desert Culture hypothesis failed to engage what seemed to be important adaptive trends that were beginning to be observed in the archaeological record. In eastern California (Bettinger 1975, 1976), for example, the intensity of resource use appeared to increase monotonically through time in a way that precluded simple climatic or technological explanation. This was especially evident in the use of nuts of the pinyon pine (*Pinus monophylla*), a major ethnographic staple, the prehistoric use of which remained quite casual until relatively late in time. It is by now quite clear from a combination of excavations and surveys that pinyon pinenuts remained relatively untapped until after A.D. 600 (Bettinger 1989; Delacorte 1990). This suggested a discontinuity between the ethnographic and archaeological records for which the Desert Archaic hypothesis offered no obvious solution. The problem is that, by all accounts, pinenuts should have been an important part of subsistence from the very beginning of human occupancy in the Great Basin. Although laborious, the intensive procurement of pinenuts is technically undemanding. Its tardy appearance in eastern California was irreconcilable with the idea of a Desert Archaic adaptation guided by efficiency, leading to the "full utilization of edible materials available" (Jennings 1957:276), unless one assumed that variables other than environment and technology were involved. Population appeared to be that missing variable (Bettinger 1975: 89–92; 1977:15–16).

Directional adaptive change in eastern California and elsewhere could be explained without resorting to such externalities as climate change or technological innovation, by arguing that geometric population growth had required increasingly intensive use of local resources, which were successively added to the subsistence repertoire in relation to the costs associated with their procurement and processing. Since intensive pinenut procurement is relatively costly, for example, it could be argued that it would not have become important in Owens Valley until other, less costly resources had been tapped to the point that their marginal cost had become excessive. That sort of demographic explanation was a clear break from the Desert Culture hypothesis, which, following Steward, had treated population in Malthusian fashion as a dependent variable and had interpreted population growth and regional colonization primarily as resulting from, rather than requiring, increased technical efficiency. In this version, population could exert its own force—although it was in some sense subject to certain techno-environmental limits.

Cultural Ecology After Demography: Optimal Foraging Theory

The presence of population as a quasi-independent variable profoundly changed the way in which Great Basin archaeologists regarded environment and technology. For population to be important as a variable *it had to be variable*—that is, it had to be viewed as changing (and sometimes growing) during the period of human occupation. This implied that the relationship between population and environment had changed in the absence of external environmental change, that the effective environment could change despite climatic/environmental stability. This further implied that technology, the means through which population interacted with environment, would have had to change as well—that is, in tandem with changes in the range of resources used. Put another way, when population was added as a variable, environment and technology ceased to be the major determinants of adaptation: they were now reduced to second-order determinants envisioned as a range of adaptive alternatives that varied in cost (e.g., high-cost vs. low-cost resources, and high-cost vs. low-cost procurement methods) and were brought into play to meet the momentary demands of population. Population increase (or resource decrease) required the deployment of more costly behaviors, often involving expanded diet breadth; population decrease (or resource increase) had the opposite effect.

Much of this logic was formalized quite elegantly through an abridged version of microeconomics known as "optimal foraging theory," most notably through the models of diet breadth and patch choice and the marginal value theorem (MacArthur and Pianka 1966; Charnov 1976; Schoener 1974). In large part, these models preserved the essential elements of the Desert Culture hypothesis, which had always held that the prehistoric inhabitants of the Great Basin had made rational (i.e., efficient) choices given the environmental and technical alternatives open to them (O'Connell, Jones, and Simms 1982; Simms 1983, 1987, 1988a, 1988b; Madsen 1986, 1988, 1989). The chief difference was that because environment and technology were now viewed as less constraining, greater variability was to be expected in local adaptive patterns through time and space. Archaic efficiency was gradually assigned meaning in terms of this anticipated variability in what can be called the "rapid adapter" theorem: according to this, Great Basin populations were capable of quick adaptive responses to a wide range of changing environmental and demographic circumstances (e.g., Madsen 1982:210; 1988:418; 1993:328; 1989:63–68; Simms 1983:828; 1988b:421–424). As in environmental determinism, technology now joined social organization as a dependent variable to be brought into play at will, to suit momentary requirements determined by the balance between population and resources.

Danger Cave provided critical support for this argument (Simms 1983: 826–827). Coprolites from the early deposits there commonly include quantities of seeds that experiments show as producing very low rates of caloric return on procurement and processing time. Since in the diet-breadth model the use of such species implies the use of all higher-ranking species, the Danger Cave coprolites were offered as evidence that the early Great Basin diet included essentially the same set of species known to have been used by Numic groups ethnographically. This further implied that the early inhabitants of Danger Cave had the requisite knowledge and technology to exploit the full range of plants and animals in the Great Basin. Since it could be assumed that all later peoples were similarly endowed, the food choice (i.e., diet breadth) throughout the Holocene was simply a matter of momentary expedience determined by the ratio of people to resources.

The Numic Problem

Despite its arguments about adaptive continuity, the Desert Culture hypothesis had nothing to say about historical (i.e., ethnic) continuity in the Great Basin. In part, this was because ethnicity was no part of Stewardian cultural ecology. Yet in retrospect it seems strange, given the materialist bent of the hypothesis, that inferred adaptive continuity was not at some point taken to imply historical continuity. From the materialist perspective, adaptive success should assure evolutionary success in the form of historical persistence—that is, continuity. Historical continuity, however, was not inferred—quite possibly because it was obvious to nearly everyone that it did not fit the facts (Jennings 1957:278–279). The distribution of Numic languages, and the lack of differentiation both within Numic and between Numic and other closely related languages, indicated very clearly that Numic speakers, the ethnographic occupants of the Great Basin, were relative newcomers. Present estimates suggest that they entered the area something like 2,000 years ago at the earliest (Lamb 1958; Miller 1983, 1984, 1986).

The problem had been noticed and was the subject of abortive attempts at explanation, none of which fit either the data or the basic assumptions of the Desert Culture hypothesis (e.g., Taylor 1961; Hopkins 1965; Gunnerson 1962). For the most part, however, adherents to the Desert Culture hypothesis, including Jennings himself, seemed content with the idea that a major ethnic spread had occurred in the absence of any major change in regional culture ecology (Jennings 1978:235–237)—never mind that this was implausible in theory (especially biological theory) and, when carefully considered, completely devastating to the basic premise of cultural ecology itself. It suggested nothing less than that ethnicity and related aspects of human behavior were altogether disconnected from ecology; that is, that ecology had nothing to say about ethnicity.

In 1982 Martin Baumhoff and I proposed a solution to the Numic problem by means of a model that differed in only minor ways from Steward's original proposal and the (by then) much-modified version of the Desert Culture hypothesis. Steward had argued that technology and environment determined the rest of the culture core, which adjusted in response to them. Adding population to that mix suggested that, given essentially the same environment and technology, groups would also adjust adaptively to meet the demands of population, which was itself ultimately regulated by (i.e., adjusted to) technology and environment. We simply argued (Bettinger and Baumhoff 1982, 1983; Bettinger 1991, 1993) that all this adjustment and regulation took time. This made it possible for ethnic replacement to occur without exceptional externalities, such as dramatic changes of environment or unusual competitive circumstances in which, by virtue of technology or population, one group enjoys an insurmountable advantage that is qualitatively beyond the means of its competitors (examples would be groups that are unfairly numerous, armed, supplied, or informed).

In keeping with the general thesis of optimal foraging theory, Baumhoff and I argued that one could envision for any given techno-environmental situation a range of hunter-gatherer strategies that could be arranged in a continuum according to population density (Bettinger and Baumhoff 1982:486–488). In the low-density case, theory suggested that diet breadth would be low (specialized) and restricted to high-ranking resources, and that settlement patterns should emphasize frequent movement to high-quality resource patches. As a result, the search for high-ranking resources and travel between high-quality patches would account for the bulk of subsistence costs. In the high-density situation it is quite the reverse: diet breadth is high (generalized) and includes, in addition to high-ranking resources, low-ranking ones that are costly to procure and process; and settlement patterns are more residentially fixed, because groups use low-quality resource patches near at hand rather than more distant, high-quality patches. In this case, procurement and processing account for the greater proportion of subsistence-related costs.

We further argued that because large game animals are among the highest-ranking resources in the Great Basin (cf. Simms 1987; Broughton and Grayson 1993), in the low-density case there males would contribute relatively more to diet than females—which would favor male-biased sex ratios (cf. Hewlett 1991; Smith and Smith 1994) and female-infanticide, which would in turn depress fertility. In the high-density case, females pay a greater fraction of subsistence costs than males because they are instrumental in the procurement and processing of high-cost resources, which in the Great Basin are mainly plants; this would discourage female-infanticide, thus leading to more-even sex ratios and greater potential growth rates. We next observed (Bettinger and Baumhoff 1982:488–489) that the high-density processor strategy enjoys a competitive advantage over the low-density traveler strategy: processors compete for all the resources of the traveler, but the traveler competes for only some of the resources of the processor. We argued that in the Great Basin case, Numic and pre-Numic groups represented the high- and low-density cases, respectively, and we explained the Numic spread in terms of these competitive advantages. In our model (Bettinger and Baumhoff 1982:489–490), replacement occurs because Numic and pre-Numic systems represent local adaptive peaks, the suboptimal valley between which prevents movement from the pre-Numic peak to the higher Numic peak by gradual readjustment. Both groups, moreover, were peak-bound, in the sense that they relied on subsistence and subsistence-related behaviors that were supported by value systems whose economic impact was neither clear nor susceptible to rapid change, and because they lacked the organizational structure needed to sustain new technologies. Pre-Numic groups were replaced, in short, because they could not adjust quickly enough to the challenges posed by local increases in population density resulting from the incursion of Numic speakers.

Great Basin Alpine Research

Although elegant in theory, the Bettinger-Baumhoff model lacked a detailed exposition of how it should be tested. We cited only limited evidence in support of our argument, most of that quite general. The model did, however, argue for important differences between Numic and pre-Numic diet breadth that would, at least in theory, be detectable archaeologically. Since archaeological data are always more messy and complex than one anticipates, the principal problem in doing this was in finding a research context that would accentuate the expected differences. The diet-breadth and patch-choice models provided useful hints in this regard. They suggest that high-quality resources and high-quality patches would be used with roughly similar intensity by both Numic and pre-Numic groups, which would blur the distinction between them (Bettinger and Baumhoff 1983:831); marginal resources and patches, on the other hand, would make the distinction much more apparent, and were the obvious place to test the hypothesis. Great Basin alpine environments are marginal in many senses, and so were ideally suited to such a test (Bettinger 1991).

A variety of geographical, environmental, and physiological constraints conspire to make alpine environments inherently marginal in comparison with their lowland counterparts (Bettinger 1993). They contain a limited range of resources, and they are difficult to reach and to work in productively. Alpine environments throughout the Great Basin are used seasonally by herds of mountain sheep, which can be exploited economically by small hunting parties. Marmots (a large, hibernating alpine rodent) can also be taken profitably. Alpine resources other than these—most notably roots, seeds, and berries—are sparse or difficult to procure (and generally both). This leads to the expectation that the low-cost, pre-Numic-type use of alpine environments should be limited to the taking of mountain sheep (*Ovis canadensis*) and marmots (*Marmota flaviventris*) by small hunting parties. A high-cost, Numic-type use would entail the harvesting of a greater range of alpine resources, including plants, by larger groups, for longer periods.

Alpine research to test the Bettinger-Baumhoff Numic-spread hypothesis was conducted across the Great Basin between 1981 and the present. In the main, the data are in keeping with the general expectations of the hypothesis. In the two highest mountain ranges in the Great Basin (figure 8.1)—the Toquima Range in central Nevada (Thomas 1982) and the White Mountains of eastern California (Bettinger 1991)—an adaptive shift has been noted, from an early specialized pattern that emphasizes short-term, hunting-related use to a later generalized pattern that features major residential villages seasonally occupied by groups engaged in intensive plant procurement as well as hunting. The shift occurs within the Christian era sometime before A.D. 1000, which is consistent with linguistic evidence. On the other hand, it clearly contradicts Steward's (1938:14) assumption that the Great Basin alpine zone could not have supported hunter-gatherers for any extended period.

Ethnicity and Ethnic Spreads

To many, the most troublesome aspect of the Numic-spread hypothesis is its implication that pre-Numic groups throughout their range were incapable of responding quickly enough to a competitive threat posed by Numic encroachment. As we have seen, in the "rapid adapter" scenario, pre-Numic groups had effectively responded to all previous environmental challenges, whether temporal (e.g., the disastrously harsh Holocene climatic optimum, or Altithermal) or spatial (e.g., Death Valley), that required the use of essentially the full range of usable Great Basin resources. It followed that they would have had no trouble in making similar adjustments to counter the reduction of resources that resulted from Numic incursion. It is obvious, however, that they *did* have trouble, did *not* adapt, and so were replaced. In reviewing the possible causes for this, it seems likely, first, that while pre-Numic groups probably tapped the full range of usable resources, they never exploited them routinely with the intensity that was characteristic of Numic exploitation. In this sense, Numic and pre-Numic groups used the same species but not the same resources—the term *resource* here referring to a species taken in a specific kind of context in a specific way, as Steward would model this. Pinyon nuts, for example, can be procured casually by picking them off the ground, which requires little labor and technology, or intensively by gathering and processing green cones, which is both laborious and technologically demanding. The species is the same, but the modes of procurement define what are effectively two resources with different ecological implications. This, as noted, was the original thrust of Steward's cultural ecology.

A second objection to the Numic-spread hypothesis regards the suggestion that pre-Numic groups who observed Numic behavior, especially exploitative patterns, were incapable of mimicking that behavior to their own advantage. That they obviously did not do so suggests that the problem was not the technology (which was relatively simple), but organization: pre-Numic groups either did not grasp the way in which Numic organization (e.g., settlement system, marriage patterns) served the intensive exploitation of Great Basin resources, or could or would not make the necessary organizational adjustments in time to avert replacement. Perhaps this is because social organization and structure have no material existence and are less subject to experimental trial than the tangible components of technology. This is consistent with ethnographic data (Jorgensen 1969) suggesting that social organization and structure are more conservative and less subject to borrowing among neighboring groups than is technology.

This sort of thinking opens the door to a rather different kind of analysis because it suggests that in order to understand the Numic spread one must understand the mechanisms that produce, maintain, and reproduce cultural behavior—which is to say, the cultural context within which individuals must make rational choices. Cultural transmission theory provides one kind of framework that is likely to be useful in pursuing research in this area. Robert Boyd and Peter Richerson (1985, 1987), for example, have adopted this kind of approach to show how symbolic markers, such as those that commonly distinguish ethnic groups, might evolve and facilitate more effective adaptive behavior by reducing the cost of social learning, or by preventing the movement of behaviors from areas where they are adaptive to areas where they are not. They suggest this without arguing (as neofunctionalists are prone to do) that such symbolic behaviors need have direct adaptive value, or that the process that leads to ethnic differentiation necessarily stabilizes (reaches equilibrium) at levels that are adaptively optimal for the groups involved. With reference to the Numic spread, it seems quite likely that there were ethnic barriers between Numic and pre-Numic peoples. If so, these barriers might well have prevented pre-Numic adoption of the organizational poses that made Numic peoples so successful.

Summary

Great Basin anthropology has always been ecological, but its conduct as ecological anthropology has changed over time. *Cultural ecology*, the initial form, became *evolutionary ecology* (optimal-foraging version), and recently *evolutionary cultural ecology* (cultural-transmission version)—or, if you will, *culture historical ecology*. The meaning and method of cultural ecology have been attended to at length. The chief differences between cultural and evolutionary ecology are the latter's use of inclusive genetic fitness as the basis for judging rational choice and its discounting of cultural processes acting over time (i.e., of culture history). The chief difference between evolutionary ecology and evolutionary cultural ecology is the latter's provision for history and historical processes as in traditional cultural ecology. Simple evolutionary ecology emphasizes adaptation as the strong force standing behind patterns of behavior. In this it closely resembles, and was in many ways anticipated by, cultural materialism (e.g., Harris 1979), neofunctionalism (e.g., Rappaport 1968), and New (currently styled "processual") archaeology (e.g., Binford 1962).

Against these, evolutionary ecology differs chiefly in its casting of hypotheses about adaptation in terms of genetic inclusive fitness. As in primitive environmental determinism, when adaptation is taken as the strong force in this way, it reduces all behavior to genes and environment, between which an equilibrium is rapidly attained—removing all vestiges of history.

In evolutionary cultural ecology, by contrast, technology and nearly all other behavioral norms are the product of evolutionary processes that are imagined to be acting in time, i.e., as *beginning* somewhere. This is important because, to avoid the banalities of teleology, the entity that is the subject of process (e.g., a cultural behavior) must assume an initial form that is not dictated by the adaptive context itself. This initial state is, in short, arbitrary with respect to (i.e., independent of) evolutionary forces and context, and must be furnished with reference to other processes. For most of human behavior, culture history (i.e., by cultural inheritance or transmission) provides the best processual account of these initial conditions. The importance of this cannot be overstated, because the outcomes of many kinds of evolutionary processes are highly contingent on initial conditions. Especially notable here are the kind of frequency-dependent processes that coordinate many useful, group-level behaviors—driving on the right side of the road in the United States, for example. In this case, techno-environmental context favors the evolution of a rule for driving on just one side of the road without specifying which side, which is largely determined by the side that happens, by chance or by history, to be most common when the process is set in motion. Further, the force of frequency-dependence in this process is often as strong as or stronger than any selective processes that might otherwise weed out suboptimal alternatives. Because this is so, the equilibrium outcome need not be optimal, as the recent triumph of VHS- over Beta-format video demonstrates. In all such analyses, history both mediates and constrains the adaptive process. In the case of Numic replacement, for example, one is inclined to assume that Numic and pre-Numic groups climbed different adaptive peaks largely because they began their adaptive hill-climbing from different points on the evolutionary landscape. The general form of the process is clearly evolutionary; the specific form and outcome, however, are just as clearly historical.

What, then, is Steward's contribution? The scope of early Great Basin ecological anthropology, specifically cultural ecology, was quite clearly confined by Steward's interests and publications. It is important to understand that his interests were not, in the main, evolutionary. He was little concerned with origins and change (although *Basin-Plateau Aboriginal Sociopolitical Groups* briefly considers postcontact developments related to the introduction of the horse and other European elements). Steward was attempting to show how one could make sense of social and political behavior in terms of ecology at a given point in time (the ethnographic present): to show how a system worked in its environmental context. His analysis was purposely static because the subject matter was static. Early attempts to extend that model into the past (the Desert Culture hypothesis) employed Steward's arguments about the static situation, and in so doing they implied that the forces of ecology and environment were stronger, and the nature of ecological relationships simpler, than Steward himself had been required to argue. As the more untenable elements of the archaeological hypothesis have become evident, archaeologists have developed less deterministic and complex models in which, recently at least, culture history exerts a force that is potentially quite strong. I think Steward would approve of this. He never argued that culture was determined by environment, any more than he argued that culture could do what it liked. Rather, he argued that culture reached out to environment in creative ways, and that the manner in which it did so explained many things about human behavior. What is missing in Steward, what I refer to as "Steward's Problem," is any attempt to reconcile the culture core with the culture non-core—i.e., to relate the things that are heavily and most directly influenced by ecology with the things that are not (e.g., language and ideology). Great Basin archaeologists are currently using evolutionary cultural ecology and cultural transmission theory to attempt this bridge. The enterprise as a whole fits quite nicely under the program of research that has come to be called historical ecology—which I think Steward would approve of as well.

References

Aikens, C. M. 1970. *Hogup Cave*. Anthropological Papers, no. 93. Salt Lake City: Department of Anthropology, University of Utah.

Antevs, E. 1948. Climatic changes and pre-white man. *University of Utah Bulletin* 38:168–191.

Baumhoff, M. A. and R. F. Heizer. 1965. Postglacial climate and archaeology in the Desert West. Pp. 697–707 in H. Wright and D. Frey (eds.), *Quaternary of the United States*. Princeton: Princeton University Press.

Bettinger, R. L. 1975. The surface archaeology of Owens Valley, eastern California: Prehistoric man-land relationships in the Great Basin. Ph.D. diss., University of California, Riverside.

———. 1976. The development of pinyon exploitation in central eastern California. *Journal of California Anthropology* 2(2): 81–95.

———. 1977. Aboriginal human ecology in Owens Valley: Prehistoric change in the Great Basin. *American Antiquity* 42(1): 3–17.

———. 1989. *The Archaeology of Pinyon House, Two Eagles, and Crater Middens: Three Residential Sites in Owens Valley, Eastern California*. Anthropological Papers of the American Museum of Natural History, no. 67. New York: American Museum of Natural History.

———. 1991. Aboriginal occupation at high altitude: Alpine villages in the White Mountains of eastern California. *American Anthropologist* 93:656–689.

———. 1993. Doing Great Basin archaeology recently: Coping with variability. *Journal of Archaeological Research* 1:43–66.

Bettinger, R. L. and M. A. Baumhoff. 1982. The Numic spread: Great Basin cultures in competition. *American Antiquity* 47:485–503.

———. 1983. Return rates and intensity of resource use in Numic and Prenumic adaptive strategies. *American Antiquity* 48:830–834.

Binford, L. R. 1962. Archaeology as anthropology. *American Antiquity* 28(2): 217–225.

Boyd, R. and P. J. Richerson. 1985. *Culture and Evolutionary Process*. Chicago: University of Chicago Press.

————. 1987. The evolution of ethnic markers. *Cultural Anthropology* 2:65–79.

Broughton, J. M. and D. K. Grayson. 1993. Diet breadth, adaptive change, and the White Mountains faunas. *Journal of Archaeological Science* 20(3): 331–336.

Charnov, E. L. 1976. Optimal foraging: The marginal value theorem. *Theoretical Population Biology* 9:129–136.

Clark, Grahame. 1953. *From Savagery to Civilization*. New York: Henry Schuman.

Cronquist, A., A. H. Holmgren, N. H. Holmgren, and J. L. Reveal. 1972. *Intermountain Flora: Vascular Plants of the Intermountain West, U.S.A.*, vol. 1. New York: Hafner.

d'Azevedo, W., W. Davis, D. D. Fowler, and W. Suttles, eds. *The Current Status of Anthropological Research in the Great Basin: 1964*. University of Nevada, Desert Research Institute, Social Sciences and Humanities Publications no. 1. Reno: University of Nevada.

Delacorte, M. G. 1990. Prehistory of Deep Springs Valley, eastern California: Adaptive variation in the western Great Basin. Ph.D. diss., University of California, Davis.

Downs, J. F. 1966. The significance of environmental manipulation in Great Basin cultural development. Pp. 39–56 in d'Azevedo et al. (eds.), *The Current Status of Anthropological Research in the Great Basin*.

Driver, H. E., and W. C. Massey. 1957. Comparative studies of North American Indians. *Transactions of the American Philosophical Society*, n.s., 47, part 2.

Fowler, C. S. 1982. Settlement patterns and subsistence systems in the Great Basin: The ethnographic record. Pp. 121–138 in D. B. Madsen and J. F. O'Connell (eds.), *Man and Environment in the Great Basin*. SAA Papers 2. Washington, D.C.: Society for American Archaeology.

————. 1990. Ethnographic perspectives on marsh-based cultures in western Nevada. Pp. 17–31 in J. C. Janetski and D. B. Madsen (eds.), *Wetland Adaptations in the Great Basin*. Occasional Papers no. 1. Provo: Museum of Peoples and Cultures, Brigham Young University.

Fowler, D. D. 1966. Great Basin social organization. Pp. 57–74 in d'Azevedo et al. (eds.), *The Current Status of Anthropological Research in the Great Basin*.

Gunnerson, J. H. 1962. Plateau Shoshonean prehistory: A suggested reconstruction. *American Antiquity* 25(3): 373–380.

Harper, K. T. 1986. Historical environments. Pp. 51–63 in W. L. d'Azevedo (ed.), *Handbook of North American Indians*, vol. 11, *Great Basin*. Washington, D.C.: Smithsonian Institution.

Harris, M. 1979. *Cultural Materialism: The Struggle for a Science of Culture*. New York: Random House.

Heizer, R. F. 1941. The direct historical approach in California archaeology. *American Antiquity* 7(2): 98–122.

————. 1956. Recent cave explorations in the Lower Humboldt Valley. *University of California Archaeological Survey Report* 33:50–57.

Heizer, R. F. and L. K. Napton. 1970. *Archaeology and the Prehistoric Great Basin Lacustrine Subsistence Regime as Seen from Lovelock Cave, Nevada*. Archaeological Research Facility Contributions, no. 10. Berkeley: University of California.

Hewlett, B. S. 1991. Demography and childcare in preindustrial societies. *Journal of Anthropological Research* 47:1–37.

Hopkins, N. A. 1965. Great Basin prehistory and Uto-Aztecan. *American Antiquity* 31(1): 48–60.

Huntington, E. 1945. *Mainsprings of Civilization*. New York: Wiley.

————. 1963. [1927]. *The Human Habitat*. New York: Norton.

Jennings, J. D. 1957. *Danger Cave*. Anthropological Papers, no. 27. Salt Lake City: Department of Anthropology, University of Utah .

————. 1964. The Desert West. Pp. 149–174 in J. D. Jennings and E. Norbeck (eds.), *Prehistoric Man in the New World*. Chicago: University of Chicago Press.

————. 1968. *Prehistory of North America*. New York: McGraw-Hill.

————. 1978. *Prehistory of Utah and the Eastern Great Basin*. Anthropological Papers, no. 98. Salt Lake City: University of Utah Press.

Jennings, J. D. and E. Norbeck. 1955. Great Basin prehistory: A review. *American Antiquity* 21:1–11.

Jones, K. T. and D. B. Madsen. 1989. Calculating the cost of resource transportation: A Great Basin example. *Current Anthropology* 30:529–534.

Jorgensen, J. G. 1969. *Salish Language and Culture: A Statistical Analysis of Internal Relationship, History, and Evolution*. Indiana University Publications, Language Science Monographs 3. Bloomington: Indiana University.

————. 1980. *Western Indians: Comparative Environments, Languages, and Cultures of 172 Western American Indian Tribes*. San Francisco: Freeman.

Kelly, Isabel T. 1932. Ethnography of the Surprise Valley Paiute. *University of California Publications in American Archaeology and Ethnology* 31(3): 67–210.

————. 1964. *Southern Paiute Ethnography*. Anthropological Papers, no. 69. Salt Lake City: Department of Anthropology, University of Utah.

Kidder, A. V. 1927. Southwestern Archaeological Conference. *Science* 66:486–491.

Kroeber, A. L. 1923. The history of native culture in California. *University of California Publications in American Archaeology and Ethnology* 20:125–142.

————. 1939. Cultural and natural areas of native North America. *University of California Publications in American Archaeology and Ethnology* 38:1–242.

————. 1959. Ethnographic interpretations: Recent ethnic spreads. *University of California Publications in American Archaeology and Ethnology* 47:259–281.

Lamb, S. M. 1958. Linguistic prehistory in the Great Basin. *International Journal of Linguistics* 24:95–100.

Lawton, Harry W., Philip J. Wilke, Mary DeDecker, and William M. Mason. 1976. Agriculture among the Paiute of Owens Valley. *Journal of California Anthropology* 3(1): 13–50.

MacArthur, R. H. and Pianka, E. R. 1966. On optimal use of a patchy environment. *American Naturalist* 100:603–609.

Madsen, D. B. 1982. Get it where the gettin's good: A variable model of Great Basin subsistence and settlement based on data from the eastern Great Basin. Pp. 207- 226 in Madsen and O'Connell (eds.), *Man and Environment in the Great Basin*.

————. 1986. Great Basin nuts: A short (sic) treatise on the distribution, productivity, and use of pinyon. Pp. 21–41 in C. Condie and D. Fowler (eds.), *Anthropology of the Desert West*. Anthropological Papers, no. 110. Salt Lake City: University of Utah Press.

————. 1988. The prehistoric use of Great Basin marshes. Pp. 414–418 in C. Raven and R. G. Elston (eds.), *Preliminary Investigations in Stillwater Marsh: Human Prehistory and Geoarchaeology*. Cultural Resource Series, no. 1. Portland, Ore.: U.S. Department of Interior, U.S. Fish and Wildlife Service (Region 1).

————. 1989. *Exploring the Fremont*. Salt Lake City: Utah Museum of Natural History, University of Utah.

————. 1993. Testing diet breadth models: Examining adaptive change in the late prehistoric Great Basin. *Journal of Archaeological Science* 20:321–329.

Madsen, D. B. and J. F. O'Connell, eds. 1982. *Man and Environment in the Great Basin.* SAA Papers 2. Washington, D.C.: Society for American Archaeology.

Miller, W. R. 1966. Anthropological linguistics in the Great Basin. Pp. 75–111 in d'Azevedo et al. (eds.), *The Current Status of Anthropological Research in the Great Basin.*

———. 1983. Uto-Aztecan languages. Pp. 113–124 in A. Ortiz (ed.), *Handbook of North American Indians,* vol. 10, *Southwest.* Washington, D.C.: Smithsonian Institution.

———. 1984. The classification of Uto-Aztecan languages based on lexical evidence. *International Journal of Linguistics* 50(1): 1–24.

———. 1986. Numic languages. Pp. 107–112 in W. L. D'Azevedo (ed.), *Handbook of North American Indians,* vol. 11, *Great Basin.* Washington, D.C.: Smithsonian Institution.

O'Connell, J. F. 1971. The archaeology and cultural ecology of Surprise Valley, northeastern California. Ph.D. diss., University of California, Berkeley.

———. 1975. *The Prehistory of Surprise Valley.* Anthropological Papers, no. 4. Ramona, Calif.: Ballena Press.

O'Connell, J. F. and P. S. Hayward. 1972. Altithermal and Medithermal human adaptations in Surprise Valley, northeast California. Pp. 25–41 in D. D. Fowler (ed.), *Great Basin Cultural Ecology: A Symposium.* University of Nevada Desert Research Institute Publications in the Social Sciences no. 8. Reno: University of Nevada.

O'Connell, J. F., K. T. Jones, and S. R. Simms. 1982. Some thoughts on prehistoric archaeology in the Great Basin. Pp. 227–240 in Madsen and O'Connell (eds.), *Man and Environment in the Great Basin.*

Rappaport, R. A. 1968. *Pigs for the Ancestors: Ritual in the Ecology of a New Guinea People.* New Haven: Yale University Press.

Schoener, T. W. 1974. The compression hypothesis and temporal resource partitioning. *Proceedings of the National Academy of Sciences* 71:4169–4172.

Semple, E. C. 1911. *Influences of Geographic Environment: On the Basis of Ratzel's System of Anthropogeography.* New York: Holt.

Simms, S. R. 1983. Comments on Bettinger and Baumhoff's explanation of the "Numic spread" in the Great Basin. *American Antiquity* 48:825–830.

———. 1987. *Behavioral Ecology and Hunter-Gatherer Foraging: An Example from the Great Basin.* BAR International Series, no. 381. Oxford: B.A.R.

———. 1988a. Conceptualizing the Paleoindian and Archaic in the Great Basin. Pp. 41–52 in J. A. Willig, C. M. Aikens, and J. L. Fagan (eds.), *Early Human Occupation in Far Western North America.* Anthropological Papers, no. 21. Carson City: Nevada State Museum.

———. 1988b. Some theoretical bases for archaeological research at Stillwater Marsh. Pp. 420–426 in C. Raven and and R. G. Elston (eds.), *Preliminary Investigations in Stillwater Marsh: Human Prehistory and Geoarchaeology.* Cultural Resource Series, no. 1. Portland, Ore.: U.S. Department of Interior, U.S. Fish and Wildlife Service (Region 1).

Smith, E. A. and S. A. Smith. 1994. Inuit sex-ratio variation: Population control, ethnographic error, or parental manipulation. *Current Anthropology* 35(4): 595–624.

Steward, J. H. 1936. The economic basis of primitive bands. Pp. 331–350 in R. Lowie (ed.), *Essays in Honor of A. L. Kroeber.* Berkeley: University of California Press.

———. 1937. Ecological aspects of Southwestern society. *Anthropos* 32:85–104.

———. 1938. *Basin-Plateau Aboriginal Sociopolitical Groups.* Bureau of American Ethnology Bulletin 120. Washington, D.C.: GPO.

———. 1940. Native cultures of the Intermontane (Great Basin) area. *Smithsonian Miscellaneous Collections* 100:445–502.

———. 1941. Culture element distributions: XIII, Nevada Shoshone. *Anthropological Records* 4(2).

———. 1942. The direct historical approach to archaeology. *American Antiquity* 7(4): 337–343.

———. 1943. Culture element distributions: XXIII, Northern and Gosiute Shoshoni. *Anthropological Records* 8(3).

———. 1955. *Theory of Culture Change.* Urbana: University of Illinois Press.

Stewart, O. C. 1941. Culture element distributions: XIV, Northern Paiute. *Anthropological Records* 4(3).

———. 1942. Culture element distributions: XVIII, Ute-Southern Paiute. *Anthropological Records* 6(4).

Taylor, W. W. 1961. Archaeology and language in western North America. *American Antiquity* 27:71–81.

Thomas, D. H. 1971. Prehistoric subsistence-settlement patterns of the Reese River Valley, central Nevada. Ph.D. diss., University of California, Davis.

———. 1973. Empirical test for Steward's model of Great Basin settlement patterns. *American Antiquity* 38(2): 155–176.

———. 1974. An archaeological perspective on Shoshonean bands. *American Anthropologist* 76(1): 11–23.

———. 1982. *The 1981 Alta Toquima Village Project: A Preliminary Report.* Desert Research Institute Social Sciences Technical Report Series, no. 27. Reno: University of Nevada.

———. 1983a. *The Archaeology of Monitor Valley: 1, Epistemology.* Anthropological Papers of the American Museum of Natural History, no. 58, part 1. New York: American Museum of Natural History.

———. 1983b. On Steward's models of Shoshonean sociopolitical organization: A great bias in the Basin? Pp. 59–68 in E. Tooker (ed.), *Development of Political Organization in Native North America.* Proceedings of the American Ethnological Society, 1979. Washington, D.C.: American Ethnological Society.

———. 1988. *The Archaeology of Monitor Valley: 3, Survey and Additional Excavations.* Anthropological Papers of the American Museum of Natural History, no. 66, part 2. New York: American Museum of Natural History.

Wauchope, Robert, ed. 1956. Seminars in archaeology: 1955. *Memoirs of the Society for American Archaeology* 11:1–158.

Wedel, W. R. 1938. The direct historical approach in Pawnee archaeology. *Smithsonian Miscellaneous Collections* 97(7).

Weide, Margaret Lyneis. 1968. Cultural ecology of lakeside adaptation in the western Great Basin. Ph.D. diss., University of California, Los Angeles.

———. 1974. North Warner subsistence network: A prehistoric band territory. *Nevada Archaeological Survey Research Paper,* no. 5, pp. 62–79.

Wilke, P. J., R. L. Bettinger, T. F. King, and J. F. O'Connell. 1972. Harvest selection and domestication in seed plants. *Antiquity* 46:203–209.

Young, D. A. and R. L. Bettinger. 1992. The Numic spread: A computer simulation. *American Antiquity* 57(1): 85–99.

Zingg, R. M. 1939. A reconstruction of Uto-Aztecan history. *University of Denver Contributions to Ethnography* 2:1–274.

CHAPTER 9

Ancient and Modern Hunter-Gatherers of Lowland South America: An Evolutionary Problem

ANNA C. ROOSEVELT

Living hunter-gatherers have long been studied for clues to human behavior and conditions of life during the evolution of our species in the Pleistocene. The human genome that formed then is thought to dominate present human sociobiology—profoundly influencing diet, land use, social organization, sexuality, war, and intelligence. Hunting-gathering peoples of Amazonia, in particular, have become central to theories about the ecology and genetics of human adaptation. For theoretical purposes, Amazonian Indian groups sometimes have been depicted as Stone Age hunter-gatherers, without acknowledgment of their broader cultural-ecological adaptations and historical backgrounds. This uncontextualized evolutionary ethnology divorces people from their societies and history and makes adaptationist assumptions that are not supported by the data. Regional trajectories of development suggest that present foragers are not the lineal descendants of prehistoric foragers. In fact, many "primitive" behavior patterns commonly attributed to a foraging past appear to be recent adaptations that took place long after primary foraging was superseded by horticulture. Rather than being linked to a foraging lifeway, such behavior patterns are associated historically with the establishment of commercial extraction and of ranching in indigenous territories by Euro-American settlers since the conquest of the New World. In this way, present native lifeways can be seen as an adaptation to life in postcolonial nations as well as to the tropical environment. Untangling the causality of human lifeways requires what Balée refers to as critical historical ecology, which follows the interaction of ecological and cultural phenomena over time and space.

When foragers first appear in the South American lowlands in the late Pleistocene and early Holocene, they exemplify a wide range of ecological adaptations—including intensive fishing, hunting, and plant collecting along rivers, and broad-spectrum collecting and hunting in the forests and savannas. The 11,200- to 10,000-year-old Paleoindian culture recently discovered at Monte Alegre in the heart of the Amazon basin, with its well-developed rock-painting, large and small, carefully chipped spear points, and economy of tropical forest and river foraging, shows that the earliest foragers were neither primitive in culture nor unable to adapt to the humid tropical environment. The early lowland cultures were not static, however, and within a few thousand years people had settled down in fishing villages along the mainstream of the Amazon and began to make pottery, the earliest in the Americas. By about 4,000 years ago people had begun to cultivate crops, which appear to have become the most important calorie source in subsistence until late prehistoric times, when crops became the main source of protein (as well as calories) in the populous agricultural chiefdoms that developed along the mainstreams and in some uplands. When the chiefdoms were destroyed during the European invasion, many people returned to a lifeway of mixed horticulture and foraging. Others, forced into a more mobile settlement pattern by the manifold disruptions of conquest, developed new forms of foraging on the abandoned crop fields of their neighbors and the orchards and free-range cattle of the European settlers.

The developmental disjunction between ancient and modern foragers means that the latter may not be theoretical proxies for the lifeways of the former. Understanding the evolutionary significance of hunting and gathering in human evolution will require more direct research on the actual history of primary hunter-gatherer adaptations. This chapter critically reviews the evidence for foragers in Greater Amazonia, focusing on the Bororo, Aché, Yora, Sirionó, Yuquí, Hiwi, and their ethnohistoric and archaeological predecessors—examining their resources, hunting and gathering practices, cultivation, biology, organization, and sociopolitical contexts. This evidence has significant implications for future research strategies in the investigation of human nature and its origins.

The Role of the Tropical Forest in Human Evolution

Some scholars who study human ecology have questioned whether the tropical lowland forests were habitable by humans before the development of slash-and-burn farming, which allowed the nutrients of the forest to be used for food production (Bailey et al. 1989; Headland 1987). The premise is that the scarce game and fruits in the poor-soil tropical forest would have been inadequate sources of subsistence, especially of calories from starch and fat.

Such theories about ecological limitations on cultural development are for the most part founded on environmental stereotypes (Goodland and Irwin 1975; Meggers 1971). In the stereotypes, the tropical region is covered with nearly uninterrupted tropical rain forest growing upon highly leached, ancient soils developed upon nutrient-poor rocks resistant to weathering. Food for animals and humans is scarce in the forest, which concentrates its nutrients in indigestible wood and leaves. Permanent agriculture with annual crop production, the putative economic base for civilization (Steward and Faron 1959:44–64), is impossible in the poor forest soils, so farming is limited to slash-and-burn cultivation, in which the nutrients of the vegetation are used by cutting and burning the vegetation and growing crops in the ashes. The high-nutrient river floodplains are considered too rare and unstable to support civilization (Meggers 1971).

The realities of lowland environments, however, are quite different from the stereotypes. Rather than primarily leached, acid soils from nutrient-poor rock, large areas of the region have nutrient-rich soils developed on heterogeneous geological substrates: recent alluvium, lifted marine sediments, extensive volcanic rock, and limestones (Franzinelli and Latrubesse 1993:123–126; Sioli 1984:15–214, 537–580). Recent alluvium alone makes up 25 percent of the area in some places, according to surface surveys and remote sensing (Rasänen et al. 1991), and volcanic deposits blanket the Amazonian region of Ecuador. The soils on such rocks are not the nutrient-poor tropical soil classes considered suitable only for shifting cultivation but include some of the most favorable agricultural soil classes in the tropics.

The forests on these soils, consequently, differ significantly in their food resources from those on very poor soils. Rather than being dispersed inaccessibly in the forests, large stands of fruit and nut trees cluster on nutrient-rich soils, attracting a wide variety of game (Balée 1989; Beckerman 1979; Sponsel 1986). In most areas rainfall is not uniformly high or evenly distributed, but moderate and seasonal (Whitmore 1990; Whitmore and Prance: figure 2.4; Sioli 1984:106), creating the possibility for scattered open habitats, numerous ecotones, and seasonal flushes of biological productivity.

Further, much of Greater Amazonia is crisscrossed by rivers, which bring in new stores of nutrients every year and harbor the fish that are the near-universal protein source for native peoples in the lowlands. Floodplain lakes in particular (Junk 1970) are foci of highly productive, annually harvestable biomass in the form of wildlife, edible herbs, and fruit trees clustered there. Some of the most common floodplain trees and palms bear large annual crops of oily and/or starchy fruits. Ubiquitous reptiles lay large clutches of oil-rich eggs, and many of the abundant fish lay down fat stores seasonally. By referring exclusively to poor-soil forest resources and neglecting productive aquatic microenvironments, ecological-barrier theorists have left out significant natural resources that would be available to foragers. The environments of the lowlands, therefore, would not necessarily have prevented either foraging or agriculture in the evolutionary trajectory of native societies.

The Clovis Migration Theory

Nevertheless, the tropical forests have long been assumed to have been an ecological barrier to the first migrants into the continent: the Paleo-Indians, traditionally pictured as specialized hunters who pursued Pleistocene megafauna through the open, temperate habitats of the Americas. Bands of Paleo-Indians are thought to have entered the Americas across the Bering Straits and settled in the high plains of western North America by 11,200 years ago, near the end of the Pleistocene epoch (Haynes 1992; Fagan 1987). Sites of the Clovis-tradition cultures in this area have fluted spear points and other tools, and bones of large, extinct animals. According to the migration theory, offshoots from these cultures descended into South America down the Andean chain, avoiding the seacoasts and tropical forests (Lynch 1978, 1983). Paleo-Indians were considered to have avoided aquatic habitats, due to their focus on large-herd game and their lack of fishing technologies and watercraft. Big-game hunting was also considered inappropriate for the dispersed, cryptic game of the resource-poor tropical forests.

The Clovis migration theory of the peopling of the New World has been the most widely accepted so far, and as a result most research on Paleo-Indians has focused on temperate open areas of South America; littoral and tropical lowland areas have not been systematically studied until recently. The migration theory, consequently, has rested primarily on evidence from the presumed center of the Paleo-Indian expansion, without corroboration from the peripheries. As negative evidence in support of the theory, specialists cited a lack of finely flaked, late Pleistocene bifacial spear points in humid tropical lowland South American sites (Lynch 1978, 1983). But since little research had been carried out on the preceramic of that area, the lack of evidence for any particular trait was hardly conclusive.

Despite the evidence problems, the idea of the tropical forest as an ecological barrier to Paleo-Indians came to have the status of fact. Preceramic cultures had been discovered in lowland South America, but most North American Paleo-Indian specialists denied that they were Paleo-Indian because they did not fit the tenets of the theory.

Pre-Clovis Adaptation in South America

The Clovis migration theory has been strongly opposed by Alan Bryan and several other scholars (Bednarik 1989; Bryan 1983, 1986; Dillehay et al. 1992), who hypothesize the existence of a pre-Clovis occupation by people with a broad-spectrum culture of plant collecting, fishing, and small-game hunting, and a simple stone tool kit lacking the finely chipped projectile points of the later Paleo-Indians. The archaeologists found evidence of this culture east of the Andes at pre-Clovis Paleo-Indian sites ranging from 300,000 to 12,000 years old. Some of the best known deposits are the lower levels of Neide Guidon's site of Pedra Furada in northeast Brazil (Guidon and Delibrias 1986), Conceição Beltrão's site of Alice Boer in southern Brazil (Beltrão et al. 1986), and Ruth Gruhn and Alan Bryan's site of Taima Taima in northern Venezuela (Bryan et al. 1978; Ochsenius and Gruhn 1979). Less known example are the basal deposits at Eurico Miller's site, Abrigo do Sol, in Mato Grosso, Brazil (Schmitz 1987), and André Prous' site, Boquete, in Minas

Gerais (Prous 1991). Their claims, however, are based on evidence of artifacts, hearths, or dating that many archaeologists find questionable (Dincauze 1984; Lynch 1990; Roosevelt et al. 1996). Disagreement over the cultures of claimed great antiquity has tended to obscure the presence east of the Andes of numerous well-documented Paleo-Indian cultures that were identical in age with Clovis tradition, but quite distinct in culture and ecological adaptation (Roosevelt et al. 1996).

At the same time, sites with problematic stratigraphy, rare and inconsistent dates, rare or absent megafauna, or questionable artifacts in temperate Andean regions were readily accepted as evidence of the arrival of Paleo-Indians migrating from North America (Lynch 1983). The El Inga site in Ecuador, and Guitarrero, Pikimachay, and Lauricocha Caves and Talara in Peru are examples of such sites. In addition, the claimed pre-Clovis site of Monte Verde (Dillehay 1989) in the Chilean Andes area has gained surprising credibility, despite its wide-ranging dates, rare, crude lithics, discontinuous stratigraphy, waterlogging, and carbon contaminants. Monte Verde's finely flaked projectile points were un-associated finds.

These and other Andean sites had as many evidence problems as preceramic lowland sites, if not more (Dillehay et al. 1992; Roosevelt et al. 1996), but the Andean sites' consonance with the theory has tended to shield them from the critical analysis directed at sites east of the Andes.

Unfortunately, the debate over the sites has been harsh and polarized, marred by rigidity and unnecessary ignorance of South American sites and complexes on the part of the "Clovists," and by credulity and unwillingness to acknowledge empirical problems on the part of the "pre-Clovists." Not all South American preceramic sites were excavated using the now-standard archaeological techniques of stratigraphic mapping, piece plotting, fine-screening, total collection, archaeobotany, and analytical dating programs, which means that the quality of the evidence is often inadequate for the interpretive problems. And influential North American Paleo-Indian experts, for their part, frequently do not bother to consult South American publications not in English, and are often not personally familiar with lowland South American sites—making their opinions rather uninformed. The upshot of the debate has been a situation in which widespread doubt about the validity of Pleistocene human sites in eastern South America has continued.

Nevertheless, despite the weakness of the claims for the earliest dates, several lowland South American preceramic human occupation sites without apparent stratigraphic or contamination problems have consistent initial dates between about 11,200 and 10,000 B.P., contemporary with the Clovis tradition (Dillehay et al. 1992; Prous 1991; Roosevelt et al. 1996; Schmitz 1987). Many of these sites contain large and small, finely flaked, triangular, tanged projectile points and broad-spectrum subsistence remains quite different from those of North American Paleo-Indians. Both the presence and the character of these cultures tend to contradict the theory that the descendants of Clovis colonized South America through the Andean chain. The existence of late Pleistocene cultures in eastern South America forces a reconsideration of the migrations and ecological adaptations of foragers in the Americas.

Early Cave Dwellers of Monte Alegre

Possible Paleo-Indian cultures in the tropical lowlands had been easy to dismiss in the past because of evidence problems. Accordingly, my recent research to investigate early hunter-gatherers in Amazonia was designed to collect more definitive evidence. My collaborators and I searched for sites with indubitable artifacts, excellent stratigraphy, abundant datable paleoecological evidence, and no possible contamination agents. We also applied a multifactorial methodology using several different dating methods: numerous dates on individual, rather than grouped, specimens, and split-sample, blind dating trials by different laboratories on a range of different types of materials (Roosevelt et al. 1996).

We focused our search for Paleo-Indians in the Santarem–Monte Alegre region, alerted to their possible existence by large, finely chipped spear points in museums, and by an abundance of rock art (Hartt 1871; Museo Paraense Emílio Goeldi 1989; Simões 1976). In this region of the confluence of the Tapajós River with the Amazon, the mainstream Amazon winds through extensive floodplains and river terraces, between high *terra firme* and ranges of sandstone hills. At Monte Alegre on the north bank, a hypothesized Pleistocene refugium,[1] we found among the painted caves first described by Alfred Russel Wallace (1889) one that lacked the evidence problems of other lowland sites. At this site, Caverna da Pedra Pintada (see figure 9.1), we found a Paleo-Indian campsite of tropical forest hunter-gatherers sealed under sterile layers at the base of a deeply stratified deposit of prehistoric cultures. The Paleo-Indian layers were full of carbonized wood and the pits and nutshells of the fruits of evergreen tropical forest and seasonal woodland trees and palms that still grow in the region today. Among them was the highly important economic species Brazil nut, *Bertholletia excelsa*, and other trees: *Sacoglottis guianensis*, *Mouriri apiranga*, *Byrsonima crispa*, numerous palms (*Attalea* spp. and *Astrocaryum* spp.), and tree legumes (*Hymenea*, c.f., *parvifolia* and *oblongifolia*).[2] The faunal remains were poorly preserved but highly diverse: taxa included fish, tortoises, turtles, toads, snakes, shellfish, small and medium-sized rodents, bats, and very rare large mammals (probably ungulates). Many of the reptiles and rodents were juveniles. Fish, the most common fauna, included very large fishes (1.5 m) to very small ones (10 cm); species ranged from *Hoplias malabaricus* to *Arapaima gigas*, and there were numerous unidentified catfishes, characins, and cichlids. The fact that none of the plant or animal species are particularly adapted to cold, desert, or grass savanna environment confirmed that this was indeed a tropical forest adaptation, as did the carbon isotope patterns.[3] The presence of several tree species considered to be adapted to human disturbance suggests that the Paleo-Indian occupation may have already begun to have an impact on the character of the forest. The camp was apparently visited at different times of the year, for the tree species found in the cave tend to fruit in the rainy season today (Cavalcante 1991) but the fish fauna and turtles are most accessible in the dry season and seasonal transitions, at least today. Today, men tend to hunt very large game singly or in small groups, but small fish, small reptiles, amphibians, mammals, and juvenile animals are frequently

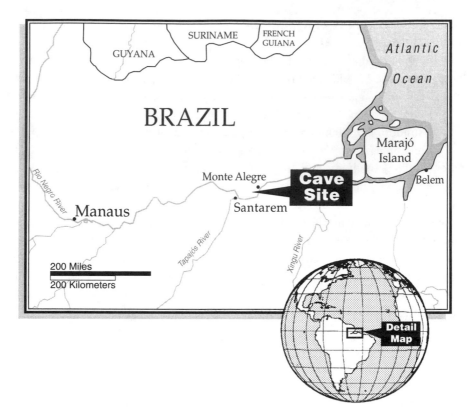

Fig. 9.1 *Location of Paleo-Indian Cave Site, Caverna da Pedra Pintada.*

caught by women and children. The wide range of both small and large species in the cave suggests a pattern of broad-spectrum tropical forest and riverine gathering similar to that of the other eastern Paleo-Indian cultures, rather than the specialized large-game hunting by males envisioned in the Clovis migration theory.

Among the pressure-flaked stone tools in the cave, there were large and small, bifacial projectile points sometimes with stems and wings that could have functioned as harpoon and spear points, and heavy scrapers and gravers that could have been used for woodworking. The abundant pigment in the Paleo-Indian levels of the cave is of the same composition as that of the cave paintings, documenting aesthetic-ceremonial activity as well as subsistence foraging. Among the designs are human handprints of adults and children.

The dates of this component of the site—which include fifty-six radiocarbon dates between 11,200 and 9,800 B.P. on wood and fruits and thirteen corroborative luminescence dates on soil and burned lithics—show that Paleo-Indians adapted to tropical habitats at the same time that they adapted to temperate habitats. Not surprisingly, the adaptation to the Amazonian habitat involved different subsistence forms, artifacts, and art forms than did the adaptation to the temperate plains of North America. The existence of a tropical South American Paleo-Indian culture the same age as Clovis but culturally distinct suggests that, rather than being the progenitor of South American Paleo-Indian cultures, Clovis was just one of several regional Paleo-Indian cultures. Research along the now-submerged coasts of the hemisphere is needed to investigate the forebears of eastern South American Paleo-Indians.

Floodplain Foragers Discover Pottery

Just as Paleo-Indians were not limited to one particular adaptation, the type of adaptation represented by Monte Alegre was not the only one established in Amazonia. In the early postglacial period, Amazonian foragers settled in large fishing villages along the floodplains and began to make pottery, the earliest in the Americas, between ca. 7,500 and 4,000 years ago. We know this culture best from the site of Taperinha, near Santarem, which we excavated in 1987 and 1993, led there by the writings and museum collections of nineteenth-century scientists (Hartt 1885; Roosevelt et al. 1991).

According to the large shell midden at the site, these peoples' subsistence heavily emphasized turtles, fish, and shellfish, with tree fruits and seeds a lesser source of food. These resources of the tropical aquatic habitat furnished the economic support for the development of pottery. Despite the orientation to the floodplain, the early pottery-age people continued to use the interfluvial forests seasonally (we find them in the middle layers at Caverna da Pedra Pintada), and a variety of broad-spectrum foragers roamed the Amazon interior away from the floodplain (Magalhães 1994). It seems that comparative sedentism and the use of pottery were floodplain phenomena, for sites far away from the large rivers usually lack pottery at this early date. Contrary to the barrier hypothesis, the development of ceramics and village life did not require agriculture—for, despite exhaustive soil screening and flotation, no trace of cultigens has been found in the early pottery sites. The foraging economy appears to have supported a relatively healthy lifestyle, for the teeth of these early people lack evidence of the growth-arrests common in populations whose young are plagued by chronic or periodic infectious disease or nutritional stress.

Although some archaeologists expected early pottery to be undecorated, this pottery bore rare incised designs on rim and sides, possibly copied from designs incised or burnt on gourd vessels. In contrast to Paleo-Indian lithics, the lithic tools of the shell-mound dwellers were rare and ill-made, consisting of unshaped stone mashers, grinders, or hearth rocks, and flakes. Bones and shells were used for tools, and we recovered a manatee bone toggle and awl, and turtle- and mollusc-shell scrapers.

The early pottery foragers have been found at many sites in eastern South America and are now dated by thirty-seven radiocarbon assays to between ca. 7500 and 4000 B.P. (uncalibrated). Their age has been verified exhaustively by radiocarbon dating done directly on pottery as well as on associated biological remains, and by TL dating of pottery also dated by radiocarbon. In addition to Taperinha, early

Fig. 9.2   *Caverna da Pedra Pintada, Pará, Brazil.*

pottery cultures have been dated at Caverna da Pedra Pintada and at seven other
sites at the mouth of the Amazon and the coasts of the Guianas (Roosevelt 1995).

   These other early pottery sites, dated in the 1970s and 1980s, are not well known
to North American anthropologists because many were published only in South
America and some only in Portuguese, a language that few North Americans read.
Also, some of the earliest pottery dates run by the radiocarbon laboratory at the
Smithsonian Institution were withheld from publication (Simões 1981) or published as preceramic (Williams 1981). The Smithsonian scholars Clifford Evans
and Betty Meggers believed that pottery had been introduced from Japan to the
west coast of South America about 5,000 years ago and spread from there through
the continent (Meggers, Evans, and Estrada 1965); when the institution's laboratory
produced dates earlier than 5,000 B.P. in eastern Amazonia, they were reluctant to
accept the dates. Through the Freedom of Information Act, the Smithsonian Anthropology Archives have made the original laboratory records and correspondence
available to us, and we have been able to publish all the Smithsonian dates along
with those from Taperinha and Pedra Pintada (Roosevelt 1995). If this had not been
possible, it would have been difficult indeed to gain recognition for the existence of
Archaic pottery cultures in eastern South America.

Prehistoric Farmers Replace Foragers

The lack of importance of farming among the early pottery cultures of Amazonia
was not due to any environmental limitation for agriculture, because it is thought

that during the next phase of occupation people began to farm (Bush and Colinvaux 1988; Denevan 1966; Lathrap 1970; Roosevelt 1980, 1989, 1993, 1995b; van der Merwe, Roosevelt, and Vogel 1981). The evidence for prehistoric cultivation consists variously of crop pollen from lake cores, human bone chemistry, carbonized maize, manioc griddles, and agricultural earthworks.

At first, between about 4000 and 2000 B.P., crops may have become an important part of the calorie supply, but the human bone chemistries show that they were supplemented with fish and game for the majority of the protein in the diet. These early horticultural cultures are known by archaeologists for their elaborate pottery and stone carvings, which depict animals, human-animal combinations, and geometric designs.

The phase of mixed horticulture and foraging lasted many millennia, and only in the last prehistoric period does maize agriculture become the predominant source of both calories and protein in some areas of Greater Amazonia. During the years between about 2000 B.P. and the sixteenth century, when the Amazon was invaded by Europeans, intensive agriculture became the main subsistence economy in several areas, not only in floodplains but in forests also. At the same time, the remaining forests were becoming more open and diverse, and rainfall may have diminished as a result. Naturalized or cultivated species, such as cashew (*Anacardium occidentale*) and passion fruit (*Passiflora edulis*), spread widely, and people moved valued tree species out of their customary microhabitats by cultivation. Acai (*Euterpe oleracea*), for example, was grown in areas where it must be watered artificially to survive. The subsistence system appears to have been quite adequate, as skeletal populations of the time show few pathologies other than minor anemic and alveolar pathologies attributable to minor nutrient deficiencies, gum abrasion, and intestinal parasites. Stature is generally tall, and pathologies of chronic growth-arrests are not common.

In some areas, the pollen evidence indicates extensive deforestation around settlements (Bush and Colinvaux 1988), which were sometimes large and sedentary (Athens 1989; Porras 1987; Roosevelt 1991; Smith 1980). At this time, population density and size appear to have increased to levels above those observed in Amazonian Indian societies today. Concurrently, changes in settlement pattern and material culture, along with conquest-period documents, indicate that some areas came to have more centralized and hierarchical political organization than earlier in prehistory (Porro 1994; Rodrigues de Oliveira 1994; Roosevelt 1993). Artifacts become even more elaborate and larger in scale than before, and long-distance trade systems expanded. Ethnohistoric accounts by the early invaders from Europe describe dense and settled agricultural populations and warring paramount chieftaincies.

In terms of timing, it is the European conquest that correlates with the disappearance from Amazonian Indian cultures of many of the traits of the late prehistoric period. The native chiefdoms were defeated, politically decapitated, and their populations decimated (Porro 1994). Gathered into missions, the survivors were deculturated and intermarried with Europeans (Rodrigues de Oliveira 1994). The best land went to the missions and to settlers, who established plantations and

ranches. In the cattle area, agriculture diminished since the unfenced cattle trampled and grazed on the crops. Many indigenous people fled to the poorer hinterlands away from the main river. This process of disorganization and displacement
of societies and the introduction of new diseases led to a less favorable physiological condition, such that aggravated pathologies of generalized anemia not found in
prehistoric skeletons are common in nineteenth-century Indian skeletons.

With lower population density and more fluid settlement forced by the pressure
from missionaries and settlers, many groups gave up intensive agriculture and
began to rely more on shifting cultivation. Fish returned to its former prominence
as the main protein supply, and forest regenerated in the abandoned agricultural
fields. The process of cultural change and displacement has only intensified during
the national period, and Indians today are primarily confined to the peripheries of
the lowlands (Clay 1988; Roosevelt 1994).

Modern Hunter-Gatherers in Theoretical Perspective

Hunter-gatherers have been identified by some scholars among modern Indian
groups of Greater Amazonia. Some of these groups are important in the research of
a group of evolutionary anthropologists called *sociobiologists*, who are particularly
interested in the ecological and biological causes of the evolution of human social
behavior. Sociobiologists believe that living hunter-gatherers of the tropical forests
represent an ancient lifeway that predominated during the long Paleolithic period
and did much to shape present human nature (Hill and Hurtado 1989). They are
interested in studying living foragers to see what life was like during the Paleolithic.
This research is, thus, a search for the primitive, the original, the primeval, and it assumes a linear evolutionism from the Paleolithic directly to modern hunter-gatherers,
often without regard for the actual history of ecological adaptations in the area.

One of the most interesting and problematic aspects of sociobiologists' theoretical approach is their projection of expected subsistence patterns and division of
labor onto Paleolithic humans (Roosevelt n.d.). Ethnocentric assumptions from
Western society lead sociobiologists to expect early hunter-gatherers to have been
male-dominated, both economically and socially. This line of thought follows from
the belief that male hunting was the main source of food during the Paleolithic, and
that this, together with male physical superiority for defense and the necessary domestic confinement of women due to child-rearing, was responsible for the evolution of gender inequality and patrilineal band organization. Of great interest in
terms of the history of ideas is the connection of such theoretical interpretations to
late-Victorian notions of the proper roles for men and women—as breadwinners
and keepers of the domestic fire, respectively.

Several scholars have pointed out, however, that the South American lowland
"foragers" are not exclusive foragers, relying purely on wild resources (Balée
1992). Many of the relevant South American groups are cultivators, like the majority of lowland Indians, and the supposition that they were foragers is based on uninformed hearsay or the presumption that the cultivators must have only recently

taken up cropping. In some cases the identification of a group as foragers is based purely on the lack of the craft of pottery-making, not the lack of farming: it is assumed that the lack of pottery indicates that these are foragers recently converted to agriculture.

Nevertheless, none of the lowland "foragers" are preceded in their areas by hunter-gatherers or preceramic cultures. In fact, the earliest sources describe the groups in question as making pottery, having relatively large populations and sedentary settlements, and practicing agriculture. Their prehistoric predecessors appear in many cases to have been ceramic-age, agricultural chiefdoms. The process by which these people subsequently lost population, increased their mobility, stopped making pottery, and lessened their reliance on farming appears to have been an adaptation to the disruptions caused by conquest, missionization, colonization, and national development. Many of the groups identified as foragers live today on missions or privately owned ranches and either work for wages or gifts, or live as beggars or virtual slaves. Some are refugees from war or harassment and live on the run in very small groups; they frequently cannot maintain gardens under these conditions and must live off gathered resources or the theft of others' produce.

Further, the division of labor and status among these lowland Indian "foragers" is quite different from sociobiologists' models for the most part, with women having comparatively strong social roles, furnishing the majority of subsistence calories in the form of crops and benefitting from a tendency to uxorilocal residence of matrilineally related relatives.

The Hiwi of Venezuela and Colombia

The Hiwi (also known as Cuiva or Guahibo) of the middle Orinoco are one of the groups studied by sociobiologists, who claim that they have always been exclusive foragers and are good models for the Paleolithic (Hurtado and Hill 1991).

The sociobiologists describe the Hiwi as a society in which men do most of the foraging, and women tend to remain in the domestic sphere, in contrast to the general lowland pattern. However, the footnotes to the anthropologists' articles and the ethnohistoric documentary background of the Hiwi reveal that a particular sociopolitical context and recent historical change are responsible for these patterns. According to Magdalena Hurtado and Kim Hill (1991), Hiwi women do not forage today because they are threatened with rape by the Venezuelan cowboys who patrol the cattle ranges that the Hiwi occupy. Further, the "game" that men capture for food, ca. 30 percent of the diet, are domestic cattle, not wild game. In addition, the fruit that men collect is mango, an Asian species that, like cattle, is a product of the intrusion of Euro-Americans into Hiwi territory. Major elements in Hiwi subsistence derive from the ranch economy of Colombia and Venezuela, and from wage labor in nearby cities.

The identification of Hiwi as pure hunter-gatherers ignores the primary literature. The earliest documentary references to these people, c. 1530, describe their large villages (six houses or more), large fields of tall, abundant maize, gold ornaments, and bead money (Federman 1958:71–84). This and numerous other early sources describe the group as clients of the powerful Achagua chiefdom (Morey 1975). Archaeological research in the middle Orinoco confirms at least two thousand years of occupation in the area by stratified farmer-fisher cultures before the conquest (Petrullo 1939; Roosevelt 1980; Spencer and Redman 1991).

The Hiwi are still cultivators and potters today (Conaway 1976), but they have become more mobile in settlement pattern since the early days of the conquest. With the division of their territory into ranches, free-range cattle have become their main source of protein, and the introduced mango trees at abandoned rancher settlements, their main collected plant food. But wages or produce from seasonal labor, not natural resources, have become their main source of support, as a consequence of their embeddedness in the national society.

The Yora of Peru

The Yora of the Manu Park in the Peruvian Amazon have also been presented by sociobiologists as primeval hunter-gatherers in the tropical forest (Hill and Kaplan 1989; Hill and Hurtado 1989). But, like the Hiwi, the Yora also get their food from agriculture, as the numerous quantitative tables from the research documentation show. Refugees from intertribal conflict, they no longer make pottery or farm their own fields, but their food nonetheless comes from fields of planted crops, and their settlements lie on ceramic archaeological sites (a fact recorded by biologists surveying in the area). Prehistoric groups in the foothills of the Upper Amazon, in fact, made elaborate pottery and had maize agriculture, and in some areas they built earth and stone terraces and stone buildings with decoration (Lathrap 1970; Roosevelt 1989).

Neither the Hiwi nor the Yora are demonstrated to have evolved directly out of prehistoric hunting-gathering societies, and their adaptations cannot be understood apart from the national societies that they lie in today.

The Aché of Paraguay

A lowland South American group very important in the hunter-gatherer literature is the Tupi-speaking Aché of the subtropical hills of Paraguay, recently studied by sociobiologists (Hill and Hawkes 1983). The researchers characterize the Aché as relict hunter-gatherers who have survived from ancient times despite the late prehistoric spread of farming in the lowlands. The research focuses on trekking behavior and the subsistence gained from foraging on trek. According to the researchers, Aché families live primarily by trekking for game. The food collected by women on trek is insignificant in the diet, in which meat from game hunted by men makes up 90 percent. This percentage of meat in the diet has reverberated in the theoretical literature on hunter-gatherers, because recent writings by other authors have shown quantitative evidence of the large contribution that plant food collected by women tends to make to forager diets. The Aché are presented as a contradiction to this pattern.

However, by looking closely at earlier literature on the Aché (also referred to as Guayaki) (Métraux and Baldus 1946), and at the details and footnotes of the many overlapping articles by the Aché research team, one finds a very different story. The oldest sources on the Aché raise serious doubts about their lack of horticulture. It seems, in fact, that their identification as hunter-gatherers may derive from the pejorative epithets of their indigenous enemies, and of missionaries and ranchers who wished to settle them on their lands in order to gain their souls and labor.

The Aché certainly still subsist mainly on agriculture today and did so when studied by the team, as quantitative consumption tables compiled by the researchers show. Why, then, do the researchers make the statement that meat from game makes up 90 percent of the Aché diet? The situation is as follows: Before the sociobiologists began their research, the Aché had already settled permanently with missionaries, for whom they worked as servants and farm laborers and from whom they received wages, goods, and crops, as well as meat from domestic animals and purchased goods such as flour and sugar. By the time of their move to the missions, the Aché had given up their traditional hunting treks—but the anthropologists persuaded them to go on trek for purposes of the research and paid them for doing so. Since the Aché did not bring with them on the trek enough food to last for the trip, they were obliged to subsist on what they caught. Hence the 90 percent meat figure, which was what they ate on the hunting trek commissioned by the anthropologists. But in the theoretical literature the anthropologists do not make it clear that, according to their own careful measurements, crops, not game, make up ca. 90 percent of Aché food over the year—a pattern that makes them similar to most other Amazonian Indians. The fact that they once trekked seasonally does not make them primeval foragers, for most Amazonian Indians regularly go trekking (Lathrap 1968), a valued type of family vacation as well as a seasonal supplement to food from crops.

There seems no valid reason to consider the pre-mission Aché different from other Amazonian horticulturalists, who supplement their starchy crops with fish and game. Their identity as primary foragers is not based on any documented evidence.

The Sirionó and Yuquí of Bolivia

The Tupi-speaking Sirionó of the floodplains and seasonal forests of lowland Bolivia were one of the first groups to be tagged as primitive hunter-gatherers. One of their first ethnographers, Alan Holmberg, regarded them as primitive and culturally depauperate (Holmberg 1969). He described their dwellings and gardens as pitiful, dirty, and crude, and characterized their psychological state as obsessed by an insufficiency of food.

Nonetheless, the Sirionó were cultivators then as now and possessed cultural items typical of Amazonian horticultural groups' material culture, such as stone axes for clearing fields and pottery for vessels. The poverty of their lifeway can be attributed to several factors in their current political environment, rather than to their primitive evolutionary status. Holmberg, himself, affected their settlement pattern greatly by asking them to stay in a particular place so that he could study them; at the time, they were running away from a forced-labor camp at a mission station in a region that had long ago been converted to ranching by white and Mestizo farm owners, for whom the Sirionó customarily worked as hands.

The immediate predecessors of the Sirionó in the area were not hunter-gatherers but the people of the sedentary, agricultural chiefdoms that for more than 1,000 years lived in the midst of large mounds and agricultural earthworks that they constructed (and that the Sirionó live on today) (Erickson 1980). The Sirionó could be the decimated descendants of such Upper Amazon floodplain chiefdoms—or they could be deculturated migrants from Tupi-Guarani Polychrome-horizon chiefdoms farther to the south, as some have speculated (Stearman 1984). Their oral history, collected by an Argentine archaeologist, includes evidence that they practiced urn burial cults of types found in the archaeological record of mounds of both areas (Fernandez-Distel 1984–85). Only careful archaeological and historical work will tell which explanation is more likely.

The Yuquí (Stearman 1984, 1988) of the Andean piedmont of the Bolivian Amazon, linguistic relatives of the Sirionó, similarly lack documentation for their hunter-gatherer status. Their ethnographer presents them as extremely primitive hunter-gatherers who do not even know the simplest crafts, such as basketry. However, the Yuquí fit better into the classification of impoverished recent refugees, like the Yora, than into primary foragers. They have knowledge of cultivation and report that they practiced farming in the past, but they barter for a living at present. Their social context when studied by the anthropologists was a Protestant mission station, where they lived and exchanged labor and hammocks for cash, guns, ammunition, clothing, and food. The anthropologists called them primary hunters but observed that fishing supplied most of their protein, as is common elsewhere in Amazonia. The majority of their calories is from store-bought food and locally produced crops.

Like the Sirionó, the Yuquí lack forager predecessors. They were preceded in the tropical forest piedmont by people who farmed maize, made elaborate pottery, and lived in substantial, permanent settlements until the eighteenth century (Bennett 1936). Why the elaborate indigenous cultures disappeared, and where the Yuquí came from, if not from there, will remain unknown until someone researches their past. But it is clear that the Yuquí do not represent ancient foragers who somehow survived in the deep forest uncontacted by civilization.

The Bororo of Central Brazil

Finally, the Bororo are perhaps the most famous hunter-gatherer group in the tropical lowlands east of the Andes. Studied by anthropologists as prestigious as Claude Lévi-Strauss (1955), the Bororo often have been considered noteworthy in part because they seemed to be foragers with an unusually elaborate social organization, religious ritual, and material culture. They were regarded as paradoxical because anthropologists have believed that foragers would necessarily have a primitive culture and social organization. (Recently, anthropologists have concluded that there are, in fact, numerous prehistoric hunter-gatherer societies characterized by cultural and/or organizational complexity [Price and Brown 1985].)

The forager status of the Bororo, however, appears to have been concluded on the basis of their supposed lack of pottery—not on a lack of cropping. They have always farmed maize and other crops since the first observations (Lowie 1946). This is also true of the entire Gê-speaking group of tribes in central Brazil, but the misidentification of the Bororo as hunter-gatherers led to the conclusion that Gê peoples had only recently converted to horticulture. Accordingly, the article on them by Lowie was included in the volume on South American "marginals" in the influential *Handbook of South American Indians* (Steward 1946).

Several aspects of Bororo society do not fit the evolutionary theories. Unlike sociobiologists' models of early forager societies as patrilineal, male-dominated hunting bands, the Bororo, like many of the Gê and other supposed Amazonian foragers, are predominantly matrilocal. In fact, the laments by young Bororo men about this fact have achieved mythic literary status (Crocker 1985). In this society, men join their wives' families at marriage and owe their in-laws bride-service. The burdens of sons-in-law are so heavy that in Amazonia the *Língua Geral* word for "son-in-law," *poitos*, is also used for "slave." Although a husband has to clear the fields and build the house, the fields and house are nonetheless considered to be owned by the women, who farm and keep the house, not by the man; and a man's children are considered to belong to and owe obedience to their maternal uncle, rather than to their biological father. Marriage bonds in such societies are weak, and young men who are unlucky or poor providers complain that they are often kicked out without ceremony and have no place to go, for their own families are not glad to have them home again.

Recently, Bororo archaeology has been investigated with their help, and a documented history has been compiled for them (Wüst 1994). As in the other cases discussed here, the history does not include the survival of primitive hunter-gatherers to the present. The sequence begins ca. 10,000 B.P. with preceramic hunter-gatherers, who are followed several thousand years later by pottery cultures who apparently farmed manioc and other crops. The Bororo entered the region only after the European conquest, sometime in the eighteenth century, displacing the manioc cultivators. At that time they made pottery, farmed maize and other crops, and foraged, as they continue to do today. The fact that they have recently abandoned pottery manufacture has more to do with the adaptation of their society to the economy of the national society and the availability of metal and plastic containers than with the exigencies of a foraging lifeway.

The misidentification of the Bororo as relict hunter-gatherers is not inconsequential scientifically. At this moment, scientists working on mapping the human genome have targeted Gê people for DNA sampling, purely because of their mistaken identification as primeval foragers. But the Bororo are not necessarily closer genetically to Paleolithic people than Quechua herders are. Whatever the results of the assays, they cannot but be misinterpreted, due to the profound misunderstanding of the evolutionary status of the Bororo and other Gê peoples.

Conclusions

The conclusion to be drawn from this human ecological history is that many modern "foragers" of the lowlands represent new adaptations, not primeval ones. In addition to being evolutionarily disjunct from ancient foragers, today's indigenous Amazonians often have poorer land than their predecessors and have to share their territories with ill-disposed others, who are more powerful in present-day social, political, and economic entities. In the course of indigenous adaptation to the changed political landscape, the people have changed their organization, subsistence, habitation technology, and art. These changes cannot legitimately be attributed solely to adaptation to the environment: they must be considered a response to the presence of other people and the consequent stress and privation that Indians have endured (Roosevelt 1990). Early prehistoric hunter-gatherers were healthier and wealthier, and had more complex and elaborate cultures. Except for groups like the Bororo, who live farther from modern urban centers, most living Amazonian Indian adults are short-statured; their children have depressed growth curves; and they are plagued by nutrient deficiencies and infections to a degree not experienced by their predecessors.

These findings encourage the consideration of changing sociopolitical and demographic factors as influences on cultural ecology. The density and sedentism of settlement affect subsistence and organization. Sedentism may make game scarce, while mobility makes permanent agriculture difficult. Population sparsity, in turn, makes intensive agriculture unnecessary. Competition with other groups affects access to resources, and new migrants' cash-cropping may force changes in indigenous peoples' subsistence. The establishment of population centers nearby may influence shifting horticulturalists to become specialist foragers, who bring forest resources to sedentary agriculturalists or city markets. The exact causality varies greatly, and our understanding of these historical ecological processes requires empirical investigation of the specific circumstances of each group in each region. However, the epistemological principle is the same: cultural and biological evolutionary explanations are no more than just-so stories until their validity is tested with historical, diachronic evidence from the time when the processes were working themselves out.

But this historical ecology of foraging in lowland South America leaves us at a loss for direct ethnographic evidence for the ancient forager lifestyle, which remains of great theoretical interest. The implication of all our findings would seem to be that scholars who are interested in learning about the lifeway of hunting and gathering will have to consider archaeology—for it is to a great extent in the archaeological record of the Pleistocene and early Holocene that the data on primary tropical hunting and gathering are to be found. At least in tropical South America, there have been few independent primary foragers since about 3,000 to 4,000 years ago. There is no reason not to study the foraging of horticulturalists, or refugees, or of recent client-foragers for agricultural societies—but food production adds a new dimension to cultural ecology, and such societies are not necessarily proxies for the behavior of our earliest ancestors. Ideally, future researchers will compare the ancient and modern foragers of a region, as Gustavo Politis and his colleagues have in the Colombian Amazon (Politis 1996). Bringing broader theoretical interests and updated methods to Paleolithic and early Holocene archaeology will do it no harm. When motivated by questions about the details of ecological adaptations, division of labor, residence patterns, and social organization, archaeologists may well need to expand their methodologies with more meticulous archaeobotany and zooarchaeology, osteological analysis of activity patterns, DNA analysis, stable isotope analysis, and so on.

The evidence for the long history of hunter-gatherers in the Amazonian tropical forest is an incentive to take another look at the role of tropical forests in human evolution. In Africa, for example, it had long been assumed that the critical steps toward the human species took place in the tropical savanna—not the tropical rain forest, which was thought an evolutionary backwater. Nevertheless, recent findings in paleoanthropology suggest the need to rethink the assumptions about the ecology of hominization. The earliest known bipedal human forebear, *Australopithecus afarensis*, turns out to have well-developed bone morphology for climbing but somewhat inefficient morphology for walking. The paleoecological context of *Australopithecus ramidus*, the new Ethiopian primate fossil hominid discovered by U.C. Berkeley scholars, is humid tropical woodland, not open savanna, and stable carbon isotopes in east African soils do not show the savannaization of habitats hypothesized to have inspired the rise of the hominids. Given these scattered indications and the fact that humans' nearest living relatives are tropical forest primates, not savanna species, the tropical forest should at least be investigated as an alternate hearth of development for the human species.

Notes

1. Monte Alegre at present bears a mosaic of secondary forest and savanna woodland created by long-term cutting and burning of the rain forest for ranching and agriculture. However, the botanical taxa and their stable carbon isotope ratios in the archaeological deposit at Caverna da Pedra Pintada indicate fully developed, closed-canopy tropical rain forest cover in the late Pleistocene (see below). Pollen and sediment studies in nearby equatorial, lowland Amazonian regions also show the persistence of tropical rain forest cover in the terminal Pleistocene (Colinvaux et al. 1996).

2. The dated and taxonomically identified burned tree fruits and wood are considered human, not animal, subsistence remains for several reasons: The plant remains were accompanied by numerous artifacts and burned fragments of faunal bones, shells, and carapaces. Layers that lacked the artifacts and faunal remains also lacked the plant remains. The palm

seeds had all been neatly cracked to remove their kernels, not gnawed or bored, and no other tree seeds showed any animal tooth marks or insect bore holes.

3. The stable carbon-isotope ratios of the fifty-six dated carbonized palm seeds and charcoal chunks from the deposit exemplified the ranges expected for closed-canopy tropical rain forest. All but two of the fifty-six showed carbon 13/12 ratios between −27 and −35 o/oo (parts per mil) after subtracting 5 o/oo for enrichment in carbon 13 due to metabolic fractionation during seed formation (Tieszen 1991). The two remaining samples had ratios of 12.3 and 16.6 o/oo, characteristic of CAM-photosynthesizing tropical forest epiphytes. No samples showed the ratios expected for either open forest or savanna.

References

Athens, S. 1989. Pumpuentsa and the Pastaza phase of southeastern lowland Ecuador. *Nawpa Pacha* 24:1–29.

Bailey, R. C., G. Head, M. Jenike, B. Owen, R. Rechtman, and E. Zechenter. 1989. Hunting and gathering in tropical rain forest: Is it possible? *American Anthropologist* 91:59–82.

Balée, W. 1989. The culture of Amazonian forests. Pp. 1–21 in D. A. Posey and W. Balée (eds.), *Resource Management in Amazonia: Indigenous and Folk Strategies*. Advances in Economic Botany, vol. 7. Bronx: New York Botanical Garden.

———. 1992. People of the fallow: A historical ecology of foraging in South America. Pp. 35–57 in K. H. Redford and C. Padoch (eds.), *Conservation of Neotropical Forests: Working from Traditional Resource Use*. New York: Columbia University Press.

Beckerman, S. 1979. The abundance of protein in Amazonia: A reply to Gross. *American Anthropologist* 81:533–560.

Bednarik, R. G. 1989. Pleistocene settlement of South America. *Antiquity* 63(238): 101–111.

Beltrão, M. C. de M. C., C. R. Enriquez, J. Danon, E. Zuleta, and G. Poupeau. 1986. Thermoluminescence dating of burnt cherts from the Alice Boer site (Brazil). Pp. 203–213 in Bryan (ed.), *New Evidence for the Pleistocene Peopling of the Americas*.

Bennett, W. 1936. Excavations in Bolivia. *Anthropological Papers of the American Museum of Natural History* 34(3): 359–494.

Bryan, A. L. 1983. South America. Pp. 137–146 in R. Shutler (ed.), *Early Man in America*. Beverly Hills: Sage.

———. 1986. Paleoamerican prehistory as seen from South America. Pp. 1–14 in A. L. Bryan (ed.), *New Evidence for the Pleistocene Peopling of the Americas*. Orono: Center for the Study of Early Man, University of Maine.

Bryan, A. L., R. M. Casamiguela, J. M. Cruxent, R. Gruhn, and C. Ochsenius. 1978. An El Jobo mastodon kill at Taima Taima, Venezuela. *Science* 200:1275–1277.

Bush, M. and P. Colinvaux. 1988. A 7000-year pollen record from the Amazon lowlands, Ecuador. *Vegetatio* 76:141–154.

Cavalcante. P. B. 1991. *Frutas comestíveis da Amazônia*. 5th ed. Belém: Edições Cegup.

Clay, J. 1988. *Indigenous Peoples and Tropical Forests: Models of Land Use and Management from Latin America*. Cambridge, Mass.: Cultural Survival.

Colinvaux, P., P. E. Oliveira, J. E. Moreno, M. C. Miller, and M. B. Bush. 1996. A long pollen record from lowland Amazonia: Forest and cooling in glacial times. *Science* 274:85–88.

Conaway, M. E. 1976. Still Guahibo, still moving: A study of circular migration and marginality in Venezuela. Ph.D. diss., University of Pittsburgh.

Crocker, C. 1985. My brother the parrot. Pp. 13–48 in G. Urton (ed.), *Animal Myths and Metaphors in South America*. Salt Lake City: University of Utah Press.

Denevan, W. 1966. *The Aboriginal Cultural Geography of the Llanos de Mojos de Bolivia*. Ibero-Americana 48. Berkeley: University of California Press.

Dillehay, T. 1989. *Monte Verde: A Late Pleistocene Settlement in Chile*. Vol. 1, *Paleoenvironment and Site Context*. Washington, D.C.: Smithsonian Institution.

Dillehay, T. D., G. Ardilla Calderon, G. Politis, and M. C. Beltrão. 1992. Earliest hunters and gatherers of South America. *Journal of World Prehistory* 6(2): 145–203.

Dincauze, D. F. 1984. An archaeological evaluation of the case for pre-Clovis occupations. *Advances in World Archaeology* 3:275–323.

Erickson, C. 1980. Sistemas agrícolas prehispánicas en los Llanos de Mojos. *América Indígena* 40(4): 731–755.

Fagan, Brian M. 1987. *The Great Journey: The Peopling of Ancient America*. London: Thames and Hudson.

Federman, N. 1958. *Historia indiana seguida del itinerario de la expedición*. Trans. Juan Friede. Madrid: ARO Artes Graficas.

Fernandez Distel, A. 1984–85. Hábitos funerarios de los Sirionó (Oriente de Bolivia): Intento de proyección hacia el pasado arqueológico de su habitat. *Acta Praehistorica Arqueologica* 16/17:159–182.

Franzinelli, E. and E. Latrubesse, eds. 1993. *Geologia Quaternária da Amazonia. Conferência internacional: Resumos*. Manaus: Universidade Federal do Amazonas.

Goodland, R. and H. Irwin. 1975. *Amazon Jungle: Green Hell to Red Desert?* New York: Elsevier.

Guidon, N. and G. Delibrias. 1986. Carbon-14 dates point to man in the Americas 32,000 years ago. *Nature* 321:769–771.

Hartt, C. F. 1871. Brazilian rock inscriptions. *American Naturalist* 5(3): 139–147.

———. 1885. Contribuições para a ethnologia do valle do Amazonas. *Arquivos do Museu Nacional do Rio de Janeiro* 6:1–174.

Haynes, C. V., Jr. 1992. Contributions of radiocarbon dating to the geochronology of the peopling of the New World. Pp. 355–374 in R. E. Taylor, A. Long, and R. S. Kra (eds.), *Radiocarbon Dating After Four Decades*. New York: Springer-Verlag.

Headland, T. 1987. The wild yam question: How well could independent hunter-gatherers live in a tropical rain forest environment? *Human Ecology* 15(4): 463–491.

Hill, K. and K. Hawkes. 1983. Neotropical hunting among the Aché of eastern Paraguay. Pp. 139–188 in Raymond Hames (ed.), *Adaptive Responses of Native Amazonians*. New York: Academic Press.

Hill, K. and A. M. Hurtado. 1989. Hunter-gatherers of the New World. *American Scientist* 77(5): 437–443.

Hill, K. and H. Kaplan. 1989. Population and dry-season subsistence strategies of the recently contacted Yora of Peru. *National Geographic Research* 5:317–334.

Holmberg, A. R. 1969. [1950]. *Nomads of the Long Bow*. Garden City, N.Y.: American Museum Science Books.

Hurtado, A. M. and K. R. Hill. 1991. Seasonality in a foraging society: Variation in diet, work effort, fertility, and sexual division of labor. *Journal of Anthropological Research* 46(3): 293–346.

Junk, W. J. 1970. Investigations on the ecology and production biology of the "floating meadows" (*Pasalum-Echinochloetum*) on the middle Amazon. *Amazoniana* 4:9–102.

Lathrap, D. W. 1968. The hunting economies of the tropical forest zone of South America: An attempt at historical perspective. Pp. 23–29 in R. B. Lee and I. DeVore (eds.), *Man the Hunter*. Chicago: Aldine.

———. 1970. *The Upper Amazon*. New York: Praeger.

Lévi-Strauss, C. 1955. *Tristes tropiques*. Paris: Librairie Plon.

Lowie, R. 1946. The Bororo. Pp. 419–434 in Steward (ed.), *The Marginal Tribes*.

Lynch, T. E. 1978. The South American Paleo-Indians. Pp. 454–489 in J. D. Jennings (ed.), *Ancient Native Americans*. San Francisco: Freeman.

———. 1983. The Paleo-Indians. Pp. 87–138 in J. D. Jennings (ed.), *Ancient South Americans*. San Francisco: Freeman.

———. 1990. Glacial-age man in South America: A critical review. *American Antiquity* 55:12–36.

Magalhães, Marcos. 1994. *Archaeology of Carajás: The Pre-historic Presence of Man in Amazonia*. Rio de Janeiro: Companhia Vale do Rio Doce.

Meggers, B. J. 1971. *Amazonia: Man and Culture in a Counterfeit Paradise*. Chicago: Aldine.

Meggers, B. J., C. Evans, and E. Estrada. 1965. *Early Formative Period of Coastal Ecuador*. Smithsonian Contributions to Anthropology, vol. ll. Washington, D.C.: Smithsonian Institution.

Métraux, A. and H. Baldus. 1946. The Guayaki. Pp. 435–444 and plates 95–96 in Steward (ed.), *The Marginal Tribes*.

Morey, N. 1975. Ethnohistory of the Colombian Llanos. Ph.D. diss., University of Utah, Salt Lake City.

Museu Paraense Emílio Goeldi. 1989. *Museu Paraense Emílio Goeldi*. São Paulo: Banco Safra and Conselho Nacional de Desenvolvimento Científico e Tecnológico.

Ochsenius, C. and R. Gruhn. 1979. *Taima-Taima: A Late Pleistocene Paleo-Indian Kill Site in Northernmost South America—Final Report of 1976 Excavations*. Coro, Venezuela: CIPICS/ South American Quaternary Documentation Program.

Petrullo, V. 1939. *Archaeology of Arauquin*. Bureau of American Ethnology Bulletin 123, Anthropological Papers 12. Washington, D.C.: Smithsonian Institution.

Politis, Gustavo G., ed. 1996. *Nukak*. Bogotá: Instituto Amazónico de Investigaciones Cientificas—SINCHI.

Porras, P. 1987. *Investigaciones archaeológicas a Las Faldas de Sangay, Provincia Morona Santiago*. Quito: Artes Gráficas Senal.

Porro, A. 1994. Social organization and political power in the Amazon floodplain: The ethnohistorical sources. Pp. 79–94 in Roosevelt (ed.), *Amazonian Indians from Prehistory to the Present*.

Price, T. D. and J. Brown, eds. 1985. *Prehistoric Hunter-Gatherers: The Emergence of Cultural Complexity*. Orlando: Academic Press.

Prous, A. 1991. Fouilles de l'abri du Boquete, Minas Gerais, Brésil. *Journal de la Société des Américanistes* 77:77–109.

Rasänen, M., J. S. Salo, and H. Jungner. 1991. Holocene floodplain lake sediments in the Amazon: 14C dating and paleoecological use. *Quaternary Science Reviews* 10:363–372.

Rodrigues de Oliveira, A. 1994. Evidence of the nature of the process of indigenous deculturation and destabilization in the Brazilian Amazon in the last three hundred years: Preliminary data. Pp. 95–122 in Roosevelt (ed.), *Amazonian Indians from Prehistory to the Present*.

Roosevelt, A. C. 1980. *Parmana: Prehistoric Maize and Manioc Along the Amazon and Orinoco*. New York: Academic Press.

———. 1989. Resource management in Amazonia before the conquest: Beyond ethnographic projection. Pp. 30–62 in D. A. Posey and W. Balée (eds.), *Resource Management in Amazonia: Indigenous and Folk Strategies*. Advances in Economic Botany, vol. 7. New York: New York Botanical Garden.

———. 1990. The historical perspective on resource use in Latin America. Pp. 29–64 in *Economic Catalysts to Ecological Change: Working Papers of the Tropical Conservation and Development Program*. Gainesville: University of Florida Latin American Studies Center.

———. 1991. *Moundbuilders of the Amazon: Geophysical Archaeology on Marajo Island, Brazil*. San Diego: Academic Press.

———. 1993. The rise and fall of the Amazon chiefdoms. *L'Homme* 126–128, 33(2–4): 255–283.

———. 1995. Early pottery in Amazonia: Twenty years of obscurity. Pp. 115–132 in W. Barnett and J. Hoopes (eds.), *The Emergence of Pottery*. Washington, D.C.: Smithsonian Institution.

———. 1997. *The Excavations at Corozal, Venezuela*. Yale University Publications in Anthropology, vol. 82.

———. N.d. Hunter-gatherers of South America (book manuscript).

———, ed. 1994. *Amazonian Indians from Prehistory to the Present: Anthropological Perspectives*. Tucson: University of Arizona Press.

Roosevelt, A. C., R. A. Housley, M. Imazio, S. Maranca, and R. Johnson. 1991. Eighth millennium pottery from a shell midden in the Brazilian Amazon. *Science* 254:1621–24.

Roosevelt, A. C., M. Lima da Costa, W. Barnett, J. Feathers, C. Machado, M. Michab, M. Imazio da Silveira, H. Valladas, A. Henderson, N. Toth, and K. Schick. 1996. Cave dwellers in the Amazon: The peopling of the Americas. *Science* 272:373–384.

Schmitz, P. I. 1987. Prehistoric hunters and gatherers of Brazil. *Journal of World Prehistory* 1(1): 53–126.

Simões, M. 1976. Nota sobre duas pontas de projetil da Bacia Tapajós (Para). *Boletim do Museu Paraense Emílio Goeldi*, n.s., 62:1–389.

———. 1981. Coletores-pescadores ceramistas do litoral do Salgado (Para): Nota preliminar. *Boletim do Museu Paraense Emílio Goeldi*, n.s., 78:1–32.

Sioli, H., ed. 1984. *The Amazon: Limnology and Landscape Ecology of a Mighty Tropical River*. Dordrecht: W. Junk.

Smith, N. J. H. 1980. Anthrosols and human carrying capacity in Amazonia. *Annals of the American Association of Geographers* 70:553–566.

Spencer, C. S. and E. M. Redmond. 1992. Prehispanic chiefdoms of the western Venezuelan llanos. *World Archaeology* 24(1): 134–157.

Sponsel, L. E. 1986. Amazon ecology and adaptation. *Annual Review of Anthropology* 15:67–97.

Stearman, A. M. 1984. The Yuquí connection: Another look at Sirionó acculturation. *American Anthropologist* 86:630–650.

———. 1989. *Yuquí: Forest Nomads in a Changing World*. New York: Holt, Rinehart, and Winston.

Steward, J. H., ed. 1946. *Handbook of South American Indians*. Vol. 1, *The Marginal Tribes*. Washington, D.C.: Smithsonian Institution.

Steward, J. H. and L. C. Faron. 1959. *Native Peoples of South America*. New York: McGraw-Hill.

Tieszen, L. L. 1991. Natural variations in the carbon isotope values of plants: Implications for archaeology, ecology, and paleoecology. *Journal of Archaeological Science* 18: 227–248.

Van der Merwe, N., A. C. Roosevelt, and J. Vogel. 1981. Isotopic evidence for prehistoric subsistence change at Parmana, Venezuela. *Nature* 292:536–538.

Wallace, A. R. 1889. *A Narrative of Travels on the Amazon and Rio Negro*. 2d ed. London: Ward, Lock.

Whitmore, T. C. 1990. *An Introduction to Tropical Rain Forests*. Oxford: Clarendon Press.

Whitmore, T. C. and G. T. Prance, eds. 1987. *Biogeography and Quaternary History in Tropical America*. Oxford Science Publications. Oxford: Clarendon Press.

Williams, D. 1981. Excavations of the Balbina shell mound, Northwest District: An interim report. *Journal of the Walter Roth Museum of Archaeology and Anthropology* 4(1–2): 13–36.

Wüst, Irmhild. 1994. The eastern Bororo from an archaeological perspective. Pp. 315–342 in Roosevelt (ed.), *Amazonian Indians from Prehistory to the Present*.

CHAPTER 10

Potential Versus Actual Vegetation: Human Behavior in a Landscape Medium

TED GRAGSON

> Historical ecology . . . seeks a synthetic understanding of human/
> environmental interactions within specific societal, biological, and re-
> gional contexts. In other words, the focus of historical ecology is a rela-
> tionship, not an organism, species, society—not a "thing."
> —WILLIAM BALÉE (1994:1)

> Whenever a plant cover does not closely resemble either a cultivated
> field or a manicured lawn or park, it is often perceived as being "natural."
> —THOMAS VALE (1982:1)

Carole Crumley argues (1994) that the role of environment in human existence is now reduced to that of a resource to be commoditized, and that values, beliefs, issues, history, and culture are the key elements in our explanatory frameworks. The impasse she presents as immanent to much of anthropology is surmountable, but not through the false dichotomization or reification of concepts and approaches. I view the purpose of this volume as furthering the growing efforts at generating the inter- and intradisciplinary scientific understanding necessary for overcoming this impasse. At the most general level, I wish to explore the question of why people use the land the way they do. To adequately address this question it is necessary to confront the problem identified by Crumley and to define the relationship identified by William Balée in the initial quotation. The particular aspect of the human/ environment relationship that I wish to explore fits under the general heading of *disturbance*, to which the quotation by Thomas Vale is pointedly addressed.

The integrative and comparative goals of anthropology are impossible to achieve either by using a simplistic equality such as "environment = commodity" or by relying on poorly operationalized explanatory elements such as "culture" (figure 10.1). The conceptual and methodological advances needed to transcend this situation are now available, and the process of translation and integration to anthropology has

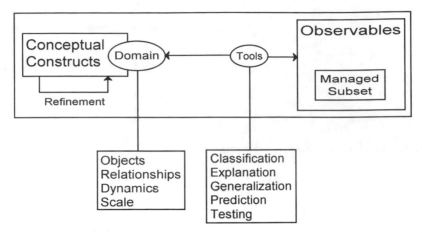

Fig. 10.1 *The Process of Generating Understanding, and the Means to Its Application.* (Adapted from Pickett, Kolasa, and Jones 1994:28.)

begun (Gunn 1994; Little 1991; Aunger 1995). Some of the most important contributions are emerging from (1) the paradigm of landscape ecology (Forman and Godron 1986); (2) the shift away from a syntactic toward a semantic view of science, and of evolutionary theory in particular (Lloyd 1994; Pickett, Kolasa, and Jones 1994; Pera 1994); and (3) the increased demand for conceptual understandings with empirical utility (SGCR 1995; Funtowicz and Ravetz 1991; Aunger 1995). These contributions to the understanding of the human/environment relationship will be evaluated from the perspective of my research among the Manxuj in the Chaco Boreal of Paraguay, by considering the contrast between potential and actual vegetation. The application takes a diachronic approach, since understanding a vegetative system requires considering how past system states affect present system expression. The application is ecological by its concern with process and interaction, and dialectical by its focus on unity and contradiction (Pera 1994; Allen and Starr 1982; Pickett, Kolasa, and Jones 1994).

The Landscape Paradigm

That humans interact with environment through the medium of a landscape was early recognized in the history of the natural and social sciences (Worster 1994; Forman and Godron 1986; Golley 1993; Marsh 1864). As research objects, however, landscapes are too data-rich to be manageable without appropriate methods and techniques. As a consequence, they have emerged as legitimate objects of study only during the last fifty years—following methodological advances in spatial and analytical geography, and technical advances in computer hardware/software such as geographic information systems and relational data-managers. The paradigmatic approach to the study of landscapes is called *landscape ecology* (Forman and Godron 1986; Pickett, Kolasa, and Jones 1994; Allen and Hoekstra 1992; Haines-

Young, Green, and Cousins 1993). Landscape ecology is more than an academic reformulation and synthesis of geographic and ecological concepts, however, for it seeks to integrate principles from areas such as ecosystem ecology, island biogeography, and plant social ecology to achieve understanding at a level largely ignored until recently.

Landscape ecology has been instrumental to land-use planning and management in Australia, Canada, Germany, the Netherlands, and Czechoslovakia, and it is beginning to be applied in the United States (Golley 1993, pers. comm. 1994; Forman and Godron 1986; Allen and Hoekstra 1992). The particular strength of the landscape paradigm is its focus on the *relationship* between entities and place; consideration is given both to the factors controlling the location and action of organisms, and to the influence of these organisms on the pattern of the landscape (Crumley and Marquardt 1987; Bridgewater 1993; Winterhalder 1994; Stafford and Hajic 1992; Lepart and Debussche 1992). A relationship exists when two or more elements are substantively connected by the content of their relation, a "jointly occupied" position filled by empirical actors and/or events (Knoke and Kuklinski 1982; Laurini and Thompson 1992; Pickett, Kolasa, and Jones 1994). The landscape paradigm is relevant to anthropology because it provides the means for merging scientifically derived conclusions about the structure and function of a particular environment with the knowledge of place that individual members of a culture have acquired from long-term experience with local areas and resources (Addicott et al. 1987; Cousins 1993; Kempton, Boster, and Hartley 1995; DeWalt 1994; Brandenburg and Carroll 1995).

The defining characteristic of a landscape is spatial contiguity, which implies a hierarchy of self-similar elements united by the significance and asymmetry of information passing between them (Winterhalder 1994; Addicott et al. 1987; Turner 1989; Cousins 1993; Allen and Hoekstra 1992). Humans conceptually organize the vertical and horizontal complexity of a landscape into prototypical units on the basis of socially and geographically contingent beliefs about the world, values about what is moral or just, and facts about the availability and distribution of resources (Turner 1989; Rayner et al. 1994; Kempton, Boster, and Hartley 1995; Hrenchuk 1993; Scott 1995; Rochcleau, Thomas-Slayter, and Edmunds 1995). The resulting *knowledge-of-place* model helps individuals interpret observations and generate novel inferences, with operational decisions about land use being made in social situations where individuals with common knowledge of local areas and resources use each other as reference points in their strategic behavior. In line with this reasoning, change in the cover of the land is the cumulative outcome of time-dependent activities carried out by individuals (figure 10.2).

Strategic behavior subverts conventional notions about directional causality by recognizing that an actor's participation in the events leading up to a particular outcome are partially determined by the outcome itself, although anchored by the envisioned constancy of the decision-making process itself (Tooby and DeVore 1987; Ostrom, Gardner, and Walker 1994; Agrawal 1994; Gragson 1993). Actors have relationships with entities in the physical environment (e.g., resources, through

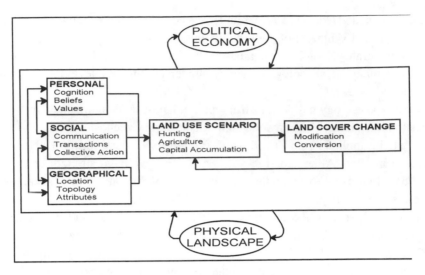

Fig. 10.2 *Graphic Model of Situational Interactions Leading to Land-Cover Change.*

extractive encounters) and the social environment (e.g., members, through a shared symbol system), which affect the focal actor's perceptions, beliefs, and actions. The "heroic assumption" that actors possess perfect information runs counter to the bounded rationality of individuals reflected in the information asymmetries recorded for members of single social systems (Kempton, Boster, and Hartley 1995; Sanderson 1994; Ostrom, Gardner, and Walker 1994; Knoke and Kuklinski 1982). Actors are fallible learners who make decisions in uncertain and complex situations on the basis of an incomplete knowledge of alternatives, outcomes, and participants. Actors choose between probability distributions rather than means, and the measured differences in their responses reflect risk-prone, risk-indifferent, and risk-averse behavior (Stephens 1990; Stephens and Krebs 1986; Tversky and Kahneman 1974; Cooke 1991; Gragson 1993).

Vegetation and Humans

While floristics as an endeavor seeks to identify the kinds of plants growing in a particular area, the study of *vegetation* takes as its object of interest the collective phenomenon that results from the interaction of multiple species (Nicolson 1987; Vale 1982). The potential or hypothetical vegetation of a given locale is predicted from the contingency of four primary environmental constraints (Vale 1982):

1. *regional climate*, which determines the availability of water and energy;
2. *topography*, which modifies moisture availability by influencing local climate and water movement over the landscape;

3. *soil* (or *substrate*), which affects the chemical relationships linking plants and environment; and

4. *biotic interactions*, plant/plant and plant/animal, which are characterized as competition, mutualism, and so on.

These environmental constraints can be ranked hierarchically with respect to each other (figure 10.3), with lower-ranked constraints being significantly more dependent on superordinate constraints (Allen and Starr 1982; Klijn and de Haes 1994; Pickett et al. 1989). This hierarchy is both structural (since it implies changes in reservoir size) and functional (since it reflects transport of energy and matter between levels). It also has dimensionality: the spatial dimensions refer to the comprehensiveness of level-effects (e.g., climate affects larger areas than does the activity of individual organisms), while the temporal dimension refers to the response characteristics of individual levels to particular challenges (e.g., individual biota respond faster than topography). Questions can be posed and explanations formulated for any level of the hierarchy (Allen and Starr 1982; O'Neill et al. 1986; Pickett et al. 1989; Urban, O'Neill, and Shugart 1987; Klijn and de Haes 1994). As a rule, the level of the hierarchy containing the interactions of interest is selected for analytical study; mechanisms for these interactions are found on the level below the analytical level, while constraints and context for the interaction are found on the level above the analytical level.

Potential vegetation becomes actual vegetation through the realization of constraint contingency in a particular time/space context when linked to disturbance events associated with the behavior of individuals using the land. A *disturbance* is

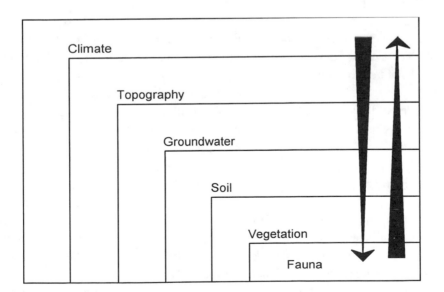

Fig. 10.3 *Hierarchy of Environmental Constraints.* (Adapted from Klijn and de Haes 1994:91.)

"any relatively discrete event in time that disrupts ecosystem, community, or population structure and changes resources, substrate availability, or the physical environment" (White and Pickett 1985:7; and see Uhl et al. 1990). Depending on both the magnitude of impact that particular human activities have on land cover and the knowledge or awareness of prior land-cover conditions, land-cover change is classified as belonging to one of two fundamental types (Turner, Moss, and Skole 1993): (1) modification, a change of condition within a land-cover type (e.g., forest thinning); and (2) conversion, a change from one type of land cover to another (e.g., forest to pasture). These fundamental land-cover outcomes are delineated by the set of all possible land-use scenarios that can potentially entail disturbance (e.g., hunting, agriculture, or capital accumulation). The concept of disturbance so defined becomes a dynamic process occurring at a variety of spatial and temporal scales. The typical spatial scales of interest range from 10^2 to 10^{10} m^2, while the temporal scales range from 10^{-2} to 10^4 days (White and Pickett 1985; Uhl et al. 1990; Pickett et al. 1989).

Humans as components of the landscape are not merely exogenous system disruptors, but one of many contributors to the various environmental constraints in operation (Vale 1982; White and Pickett 1985; Uhl et al. 1990). Two generalizations can be advanced regarding human contributions to the realization of vegetative pattern:

1. Human activity is an environmental constraint influencing the dynamics of plant species.
2. The effects of human activities replicate in varying degrees the effects of "natural" processes.

The product of these effects may lead to the elimination of vegetative cover, yield a new equilibrium, initiate a directional change, or result in vegetation with reduced stature, diversity, or productivity.

Land-use scenarios, as used in figure 10.2, are unitary concatenations of temporally ordered decision events, each associated with an authoritative actor and each conceived as an input, throughput, or output stage of a particular resource-production process. The unitary character of land-use scenarios derives from the consequential nature of strategic behavior and the flow of information between functional events (Allen and Starr 1982; Borgerhoff-Mulder and Caro 1985; Ostrom, Gardner, and Walker 1994; Laumann and Knoke 1987; Kempton, Boster, and Hartley 1995). Authoritative actors render a discretionary decision at a particular moment in time, with the consequence that each *decision event* is a historically contingent occurrence associated with other antecedent, concurrent, and impending decisions about land use. In order to take individual decision events out of the narrative stream of history for the purpose of determining causality, they are operationally defined as stimuli eliciting actor participation.

Events represent a particular type of good, and the strategic behavior of actors in particular land-use scenarios represents their individual or collective demand relative to the supply of this good. "Hunting" or "capital accumulation," for example, comprises many subsidiary functional events (which can be further decomposed into structural events) that attract or demand particular types of actors based on the event's intrinsic, observable attributes: time, place, resource, technology, and visibility (Borgerhoff-Mulder and Caro 1985; Ostrom, Gardner, and Walker 1994; Laumann and Knoke 1987; Gragson 1992, 1993). *Time* specifies the precise moment an event occurs; *place* identifies the location where an event occurs; *resource* specifies the "nature" and "quantity" of real (e.g., meat) or symbolic (e.g., authority) items involved; *technology* specifies the tools needed to accomplish the event; and *visibility* is indexed by the number and variety of actors interested in the substantive content of the event. A comprehensive understanding of human decision-making depends on clarifying actions as much as the perceptions and categorization of the landscape they occupy, and in recent work among the Manxuj I took the first step in addressing the envisioned human/landscape relationship.

Manxuj and Their Landscape

The Manxuj live in west-central Paraguay (Chaco Boreal) and represent a subgroup of the northern Chorote dialect group within the Matacoan language family spoken by people living in the three-corners area of Argentina, Paraguay, and Bolivia. There are between 200 and 500 Manxuj, and a total Chorote population between 1,000 and 2,000 (Gerzenstein 1983; Chase-Sardi, Brun, and Enciso 1990). Members of the southern dialect group generally referred to in the literature as Chorote (Yoxwáha) are concentrated near La Paz Mission (Salesian) on the right bank of the Upper Pilcomayo River in Argentina, but some reside on the left bank of the same river in Paraguay near the Immaculate Heart of Mary Mission (Oblates). Members of the northern dialect group referred to as Yowúxua are settled in the vicinity of the Immaculate Heart of Mary Mission, while members of the northernmost dialect group calling themselves Xuikína-wo, or more commonly Manxuj, live in the vicinity of the Santa Rosa Mission (New Tribes). There may also be some Chorote living on the Bolivian side of the Pilcomayo River, as reported at the turn of the century, but the evidence at present is unclear. Historically, the northern and southern dialect groups were hostile toward each other and practiced internecine fighting for the purpose of gaining prestige and scalps. At the present time, however, all Chorote—with the exception of the Manxuj—live in mixed communities that also include Mataco, Nivaclé, and Toba (Gerzenstein 1983; Wilbert and Simoneau 1985; Chase-Sardi, Brun, and Enciso 1990).

The area of Paraguay occupied by the Chorote remains one of the least known areas of the country. While explorers and prospectors have occasionally passed through, the major influence historically and at the present remains that of missionaries. Ethnohistorical evidence suggests that the Chorote came to occupy the three-corners area as a consequence of general population displacement caused by the intrusion of the Chiriguano just prior to Spanish contact. There were also considerable readjustments of tribal territories as a result of Spanish military expeditions against Indian groups of the Chaco in the early part of the colonial period. The Chorote are first mentioned by name in Pedro Lozano's 1733 account of the punitive expedition of 1673 out of Tarija. In 1673 they occupied the headwaters of the Bermejo River (south of the Pilcomayo River); however, under the pressure of this foray and subsequent interethnic feuding with Toba and Mocovi, they moved north to the Pilcomayo during the seventeenth and eighteenth centuries and possibly into what is now Paraguay (Kersten 1968; Chase-Sardi, Brun, and Enciso 1990; Susnik 1983; Métraux 1946).

The Anglican Church began missionary work in the Paraguayan Chaco in 1889 along the Lower Paraguay River, and then moved out to Moclhavaya in the central Paraguayan Chaco about 1910 (Grubb 1993; Kidd 1995). Here they came into contact with Lengua (Enxet), Sanapana, and Angaite, as well as Nivaclé and Mak'a. During this period, the Chorote were infamous for their reported involvement with the Toba in the massacre of Jules Crevaux and the members of his 1882 expedition. Despite this reputation, Count Eric von Rosen, a member of Baron Nils Erland Herbert Nordenskiöld's 1901–1902 Gran Chaco expedition, managed to spend some time with the Chorote and to record a fair amount of cultural information (Rosen 1924).

The process of contact and change picked up in tempo with the arrival in the central Paraguayan Chaco (Filadelfia area) of the Russian and Canadian Mennonites in 1927 and 1930. Most Mennonite missionary work has been with Nivaclé and Lengua, but in some measure they have economically affected all indigenous Chacoan groups since so many have migrated to the areas around Filadelfia in search of work. The Chaco War (1932–1935) had a severe impact on many groups including the Chorote, particularly in terms of population displacement (Kidd 1995; Plett 1979; Kleinpenning 1987).

After the Chaco War, the major outside influences on the Chacoan groups have been petroleum exploration, the extraction of tannin and wild animal skins, and a renewed interest in proselytization by Oblates, Salesians, and New Tribes missionaries (Stahl 1986; Kleinpenning 1987; Renshaw 1986; Kidd 1995). The Manxuj, while affected in some measure by all these events, were the last group of Chorote to come into sustained contact with missionaries. New Tribes made contact with them in 1973, and at present there are some 200 Manxuj living at Santa Rosa Mission (Gordon Hunt pers. comm. 1993; Chase-Sardi, Brun, and Enciso 1990; Chase-Sardi 1972).

The Manxuj, unlike other Chacoan groups at the present time, continue to occupy and use their traditional ecological neighborhood, which has yet to be seriously affected by outside forces. Some other Chacoan groups continue to live in areas they occupied prior to contact (e.g., Lengua), but these areas suffer the effects of drastic, and in some cases nearly catastrophic, environmental change; a more common situation during the last twenty to thirty years is represented by groups that have migrated (e.g., Ayoreo) or have been forcefully removed to new areas (e.g., Mak'a). While some members of all Chacoan ethnic groups practice some agriculture, most groups have historically been considered foragers (Renshaw 1986; Métraux 1946). They continue foraging in some measure for resources, but the forces of acculturation have been extremely rapid during the last twenty years and no group survives solely from this activity at the present time (Renshaw 1986; Chase-Sardi 1990; Kidd 1995).

The area occupied by the Manxuj has a median elevation of 220 m above mean sea level and a semiarid climate (figure 10.4). The vegetation is a dense, thorny, and difficult-to-traverse xerophytic forest; the soils are slightly acid to slightly alkaline, with sand to sandy-clay textures (Sprichiger and Ramella 1989; Prado 1993; USAID 1985; CIF 1991; Gragson 1994a). My ethnographic fieldwork conducted in 1993–1994 focused on Manxuj perceptual discrimination of vegetative discontinuities reflecting the equilibrium of time and space (Gragson 1994a,b). While this information is still in the process of analysis, I have developed a landscape classification that combines standard physical and biotic observations with the knowledge of place that Manxuj individuals have acquired from long-term experience with local areas and resources (table 10.1). This is the first step toward evaluating the causal connection between the pattern generated by self-similar landscape elements and the decisions of consequential actors associated with extractive activities.

Three points can be made about the taxonomic process applied by the Manxuj to the surrounding landscape: (1) the Manxuj conceptually recognize the biologically

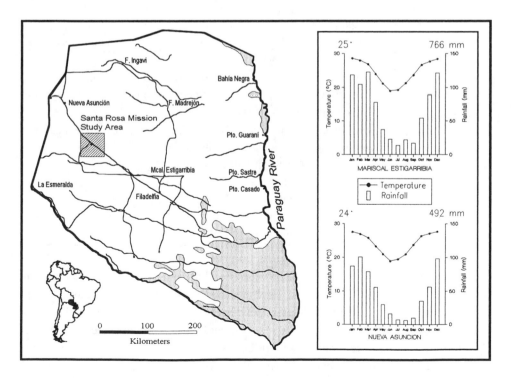

Fig. 10.4 *The Santa Rosa Mission Research Area, and Thirty-Year Climate Profiles for Two Regional Meteorological Stations.*

Table 10. 1 Manxuj Landscape-Element Classification

axná'ˣlayi	Area with clay soils flooded seasonally. Elfin-like forest with most species under 2.5 m tall (even trees that are normally taller in other settings) and virtually no undercover. The area is used sporadically for hunting.
kiˣléta	Area with loose sand to loamy sand soils and a mixed-species forest with many open glades. The area can be used for hunting, but primarily serves to collect fruits, seeds, and tubers.
kishí'	Area with loam to clay loam soils and a mixed-species forest with dense bromeliad ground cover. Large numbers of termite mounds are common. The area is important for hunting as well as harvesting plants used in manufacturing.
sé'nkita	Area with clay loam soils flooded seasonally. Coer is a mixed-species forest less than 2 m tall with little ground cover. The area is used irregularly, but occasional hunting reported.
ˣlopuushá't	Area with silt loam soils and a parkland forest characterized by a particular species of tussock grass. The area is important for hunting armadillos and Plains Vizcacha.
xótiki	Area with sandy loam soils and a mixed-species forest referred to as "home of the honey." In addition to honey, various types of edible fruits are harvested, and rocket deer and Tagua Peccary are hunted.
xot	Area with sand loam soils. Cover is a mixture of low brush (1.25–1.5 m), large expanses of a tall grass species (75–100 cm), and solitary trees (<8 m). The area is used for growing watermelon and maioc, and harvesting some types of edible seeds and fruites; some hunting and honey harvesting take place.
aloxwishúk	Area probably with gley soils flooded more or less permanently and characterized by high stem-counts of palms. Use of this area is unknown, but the palm has an edible meristem.
xajipú'	Area with loam soils and a mixed-species forest with a ground cover of moss. The area is important for hunting Plains Vizcacha, reported as responsible for inducing and maintaining the moss ground cover.
xué'	Area with gley soils flooded seasonally and with a natural "campo" ($\approx$ savanna) vegetation. The area is important for its association with permanent water sources (ˣlawa) and is used for harvesting certain edible fruits and hunting brocket deer and armadillos.

most distinctive or salient subset of the floral (and apparently faunal) species characterizing their habitat; (2) the Manxuj ethnobotanical classification is based on the affinities observed among the taxa independent of their actual or potential significance; and (3) the Manxuj ethnobotanical and landscape-element systems of classification are organized into shallow hierarchic structures. Manxuj extractive activities are guided by their categorical partitioning of the landscape into prototypical landscape elements that are differentially exploited on the basis of botanical species that (a) directly provide edible, medicinal, and other products; (b) facilitate the manipulation and extraction of nonbotanical products such as honey; or (c) serve as in-

Table 10.2 Detailed Characterization of Two *manxuj* Landscape Elements

name: *kixléta*	
Soil:	Sand to loamy sand spoils; pH of 6.7; 0.13 kg/ha N_{tot}; 21.462 kg/ha P.
Vegetation:	A forest association with eighteen woody fold-taxa in a density of 92 indiv./ha. Most trees are around 3 m in height, but cacti are 3–5 m tall and a few emergent trees reach 10–12 m. Sixteen out of the eighteen woody folk-taxa (88 percent) are "useful" in some way to the Manxuj. Important folk-taxa include *ištá'k* (cactus: fruit, construction, honey), *á'iyik* (tree: dye), *i 'tsá'jituk* (tree: fruit), *šimták* (tree: dye, species indicator), *nájosuk* (vine: tuber), *sé'm^{x}lak* (tree: manufacture, beekeeping, emergency water), and *istéine'* (tree: medicine, honey).
animals:	White-lipped, collared, and Tagua peccary; doves, charata, brocket deer, tapir, mnkeys (night and howler), plains vizcacha
Comments:	"This forest contains lots of fruit." While animals are known to occur inthis association, little hunting is actually performed; the association is primarily used to collect edible fruits, seeds, tubers, and additional resources.
Name: *kishí*	
Soil:	Loam to clay loam soil; pH of 6.4; .027 kg/ha N_{tot}; 27.667 kg/ha P.
Vegetation:	A forest association with eighteen woody folk-taxa in a density of 275 indiv./ha. Most species reach a height of approx. 2.5 m, but occasional emergents grow to 4 m. Fifteen out of the eighteen woody folk-taxa (83 %) are "useful" to the Manxuj in some way. Important species are *šinúk* (tree: manufacture, tobacco substitute, species indicator), *óxneyik* (tree: seed), *néinuk* (tree: fruit), *lótejik* (bush-firewood, manufacture), *ká'* (bush: dye), *isáx* (bromeliad: food), *álesa* (bromeliad: fiber), *náwo* (cactus: emergency water), and *axájik* (tree: fruit).
Animals:	Brocket deer, rabbit, various armadillo species, doves, tapir, palmate anteater
Comments:	This is an important association for hunting—epecially during the dry season, as this association tends to surround permanent water holes and small, seasonal water holes are also found within the association. Manxuj dry-season camps are frequently located in this association.

dicators for the likely presence of vertebrate species that feed on the botanical species. Preliminary results from a comparison of two of the landscape elements identified in table 10.1 along the noted dimensions are presented in tables 10.2, 10.3, and 10.4.

Further research among the Manxuj on the spatial relation and temporal linkage between landscape elements of the Chaco landscape should eventually lead to a determination of the structural and functional nature of Manxuj foraging activities. The conclusions so far generated nevertheless contribute to the search for balance between the doctrines of "*Homo devastans*" and the "Ecologically Noble Savage" (see Balée, chapter 1), two extreme positions reflecting the broader debate over reductionism versus holism.

Table 10.3   Ranked Frequency of Woody Fold-Taxa in Two Manxuj Landscape Elements

Rank	kixéta		kishí	
1	*šinúk*	(1)	*šinúk*	(1)
2	*aí'tsa'mak*	(2)	*óxneyik*	(2)
3	*óxneyik*	(3)	*kí'tu xunoé*	(3)
4	*ištá'k*	(4.5)	*ištá'k*	(4)
5	*á 'iyik*	(4.5)	*néinuk*	(5)
6	*i 'tsá'jituk*	(6.5)	*xálek*	(6)
7	*itíšinik*	(6.5)	*kéjašuk*	(7)
8	*kíxilek*	(10.5)	*kijútunuk*	(9)
9	*šimták*	(10.5)	*lótejik*	(9)
10	*néinuk*	(10.5)	*i 'tsá'jituk*	(12)
11	*kixé'tok*	(10.5)	*náwo*	(11)
12	*ištéine'*	(10.50	*i 'tsá'jituk*	(12)
13	*axuínuk*	(10.50	*axájik*	(13.5)
14	*istáfʷe kawajik*	(16)	*kixé'tok*	(13.5)
15	*nájosuk*	(16)	*ištéine'*	(16.5)
16	*kéjašuk*	(16)	*lináaˣlak*	(16.5)
17	*sé'mˣlak*	(16)	*pówiyik*	(16.5)
18	*xuétsekak*	(16)	*xuétsekak*	(16.5)

• nine out of 18 folk-taxa are common to both units (50% similarity).

• fourteen out of 18 folk-taxa in both uits are "useful" (78%); 7 of these folk-taxa are unique to each unit.

• in kixléta, 2 additional folk-taxa are "bio-indicators."

• in kishí, one additinal folk-taxon is a "bio-indicator."

Reductionism Versus Holism

"Natural" vegetation and "natural" ecosystems are frequently viewed as preferable to the altered systems that result from human activities, without realizing the continuum of change that is associated with the establishment of these systems. The vegetative equilibrium that we detect by observation as a localized discontinuity implies (1) a timescale, (2) a spatial scale, and (3) a degree of analytical resolution (Allen and Hoekstra 1992; Pickett, Kolasa, and Jones 1994; Vale 1982). From a temporal standpoint, a clear-cut forest may represent a finite disturbance event in a continuing equilibrium; from a spatial standpoint, a strip of trampled ground beside a trail may influence diversity no more than the "naturally" disturbed zone along a dry ravine. Analytical resolution, which subsumes the aspects of prototypicality and mutability, relates directly to whether human activity is interpreted as causing vegetation change or as creating a new equilibrium (Vale 1982; Berlin 1992; Thompson 1989; Turner, Moss, and Skole 1993). As the intensity of grazing on a grassland increases, the relative importance of different grass species may change; if the species composition of the pregrazed vegetation is ignored and the area is simply recognized as a "grassland" (prototypical category), only a severely altered plant cover would appear as a change in vegetation. To the extent that change (mu-

Table 10.4   Likely Taxonomic Affiliation of Vertebrate Species Mentioned in Tables 1, 2, and 3

Common Name	Species Name
armadillos:	
naked-tailed	*Cabassous chacoensis*
small hariy	*Chaetophractus vellerosus*
larger hairy	*Chaetophractus villosus*
chacoan hairy	*Chlamyphorus retusus*
common long-nosed	*Dasypus novemcinctus*
yellow	*Euphractus sexcinctus*
giant	*Priodontes maximus*
southern three-banded	*Tolypeutes matacus*
howler monkey (black)	*Alouatta* spp.
brocket deer (gray)	*Mazama gouazoubira*
charata	*Chauna torquata* (?)
collared peccary	*Tayassu tajacu*
doves	*Columba* spp.
night monkey	*Aotus azarae*
palmate anteater	*Myrmecophaga tridactyla*
plains vizcacha	*Lagostomus maximus*
rabbit	*Sylvilagus brasiliensis*
tagua peccary	*Catagonus wagneri*
tapir	*Tapirus terrestris*
white-lipped peccary	*Tayassu pecari*

tability) is part of the definition of a vegetative system, human activities will be viewed as either system components or exogenous system disruptors (Hoekstra, Allen, and Flather 1991; Worster 1994; Pickett, Kolasa, and Jones 1994; Durham 1995).

Time, space, and analytical resolution are as closely connected to the changing perspective on what constitutes a vegetative equilibrium as they are to the debate over reduction versus holism. Frederic Clements (1936) viewed disturbance as initiating a predictable sequence of change (succession) by which vegetation returned to the equilibrium of its predisturbance condition (climax). The climax was a stable, undisturbed state of vegetation characterized by great internal organization, similar in many respects to an individual organism. Alex Watt (1947) placed disturbance phenomena on a par with other environmental factors influencing vegetation: instead of describing disturbance as destructive of equilibria, it was portrayed as part of the equilibrium condition. Contemporary usage views nonclimactic factors, such as soils and herbivores, as contributing to the generation of equilibrium structure and community composition; more importantly, however, rather than emphasizing the end point of a process, most contemporary studies attempt to understand patterns of community dynamics and to evaluate their causes (Hoekstra, Allen, and Flather 1991; Winterhalder 1994; Pickett, Kolasa, and Jones 1994; Pickett and Mc-Donnell 1989).

Classical reductionism as presented by Ernest Nagel in *The Structure of Science* (1961) seeks explanatory factors for specific phenomena among the entities and processes associated with a lower level. This reductionistic approach frequently begins by simplifying the context of system-existence, rather than the system itself; but isolating the phenomenon from its environmental context leads to causal explanations that are unrealistic, or even irrelevant to the historical or evolutionary development of the phenomenon. Reductionism establishes explanatory precision by sacrificing explanatory realism, knowledge about action boundaries, and the evolutionary significance of the phenomena under investigation (Inchausti 1994; Pera 1994; Lloyd 1994; Aunger 1995). The generally advocated alternative approach to reductionism—holism—in many ways is not much better, since it frequently suffers from the legacy of Herbert Spencer.

Herbert Spencer had a direct and obvious impact on anthropology, but he has been equally influential in ecology, as evidenced by the use of the "volitional organism" analogy (Worster 1994; Winterhalder 1994; Stoddart 1966). Well-known generalizations such as "quantity production characterizes the young ecosystem while quality production and feedback control are the trademarks of the mature system" (Odum 1969:266) are favorites of both anthropologists and ecologists. However, just how to establish what these generalizations mean in practice, and then to determine how they will be measured, has not been as easily agreed upon (Stoddart 1966; Worster 1994). Reification of the organism to scalar levels beyond those for which the concept was designed is just as much a disservice to the process of building explanatory frameworks as the appeal to values, beliefs, issues, and culture is to the exclusion of environment.

Conclusion

Historical ecology, landscape ecology, and evolutionary ecology undertake diachronic analyses of systems of relationship, which can be contrasted to instantaneous approaches built on the description of entity form (Winterhalder 1994; Balée 1994; Pickett, Kolasa, and Jones 1994; Crumley 1994; Allen and Hoekstra 1992). The three diachronic approaches are united by the central belief in the mutability of entities (E_n): given time, $E_1 \rightarrow E_2$ (Thompson 1989; Lloyd 1994; Pickett, Kolasa, and Jones 1994). They differ, fundamentally, in the analytical resolution used in explaining interactions of interest: society (historical ecology), landscape (landscape ecology), and individuals (evolutionary ecology).

This difference in the level selected for analysis should not be allowed to confuse the central premise underlying diachronic analysis: past states affect current states of a system. Furthermore, the ecological orientation of these diachronic approaches means that the central objective of analysis is not to understand an entity in itself, but to understand its relationship to environment. Ecological systems are hierarchical by nature, and the first requirement in any paradigmatic approach is the identification of the level of interest. This must be followed by definition of the mechanisms, constraints, and context serving to bound the level of interest for analytical purposes. Only then can the process of understanding proceed without falling into the traps of simple equalities or poorly operationalized explanatory elements.

Acknowledgments

My research was carried out with support from a Fulbright Research/Teaching grant, the L. S. B. Leakey Foundation (#1247-92), and the National Geographic Society (#4896-92). Logistical support while in Paraguay was provided by the staff of the Centro de Estudios Antropológicos of the Universidad Católica de Asunción and the United States Information Service at the American Embassy. Miguel "Gato" Chase-Sardi smoothed my initial field contacts in the Chaco and was an interesting companion on my first trip to the field. William Balée provided insightful and useful commentary on earlier versions of this paper, but any errors of omission or commission remain mine alone.

References

Addicott, John F., John M. Aho, Michael F. Antonin, Dianna K. Padilla, John S. Richardson, and Daniel A. Soluk. 1987. Ecological neighborhoods: Scaling environmental patterns. *Oikos* 49:340–346.

Agrawal, Arun. 1994. Mobility and control among nomadic shepherds: The case of the Raikas. II. *Human Ecology* 22(2): 131–144.

Allen, Timothy F. H. and Thomas W. Hoekstra. 1992. *Toward a Unified Ecology*. New York: Columbia University Press.

Allen, Timothy F. H. and Thomas B. Starr. 1982. *Hierarchy: Perspectives for Ecological Complexity*. Chicago: University of Chicago Press.

Aunger, R. 1995. On ethnography: Storytelling or science? *Current Anthropology* 36: 97–130.

Balée, William. 1994. *Footprints of the Forest: Ka'apor Ethnobotany—The Historical Ecology of Plant Utilization by an Amazonian People*. New York: Columbia University Press.

Berlin, Brent. 1992. *Ethnobiological Classification: Principles of Categorization of Plants and Animals in Traditional Societies*. Princeton: Princeton University Press.

Borgerhoff-Mulder, Monique and T. M. Caro. 1985. The use of quantitative observational techniques in anthropology. *Current Anthropology* 26:323–335.

Brandenburg, A. M. and M. S. Carroll. 1995. Your place or mine? The effect of place creation on environmental values and landscape meanings. *Society and Natural Resources* 8:381–398.

Bridgewater, P. B. 1993. Landscape ecology, geographic information systems, and nature conservation. Pp. 23–36 in Haines-Young, Green, and Cousins (eds.), *Landscape Ecology and Geographic Information Systems*.

Chase-Sardi, Miguel. 1972. *La situación actual de los indígenas en el Paraguay*. Asunción: Centro de Estudios Antropológicos, Universidad Católica.

Chase-Sardi, Miguel, Augusto Brun, and Miguel Angel Enciso. 1990. *Situación sociocultural, económica, jurídico-política actual de las comunidades indígenas en el Paraguay*. Asunción: CIDSEP.

CIF (Carrera de Ingeniería Forestal). 1991. *Vegetación y uso de la tierra de la región occidental del Paraguay (chaco): Años 1986–1987*. San Lorenzo, Paraguay: Universidad Nacional de Asunción.

Clements, Frederic E. 1936. Nature and structure of the climax. *Journal of Ecology* 24:252–284. (Also in Real and Brown [eds.], *Foundations of Ecology*, pp. 59—97.)

Cooke, R. M. 1991. *Experts in Uncertainty: Opinion and Subjective Probability in Science*. New York: Oxford University Press.

Cousins, S. 1993. Hierarchy in ecology: Its relevance to landscape ecology and geographic information systems. Pp. 75–86 in Haines-Young, Green, and Cousins (eds.), *Landscape Ecology and Geographic Information Systems*.

Crumley, Carole L. 1994. Historical ecology: A multidimensional ecological orientation. Pp. 1–16 in Carole L. Crumley (ed.), *Historical Ecology: Cultural Knowledge and Changing Landscapes*. Santa Fe: School of American Research.

Crumley, Carole L. and William H. Marquardt, eds. 1987. *Regional Dynamics: Burgundian Landscapes in Historical Perspective*. San Diego: Academic Press.

DeWalt, Billie R. 1994. Using indigenous knowledge to improve agriculture and natural resource management. *Human Organization* 53(2): 123–131.

Durham, William H. 1995. Political ecology and environmental destruction in Latin America. Pp. 249–264 in Michael Painter and William H. Durham (eds.), *The Social Causes of Environmental Destruction in Latin America*. Ann Arbor: University of Michigan Press.

Forman, Richard T. T. and Michel Godron. 1986. *Landscape Ecology*. New York: Wiley.

Funtowicz, Silvio O. and Jerome R. Ravetz. 1991. A new scientific methodology for global environmental issues. Pp. 137–152 in Robert Costanza (ed.), *Ecological Economics: The Science and Management of Sustainability*. New York: Columbia University Press.

Gerzenstein, Ana. 1983. *Lengua Chorote: Variedad no. 2, estudio descriptivo-comparativo y vocabulario*. Archivo de Lenguas Precolombinas, vol. 4. Buenos Aires: Instituto de Lingüística.

Golley, Frank Benjamin. 1993. *A History of the Ecosystem Concept in Ecology: More Than the Sum of the Parts*. New Haven: Yale University Press.

Gragson, Ted L. 1992. Strategic procurement of fish by the Pumé: A South American "fishing culture." *Human Ecology* 20(1): 109–130.

———. 1993. Human foraging in lowland South America: Pattern and process of resource procurement. *Research in Economic Anthropology* 14:107–138.

———. 1994a. *Final Report to the National Geographic Society on Grant #4896-92*. Athens, Ga.: author.

———. 1994b. *Preliminary Manxuj Ethnobotanical Listing*. Athens, Ga.: author.

Grubb, W. Barbrooke. 1993. *Un pueblo desconocido en tierra desconocida*. Asunción: OMEGA.

Gunn, Joel D. 1994. Introduction: A perspective from the humanities-science boundary. *Human Ecology* 22(1): 1–22.

Haines-Young, R., D. R. Green, and S. Cousins. 1993. Landscape ecology and spatial information systems. Pp. 3–8 in R. Haines-Young, D. R. Green, and S. Cousins (eds.), *Landscape Ecology and Geographic Information Systems*. London: Taylor and Francis.

Hoekstra, Thomas W., Timothy F. H. Allen, and Curtis H. Flather. 1991. Implicit scaling in ecological research. *BioScience* 41(3): 148–154.

Hrenchuk, Carl. 1993. Native land use and common property: Whose common? Pp. 69–86 in Julian T. Inglis (ed.), *Traditional Ecological Knowledge: Concepts and Cases*. Ottawa: Canadian Museum of Nature.

Inchausti, Pablo. 1994. Reductionist approaches in community ecology. *American Naturalist* 143(2): 201–221.

Kempton, Willett, James S. Boster, and Jennifer A. Hartley. 1995. *Environmental Values in American Culture*. Cambridge: MIT Press.

Kersten, Kudwig. 1968. *Las tribus indígenas del gran chaco hasta fines del siglo XVIII.* Resistencia: Universidad Nacional del Nordeste.

Kidd, Stephen W. 1995. Land, politics, and benevolent shamanism: The Enxet Indians in a democratic Paraguay. *Journal of Latin American Studies* 27:43–75.

Kleinpenning, J. M. G. 1987. *Man and Land in Paraguay.* Providence: Foris Publications.

Klijn, Frans and Helias A. Udo de Haes. 1994. A hierarchical approach to ecosystems and its implications for ecological land classification. *Landscape Ecology* 9(2): 89–104.

Knoke, David and James H. Kuklinski. 1982. *Network Analysis.* Beverly Hills: Sage.

Laumann, Edward O. and David Knoke. 1987. *The Organizational State: Social Choice in National Policy Domains.* Madison: University of Wisconsin Press.

Laurini, Robert and Derek Thompson. 1992. *Fundamentals of Spatial Information Systems.* New York: Academic Press.

Lepart, Jacques and Max Debussche. 1992. Human impact on landscape patterning: Mediterranean examples. Pp. 76–106 in Andrew J. Hansen and Francesco di Castri (eds.), *Landscape Boundaries: Consequences for Biotic Diversity and Ecological Flows.* New York: Springer-Verlag.

Little, Daniel. 1991. *Varieties of Social Explanation: An Introduction to the Philosophy of Social Science.* Boulder, Colo.: Westview Press.

Lloyd, Elisabeth A. 1994. *The Structure and Confirmation of Evolutionary Theory.* Princeton: Princeton University Press.

Marsh, G. P. 1864. *Man and Nature, or Physical Geography as Modified by Human Action.* New York: Scribner.

Métraux, A. 1946. Ethnography of the Chaco. Pp. 197–370 in J. H. Steward (ed.), *Handbook of South American Indians*, vol. 1, *The Marginal Tribes.* Washington, D.C.: U.S. Government Printing Office.

Nagel, E. 1961. *The Structure of Science: Problems in the Logic of Scientific Explanation.* New York: Harcourt, Brace and World.

Nicolson, M. 1987. Alexander von Humboldt, Humboldtian science, and the origins of the study of vegetation. *History of Science* 25:167–194.

Odum, Eugene P. 1969. The strategy of ecosystem development. *Science* 164:262–270. (Also in Real and Brown [eds.], *Foundations of Ecology*, pp. 596–604.)

O'Neill, R. V., D. L. DeAngelis, J. B. Waide, and T. F. H. Allen. 1986. *A Hierarchical Concept of Ecosystems.* Princeton: Princeton University Press.

Ostrom, Elinor, Roy Gardner, and James Walker. 1994. *Rules, Games, and Common-Pool Resources.* Ann Arbor: University of Michigan Press.

Pera, Marcello. 1994. *The Discourses of Science.* Chicago: University of Chicago Press.

Pickett, S. T. A., J. Kolasa, J. J. Armesto, and S. L. Collins. 1989. The ecological concept of disturbance and its expression at various hierarchical levels. *Oikos* 54(2): 129–136.

Pickett, Steward T. A., Jurek Kolasa, and Clive G. Jones. 1994. *Ecological Understanding: The Nature of Theory and the Theory of Nature.* San Diego: Academic Press.

Pickett, S. T. A. and M. J. McDonnell. 1989. Changing perspectives in community dynamics: A theory of successional forces. *Tree* 4(8): 241–245.

Plett, Rudolf. 1979. *Presencia menonita en el Paraguay.* Asunción: Instituto Bíblico Asunción.

Prado, Darién E. 1993. What is the Gran Chaco vegetation in South America? I. A review. (Contribution to the study of flora and vegetation of the Chaco, V.) *Candollea* 48(1): 145–172.

Rayner, Steve et al. 1994. A wiring diagram for the study of land-use/cover change: Report of working group A. Pp. 13–53 in William B. Meyer and B. L. Turner II (eds.), *Changes in Land Use and Land Cover: A Global Perspective*. Cambridge: Cambridge University Press.

Real, Leslie A. and James H. Brown, eds. 1991. *Foundations of Ecology: Classic Papers with Commentaries*. Chicago: University of Chicago Press.

Renshaw, J. 1986. The economy and economic morality of the Indians of the Paraguayan Chaco. Ph.D. diss., London School of Economics and Political Science, University of London.

Rocheleau, Dianne, Barbara Thomas-Slayter, and David Edmunds. 1995. Gendered resource mapping: Focusing on women's spaces in the landscape. *Cultural Survival Quarterly* 18(4): 62–68.

Rosen, Eric von. 1924. *Ethnographical Research Work During the Swedish Chaco-Cordillera-Expedition, 1901–1902*. Stockholm: C. E. Fritze.

Sanderson, Steven. 1994. Political-economic institutions. Pp. 329–355 in William B. Meyer and B. L. Turner II (eds.), *Changes in Land Use and Land Cover: A Global Perspective*. Cambridge: Cambridge University Press.

Scott, David. 1995. Habitation sites and culturally modified trees: Using predictive models in Ditidaht territory. *Cultural Survival Quarterly* 18(4): 19–20.

SGCR (Subcomittee on Global Change Research). 1995. *Our Changing Planet: The FY 1996 U.S. Global Change Research Program*. Washington, D.C.: Office of Science and Technology Policy.

Sprichiger, R. and L. Ramella. 1989. The forests of the Paraguayan Chaco. Pp. 259–270 in L. B. Holm-Nielsen, I. C. Nielsen, and H. Balslev (eds.), *Tropical Forests: Botanical Dynamics, Speciation, and Diversity*. New York: Academic Press.

Stafford, C. Russell and Edwin R. Hajic. 1992. Landscape scale: Geoenvironmental approaches to prehistoric settlement strategies. Pp. 137–161 in Jacqueline Rossignol and LuAnn Wandsnider (eds.), *Space, Time, and Archaeological Landscapes*. New York: Plenum Press.

Stahl, Wilmar. 1986. *Escenario indígena chaqueño pasado y presente*. Filadelfia, C.R.: ASCIM.

Stephens, D. W. 1990. Risk and incomplete information in behavioral ecology. Pp. 19–46 in Elisabeth Cashdan (ed.), *Risk and Uncertainty in Tribal and Peasant Economies*. Boulder, Colo.: Westview Press.

Stephens, D. W. and J. R. Krebs. 1986. *Foraging Theory*. Princeton: Princeton University Press.

Stoddart, D. R. 1966. Darwin's impact on geography. *Annals of the Association of American Geographers* 56:683–698.

Susnik, Branislava. 1983. *El rol de los indígenas en la formación y en la vivencia del Paraguay*, vol. 2. Asunción: Editorial UNIVERSO.

Thompson, Paul. 1989. *The Structure of Biological Theories*. Albany: State University of New York Press.

Tooby, J. and I. DeVore. 1987. The reconstruction of hominid behavioral evolution through strategic modeling. Pp. 183–237 in W. G. Kinzey (ed.), *The Evolution of Human Behavior: Primate Models*. Albany: State University of New York Press.

Turner, B. L. II, R. H. Moss, and D. L. Skole. 1993. *Relating Land Use and Global Land-Cover Change: A Proposal for an IGBP-HDP Core Project*. Stockholm: International Geosphere-Biosphere Programme.

Turner, Monica G. 1989. Landscape ecology: The effect of pattern on process. *Annual Review of Ecology and Systematics* 20:171–197.

Tversky, A. and D. Kahneman. 1974. Judgment under uncertainty: Heuristics and biases. *Science* 185:1124–1131.

Uhl, Christopher, Daniel Nepstad, Robert Buschbacher, Kathleen Clark, Boone Kauffman, and Scott Subler. 1990. Studies of ecosystem response to natural and anthropogenic disturbances provide guidelines for designing sustainable land-use systems in Amazonia. Pp. 24–42 in Anthony B. Anderson (ed.), *Alternatives to Deforestation: Steps Toward Sustainable Use of the Amazon Rain Forest.* New York: Columbia University Press.

Urban, D. L., R. V. O'Neill, and H. H. Shugart, Jr. 1987. Landscape ecology: A hierarchical perspective can help scientists understand spatial patterns. *BioScience* 37(2): 119–127.

USAID. 1985. *Environmental Profile of Paraguay.* Washington, D.C.: United States Agency for International Development.

Vale, Thomas R. 1982. *Plants and People: Vegetation Change in North America.* Washington, D.C.: Association of American Geographers.

Watt, Alex S. 1947. Pattern and process in the plant community. *Journal of Ecology* 35:1–22. (Also in Real and Brown [eds.], *Foundations of Ecology*, pp. 664–686.)

White, P. S. and S. T. A. Pickett. 1985. Natural disturbance and patch dynamics: An introduction. Pp. 3–13 in S. T. A. Pickett and P. S. White (eds.), *The Ecology of Natural Disturbance and Patch Dynamics.* Orlando: Academic Press.

Wilbert, Johannes and Karin Simoneau. 1985. *Folk Literature of the Chorote Indians.* Los Angeles: UCLA Latin American Center Publications.

Winterhalder, Bruce P. 1994. Concepts in historical ecology: The view from evolutionary ecology. Pp. 17–41 in Crumley (ed.), *Historical Ecology.*

Worster, Donald. 1994. *Nature's Economy: A History of Ecological Ideas.* 2d ed. Cambridge: Cambridge University Press.

CHAPTER 11

Domestication as a Historical and Symbolic Process: Wild Gardens and Cultivated Forests in the Ecuadorian Amazon

LAURA RIVAL

Accounting for the Nonpractice of Agriculture in Amazonia

Most of the authors who have discussed the existence of nomadic bands subsisting with few or no domesticates in lowland South America see these marginal groups not as genuine hunters and gatherers, but as "deculturated" agriculturalists. These cases of devolution are believed to have resulted from a combination of environmental and/or historical causes. Betty Meggers's famous *Amazonia: Man and Culture in a Counterfeit Paradise* (1971) was the first in a long series of works to argue that low population density, incipient warfare, transient slash-and-burn horticulture, and food taboos are all manifestations of human adaptation to environmental limiting factors, particularly to the depletion of critical natural resources. For Meggers, the more complex societies of the Lower Amazon described in Spanish chronicles were originally formed by Andean colonists unable to maintain the same degree of social and cultural sophistication in an environment too poor to sustain intensive maize cultivation. Her model of cultural and social devolution was the first comprehensive formulation of a thesis that still has some popularity.[1] Other authors (most notably Gross 1975; Ross 1978; Harris 1984) have departed from her narrow focus on soil fertility to look for other limiting factors in the environment. In particular, they have interpreted the form of Amazonian indigenous settlements—which are typically small, widely scattered, and often deserted for months on end by residents who have gone on long treks and foraging expeditions—as clear evidence of cultural adaptation to game scarcity. Like Meggers, they attribute the lack of complex and hierarchical sociopolitical systems to a lack of resource potential. There is no need to counter here these naturalistic arguments, which have attracted numerous critical reviews and discussions.[2]

In their two papers given at the 1966 Chicago Conference "Man the Hunter," Claude Lévi-Strauss (1968) and Donald Lathrap (1973) both rejected the existence of hunter-gatherer societies in the Amazon, arguing that the marginal nomadic bands found in this region were in fact devolved agriculturalists who had adopted a hunting and gathering mode of existence only recently. Lévi-Strauss (1968:350) warned that the groups classified in the *Handbook of South American Indians* as marginal tribes were "regressive rather than primitive." Lathrap, for his part, claimed that the highly mobile foraging peoples of the northwestern Amazon inter-fluvial areas were the descendants of riverine peoples who had been forced off the more favorable riverfronts by more powerful and developed populations from the Amazon flood plains. Lathrap, like Anna Roosevelt, makes the flood plains of the Lower Amazon (with their late prehistoric chiefdoms) the source of all cultural innovation and complexity. Although basically similar in their conclusions, Lévi-Strauss's and Lathrap's arguments emphasize different factors. While Lathrap stresses, in the Stewardian tradition, the link between environmental and historical factors, Lévi-Strauss notes the disharmony between rudimentary technological achievements, on the one hand, and complex kinship systems and sophisticated cosmologies, on the other—hence arguing that cultural devolution has first and foremost affected productive practices, not the representation of social relations as encoded in kinship systems and myths. In other terms, if Lathrap's argument is, in the end, an environmentalist one (history displaces populations to environments less favorable to cultural development), Lévi-Strauss's is fundamentally culturalist (environmental conditions affect units of meaning, not the structural relations between units).

For William Balée, it is by focusing on the dynamic history of plant/human interaction that one can best account for the existence of Amazonian foraging bands. Balée (1988a, 1989, 1992, 1993) and a number of ethnobotanists after him (Posey and Balée 1989; Posey 1985; Sponsel 1986; Eden 1990), have contended that Amazonian forests are in part cultural artifacts.[3] Balée argues that far from having been limited by scarce resources, the indigenous peoples of the Amazon have created biotic niches since prehistoric times. His hypothesis that these biotic niches have become "anthropogenic forests" rests upon a number of observations of contemporary gardening activities, as well as inductions about the long-lasting effects of past human interference. Moreover, his work with a number of Brazilian marginal groups has led him to conclude that foragers can survive without cultivated crops thanks to a few essential nondomesticated resources (palms and other fruit trees), which are in fact the product of the activities of ancient populations. Nomadic bands do not wander at random in the forest, but move their camps between palm forests, bamboo forests, or Brazil nut forests, which all are "cultural forests"—that is, ancient dwelling sites. The existence of anthropogenic forests, the product of a close and long-term association between certain plant species and humans, is further supported by two observations: the wide occurrence of charcoal and numerous potsherds in the forest soil, and the greater concentration of palms, lianas, fruit trees, and other heavily used forest resources on archaeological sites. The twin propositions that species distribution is a good indicator of human disturbance and that foraging bands have adapted to disturbed forests are particularly well exposed in three of Balée's articles: the article on Ka'apor warfare (1988b), the one on the progressive devolution from agriculture to foraging activities (1992), and the one on the distinction between old fallows and high forest (1993).

In his historical reconstruction of the colonization of the Amazon's lower course between the seventeenth and the nineteenth centuries, Balée (1988b:158–159) argues that the Indian populations of Brazil responded to political domination with five basic strategies. Those living along important rivers allied with the Brazilian military, whom they helped to capture slaves from rebellious tribes. Less powerful groups fled; some adopted a wandering, nonhorticulturist way of life, while others continued to cultivate fast-growing crops, such as sweet manioc and maize. The two other strategies were to either resist domination violently—and risk extermination—or migrate to remote forested areas where settled villages, organized around the production of bitter manioc, could be maintained. On the basis of this historical reconstruction Balée (1992), not unlike Lathrap (1970, 1973), Roosevelt (1993), and Roosevelt et al. (1991), proposes a model to account for the progressive loss of cultivation by wandering marginal tribes through disease, depopulation, and warfare. Finally, his study (1993) of the indigenous agroforestry complex of the Ka'apor people of Brazil defends convincingly the idea that forests of biocultural origin can be treated as objective records of past human interactions with plants, even if the local population does not have any social memory of such history and cannot differentiate old fallows from patches of undisturbed forest. The presence of surface pottery and charcoal in the soil, the distribution of species, the size of trunks, oral history, and native classifications of forest and swidden types can all be used to differentiate old fallows from high forests.

To summarize, Balée uses historical ecology to counter the ahistorical explanations offered by cultural ecology and evolutionary ecology. If Amazonian foragers exploit "wild" resources, they are not preagriculturalists; and their agricultural regression, which follows a recurrent pattern, can be documented. There are four important points in this argument. First, it is argued that the environment does not limit cultural development as much as has previously been believed. An ahistorical view of the environment blinds us to the fact that what we take as a "pristine" environment may actually be an ancient agricultural site. Second, better explanations can be offered by taking into consideration nonenvironmental factors, particularly historical ones, and stressing sociopolitical dynamics. Third, the historical evidence of past agriculture is twofold: it is both linguistic and botanical. Living foragers do not remember that their forebears cultivated, but their languages possess cognates for cultigens. In other words, amnesia affects two types of knowledge: the cultural past of the group, and technical savoir faire. The only cultural transmission that seems to have worked and continued through time is unconscious linguistic knowledge.[4] Fourth, the process of agricultural regression—and of regression from sedentarism to nomadism—is progressive: at each stage, one or more cultigens is lost and the dependence on uncultivated plants increases. If the argument for the loss of cultigens is essentially similar to that of Roosevelt, the great merit and originality of Balée's work is to have shown that the increased reliance on uncultivated plants is not a return to nature, but an adaptation to "vegetational artifacts of another society" (Balée 1988a:48).

Are the Huaorani Devolved Agriculturalists?
Trekking Through History

The Huaorani, like the Brazilian groups discussed by Balée (1992), chose to flee from coercive powers and to adopt a wandering way of life. We will probably never know whether they are the descendants of more sedentarized and sophisticated cultivators—but there is little doubt that their way of life has, for centuries, depended more on foraging activities than on agriculture. To this day, Huaorani people invest considerable time and show great interest in hunting and gathering activities. Although they cultivate fast-growing food plants in a rudimental fashion, they subsist mainly on resources that are the product of ongoing forest management activities. Their rejection of elaborate gardening corresponds to a specific historical experience, and a particular type of social organization.

The history of the Huaorani is still poorly known. They form a very isolated group, whose language is not attached to any known phylum, and whose borrowing of non-Huaorani cultural traits was literally nil at the time of contact in the early 1960s. To the best of our knowledge, they have lived for centuries in the interstices between the great Zaparo, Shuar, and Tukanoan nations of the Upper Marañon, constituting nomadic and autarchic enclaves that fiercely refused contact, trade, or exchange with their powerful neighbors. The core of their ancestral territory seems to have been the Tiputini River, from where they appear to have expanded east, west, and southward until occupying most of the hinterlands between the Napo and Curaray Rivers (see figure 11.1) in the aftermath of the rubber boom, which caused the disappearance of most Zaparo communities (Rival 1992).

Isolationism is still an important means for maintaining ethnic boundaries. Only the Huaorani, the speakers of *huao terero*—the Huaorani language—are truly "humans"; all other people are *cohuori*, strangers coming from the other side of the Napo River, or *quehueri,* cannibals feeding on Huaorani children. The last group of uncontacted Huaorani, the Tagaeri, still maintains a complete state of isolation. They refuse all communication with outsiders, and with their relatives who have accepted peaceful contact and exchange with non-Huaorani. The Tagaeri live in hiding, with no cultivated crops, their fires burning only at night. They refuse marriage alliances outside their group. Each year, despite the danger of being seen by the oil crews who are now occupying their land, they try to go back to their palm groves for the fruiting season.

I have shown elsewhere (Rival 1992, 1993) that the Huaorani are not only autarchic, but also highly endogamous. Their kinship system is flexible enough, however, to accommodate demographic variations. The population is divided into

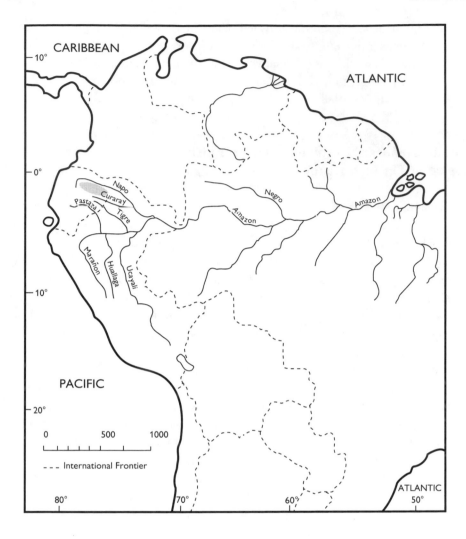

Fig. 11.1 *Huaorani Study Area.*

dispersed networks of intermarrying longhouses separated by vast stretches of un-
occupied forest. These intermarrying longhouses form regional groups (called
huaomoni; literally, "we-people"), who maintain relations of hostility with each
other. A *huaomoni* group calls all other groups *huarani*, that is, "others" or "enemies." The core relation within a *huaomoni* group is the relation between a brother
and a sister, who, once married in the same nexus, exchange their children in marriage. Marriage tends to be uxorilocal, with men going to live with their wives'
kin.[5] Despite the prevalence of hostility and no-contact, *huarani* groups are loosely
linked by personal ties between individual relatives who, for one reason or another,
do not belong to the same *huaomoni* group. These privileged relations, mainly used
in case of spouse scarcity within the endogamous nexus, secure the renewal of alliances without which the group could not socially reproduce.

Making the Forest Anthropogenic

The traditional system of social alliances, based on a strict closure of the Huaorani social world onto itself, as well as on the partial isolation and mutual avoidance of the regional groups, corresponds to a particular mode of subsistence and use of the forest. Lévi-Strauss (1950:465, 467, 468) noted, more than forty years ago, that in South America "there are many intermediate stages between the utilization of plants in their wild state and their true cultivation," and that "farming always accompanies, and is never a substitute for, the exploitation of wild resources."[6] This remark applies particularly well to the Huaorani context, where numerous plant species are encouraged to grow outside of cultivated areas as people engage in numerous daily actions (planting, selecting, transplanting, protecting, using, and discarding) that have a direct or indirect effect on the distribution of species, be they fully domesticated or not.

Huaorani people daily consume a great number of cultigens that are not planted in gardens. They see in their forested land the historical record of the activities of past generations. They are quite explicit about the inseparability of people and the forest, which they describe as a succession of fallows. Most of the western part of Huaorani land is said to be *ahuene*—that is, secondary forest. Only in the Yasuni, they tell me, are there pristine forests, *omere*, with really high and old trees.[7] Secondary forests are further divided into *huiyencore* (four-to-ten-year-old clearings characterized by the frequency of balsa trees), *huyenco* (ten-to-twenty-year-old clearings), *huiñeme* (twenty-to-forty-year-old clearings characterized by the high incidence of adult palms), and *durani ahuè* (forty-to-a-hundred-year-old clearings, remarkable for their big trees). Before the arrival of missions, *huiñeme* forests were the preferred sites to establish main residences. However, all types of forest were— and still are—continuously visited and lived in for longer or shorter stays.

Cultivars are found—discovered—throughout the forest. This further indicates an evident strategy of resource dispersion within specific regions. Fish-poison vines are found along the creeks where people fish, semiwild fruit trees near hunting camps, and numerous useful palms (such as *Astrocaryum chambira*; in Huaorani, *oönempa*) along trails. The regional groups (*huaomoni*) are constantly moving through their vast and relatively stable territories. Hilltop longhouses are regularly left for hunting and foraging trips, during which forest-management activities take place. Wherever a Huaorani finds herself in the forest, she chances upon needed plants. Informants are vague as to whether these strategic and handy resources were planted by someone,[8] or just happened to grow there. What matters to them is that their occurrence can be related either to individuals known for using a particular area regularly or to a house-group who lived in the area, sometime in the past. For instance, when young Huaorani unexpectedly discover useful plants in a part of the forest they are not familiar with, they often attribute them, with noticeable pleasure, to the activities of past people. If they decide that these cultigens were left by dead forebears—usually great-grandparents—they may see the plants as an invitation to move permanently and legitimately into this part of the forest, and to create a new longhouse. When no certain link with past or present human activity is established, the wide occurrence of cultigens is linked to animal activity. For example, semiwild manioc[9] is said to "belong" to the tapir.

Many useful plants, however, are not connected to any human or animal activity, even when their distribution affects human distribution. For instance, an informant who once told me, "We remain within the limits of the *oonta* [*Curarea tecunarum*] territory," was nevertheless adamant that the vine, which he gathered to prepare his hunting poison, just happened to be where we found it. However, given the cultural importance of curare poison, one wonders whether the *Curarea tecunarum* vine has not been subjected to human management. Although this cannot be solved before thorough botanical research is undertaken, the denial of plant management is interesting in itself. Another species that does not seem to be managed in any intentional way, but whose spatial distribution greatly influences the Huaorani's movements and choice of residence, is the ungurahua palm (*Jessenia bataua*; in Huaorani, *peto-huè*). A number of informants have told me that one of the reasons why longhouses are built on hilltops is that this is where ungurahua palms grow. The ungurahua palm provides rich food,[10] building materials, and raw materials for the making of a wide range of artifacts and remedies. Besides being an extremely useful plant resource, the ungurahua palm offers protection: Its wood makes good fire, even under the wettest conditions. The safest place to spend the night when lost in the forest is under an ungurahua palm. People say that ungurahua palms, which have deep roots and grow in fertile soils, can stop violent winds from felling emergent canopy trees. Finally, informants stress over and over again that those who flee from wars and spearing raids would not survive without the ungurahua fruit, which is rich in fats and proteins, and ripens throughout the year. The fruit is also appreciated by woolly monkeys (*Lagothrix lagotricha*), a favored game animal. The ungurahua palm is never planted, but grows along ridge tops, where people collect the fruit during their gathering expeditions. It is brought back to a camp or longhouse hearth to be simmered. People are perfectly aware of the fact that these cooking activities encourage the germination of ungurahua seeds, hence facilitating its propagation.

A large number of other food plants are propagated through human consumption, rather than direct planting.[11] For example, the much-appreciated *daboca* fruit (one of the Solanaceae, perhaps *Solanum* sp. or *Solanum sessiliflorum*) grows where it has been discarded. A very sour fruit, it is never completely eaten, and the seeds remain on the forest floor until the proper conditions of heat and light favor germination. There are numerous *daboca* bushes in manioc gardens, around houses, and along rivers, but, according to my informants, none of them are planted.

More generally, I would like to suggest that we should distinguish cultivation from domestication (Chase 1989; Yen 1989; Groube 1989). Some plants, particularly trees, are cultivated without having been domesticated—i.e., without showing morphological and genetic modification. Cultivation, which refers to the human activity of encouraging the growth of a particular plant—by, for instance, protecting and weeding—does not imply any genetic response in plants. It is therefore perfectly conceivable to cultivate wild plants, as it is to manage domesticated species in the wild. And the Huaorani manage, rather than cultivate, two common Amazonian domesticates, the peach palm (*Bactris gasipaes*) and sweet manioc (*Manihot esculenta*).

The Political Use of Cultigens

Charles Clement (1988, 1992) has demonstrated that the peach palm is a fully domesticated palm. The Huaorani, like many Amazonian Indians, grow peach palm seedlings (usually from the seed, rarely from a basal sucker), which they replant later in the clearing surrounding the longhouse. It is likely that most peach palm groves start in this fashion. But, because forest regrowth would overtop the palms a few decades after the abandonment of a dwelling site, the groves would not endure without human intervention. Every year, at the beginning of the fruiting season (which generally starts in January and lasts until April), the intermarrying *huaomoni* groups converge toward their groves, generally at two to three days' walking distance from their main residences. They spend the whole season collecting and preparing the fruit for their daily consumption and, more importantly, for drinking ceremonies and marriage celebrations. These groves are in fact old dwelling sites: they exhibit scattered potsherds and broken stone axes, which are proudly excavated and kept as the secure signs that "the grandparents lived there."

As people prepare and consume vast quantities of fruit in season, new seedlings develop around the temporary hearths, year after year. The peach palm fruit is not edible unless it has been simmered in water for a few hours; fruits at the top of the pot, which are not properly heated, and therefore not completely freed from proteolytic enzyme inhibitors or calcium oxalate crystals, are discarded. Some of these fruits are eaten by animals, but a substantial number are left to germinate on-site. Young saplings of *macahuè*[12] are planted at the side of the thorny peach palm trunks to provide easier access to fruit bunches. Old trees are felled for the quality of their hard wood, *tehuè* (from *tey*, "hard," and *ahuè*, "wood"), which is used to make spears and a whole range of smaller piercing or cutting tools. In wartime, palm groves are destroyed to make spears. Enemy groups destroy each other's groves as a means not only to increase their stock of precious hard wood, but also to suppress social memory: without these landmarks, a group loses its sense of continuity and its claim to a particular part of the forest.

Although peach palm groves could not persist without human intervention, they are not, properly speaking, cultivated. Maintained through activities of consumption, they are the products of the activities of past generations of Huaorani—and, more explicitly, of the deceased grandparents or great-grandparents of those who come to feed on the trees. Peach palms and their seasonal harvests are taken to be gifts from deceased relatives. The house-groups, who rarely see each other during the rest of the year, spend the fruiting season together on the sites where their forebears lived and died, remembering them, enjoying one another's company, chanting to the bounty of the forest, and celebrating the marriage of those mature enough to have children of their own. The peach palm fruit, the fruit produced by past life activities, is food to the living who, through their present consumption activities, ensure the feeding of the generations to come. The peach palm materializes a crucial link between past, present, and future generations of endogamous *huaomoni* people. And when social dynamics lead to the disappearance of a particular *huaomoni* group, its peach palm grove, no longer maintained, disappears as well. Lasting longer than human lives, these groves are a source of pride, security, and rejoicing, the concrete and material sign of continuity.[13]

Huaorani manioc gardening (manioc has always been cultivated marginally for marriage ceremonies) is not "swidden horticulture" any more than peach palm management is. Now living for the most part in sedentarized villages, Huaorani people grow manioc on a larger scale and use it for daily consumption. However, they are not, by any standard, horticulturists—especially when compared with the Shuar and Quichua Indians, whose women are accomplished gardeners, and whose manioc beer is the symbol of gender complementarity and good living (Descola 1986; Whitten 1985). A few of the most striking differences can be cited for illustration. The Huaorani manioc gardens are small in size, poor in crop variety, and abandoned after only one harvest. The soil, hardly weeded, is not cleared of all its vegetation cover, nor is it burned. There is no strict gender division of labor (gardening does not represent the secret domain of female knowledge), and no belief in the need to combine garden technology with magic.[14] Given a general lack of planning and concern for securing regular and continuous supplies of garden crops, households can spend months without any. Finally, manioc roots are never brewed into beer. Quichua Indians, for whom manioc beer is not merely a staple food, but the sacred mark of social and cultural identity, profoundly despise the Huaorani way of growing and preparing manioc. They regard it as closer to animal than to human behavior. For example, they say that the Huaorani, who sometimes eat young and tender manioc roots raw in their gardens, behave like wild pigs—disregarding the fact that they always remove the skin and clean off the dirt before eating.

Additional facts can be cited to support the contention that garden produce is primarily food for visitors, rather than food for daily subsistence. First, the level of gardening varies according to the degree of peace and the size of regional alliance networks. When a house-group wants to renew its alliance with an enemy group, it uses manioc to prepare the feast drinks. In times of warfare and feuding, longhouses disperse in the forest and live without gardens for months. The wider the alliances, the more house-groups visit each other, the more they organize feasts, and the more they plant manioc to prepare nonfermented manioc drinks, *têpê*. That planting manioc implies first and foremost hosting is further demonstrated by the expression used for rejoicing, *huatapè*, which literally means "give me another bowl of manioc drink" and elliptically entails "I laugh away with you, my visitor." It is also this main function of serving as food-drink for hosts that explains why gardens are basically monocultures. If forest products offer a rich and varied diet, gardens bring little more than manioc, plantains, or bananas. Finally, because planting manioc is an invitation to visit and feast, a house-group never consumes the harvest of its garden alone: as soon as the manioc is ripe, formally invited visitors from allied longhouses or unexpected touring relatives join the hosts for a stay that may last until the manioc crop is exhausted.[15] Manioc gardens, therefore, form short-lived plantations that exist only through the labor of the living.

To sum up, both manioc and peach palm are domesticates used for ceremonial purposes. Both the cultivation of manioc gardens and the management of peach palm groves require little investment in time or energy. Manioc cultivation is extremely basic; it hardly transforms the forest cover. Peach palm groves are old dwelling sites managed with a view to encouraging the continuous growth of a certain palm species. Both plants produce food in sufficient quantity to allow for the renewal of social ties between allied longhouses. The amphitryonic function of peach palm groves is in many ways similar to that of manioc gardens. But whereas peach palm fruit celebrates the seasonal encounters of endogamous regional house-groups, manioc is used to forge new political alliances.

I would like to propose that this difference in use is related to the fact that manioc and peach palm grow at different rates. Manioc, like all garden crops, is a fast-growing, short-lived crop unfit for daily consumption. Peach palm, like most tree fruit, comes from a slow-growing plant whose bounty makes the forest into a giving environment.[16] The sweet varieties of manioc found in Huaorani territory grow so fast that the roots can be dug out as early as five months after planting. Full of vital energy, manioc fails to reproduce in situ. Never planted twice in the same place, it migrates throughout the forest, at the mercy of human alliances. Peach palm groves, on the other hand, grow very slowly and continue to give fruit in the same place year after year as long as house-groups care for them. Manioc, a "migratory" fast-growing plant, is particularly fitted for the organization of "diplomatic" feasts. When a house-group wants to renew its alliance with an enemy group, it uses manioc to prepare the feast drinks. Given that manioc is much more productive than the peach palm, and that it can be cultivated at any time of the year, almost anywhere, it allows for the organization of large feasts to which *huarani* guests can be invited. Whereas manioc is the ideal plant for feasting with the "enemy," the peach palm, the slow-growing legacy from past generations, gives the perfect fruit for celebrating *entre nous*. The two plants, with their contrastive practical and symbolic qualities, enable the formation, or the renewal, of very different types of alliances.

The Management Strategies of Marginal Tribes: A Form of Cultural Loss?

So far, I have tried to show the weaknesses of the devolution thesis: its overemphasis of the significance of species domestication, and its treatment of "swidden horticulture" as homogeneous and unproblematic. Huaorani ethnography reminds us that the food quest is embedded in sociocultural processes, and that the "regressed agriculturalists" of Amazonia, in giving up more intensive forms of horticulture, have exercised a *political choice*. I would like to suggest in this section that the

Huaorani's use of domesticates for feasting—rather than for daily consumption—
might be of widespread occurrence in Amazonia. I would like to add that the failure
to apprehend this phenomenon correctly has led to the misinterpretation of the cultural differences between "foragers" and "cultivators."

Throughout Amazonia, gardening and hunting appear to be culturally more valued than gathering. Because it is generally true that gardening and hunting activities give rise to more elaborate representations than do gathering activities, they are
considered richer in symbols and meanings. This has led Amazonian ethnographers
to underplay the economic, social, and cultural importance of foraging and trekking. We know, however, that a number of Amazonian groups subsist primarily on
hunting and gathering, with agriculture providing no more than a bonus. This is
true, for example, of the Shavante of central Brazil (Maybury-Lewis 1979:303). An
even larger number of groups share their existence between a more sedentary period, dedicated to horticulture and ritual celebrations, and a more mobile one, spent
in foraging activities. A recent account of such a dual way of life is offered by Eduardo Viveiros de Castro (1992:92), who observes that, among the Araweté, "one
lives in a village because of maize. . . . Maize concentrates, it is practically the only
force that does." The Araweté live most of the year dispersed throughout the forest,
coming together in order to cultivate maize gardens[17] and produce ceremonial
drinks. Viveiros de Castro (1992:47) adds that the Araweté see maize consumption
as a sign of civilization: they would feel like pure savages—like some of their enemies—if they had to spend their whole lives solely in nomadic hunting.

This points to another weakness of Amazonian ethnography, which has tended to
neglect the importance of the mode of subsistence in defining group identity. Beside the type of alternation described above—between a more sedentary life organized around gardening and ritual activities, and a more nomadic one led through
foraging activities—we also find groups of markedly sedentary gardeners living
next to nomadic foragers. Conscious of their cultural distinctiveness, these two
types of people stand in opposition to each other. The "cultivators" find themselves
culturally superior to the "foragers," who scorn the former's ignorance of the forest
as well as their lack of freedom. The cultivators acknowledge the hunting skills of
the foragers, and the foragers recognize that the cultivators have a greater knowledge of domestic plants. Balée (1993:249), for instance, hints at the cultural importance of the mode of subsistence, when he mentions that both the Ka'apor and the
Guajá are very conscious of the differences in their lifestyles, and that they distinguish each other "on the basis of their radically different means of associating with
plants."

Such cultural choices do not signal mere differences in technical knowledge.
They correspond to different identities, different historical experiences, and different forms of cultural knowledge sustaining different types of social organization.
Several of these "oppositional pairs" have been mentioned in the literature.[18] The
relationship between the two groups is generally one of hostility—a mixture of disdain and fear. The more mobile group is considered less developed by the sedentary
one (for example, the Barasana call the Maku their slaves). There are (reciprocal)

accusations of cannibalism. The ritual practices of the more sedentary group are more elaborate, but are not always considered more potent than those of the more mobile one. The more mobile group often relies in part on the food cultivated by the more sedentary one—food to which it gains access through begging, stealing, or trading forest products and labor. This cultural opposition is made even more complex by the fact that, since the conquest, the cultivators are viewed—and sometimes view themselves—as "civilized," while the foragers are dismissed as "savages."[19]

Generalizing from the Huaorani case, I can see two ways in which such a cultural differentiation is historically created. On the one hand, it leads to the separation of gardening and gathering as complementary, seasonal activities. As some groups choose to widen their exchange networks, they become increasingly dependent on horticulture and trade, and neglect trekking. Villages cease to be primarily ritual centers. Plant domestication becomes the necessary part of a new domestic economy.[20] Those who continue to trek and rely on wild food for their daily consumption develop forest-management practices that lead to a greater concentration of their favored resources within the areas they inhabit. By using Tommy Carlstein's (1982) concept of "path-allocations" and "time-space regions," Jeremy Gibson's (1986) concept of "affordances," and Tim Ingold's (1993) "dwelling" perspective, we can easily understand how specialization in either gardening or trekking leads to alternative ways of being in the world and of knowing it. Cultivators dedicate their time and attention to particular species, which they try to alter through, for example, hybridization. Foragers are more interested in forest ecology, and in intraspecies relationships.

I have shown that the Huaorani (and other indigenous tribes who once lived between the Napo and Tigre Rivers) have, through their political choices and practical engagement with the forest, transformed it and, to some extent, anthropomorphized it. The forest as it is today is partly the result of this historical process of transformation. History, which records both the transformation of social relations and the transformation of nature by humans, also transforms knowledge and representations (Godelier 1973:292). Although my analysis of fallow classification is still in progress and far from conclusive, I would like to mention, for instance, that the term used for secondary forests, *ahuene*, is also used to refer to a ceremonial host (Rival 1993). In fact, the term seems appropriate for naming any source of abundance, be it "the owner of the feast," a generous leader, an oil engineer who benevolently gives away manufactured goods, or a secondary forest with its greater concentration of fruit trees. This is to me a clear indication of a hunting-and-gathering view of the environment. Nurit Bird-David (1992:39) has shown that the sharing economy of hunters and gatherers (which she calls "demand-sharing," the fact of giving without expecting an equivalent return) derives from their particular view of the environment as a sharing parent that gives unilaterally and provides for the needs of its human children. Of course, the Huaorani's anthropomorphization of peach palm groves is not metaphorical in the same sense: the plants they manage in the wild, and on which they depend heavily—such as the peach palm—do result from the activities of previous generations. Or, to put it differently, there does not seem to be in this context a metaphorical projection of society upon nature. It is the link between successive human generations that makes the peach palm grove a *gift* from the dead, an *inherited* heirloom. And it is through metaphorical extension that all those who provide abundant food and consumer goods are seen as generous ancestors.

In conclusion, underplaying the importance of cultural differences between cultivators and foragers amounts to underplaying the fact that both the experience and the representation of the world are historically constituted. Whatever their cultural past, hinterland marginal groups like the Huaorani have adopted a basically foraging way of life for many generations and have gained a unique knowledge of their forest environment. They have developed marked patterns of social relationships and cultural values, and a unique system of resource management that drastically diminishes their dependence on cultivated crops. Given that such developments represent the constitution of a different body of knowledge and an alternative way of life, the Huaorani are definitely *not* devolved agriculturists suffering from cultural loss.

The Place of Culture in the Historical Ecology Paradigm

I have tried to show that the Huaorani traditional system of social alliances, which leads to great insularity, corresponds to a particular mode of subsistence and use of the forest. Horticulture is rudimental and peripheral to the main subsistence activities. Domesticated plants are mainly used for feasting. A fast-growing and short-lived domesticate such as manioc is typically used to feast with the enemy. A slow-growing and long-living domesticated fruit tree such as the peach palm—which is encouraged, through a number of direct or indirect management practices, to continuously reproduce in the same forest patches—consolidates endogamous ties. The social practices associated with consumption, and the cultural representations arising from these practices, correspond to a view of the environment that does not discriminate between what is wild, tame, or domesticated, but, rather, between what grows fast and what grows slowly.

Historical ecology, with its stress on social history and biocultural dynamics, provides the proper methodology for studying empirically the dynamics between old fallows and high forest—an essential part of understanding the gradual transformation of natural environments into landscapes (Balée, chapter 1). It also allows for the consideration of the social relations and the cultural representations that inform material practices, particularly in relation to plants. These practices too are historical products, products that are ill analyzed by the notions of "devolution" and "deculturation." Historical ecology does not presuppose any deterministic causal order between social, cultural, and environmental factors. Societies are not entirely subjected to environmental constraints. Humans are not detached observers busy representing the natural world in a completely arbitrary fashion (Ingold 1992). Societies progressively develop cultures within specific environments, a process that is eminently social and historical. As humans are at once socially organized and socially related to a number of living forms, this process is not limited to interactions between human groups. History, which is about the production and reproduction of collectivities, must therefore also be about the social relations that have developed between human collectivities and other living organisms. As such, history is inscribed in the environment (for example, in biocultural phenomena such as anthropogenic forests), in the knowledge of the environment, and in its symbolic representation.

Historical explanations must therefore include an account of the process by which social relationships have developed between human groups and living organisms, as well as of the process by which such relationships have been cognized, represented, and imagined. For this we need a concept of culture that emphasizes the practical engagement of people with the world. This concept must not only provide psychologically tenable explanations of how representations relate to practices (Boyer 1993; Bloch 1992), but also offer accurate interpretations of the interplay between sociopolitical forces and the meanings that social actors attribute to them (Comaroff and Comaroff 1992).

Notes

1. See Meggers (1995) for a new formulation of her original thesis. She now sees climatic instability as a primary environmental constraint limiting the development of Amazonian societies.

2. See Sponsel (1986) for a concise summary; Roosevelt (1989, 1993) for archaeological findings on the relationship between resource management, demography, and settlement patterns; and Whitehead (1993a) for a historical perspective on native Amazonian societies.

3. Some researchers have even argued that many soil features underlying these forests are also the outcome of human intervention (Hecht and Posey 1989).

4. However, even linguistic knowledge may be erased over time. Balée (1992) mentions, for instance, that the Guajá, who still have a cognate for maize, have lost the term for bitter manioc.

5. Men often spend all their lives in the houses of their wives, at least until the death of their wives' parents. They may then decide to create a new nexus by allying with their younger brothers after the death of *their* wives' parents. The privileged relationship between a boy and one of his sisters starts very early during childhood, and marriage often takes place sequentially between two pairs of brothers and sisters—that is, the first marriage of two cross-cousins is soon followed by the marriage of the husband's sister with the wife's brother.

6. Lévi-Strauss's early notice of the importance of the "developed exploitation of wild resources" is now given full recognition. Irvine (1989) has recently argued that, given the extent of indigenes' manipulations of wild, semidomesticated, or domesticated plants found in their environment, swidden agriculture is best seen as the first stage of a larger agroforestry complex in which farming strategies (the selection and breeding of domesticated species in order to enhance their yields) are not differentiable from manipulations, which can be very deliberate or almost unconscious. See also Posey (1983) who, following a similar

line of argument in his analysis of Kayapó resource management, notes that no clear-cut demarcation exists between field and forest. Rather, the more general reforestation process is reflected by a continuum between undisturbed and disturbed forest.

7. I do not know whether these forests are pristine from the archaeological and paleo-botanical perspectives. I have yet to collect field data on old fallows among the Yasuni communities. However, it is my guess that they would describe at least some areas of the Yasuni as *ahuene* rather than *omere*, given that they have lived in the area for at least five generations.

8. I have seen women plant part of the vine they had brought for stunning fish near the stream before going home with their catch. One threw the seeds of *cuñi* (a bush whose leaves are mashed and mixed with clay to produce a stunning poison) along the stream where she had fished. She had also thrown some of the same seeds in her manioc plantation the previous day.

9. I could not establish whether it was truly wild manioc, or some domesticated variety still growing in an old fallow.

10. Balick and Gershoff (1981) and Schultes (1989) discuss the high quality of this palm's seed oil.

11. There is no single term to translate the verb "to plant." To insert a manioc stalk into the ground is *gay gati huiyeng*, a term that describes the action of digging an oblique hole and thrusting something into it. To plant a banana shoot is *penenca huote mi*—literally, "to fit the banana shoot in the hole with will." To sow (seeds of domesticated plants such as corn or of semiwild plants) is *yamöi gaqui*, "to spread in the open."

12. Unidentified species, probably of the Bombacaceae family. When a patch of forest is cleared for a house or a garden site, this slender, smooth, easy-to-climb, and fast-growing tree (it grows much faster than the peach palm) is protected. Young saplings are replanted at the side of old peach palm trees, or germinated *Bactris gasipaes* are planted next to *macahuè* saplings.

13. Traditionally, and as just described, peach palm groves are not planted, but result from symbiotic relations perpetuated through consumption. Despite the fact that the current practice in sedentarized villages, as in many other Amazonian societies, is to plant peach palms in swiddens and backdoor yards, the old cultural meanings have not completely died out. When, for example, families leave a village community after a dispute with its leader (a rare and dangerous undertaking), they never abandon their gardens without felling all their peach palms,— precaution never taken for other crops (large banana and manioc plantations, cacao, coffee, and groves of citrus trees are left behind). This practice indicates that peach palms do still stand for social continuity. Moreover, planted peach palms, which are like any introduced food crop, are still distinguished from the ancestral groves to which people continue to go every year.

14. I have interpreted (Rival 1993) the ritual beating of manioc stalks with large balsa leaves (mainly of *Ochroma pyramidale*, *Cecropia sciadophylla*, or *Cecropia* spp.) before planting as corresponding to a transfer of energy between two categorically similar fast-growing species. People know that manioc grows well whether this ritual beating occurs or not. Ritual beating is therefore performed for ceremonial purposes, to ensure the symbolic transformation of manioc into balsa.

15. Today, pushed (under the pressures of missionaries and non-Huaorani teachers, and because of the constraints of sedentarism) to cultivate more intensively, people have maintained a system of host/guest relations by which only one family out of four cultivates a garden, sharing its production with its "guest" relatives. A new strategy has also been developed

by young educated men, who cope with the need for more garden products and the Huaorani women's resistance to the work of cultivation by marrying Quichua women.

16. Whenever the fruit is in season, it becomes the main staple. During the peach palm fruiting season, even hunting is discontinued.

17. With less than 20 percent of garden sites planted with crops other than maize, the Araweté gardens are essentially maize plantations (Viveiros de Castro 1992:41).

18. For instance, the Barasana (S. Hugh-Jones 1979; C. Hugh-Jones 1982) and the Maku (Silverwood-Cope 1972; Reid 1979); the Yekwana and the Yanomami (Heinen 1991); the Canelos Quichua (Whitten 1985) and the Huaorani (Rival 1992); the Shuar (Pellizzaro 1983) and the Achuar (Descola 1986); Venezuelan mestizo settlers and the Cuiva (Arcand 1981); Bolivian mestizo farmers and the Sirionó (Stearman 1987).

19. The opposition between civilized (*alli* in lowland Quichua) and savage (*sacha* in lowland Quichua) can operate within the same group to differentiate those who have accepted Christianity, a pacific cohabitation with the nation-state, and a more sedentary life, from those who still refuse contact and conversion, and maintain their distance from outsiders. This has happened among the Huaorani since the first Christian conversions by the Summer Institute of Linguistics in the late 1950s, and it is currently occurring between the Tagaeri and the rest of the Huaorani.

20. Even if horticulture is primarily a male occupation in some groups (Johnson 1983), in most groups it is part of a strongly gendered economy, with men specializing in hunting and trade (and, more generally, dealing with the outside), and women specializing in horticulture and domestication (Descola 1986; Rivière 1987). For an excellent review of the European impact on Amazonian regional trade networks and agricultural development, see Whitehead (1993a,b).

References

Arcand, Bernard. 1981. The Negritos and the Penan will never be Cuiva. *Folk* 23:37–43.

Balée, William. 1988a. Indigenous adaptation to Amazonian palm forests. *Principes* 32(2): 47–54.

———. 1988b. The Ka'apor Indian wars of lower Amazonia, ca. 1825–1928. Pp. 155–169 in R. R. Randolph, D. Schneider, and M. N. Diaz (eds.), *Dialectics and Gender: Anthropological Approaches*. Boulder, Colo.: Westview Press.

———. 1989. The culture of Amazonian forests. Pp. 1–21 in Posey and Balée (eds.), *Resource Management in Amazonia*.

———. 1992. People of the fallow: A historical ecology of foraging in lowland South America. Pp. 35–57 in K. Redford and C. Padoch (eds.), *Conservation of Neotropical Forests: Working from Traditional Resource Use*. New York: Columbia University Press.

———. 1993. Indigenous transformation of Amazonian forests: An example from Maranhão, Brazil. *L'Homme* 33(2–4): 231–254.

Balick, Michael and S. N. Gershoff. 1981. Nutritional evaluation of the *Jessenia bataua* palm: Source of high-quality protein and oil from tropical America. *Economic Botany* 35:261–271.

Bird-David, Nurit. 1992. Beyond "The original affluent society": A culturalist reformulation. *Current Anthropology* 33(1): 25–47.

Bloch, Maurice. 1992. What goes without saying: The conceptualization of Zafimaniry society. Pp. 127–146 in A. Kuper (ed.), *Conceptualizing Society*. London: Routledge.

Boyer, Pascal. 1993. Cognitive processes and cultural representations. Pp. 1–47 in P. Boyer (ed.), *Cognitive Aspects of Religious Symbolism*. Cambridge: Cambridge University Press.

Carlstein, Tommy. 1982. *Time Resources, Society, and Ecology*. Vol. 1 of *On the Capacity for Human Interaction in Space and Time*. London: Allen and Unwin.

Chase, A. K. 1989. Domestication and domiculture in northern Australia: A social perspective. Pp. 42–54 in Harris and Hillman (eds.), *Foraging and Farming*.

Clement, Charles. 1988. Domestication of the pejibaye (*Bactris gasipaes*): Past and present. *Advances in Economic Botany* 6:155–174.

———. 1992. Domesticated palms. *Principes* 36(2): 70–78.

Comaroff, Jean and John Comaroff. 1992. *Ethnography and the Historical Imagination*. Chicago: University of Chicago Press.

Descola, Philippe. 1986. *La nature domestique: Symbolisme et praxis dans l'écologie des Achuars*. Paris: Editions de la Maison des Sciences de l'Homme.

Eden, Michael. 1990. *Ecology and Land Management in Amazonia*. London: Belhaven Press.

Gibson, Jeremy. 1986. *The Ecological Approach to Visual Perception*. Hillsdale, N.J.: Erlbaum.

Godelier, Maurice. 1973. Mythe et histoire: Réflexions sur les fondements de la pensée sauvage. Pp. 271–302 in Maurice Godelier, *Horizon et trajets marxistes en anthropologie*, vol. 2. Paris: Maspéro.

Gross, Daniel. 1975. Protein capture and cultural development in the Amazon Basin. *American Anthropology* 77(3): 526–549.

Groube, Les. 1989. The taming of the rain forest: A model for late Pleistocene forest exploitation in New Guinea. Pp. 292–304 in Harris and Hillman (eds.), *Foraging and Farming*.

Harris, David R. and Gordon C. Hillman, eds. 1989. *Foraging and Farming: The Evolution of Plant Exploitation*. London: Unwin Hyman.

Harris, Marvin. 1984. Animal capture and Yanomamo warfare: Retrospect and new evidence. *Journal of Anthropological Research* 40(1): 183–201.

Hecht, Suzanne and D. Posey. 1989. Preliminary results on soil management techniques of the Kayapó Indians. Pp. 174–188 in Posey and Balée (eds.), *Resource Management in Amazonia*.

Heinen, Dieter. 1991. Lathrap's concept of "interfluvial zones" in the analysis of indigenous groups in the Venezuelan Amazon. *Antropológica* 76:61–92.

Hugh-Jones, Christine. 1982. *From the Milk River: Spatial and Temporal Processes in North-West Amazonia*. Cambridge: Cambridge University Press.

Hugh-Jones, Stephen. 1979. *The Palm and the Pleiades: Initiation and Cosmology in North-West Amazonia*. Cambridge: Cambridge University Press.

Ingold, Tim. 1992. Culture and the perception of the environment. Pp. 39–56 in E. Croll and D. Parkin (eds.), *Bush Base, Forest Farm: Culture, Environment, and Development*. London: Routledge.

———. 1993. Building, dwelling, living: How animals and people make themselves at home in the world. Paper presented at the IV Decennial Conference on The Uses of Knowledge: Global and Local Relations.

Irvine, Dominique. 1989. Succession management and resource distribution in an Amazonian rain forest. Pp. 223–237 in Posey and Balée (eds.), *Resource Management in Amazonia*.

Johnson, Allen. 1983. Machiguenga gardens. Pp. 29–65 in R. B. Hames and W. T. Vickers (eds.), *Adaptive Responses of Native Amazonians*. New York: Academic Press.

Lathrap, Donald. 1970. *The Upper Amazon*. London: Thames and Hudson.

———. 1973. The "hunting" economy of the tropical forest zone of South America: An attempt at historical perspective. Pp. 83–95 in D. R. Gross (ed.), *People and Cultures of Native South America*. New York: Doubleday.

Lévi-Strauss, Claude. 1950. The use of wild plants in tropical South America. Pp. 465–486 in J. Steward (ed.), *Handbook of South American Indians*, vol. 6, *Physical Anthropology, Linguistics, and Cultural Geography of South American Indians*. Washington, D.C.: Smithsonian Institution Press.

———. 1968. The concept of primitiveness. Pp. 349–352 in R. Lee and I. DeVore (eds.), *Man the Hunter*. Chicago: Aldine.

Maybury-Lewis, David. 1979. *Dialectical Societies: The Gê and Bororo of Central Brazil*. Oxford: Oxford University Press.

Meggers, Betty. 1971. *Amazonia: Man and Culture in a Counterfeit Paradise*. Chicago: Aldine-Altherton.

———. 1995. Judging the future by the past. Pp. 15–43 in L. Sponsel (ed.), *Indigenous Peoples and the Future of Amazonia: An Ecological Anthropology of an Endangered World*. Tucson: University of Arizona Press.

Pellizzaro, Siro. 1983. *Celebración de la vida y de la fecundidad*. Mundo Shuar 11. Quito: Abya-Yala.

Posey, Darrell. 1983. Indigenous ecological knowledge and development in the Amazon. Pp. 225–257 in E. Moran (ed.), *The Dilemma of Amazonian Development*. Boulder, Colo.: Westview Press.

———. 1985. Indigenous management of tropical forest ecosystems: The case of the Kayapó Indians of the Brazilian Amazon. *Agroforestry Systems* 3:139–158.

Posey, Darrell and William Balée (eds.). 1989. *Resource Management in Amazonia: Indigenous and Folk Strategies*. Advances in Economic Botany, vol. 7. Bronx: New York Botanical Garden.

Reid, Howard. 1979. Some aspects of movement, growth, and change among the Hupdu Maku Indians of Brazil. Ph.D. diss., Cambridge University.

Rival, Laura. 1992. Social transformations and the impact of formal schooling on the Huaorani of Amazonian Ecuador. Ph.D. diss., University of London.

———. 1993. The growth of family trees: Understanding Huaorani perceptions of the forest. *Man* 28(4): 635–652.

Rivière, Peter. 1987. Of women, men, and manioc. Pp. 178–201 in H. O. Skar and F. Salomon (eds.), *Natives and Neighbours in South America*. Gothenburg, Sweden: Gothenburg Museum Publications.

Roosevelt, Anna. 1989. Resource management in Amazonia before the conquest: Beyond ethnographic projection. Pp. 30–62 in Posey and Balée (eds.), *Resource Management in Amazonia*.

———. 1993. The rise and fall of the Amazon chiefdoms. *L'Homme* 33(2–4): 255–284.

Roosevelt, Anna C., R. A. Housley, M. Imazio, S. Maranca, and R. Johnson. 1991. Eighth-millennium pottery from a shell midden in the Brazilian Amazon. *Science* 254: 1621–1624.

Ross, Eric. 1978. Food taboos, diet, and hunting strategy: The adaptation to animals in Amazonian cultural ecology. *Current Anthropology* 19(1): 1–36.

Schultes, R. E. 1989. Seje: An oil-rich palm for domestication. *Elaeis* 1(2): 126–131.

Silverwood-Cope, Peter. 1972. A contribution to the ethnography of the Colombian Maku. Ph.D. diss., Cambridge University.

Sponsel, Leslie. 1986. Amazon ecology and adaptation. *Annual Review of Anthropology* 15:67–97.

Stearman, McLean Allyn. 1987. *No Longer Nomads: The Sirionó Revisited.* Lanham, Md.: Hamilton Press.

Viveiros de Castro, Eduardo. 1992. *From the Enemy's Point of View: Humanity and Divinity in an Amazonian Society.* Chicago: University of Chicago Press.

Whitehead, Neil. 1993a. Ethnic transformation and historical discontinuity in native Amazonia and Guyana, 1500–1900. *L'Homme* 33(2–4): 285–306.

———. 1993b. Recent research on the native history of Amazonia and Guyana. *L'Homme* 33(2–4): 495–506.

Whitten, Norman. 1985. *Sicuanga Runa: The Other Side of Development in Amazonian Ecuador.* Urbana: University of Illinois Press.

Yen, D. E. 1989. The domestication of environment. Pp. 55–78 in Harris and Hillman (eds.), *Foraging and Farming.*

CHAPTER 12

Independent Yet Interdependent "Isode": The Historical Ecology of Traditional Piaroa Settlement Pattern

STANFORD ZENT

The aboriginal human groups inhabiting the interfluvial regions of lowland South America are well known in the ethnographic literature for their small, diffuse populations and simple sociopolitical organizations. Most ecologically informed theories that have been proposed to explain interfluvial sociodemographic underdevelopment invoke some form of environmental determinist argument—to wit, population and social structures are severely constrained by a stringent natural resource base (Steward and Faron 1959; Lathrap 1968; Meggers 1971; Gross 1975; Roosevelt 1980). The corollary of this argument is that interfluvial populations are maintained at or near environmental carrying capacity, with no margin of underutilized resources to support further population growth and sociocultural development. Another common perception of the interfluvial peoples is that they have been relatively isolated from contact with outsiders and therefore survived the conquest of America by Europeans without major modifications of aboriginal culture (Carneiro 1964:9; Lathrap 1968:24; Holmberg 1969; Smole 1976; but see Martin 1969; Isaac 1977; Balée 1992). In this light, their perceived primitiveness is traced to a history of intercultural isolation. The environmental determinist and the cultural isolationist interpretations of interfluvial cultural stagnation tend to support each other. On one hand, the perception that the interfluvial environment is capable of supporting only rudimentary levels of population and culture tends to negate whatever influences, whether downgrading or uplifting, biological and cultural contacts with foreigners might have had. On the other hand, the belief that interfluvial groups escaped significant impact from the colonization process clears the way for assuming that the natural environment is the major determinant of the observed sparse populations and base organizational forms. But the history and ecology of the Piaroa Indians of Venezuela challenge these well-established understandings.

The Piaroa (also known as *Wõthihã*) are a native tropical forest people, number-ing approximately 11,500 individuals, who live predominantly in Amazonas and Bolivar States of Venezuela (OCEI 1993). Linguists classify the Piaroa language as a member of the Salivan family, thought to be one of the original language groups of the Middle Orinoco valley (Migliazza 1980). Throughout most of the historical period the bulk of the Piaroa population were settled in the interfluvial mountain forests rising between the Orinoco and Ventuari Rivers, and they had few direct contacts with non-Indian colonizers whose operations were based in the fluvial lowlands. In fact, the Piaroa had earned a reputation for being extremely aloof and fearful of whites or criollos (i.e., mestizos) and for stubbornly resisting cultural in-novation. About thirty years ago, however, a period of downriver migration began, and today most of the Piaroa reside along the more accessible margins of their for-mer territory, in fluvial or savanna habitats. Accompanying this radical geodemo-graphic shift has been an unprecedented culture change and integration into West-ern society (Mansutti 1988; Zent 1992:71–79).

Although to many observers it would appear that culture change is a fairly recent phenomenon among the Piaroa, I propose in this chapter that certain elements pre-sumed to be emblematic of traditional Piaroa culture—namely, the settlement pat-tern—are really derived from historical (i.e., postcontact) rather than aboriginal (i.e., precontact) roots, being the product of the burgeoning contact situation be-tween neo-European and Amerindian populations. The colonials introduced com-pletely new technologies, economic opportunities, power relationships, and disease organisms, thereby bringing about changes in the material and cultural environment to which native peoples were compelled to adapt or die. Although direct contacts between Piaroa and foreigners were kept to a minimum for a long time, I argue that the regional confrontation between Western and indigenous worlds nonetheless touched off a series of cultural and biological processes that exerted powerful se-lective pressures, however indirect, on the Piaroa. The key postcontact historical-ecological processes were regional population decline and increased productive ca-pacity due to the arrival of iron or steel technology. These processes led to adjustments in the relationships between local groups and natural resources, as well as between local groups and other social groups; the characteristic settlement mode reflects, and indeed arose as an adaptive response to, these changing relationships.

Settlement pattern may be defined as the spatial and temporal distribution of communities over the landscape. The component parts that make up a settlement pattern include the location, duration, size, and integration of communities (Hames 1983:394). Bruce Trigger (1968:74–75) has proposed that the determinants of set-tlement pattern fall into two broad classes: synchronic-functional and diachronic-developmental. Although Trigger treats the two types of determinants as logically and formally separate, the position expressed in the present analysis is that empiri-cally they are interactive. I begin my argument by outlining the basic features of the traditional Piaroa settlement pattern as may be culled from the previous ethnographic literature. I then review their contact history with neo-Europeans, focusing on the ecological consequences of this contact. The conclusions of the historical analysis are bolstered by ethnographic data on resource appropriation and settlement patterns from relatively unacculturated Piaroa living in the Upper Cuao River region.

Traditional Piaroa Settlement Pattern

The written historical record and the native oral history agree that the traditional home territory of the Piaroa is located in the vast upland forests on the right bank of the Orinoco River between the major tributaries, the Parguaza and the Ventuari (see figure 12.1). This topographically rugged region, heavily dissected by numerous rivers and streams, is referred to by geologists as the Cuao-Sipapo massif, while the Piaroa refer to it as *hut^{h}okiyu*, meaning "highland." The earliest unambiguous historical references to the Piaroa ethnic group and their locality date to the mid-eighteenth century. These sources place them in several river basins in or near the *hut^{h}okiyu* heartland: the Upper Cataniapo (Gilij 1965, 1:59; Caulín 1966:120; Ramos Pérez 1946:448; Aguirre Elorriaga 1941:91), the Upper Parguaza (Ramos Pérez 1946:448), the Upper Suapure (Ramos Pérez 1946:448), the Autana (Gilij 1965, 1:59), the Sipapo (Tavera-Acosta 1907:230), west of the Manapiare (i.e., the Upper Cuao and Marieta) (Gilij 1965, 1:130), and the Ventuari (Gilij 1965, 3:105–106; Ramos Pérez 1946:448). Meanwhile, the mountain motif is ubiquitous in Piaroa mythology and cosmology; most sacred sites are mountains firmly situated within *hut^{h}okiyu* (Zent 1993b).

The Piaroa habit of occupying the river headwaters and upriver valleys, above the major rapids, afforded them a natural defense against the incursions of colonials who were loath to stray far from the main waterways. Throughout the nineteenth and twentieth centuries, we find several accounts of small groups of Piaroa settlers descending to the surrounding downriver zones (Mansutti 1990), but the same reports leave little doubt that the bulk of the Piaroa population remained in the upriver region until the recent decades. Documentary exploration of the interior uplands by outsiders did not begin until well into the present century. Actual descriptions of traditional Piaroa settlement pattern date mostly from the middle decades of the twentieth century (1940–1970); therefore, the following ethnographic description should be understood as referring to this time period.

Ethnographic Description of the Settlement Pattern

Most writers emphasize the interfluvial forest orientation of traditional Piaroa settlements. Settlements were typically located in the headwaters or smaller tributary creeks; the larger, navigable rivers were scrupulously avoided (Chaffanjon 1986:183; Anduze 1963:157; Smole 1966:120; Boglar 1971:331). A major reason

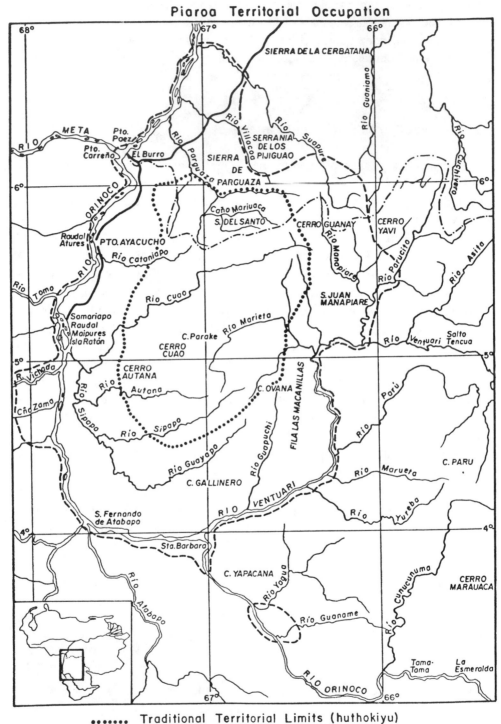

Fig. 12.1 *Piaroa Territorial Occupation. Dotted lines*: traditional territorial limits (*hutʰokiyu*). *Dashed lines*: contemporary territorial limits (1995).

given for the backwoods orientation is the fear of contact with the white man and his virulent diseases (Crevaux 1883:542–546; Deferrari 1945:9; Wavrin 1948:59; Tesch 1953:69; Zent in press). Smaller-scale factors entering into settlement location decisions include the presence of a flat space of well-drained land for the house site, close proximity to a perennial stream (< 100 m), the relative absence of vermin such as mosquitoes and nigua fleas, an abundance of arable land, and the avoidance of swampy land (Anduze 1963:157; Smole 1966:123; Boglar 1969:63; Overing and Kaplan 1988:335). Walter Dupuoy (1952:17) and José María Cruxent and Mauricio Kamen-Kaye (1950) note that some Piaroa subgroups have adopted a more riverine lifestyle—especially where seasonally flooded forests are extensive, such as in the Autana River basin. Enrique Deferrari (1945:9) and Joseph Grelier (1955:47) tell of groups inhabiting the forest-savanna fringe, apparently alluding to the vanguard of Piaroa communities who began migrating out of the deep forests ca. 1940–1950, seeking more stable contacts with the criollo world.

Another common feature of Piaroa settlement is the small size of local groups. Typical Piaroa villages were composed of two or three extended families, which may be further divided into five to ten nuclear family units (Tesch 1953:68; Grelier 1955:48; Monod 1970:6; Boglar 1971:332). The population of the average village is 25–30 inhabitants (Monod 1970:6; Boglar 1971:332), with a normal range of 16–50 (Overing 1975:349). Lajos Boglar (1969:64) observes that village membership is constantly fluctuating due to frequent dislocations to pursue subsistence, trading, or visiting interests. Joanna Overing (1975:29) reports that the village of a regional or neighborhood leader may swell beyond normal ranges during intervillage feasts, or according to the political evolution of his region of influence.

In physical terms, the typical village consisted of a single communal house (generically known as *isode*) in which all the resident families slept. According to Overing and M. R. Kaplan (1988:334; see also Overing 1975:30), the house size and style represent an "overt statement" of the sociopolitical status of the homeowner. Three basic architectural styles are noted: conical-pointed (*uc̆uhode*), conical-elliptical (*hare'bahode*), and rectangular (*tæ'bihode*). The first house design is the most valued and is usually built by politically prominent headmen; it takes the most time and labor to make, lasts longer, and typically is built on a larger scale than the other house types. The Piaroa *uc̆uhode* has been measured up to 20 m in diameter and 12 m in height (Grelier 1955:49). The structural design may vary, but the building materials are universally the same: wooden poles and beams for structural support, palm leaves for thatching, and vine for lashing together the poles and leaves.

Traditional houses are completely thatched over by palm leaves from tip to base without intervening walls, a construction style referred to as "roof-in-ground" (Dupouy 1952:20). The house is windowless and has but a single, short doorway, often covered over by a palm leaf—creating a very dark, closed dwelling space that is well protected from insect swarms as well as the midday heat (Deferrari 1945:9; Smole 1966:122; Monod 1970:6). The interior dwelling space is not partitioned by inner walls but is conceptually and behaviorally divided into two areas: (a) a central area, used as a communal work space, or where bachelors or visitors hang their hammocks; and (b) a peripheral area along the outer edge, where resident family hearths and sleeping quarters are allocated (Wavrin 1948:43, 48, 86; Grelier 1955:49–50; Smole 1966:122; Overing and Kaplan 1988:335). The village head-man occupies a privileged dwelling space, normally in the center periphery to the right or left of the entrance (Tesch 1953:69). Surrounding the house outside is a well-groomed, open clearing that serves as a commons area where children play, adults sometimes work, and ritual dances may be enacted. The house clearing is often located in the middle of a garden, and numerous footpaths lead outward to more distant gardens and hunting grounds. Another type of dwelling mentioned in the literature is the ceremonial men's house (*ruode*), where ceremonial masks and sacred musical instruments are guarded. The *ruode* is a small, beehive-like hut with a distinctive overhang on top; it is built only when the important *wærime/sāri̵* dancing and drinking ceremony, the maximum religious event of the Piaroa, is being held (Overing and Kaplan 1988:336).

The great dispersion of Piaroa settlements is another basic feature (Deferrari 1945:9; Cruxent 1947:12; Rottmayr 1949:17; Anduze 1963:157), and related to the considerable physical distances between settlement sites is the comparative political-economic autonomy of the local settlement group. The normal distance between houses in a particular river basin or district varies from several hours' to more than a full day's march away (Tesch 1953:70; Smole 1966:122; Monod 1970:6; Overing and Kaplan 1988:348). Local groups exercise independence in most economic, political, and social affairs under the authority of the household headman, the *isoderua* ("house owner"), who serves as both political and religious leader to his coresidents (Smole 1966:121; Monod 1970:6; Boglar 1971). Several villages within a contiguous area may form a temporary alliance or neighborhood, marked by intensive kinship and marriage ties, trading relationships, coparticipa-tion in religious ceremonies, and recognition of a common religious leader (*tha tʰīrua*, or simply *tʰīrua*), whose spiritual power is highly respected throughout the neighborhood and who sponsors the neighborhood intervillage rituals. Jean Monod (1970:6) estimates that the neighborhoods comprise an average of 10 houses within a 100 km^2 area, while Overing and Kaplan (1988:348) specify a mode of 6–7 and a range of 2–10 settlements per territory (neighborhood), the number of member groups fluctuating according to local population densities.

Sometime during the wet season, the neighborhood leader sponsors the *wærime/sāri̵* feast, which all the households of the neighborhood, and close rela-tives beyond, are invited to attend. This intercommunity event may last up to sev-eral months, and thus the wet season is theoretically a time of aggregation of the neighborhood population into one community (Overing and Kaplan 1988:349; cf. Tesch 1953:70; Wavrin 1948:34–35). The dry season, by contrast, corresponds to the season of settlement dispersion into smaller nuclear and extended family groups, camping and visiting expeditions being more common at this time of year (Wilbert 1966; Anduze 1974).

Relations beyond the neighborhood are pacific but still hostile, as most people fear the black magic of putatively enemy shamans outside the home neighborhood (Mansutti 1986). Indeed, the Piaroa themselves directly attribute the long distances between traditional settlements to the fallout of magical warfare between rival shamans (Zent in press). Although the neighborhoods no doubt indicate a supralocal level of social structure, both neighborhood configuration and the political position of the neighborhood leader are considered to be very dynamic and shifting (Zent 1992:392–394; Overing 1975:45–65). Lasting regional integration appears to be undermined by the relative autonomy of each separate community led by a headman who is a shaman-curer in his own right, and who competes with other such men for political influence in the neighborhood.

The short duration and attendant mobility of traditional settlements are widely mentioned in the literature. Johannes Wilbert (1966:49) and William Smole (1966) classify Piaroa settlement as "seminomadic," while others (Cruxent and Kamen-Kaye 1950:17; Grelier 1953:256) prefer the closely related term "semisedentary." The average settlement life is put at three to four years by Boglar (1969:63), while Overing (1975:58) sets the upper limit at ten years. The matter of distance between successive settlements is not well described in the literature, but in some cases it is as little as 300 m (Grelier 1953:258). Smole (1966:124) remarks that Piaroa prefer short-distance relocations for convenience's sake. M. de Wavrin (1948:286) and Alexander Mansutti (1988:6) induce a multiple-residence pattern in which a particular local group maintains more than one settlement at any one time and rotates stays between them throughout the year. The proclivity of the Piaroa to change occupation sites is reflected in the great variety of reasons motivating the abandonment of a settlement, ranging from the very mundane to the very sacred: crop failure, insect infestation, progressive distancing of mature gardens, movements of neighboring communities, discovery by outsiders or enemies, disputes among community members, death of the headman or other prominent member(s) of the community, arrival of disease epidemics, traumatic encounter with malevolent spirits, signs of tapir (a sacred animal) anywhere around the house or gardens, and fear of magical attack (Chaffanjon 1986:183; Wavrin 1948:286; Velez Boza and Baumgartner 1962:153–157; Anduze 1974:34; Mansutti 1988:6; Zent 1992:342, 427–428).

In addition to the mobility encompassed in settlement shifts, short-term expeditions into the forest to collect, hunt, or fish are commonplace. Temporary shelters are built during these trips, and thus abandoned camping lodges dot the Piaroa landscape (Wilbert 1966:49; Cruxent and Kaymen-Kaye 1950:17; Grelier 1959:53). Individuals or family groups may also go on trading or visiting trips to distant communities where relatives live; such "visits" may augur more long-term realignments of local group affiliation (cf. Grelier 1953:263; Boglar 1969:64). The frequent relocation of settlement sites, excursions of small groups to exploit distant resource areas, visitation or trading trips to other communities, and residence shifts of individuals or family groups add up to give Piaroa settlement the remarkably fluid appearance that has captivated different writers (Grelier 1953:259–260; Anduze 1963:157; Boglar 1969:64; Overing 1975:115–119).

Summarizing this section, the traditional Piaroa settlement pattern is distinguished by several key features: interfluvial orientation, small size, dispersed distribution, and frequent mobility. A brief look at Piaroa social organization reveals that these physico-spatial properties of the settlement pattern are expressed on a higher social plane in the form of certain kinship, marriage, and political relationships that interact with residence behavior.

Social Aspects of the Settlement Pattern

A cognatic kinship system prevails, which determines that the elementary kinship structures are idiosyncratic, egocentric kindreds. The *kindred*, formally codified in the term *čawaruæ*, "my kinsmen" (see also Overing 1975:69–87), is a highly unstable unit due to the individual point of reference and the continual makeover of a person's interpersonal relationships (Schusky 1972:71). Genealogical and residential relationships frequently overlap, and hence the coresidential group may be conceived as a localized "node" or "core" of multiple overlapping cognatic kindreds (Butt Colson 1983–84:103; Urbina 1983–84:194). The temporary and evolving nature of the kindred gives Piaroa society its characteristic fluid and amorphous look (Overing 1984:127). Residential groups coalesce and evanesce according to the shape and flow of individual kindreds within the context of the larger kindred nodes. Some degree of continuity of the settlement group is achieved, however, through the institutionalized preference for endogamous marriage, which has the effect of enhancing and perpetuating kin-relatedness within the coresidential unit of cognates (Overing 1983–84:332). The endogamous marriage principle provides the social rationale for the atomistic self-sufficiency of residence groups—but, in fact, the ideal of endogamy is rarely realized, due to the very small size of settlements and the intervention of political considerations. The nonendogamous marriage pact extends social networks beyond the small circle of the residence group and indeed serves as the key political instrument for building up the settlement and neighborhood groups, a main prerequisite for political ascendancy as the neighborhood leader.

Overing (1975) and Overing and Kaplan (1988), who performed the most in-depth research on traditional Piaroa social structure, depict Piaroa residence behavior as operating at two levels of organization: "ideal" and "real." The ideal level of residence consists of the local residence units (*isode*), which are conceptualized as atomistically structured and isolated from one another. In kinship terms, the *isode* embodies the close kindred of all the coresident members. At the real level of residence, by contrast, we find great variability, flexibility, fluidity, and contact among local residence groups: marriages between *isode* are in fact more common than *isode* endogamous marriages; people change residential affiliation at not-infrequent intervals; and *isode* undergo fission and fusion with apparent regularity. The relevant context here is the neighborhood of associated settlements, which transcends the level of the local group and was found by Overing and Kaplan to be more or-dered by politics than by kinship.

Overing (1975) contends that the ideal and real levels of residential behavior are not separate, but in fact are systematically interconnected by way of a complex kin-ship terminological system consisting of three parts. The first part is the term used to express the kindred (*čawaruæ*), which groups together everyone cognatically re-lated to ego. The second part is a Dravidian terminology that makes a clear distinc-tion between consanguine and affine.[1] The third part is a system of teknonymy that has the effect of reclassifying an affine as a kinsman, thus expressing the common relationship of all coresidents. Overing argues that it is the leeway and flexibility built into the terminological system, allowing a person to redefine his or her kinship universe, that permits the easy movement of people among local residence units while maintaining the ideological fiction of the close-knit, entirely autonomous local unit. Thus, according to this model, the dispersed, somewhat independent settlement and the neighborhood of somewhat interdependent, fluid and interacting settlements are complementary components of the same, unitary residential system.

A Historical-Ecological Hypothesis of Traditional Settlement Pattern

The high levels of atomism and fluidity displayed by traditional Piaroa settlement groups stand out as a key investigative problem in the social ethnography of this group. Previous theoretical discussions of the traditional social-spatial organization have concentrated on single-factor explanations, although choosing different foci: wild-resource distributions (Rottmayr 1949; Cruxent and Kamen-Kaye 1950; Tesch 1953; Wilbert 1966; Pruneti 1968), symbolic-logical gymnastics (Overing 1981, 1983–84, 1984), or historical political-economic forces (Grelier 1959; An-duze 1974; Tavera-Acosta 1907; Eden 1974; Mansutti 1988, 1994). In the follow-ing pages, I expound a historical-ecological interpretation of the traditional settle-ment pattern. This combined perspective permits a more integrative analytical treatment, taking into account micro and macro levels of settlement organization as well as the physico-spatial and socio-spatial properties. The central issue, as I see it, is to explain the seemingly contradictory tendencies of settlement dispersion (atom-ism) and mobility (fluidity). My argument consists of two interacting hypotheses:

1. The ecological effect of Western contact on the Piaroa, through depopulation and the introduction of iron technology, was to lessen population pressure on environmental resources.
2. Given this low population density, the subsistence system and defense con-siderations exert pressure toward settlement dispersion and independence, while other survival requirements promote intersettlement closeness and in-terdependence. The conflicting demands of the adaptive-ecological system account for the observed pattern of very dispersed yet fluid settlements.

Historical Process

Very little is known about the specific prehistory of the Piaroa, but we do have a number of useful data pertinent to a broad reconstruction of the regional Middle Orinoco cultural baseline. The archaeological record of settlement and subsistence patterns (Roosevelt 1980), ceramics (Rouse and Cruxent 1963; Zucchi, Tarble, and Vaz 1984), cemeteries (Humboldt and Bonpland 1876; Crevaux 1883; Chaffanjon 1986; Dickey 1932), and rock art (Scaramelli 1993), along with early ethnohistorical accounts (Gumilla 1963; Gilij 1965), suggests larger populations, more intensive food technologies (including subsistence specialization), more extensive trade networks, more complex and comprehensive polities, and higher craft and artistic development in the floodplain regions during pre- and early conquest times than survived into the modern era. Several writers (Zucchi, Tarble, and Vaz 1984; Biord Castillo 1985; Arvelo-Jiménez, Morales Méndez, and Biord Castillo 1989) hypothesize the former existence of a regional multiethnic society of diverse ethnolinguistic groups integrated by complex interdependent ties based on intermarriage, trade, and ideological sharing.

The European presence in the Middle Orinoco began with the fledgling efforts of Jesuit missionaries in the latter seventeenth century, but sustained contact between Europeans and native Orinoco peoples was not achieved until the 1730s with the founding of various missions between Cabruta downstream and the Atures rapids upstream. The ethnic groups most attracted and affected by the missions were the sedentary, agrarian-oriented peoples who occupied the accessible fluvial locations— e.g., the Saliva, Achagua, Maipure, Otomac, and Yaruro. The interfluvial-dwelling Piaroa were regarded as difficult targets for missionization and generally shunned direct contact with the missions. The Jesuit Felipe Salvador Gilij portrayed the Piaroa in the eighteenth century as "untamed and evasive" (1965, 2:154) and "coarse, lovers of the forest darkness, and having little affection for foreigners" (2:58). The Franciscan Ramón Bueno (1933:68) draws a similar character sketch, depicting them as "very inconstant, timid, and fugitive." Although the Piaroa kept their distance from the missions, events whose point of action was centered in the fluvial frontlands reverberated far away into the hinterland, deeply affecting the people who lived there. An appreciation of the impact of European contact on Piaroa culture therefore depends on a proper understanding of the relationships between regional and local (Piaroa) human systems, and between fluvial and interfluvial ecological contexts. The following narrative and analysis of Piaroa postcontact history focuses on the development and significance of two major types of change in the human biocultural environment since contact—demography and technology— which, I argue, exerted major influences on the cultural ecology of the Piaroa.

Demographic Impacts

The most striking and probably most far-reaching demographic development was a major depopulation throughout much of the Orinoco basin, although this impact was most rapid and complete in the riverfront districts. The exact rate and extent of native depopulation in the Middle Orinoco is impossible to gauge because documentation is nonexistent or inadequate for most of the historical period. However, many scholars agree that the losses were probably devastating; estimates of the toll of depopulation during the first one hundred years of contact range from 60 percent (Perera 1982:114) to 95 percent (Dobyns 1966). Assuming a strong correlation between levels of population loss and contact with foreigners, it is supposed that the eighteenth century was the most significant period of demographic collapse (cf. Whitehead 1988:21–41, 104). A rough measure of the degree of depopulation may be inferred from the number of extinctions of ethnic groups from first contact to present. Gilij named thirty linguistically distinct groups (including the Piaroa) inhabiting the Middle Orinoco region in the mid-eighteenth century (Biord Castillo 1985). Less than a hundred years later, in the Codazzi census of 1841, only ten of these groups were still present (Codazzi 1940). By 1982, viable populations of only five of these groups had survived (OCEI 1985).

Infectious disease imported by the European and African immigrants was undoubtedly the number one cause of aboriginal mortality in the Orinoco, as was the case throughout the Americas (Crosby 1972; Dobyns 1983).[2] Native population was more concentrated and settled in the more accessible fluvial regions, and these groups, who had been numerically and politically dominant, bore the brunt of early contact with foreign pathogens. The policy of the colonial missions was to gather the population into large settlements, appropriately named *reducciones* ("reductions"), where disease epidemics often broke out and spread. The Orinoco missions, José Gumilla (1963:478) writes, suffered "repeated contagions and epidemics," and in 1741 a "great wave of smallpox" swept from "nation to nation" and obliterated nearly the entire child cohort of the missions. In Gilij (1965, 2:75–76), we read that whooping cough arrived in the Orinoco with such abrupt ferocity that it "took many to meet the creator," and that the once-numerous Saliva nation was decimated by smallpox and scarlet fever (see also Román, in Aguirre Elorriaga 1941:36).

The high attrition of the reduced native population, lost through mortality or escape, engendered a constant need to recruit fresh converts. The so-called flying missionaries marched into the interior supported by a detail of Spanish soldiers, looking for heathen souls to herd back to the missions. The ongoing process of "reducing" new ranks of natives must have extended and prolonged the exposure of the Indian population to lethal foreign diseases.

The Piaroa successfully resisted most attempts at reduction in the eighteenth century, even though the Piaroa of the Cataniapo and Ventuari Rivers were persistent targets of Gilij's flying missionaries (Gilij 1965, 3:104–106; Del Rey Fajardo 1977:155–160). A small number of Piaroa were brought to live in a reduction called San Estanislao de Patura in 1751, but five years later it was defunct and all the "reduced" Piaroa had fled back to their former lands. Attempts to lure and keep Piaroa in other missions failed as well. Gilij relates his frustrating experience when trying to bring Piaroa into the fold: "With infinite efforts, having taken the Piaroa out of

the forests . . . they returned repeatedly, in the manner of indomitable wild animals"
(1965, 2:155; my translation). Beyond very brief contacts, then, the Piaroa maintained their distance from the missions, which probably explains why they escaped
the holocaust suffered by virtually all of the fluvial tribes of that time period.

The slave business, which reached a peak in the first half of the eighteenth century, also contributed to the depletion of the regional population—by exporting
people to the coasts, where many of them then succumbed to illness, and by forcing
vulnerable groups to resettle at the militarily more secure missions, exposing them
to the rampant disease vectors found there. Slave raiding was practiced by Carib Indians allied with the Dutch from downstream, and by Arawakan raiders who
worked for Portuguese brokers from upstream (Del Rey Fajardo 1977:140–145).
Historical and ethnographic evidence point to the Piaroa's being a favorite prey
population for slave raiders. The close association, sometimes substitution, of the
tribal name Piaroa with *Mako*, and the ethnolinguistic significance of the term
Mako as "slave," are suggestive of a status of slave population (Overing and Kaplan 1988:321; cf. Rivero 1956:47; Acosta Saignes 1954:71–79). The most vivid
and convincing evidence of Piaroa exposure to organized kidnappers, however,
comes from their own oral tradition. The trauma of slave predation is deeply etched
in the Piaroa collective memory in the form of tales about the depredations and
eventual defeat of the cannibalistic *kæriminæ*. The *kæriminæ* are described as a tall,
white, and bearded race of men who wore Western garb and carried firearms, and
who captured and allegedly ate the Piaroa's ancestors. The Piaroa commonly associate white people with the hated *kæriminæ* (Monod 1972), and several informants
acknowledged that the generalized fear of outsiders stems from this bloody history.
The name of *kæriminæ* appears to be a derivation of "Carib," but regardless of who
the *kæriminæ* really were, their described physical, cultural, and behavioral traits
suggest rather strongly that they were slave hunters in league with, if not in fact
themselves, Europeans.

The colonial slave business declined by the latter eighteenth century as the Spanish consolidated political-military control over most of the Orinoco River. Religious reduction of the native population fell off as well with the expulsion of the Jesuits in 1767, who were followed by less active Capuchin and Franciscan mission
operations. With the onset of the republican period (1830), all missionaries were
banished from the Orinoco territories, beginning a period of ecclesiastic exile that
lasted for more than a hundred years, and colonization activity in general lapsed.
By this time, many of the original riverine groups had been extinguished, while
small contingents of Piaroa occasionally voyaged to and even took up residence in
the criollo river towns of Atures and Maipures (Codazzi 1940:23–24, 46–47;
Michelena y Rojas 1989:294; Ayres 1967:325–326). There are also signs of north-
ward expansion of the Piaroa about this time, down the Parguaza, Villacoa, and
Suapure Rivers, facilitated by coresidence and admixture with the savanna-riverine
dwelling Mapoyo who were in serious decline by this time (Bueno 1933:70; Codazzi 1940:49; Michelena y Rojas 1989:273).

An extract-export economy geared up in the latter nineteenth century, dominated by the commercial exploitation of the natural resins or fibers of a few wild jungle plants: rubber (*Hevea brasilensis* and *H. guianensis*), balatá (*Manilkara bidentata*), chicle (*Couma* spp.), *chiquichiqui* (*Leopoldina piassaba*), and tonka bean (*Coumarouna polyphylla*). In order to meet the rising demand for Indian labor, covert forms of forced labor such as *avance* or debt peonage were instituted. The Piaroa appear to have been only marginal participants in the Amerindian labor market that fueled the extraction industry, largely because most of the commercial plant species were absent from or sparse in their traditional upland habitat. However, we do get some reports of the Piaroa being employed in this business along the southern rim of their territory, where ample stands of chicle, rubber, and *balatá* apparently are found (Chaffanjon 1986:190, 213–214; Crevaux 1883:542–546; Koch-Grünberg 1979, 1:261, 370; Eden 1974:29; Iribertegui 1987). Contemporary Piaroa universally recall the boom years as a time of extreme danger and hardship, when their ancestors were recruited by coercive means to work as jungle extractors. Oral testimony also holds that many of their people responded to such mistreatment by retreating deeper into the hills and intensifying settlement mobility in order to escape detection by the violent rubber gangs. As if to confirm etically this emic account, Miguel Perera (1982:138) describes an eastward demographic shift of the Piaroa during the boom years, away from the collecting and transit lanes of the Lower Sipapo and Orinoco and into the uplands of the Upper Sipapo (see also Eden 1974:44–45). Jean Chaffanjon (1987:222) observed that much of the Ventuari was nearly depopulated of its native inhabitants at the time of his trip (1886–1887) due to the depradations caused by the "civilized" extractors and traders. Similarly, Theodor Koch-Grünberg (1979, 1:352, 367–368) remarked that the Piaroa, among other tribes of the Ventuari and Manapiari Rivers, lived far inland from the main shores, avoiding contact with the whites. This avoidance was maintained not only because of the violence with which the chicle and rubber traders commanded Indian labor, but also because of fear of epidemic diseases carried by the transient labor force (cf. Tavera-Acosta 1927:32; Koch-Grünberg 1979, 1:371; Iribertegui 1987:214, 309). In this sense, the main effect of the extractive business on the Piaroa population appears to have been to reinforce or intensify the interfluvial settlement orientation.

The extractive economy entered a bust period after 1913, marked by pervasive economic regression and population decline throughout the region. As the criollo population shrank in the downriver areas, the Indian population gradually began to repopulate the riverbanks it had abandoned earlier to escape servitude in the extraction business (Hanson 1933:584). The violence of the rubber boom now lifted, conditions were propitious for interior groups like the Piaroa to begin moving downriver. Several factors appear to have been key in setting the stage for the great wave of downriver migration of the 1960s and 1970s. In 1924, the territorial capital was moved to the northwestern town of Puerto Ayacucho, and this part of the territory, relatively close to Piaroa land, was subsequently transformed from a backwater to a hub of regional colonization and development. Religious organizations were allowed to return; the Catholic Salesian Order, arriving in 1937, and the protestant New Tribes Mission, starting in 1947, established increasingly vigorous missionizing programs among the native Amazonian peoples. Following World War II, the Venezuelan government began an active policy of attracting native groups into the fluvial borderlands through grants of social services and economic development programs. An important component of this Indian colonization strategy has been the provision of modern medical services (e.g., hospitals and clinics, community-based medical dispensaries, resident native nurses), which has had a fundamental impact on Piaroa geodemographic dynamics. Whereas previously the Piaroa found it biomedically adaptive to retire from the fronts of interethnic contact, away from deadly diseases, the development of a modern medical infrastructure in the down-river zones and the advancing penetration of epidemic diseases upriver at some point reversed this advantage. The consequent situation of biomedical inequality was an important factor leading to downriver migration and created a gap in demographic performance between peripheral and interior groups that is still evident today (Zent 1993a).

The demographic history of the Piaroa over the past three centuries emphasizes several important population processes: regional population reduction, decline of aboriginal ethnic diversity, greater threats to population in the fluvial zones until the modern period, oscillating intensities of colonization pressures emanating from the fluvial zones, reinforcement of interfluvial orientation during high-risk periods, and relaxation of interfluvial settlement during lower-risk periods. A useful framework for discussing the relevant population processes going on here is provided by the source and sink habitat/population concepts. A *source habitat* is an area where local reproductive success is greater than local mortality, whereas a *sink habitat* is one in which within-habitat mortality exceeds reproduction. The two habitats may be linked for purposes of population regulation in the sense that population in the sink is maintained by immigration from the source (Pulliam 1988). These concepts have been applied to aboriginal Amazonia by Brian Ferguson (1989), who proposes that in pre-Columbian times the resource-rich floodplains served as demographic pumps (i.e., sources), sloughing off excess population to the less-endowed interfluvial regions, the sinks—where the incoming population suffered losses and was eventually leveled in line with lower resource capacities. My idea is that source and sink habitats of the Middle Orinoco region were reversed following contact with European populations: the floodplain, previously a source, became a sink where population was drained by the predation of foreign disease organisms or slave merchants; the interfluve, the former sink, became a source of replacement population for the sinking floodplain. The interfluvial-based Piaroa, saved from the fate of genetic and cultural extinction suffered by the fluvial groups, moved cautiously into the territorial voids left behind by the defunct floodplain peoples. A likely result of this population-spread was overall lowered population densities; hence, population did not build up within the interfluve. It would probably be more accurate to say that the Piaroa were pulled into, rather than pushed out toward, the fluvial regions.

To understand why, it is necessary to turn to a consideration of the impact of Western trade goods on the Piaroa.

Technological Impacts

Interethnic exchange was a prominent feature of the aboriginal Middle Orinoco cultural landscape. The food production systems of some groups depended on mutualistic food-food or food-labor exchanges between groups occupying different ecological niches (Morey and Morey 1973; Morey 1975:232; Henley 1983:235). For example, the Atures, who specialized in fishing, traded dried fish for agricultural produce with neighboring peoples (Gumilla 1963:228–229). According to Piaroa oral tradition, the inland Piaroa were trade partners of the riverine Atures (see also Mansutti 1990:15). Mansutti (1990:16) alludes to a subregional exchange circuit involving the Piaroa, Kiruva, and Maipure groups, who were arranged along a biotopic gradient, with Piaroa in the headwater forests, Kiruva at the savanna-forest ecotones, and Maipures in the floodplains. Beyond food and labor exchanges, a far-flung trade network of raw materials and manufactured items flourished at the time of European contact. The hallmark institutions of this regional market were the *quiripa* strings (necklaces made out of freshwater shell disk), used as a standard unit of exchange; the mirray ceremony, involving feasting and ritual discourse preceding important transactions; and the interethnic trading festivals (Morey and Morey 1975).

The cultural and demographic upheavals wrought by European colonization disrupted the structure and function of the aboriginal exchange system. Subsistence specialization and exchange disappear as population declines. The trade once controlled by floodplain groups was increasingly taken over by the missionaries, European entrepreneurs, or Indian middlemen. Western goods, mainly cloth and metal, became dominant trade items and were quickly incorporated into native exchange systems. Meanwhile, the exchange of native goods was attenuated; it was conserved mostly among interfluvial groups (Arvelo-Jiménez, Morales Méndez, and Biord Castillo 1989).

Iron was an early imported commodity of the colonial period that found its way into native trade circuits. The strong appeal of iron implements was noticed by the missionaries, and they intentionally dangled iron and other Western trade goods as a lure to entice Indians to come to the reductions (Gilij 1965, 3:60). The iron trade reached beyond the missions, however, for missionized natives commonly maintained trading partners in the interior and passed along the exotic Western goods (Gilij 1965, 2:267; Bueno 1933:62). Evidence of the early extensive distribution of iron tools is found in Gilij, who comments that, "except perhaps for the most remote savages, everyone utilizes the axes and cutting tools that the missionaries have given to them or that they have obtained by trade with the Europeans" (1965, 2:275; my translation). One of the main ways that natives gained access to European iron was by provisioning the missions and associated military garrisons with foodstuffs in exchange for iron, and some missions even had iron foundries expressly for this purpose (Alvarado 1966:244–246). Besides food, the Indians sold a number of forest and river products that were prominent in the colonial economy, including tortoise oil, cinnamon, annatto dye, calaba oil, copaiba balsam, and tropical birds (Gilij 1965, 2:266; Aguirre Elorriaga 1941:21).

Although the date when the Piaroa were first introduced to iron is unknown, their relative proximity to the Middle Orinoco missions and their active role in the regional trade system suggest that they had access to Western goods from at least Jesuit times. Gilij (1965, 2:284) relates that the Piaroa were notable producers and traders of curare hunting poison, peraman wax (from *Symphonia globulifera*), and chica dye (from *Arrabidaea chica*). Most important from a commercial standpoint, they were widely regarded as the source of the most potent curare, a distinction that has lasted until recent times (Grelier 1959).

The first contacts with Europeans and the precise introduction of iron are not well preserved in Piaroa oral history. However, some informants mentioned that the first knives and axes were obtained from other native groups who lived near the Orinoco shore and had regular contact with whites. By these accounts, ancestral Piaroa traded body oils, torch resin, wax, dogs, braided string, and manioc meal to floodplain peoples in exchange for knives, machetes, axes, and fishhooks. The antiquity of iron tools is reflected in Piaroa beliefs about stone tools. All informants questioned about it responded that their ancestors did not make stone tools but that they had used them in the remote past. Notions of the origin of stone tools vary somewhat, but several informants expressed the belief that they were made by the forest animals in a mythic time when animals lived like men and made gardens and tools. The ancestors gathered the stone tools left in the ground and employed them to fell their gardens (Zent field notes).

The first accounts of direct economic contact between Piaroa and Europeans do not appear until the beginning of the nineteenth century:

> The Piaroa . . . often come voluntarily to the reductions, promising to live peacefully, and once they obtain steel tools, they run away. (Bueno 1933:68; my translation)

> The independent Macos [Piaroa] . . . came and established themselves some time ago in the mission. They asked eagerly for knives, fishing hooks, and those coloured glass-beads. Having obtained what they sought, they returned to the woods, weary of the regulations of the mission. (Humboldt and Bonpland 1876:241)

Although the Piaroa aversion to mission life had changed little, these passages at least suggest that they had cultivated an intermittent relationship with the missions as a way to obtain the Western hardwares. Looking over the historical records of this period, one can detect a hesitant yet gradual expansion of economic contacts between Piaroa and neo-Europeans over the course of the nineteenth century. Agustin

Codazzi (1940:23–24, 46–47), who traversed the region in 1841, portrays the Macos and Piaroas as "independent" yet "used to trafficking with the criollo towns of Atures and Maipures." Francisco Michelena y Rojas (1989:294) describes the Piaroa of the Atures region ca. 1856–1858 as available and willing to work and trade with the colonists, although they refused to live in the town due to the threat of infectious disease and their distaste for the forced labor and authority imposed by the Europeans. Jules Nicolas Crevaux (1883:542–546) and Jean Chaffanjon (1986: 183–190) report trade encounters with Piaroa living very close to the Orinoco shore in the vicinity of the Mataveni River in the 1880s. As these reports make clear, while the Piaroa were expanding their trade contacts, for the most part they limited these contacts to brief encounters and kept their communities at considerable distances from the colonizing population. A main motivation for maintaining such distance seems to have been their fear of dangerous diseases carried by the (neo-) Europeans.

With the growth of an extraction business during the period from 1880 to 1945, the acquisition of Western trade goods by the Piaroa was further expanded, notwithstanding their marginal position noted earlier. The principal Piaroa commercial activity during this time was extraction of the raw forest materials, although some Piaroa may have sold manioc foodstuffs to the manioc vendors (*mañoqueros*) who in turn supplied the toasted flour (*mañoco*) to the work camps (Iribertegui 1987). Most affected were the southern Piaroa (Upper Orinoco and Sipapo), who engaged in rubber, *balatá*, and chicle collection, and those of the extreme northern fringe (Lower Parguaza and Suapure), where tonka bean grows naturally (Deferrari 1945:14; Wavrin 1948:348).

After the Second World War, the economic relationship between Piaroa and criollo populations entered a new phase. The main form of interethnic economic transaction shifted from the collection of natural forest products to cash cropping and wage labor, coinciding with the settlement migration from upriver to downriver. The first Piaroa to establish settlements within the contact zones became important economic middlemen between the criollo businessmen and their hinterland relatives (Grelier 1953). During the 1960s and 1970s, as downriver migration surged from a trickle to a flood, the regional economic importance of the Piaroa grew considerably. By the early 1970s, the Piaroa were responsible for producing 40 percent of the *mañoco* sold in the Amazonas territory (Anduze 1974:53).

Virtually all ethnographic accounts of the Piaroa in the twentieth century mention their ownership of or access to steel tools (Dalton 1912:227; Deferrari 1945:11, 14; Cruxent 1947:13; Cruxent and Kamen-Kaye 1950:20–21; Tesch 1953:68; Grelier 1959:30, 62; Wilbert 1966:50–51, 56; Smole 1966:126; Anduze 1974:62). By the time the first intensive anthropological studies of them were realized in 1968–1969, Western trade goods were extremely common, including steel tools, cooking pots and utensils, tin *budares* (griddles), and even shotguns (Boglar 1969:63; Monod 1970:5; Anduze 1974:61; Overing and Kaplan 1988:336, 339).

What was the impact of iron on the Piaroa economy? The lack of detailed descriptions of pre-iron Piaroa economy prevents a definite answer. The following reconstruction is therefore necessarily hypothetical, but it is based on lines of evidence that include extrapolation from similar ethnographic situations, variations in Piaroa cultural patterns described by ethnographers in the present century, and Piaroa oral history. I propose that iron technology stimulated a shift toward greater dependence on agricultural production (cf. Denevan 1992). There are three reasons for believing so:

(1) Studies comparing the energetic efficiency of iron versus stone verify the considerable advantage of the former, especially in the vital task of chopping down trees for swidden gardens (Carneiro 1979). The greater work capacity of the new technology would have enhanced the attractiveness of agriculture as a resource option. Greater dependence on agriculture, in turn, would have induced greater sedentism in the settlement pattern. Marcus Colchester (1984) reconstructs exactly this constellation of cultural changes—possession of iron technology, greater dependence on agriculture, more sedentary settlement—among the Yanoama, the southern neighbors of the Piaroa.

(2) Ethnographic descriptions of the Piaroa in the present century mention intratribal variation in resource, settlement, and contact patterns: subgroups inhabiting the interior mountains were observed to be more nomadic, more reliant on a foraging economy, and more isolated from the interethnic contact zones; meanwhile, other subgroups displayed more stable settlements, higher devotion to agricultural pursuits, and greater contact with the surrounding criollo society (Deferarri 1945:11, 14; Baumgartner 1954:116; Grelier 1953:259; Grelier 1959:105–110; Velez Boza and Baumgartner 1962:183, 196–197; Wilbert 1966:49–50). Thus the ethnographic record confirms the positive correlation of sedentism, agricultural dependence, and access to Western trade goods (iron in particular).

(3) Piaroa oral tradition corroborates the proposed scenario. The tree-felling technique employed by *tæbotɨhaminæ* ("our ancestors") under a regime of stone axes entailed girdling the bark, waiting for the tree to dry out and the leaves to fall, and burning the base with bonfires. The entire process took three to four months to accomplish, considerably longer than it does today. The gardens were generally smaller and more scattered in those days but contained the same basic inventory of cultivated plants; natural tree-fall gaps were even used to plant fast-growing plants like maize. Regarding former subgroups who followed a more nomadic hunting-gathering lifestyle, there is an interesting recollection about the *hurækætɨ-kʷohurime* subgroup. These people inhabited exclusively headwater regions of the Upper Cuao, Upper Parguaza, and Upper Marieta Rivers. They were frequent trekkers who lived in caves and under small temporary shelters. Some of them had no gardens, while others planted only maize in small clearings, and they were known to pilfer crops from the gardens of more settled people. They obtained Western trade goods from other hinterland dwellers, indicating that they probably occupied the figurative end of the iron pipeline. They wore bark loincloths (instead of the more typical ones made of cultivated cotton) and seed necklaces (rather than the imported glass beads), and they did not prepare the ritually important fermented manioc beverages (*sãrĩ*). The *hurækætɨ-kʷohurime* were eventually extin-

guished by tuberculosis and other diseases, or intermarried with other Piaroa (Zent field notes).

Ecological Significance of Demographic and Technological Impacts

What is the analytical interpretation of the demographic and technological impacts recounted here, and what significance do they hold for the traditional settlement pattern? I argue that the ecological effect of regional depopulation was the lowering of population density, and hence of population pressure on the existing resource base. The introduction of iron technology reinforced this pattern by siphoning off surviving interfluvial peoples, who were motivated to approach the contact zones where steel was available. Thus the interfluvial population was subject to geographical spreading following the initial shock of population decimation in the fluvial zones. However, the contact zones also presented higher mortality risks—a demographic sink phenomenon—until recently, holding the emigration of interfluvial peoples somewhat in check. In the modern period the twin forces of demography and technology are mutually supportive, and large-scale downriver migration has been the result.

At the same time, the economic shift to greater dependence on agricultural production raised environmental carrying capacity by increasing food energy production per unit land area. Smaller quantities of wild forest resources per area occupied were then needed, since their main dietary significance changed from caloric to macronutrient (protein, fat) and micronutrient (vitamins, trace elements) support. Although it is impossible to estimate in numeric terms the extent of the transition from more to less hunting-gathering and from less to more agriculture, the important point is simply to show that a shift is likely to have occurred, leading to an increase in overall environmental carrying capacity. The ecological consequence of such increase was to lessen the pressure of people on land, since the increased capacity for population growth was neutralized by the economic attraction (yet demographic sink phenomenon) operating in the contact areas.

Subsistence and Settlement Ecology in the Upper Cuao River

Given the proposed demographic and technoeconomic changes, what effect would this have had on Piaroa resource and settlement patterns? Unfortunately, the answer to this question is obscured by about three hundred years of largely unrecorded history of cultural responses to the presence of Western colonizers. Because very little information is available in the historical record, and also because the more recent ethnographic descriptions of the cultural-ecological system are too sketchy, it is necessary to base our interpretations on observations of the behavior of a contemporary

Piaroa subgroup that has remained in the traditional habitat and that most retains the traditional culture.

Despite the vast social and economic changes that have affected the bulk of the Piaroa population in recent decades (cf. Zent 1992:71–79), traditional settlement and subsistence practices still survive in a few remote areas within the hut^hokiju heartland. One of these areas is the Upper Cuao River, where I did fieldwork in 1985–1987. Here one can still find the characteristic small, dispersed, and mobile settlements described earlier, a subsistence-oriented economy, barter trade with neighboring villages, vitality of the aboriginal religion, and native forms of politico-religious leadership. Although the Upper Cuao Piaroa are among the least acculturated of the tribe living today, they are not entirely unaffected by the vast demographic and cultural changes going on around them. According to adult residents, the Upper Cuao has suffered marked depopulation in their lifetimes due to high mortality and downriver migrations. Regional population density at the time of my fieldwork was calculated to be .087 persons/km^2 (Zent 1992:95)—a population level probably lower than it was before the big downriver exodus, and also well below the .2 persons/km^2 estimate made by William Denevan (1976:82) for Amazonian *terra firme* groups in general. Although Upper Cuao subsistence and settlement forms are not necessarily equivalent to those of fifty or a hundred years ago, they still offer the best window we can get on the ecological context of traditional Piaroa settlement pattern. Given the previously stated hypothesis that historical demographic and technological impacts lessened population pressure on available resources, I particularly wish to examine the relationships between subsistence and settlement ecology under low-density conditions.

Subsistence Ecology

The Upper Cuao subsistence economy conforms generally to the typical mixed pattern found throughout native lowland South America, but quantitatively we find a clear bias toward horticulture and hunting. Approximately 90% of vegetable foods come from cultivated plant sources, while nearly 80% of animal foods are captured by hunters. A breakdown of total food production into the major components reveals that cultivated vegetables make up 77.7% of the gross food weight, wild animals 13.7%, and wild vegetables 8.6% (see figure 12.2). The clear dominance of cultivated vegetables confirms that these resources constitute the primary source of food energy in the Upper Cuao diet, whereas the main dietary significance of wild plant and animal foods appears to be the provision of nutrients (protein, fat, minerals, and vitamins) that are lacking in cultivated foods.

Although 165 edible plant types are known to Upper Cuao people (Zent field notes), actual harvests concentrate on a few important types. Six plant species (all but one of them domesticated) account for more than 90% of the total weight of the dietary vegetable component (see table 12.1). Moreover, the two top cultivated foods, manioc and maize, comprise 79% (66% and 13%, respectively) of all vegetable mass harvested. The heavy emphasis on manioc and maize production is

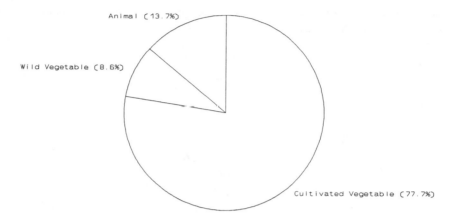

Fig. 12.2 *Major Components of Food Production (Gross Weight Contribution).*

reflected in the cultivation cycle, which consists of a very brief (2–2.5 yrs) cultivation phase—long enough for one maize crop and for one or two manioc crops—succeeded by a very long (> 20 yrs) fallow period. Manioc stands out as the dietary staple, while maize is nearly a staple during the maize harvest months (August–October). However, the Piaroa cultivate mostly bitter manioc varieties, which require extensive processing to render them edible. Upper Cuao women devote more than 1.5 hours daily (21.6% of their labor time budget) to manioc preparation. Thus the vegetable production system is efficient and reliable but also depends on substantial labor inputs, provided mostly by women.

A diverse array of wild animal species are considered edible, and many of them are in fact harvested throughout the year (see table 12.2). The most notable features of the animal harvest inventory are the great *diversity* of animal species exploited and the relative *dispersion* of weight contribution made by the individual species.

Table 12.1 **Upper Cuao Vegetable Harvest Record (Gross Weight), 07/01/85 to 06/30/86**

Piaroa Name	Common Name	Latin Name	Kg.	% of Total
ire	manioc	*Manihot esculenta*	8,203.84	66.41
yami	maize	*Zea mays*	1,568.21	12.69
uphæ	pendare	*Couma macrocarpa*	610.46	4.94
pæruru	banana	*Musa* sp.	365.22	2.96
wiriya	sweet potato	*Ipomoea batatas*	300.94	2.44
pæhærĩ	peach palm	*Bactris gasipaes*	202.95	1.64
remaining vegetable foods			1,101.46	8.92
Totals			12,353.08	100.00

Table 12.2   Upper Cuao Animal Harvest Record (Fresh Weight), 07/01/85 to 06/30/86

Piaroa Name	Common Name	Latin Name	No.	Kg	% of Total
ime	white-lipped peccary	*Tayassu pecari*	11	265.24	13.56
tuwa	paca	*Agouti paca*	38	247.12	12.63
ihure	black currasow	*Crax alector*	55	164.90	8.43
akʷa	great long-nose armadillo	*Dasypus kappleri*	13	135.41	6.92
ruæñia	caterpillar	Lepidoptera	—	92.08	4.71
akʷi	nine-banded long-nose armadillo	*Dasypus novemcinctus*	20	85.69	4.38
ahe poisa	assorted fish[a]	—	—	82.00	4.19
ækuri	red-rumped agouti	*Dasyprocta leporina*	25	77.35	3.95
tæbĩ	Spix's guan	*Penelope jacquacu*	52	64.73	3.31
mæk'iræ	collared peccary	*Tayassu tajacu*	3	61.83	3.16
hiuc̃	weeping capuchin monkey	*Cebus olivaceus*	18	56.15	2.87
wiæ	earthworm	Annelida	—	54.90	2.81
bure	frog	*Leptodactylus knudzeni*	202	54.87	2.80
niuæ	catfish	*Brachyplatystoma juruense*	83	30.49	1.56
tuwæ yæmæ	red brocket deer	*Mazama americana*	1	28.00	1.43
kuyui	blue-throated guan	*Pipile pipile*	22	27.83	1.42
wʰiᵗʰæ	brown-bearded saki	*Chiropetes satanus*	10	27.15	1.39
wʰæᵗʰæ	gray-winged trumpeter	*Psophia crepitans*	22	22.73	1.16
pærewa	"viejita" fish	Cichlidae	309	22.21	1.14
tuænisa	small minnow	*Bryconamericus* sp.	—	22.01	1.12
ñuhu	southern tamandua	*Tamandua tetradactylus*	5	20.50	1.05
wewa	great tinamou	*Tinamus major*	19	20.32	1.04
teuræ yæmæ	gray brocket deer	*Mazama gouazubira*	2	18.70	0.96
weuæ	unidentified frog	?	—	17.95	0.92
ækua	winged termite	Isoptera	—	17.43	0.89
yæho	Cuvier's toucan	*Ramphastos cuvieri*	32	16.80	0.86
kædi	channel-billed toucan	*Ramphastos vitellinus*	47	15.69	0.80
tuwæræ kua	cayman	*Paleosuchus trigonatus*	1	15.00	0.77
remaining animal foods				191.42	9.78
Totals				1,956.50	100.00

[a]Includes *ruæhu* (*Gymnotus carapo*), *tuæka* (*Hoplerythrinus* sp.), *soki* (*Callichthys callichthys*), *wæmwia* (Pimelodidae), *wʰæo* (*Crenicichla* sp.), *meretɨ* (*Hemibrycon* sp.), and *tuænisa* (*Bryconamericus* sp.).

The first 90% of the total animal weight harvested is composed of 28 different animal types, one of which is a composite category of different species (assorted fish), whereas only 6 plant items fulfill a similar proportion of the total vegetable food weight (see figure 12.3). These statistics imply that animal production is much

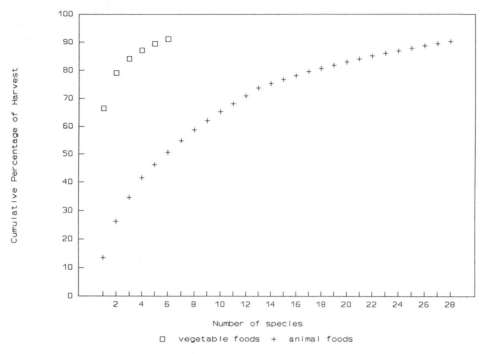

Fig. 12.3 *Cumulative Percentage of Harvest by Number of Species. Squares*: vegetable foods. *Crosses*: animal foods.

more generalized than vegetable production. Whereas the strategy of vegetable production appears to be specialized toward a few high-yielding crops, the strategy of animal production seems to depend on the exploitation of a high amount of diversity.

Another important feature of the hunting pattern is the small-game focus: animal species averaging less than 5 kg per individual (the arbitrary standard of small game according to Bourlière 1975) constitute 59% of the total animal weight harvested, while medium-sized species (≥ 5 kg, < 10 kg) account for another 14%. The heavy reliance on small- and medium-sized game interacts with the variables of high diversity and dispersion in the game harvest record, and may reflect a relatively impoverished resource base. Elsewhere, wide diet breadth has been associated theoretically and empirically with low resource productivity (Hames and Vickers 1982). Accordingly, the overall animal harvest rate in the Upper Cuao was .39 kg/manhour, a figure that ranks rather low compared with the rates recorded among various other native Amazonian groups (Hames 1989:64).[3]

The finding of inferior harvest rates, while available land resources do not appear to be limiting, is consistent with the hypothesis that game resources are rather poor in the Upper Cuao, even by Amazonian standards. Yet even though harvest rates are generally low, the high diversity enables a relatively steady output throughout the year. Animal harvest return rates varied between .25 kg/hr and .40 kg/hr during most months of the year of my study (see figure 12.4), the only exceptions being March and April, when they exceeded .6 kg/hr—due primarily to *three* very successful white-lipped peccary hunts, which are infrequent and unpredictable

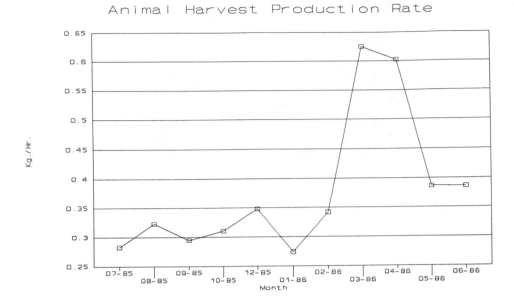

Fig. 12.4 *Animal Harvest Production Rate.*

events. Excluding these essentially random windfall harvests, animal productivity
is maintained at relatively low but steady levels throughout the year.

The effective range of most foraging expeditions can be mapped as a 10 km radius from the house. Most overnight camping trips also take place within this
range—in fact, about half are within 5 km. Since walking is the main form of transport in this hilly region, the distance of foraging expeditions may be held down
by the high costs of traveling to and from new resource patches. Travel constitutes
a major constraint here, to the extent that the high-diversity–small-game–low-
efficiency hunting focus is a function of the limited hunting territory. Camping implies movement within this territory and results in superior game yields, as verified
by comparing the returns from main house and camping locations in similar seasons (see table 12.3). Even though the distance between the two locations studied
here is only 3.3 km, hunting returns at the latter site were up to three times greater.

In sum, the Upper Cuao economic pattern may be characterized as an essentially
horticultural-hunting subsistence system. Horticulture supplies the greater part of
the caloric component of the diet and displays a specialized focus, dominated by
bitter manioc and maize. The staple manioc is a highly efficient crop and provides a
stable food energy supply year-round, but the bitter varieties require considerable
labor input in the processing phase. Hunting provides the vast bulk of dietary protein and fat, and shows a generalized resource focus, in which a large variety of
wild animal species contribute fairly evenly to total intake. The return rate on hunting is fairly low but steady throughout the year, the major exception to this pattern
being the infrequent and unpredictable but enormously productive peccary hunts.
The two facets of Upper Cuao subsistence imply distinct spatial pressures in the
quest to achieve maximum input-output efficiency: the cultivation of fixed gardens
exerts a sedentary pressure, while the hunting of dispersed wild fauna promotes no-

Table 12.3 Comparison of Animal Harvest Productivities at Main Settlement and Camping Locations (1985 and 1987)

Month/Year	Output (kg)	Input (hr)	Rate (kg/hr)
Main settlement:			
07/85	61.48	274.0	.22
08/85	67.95	191.8	.35
09/85	99.27	376.3	.26
Camping location:			
09/87	88.09	121.0	.73

madism. The two opposing tendencies are balanced on the level of subsistence by the rapid-shift, low-intensity cultivation pattern and the high-diversity, low-efficiency animal capture system. New gardens are opened up in new areas rather than extending production in existing gardens, while long-distance hunting is eschewed for the sake of staying near to one's gardens.

Settlement Ecology

The precarious balance between horticulture and hunting is also articulated at the level of settlement behavior. Traditional Upper Cuao settlements are small, ranging from five to thirty-five people, and consist physically of one or two houses (*isode*) and the surrounding gardens. The distance between nearest neighbor communities varies from as little as 2 km (50 minutes' walk) to as much as 15 km (8–9 hours' hike). All settlements observed during the fieldwork were located in fluvial valleys, between 300 and 650 m altitude. Although the distance between house and garden is to some extent a function of the age and size of the settlement, the maximum observed distance from house to garden was 1.5 km.

An important feature of the settlement landscape is the prominent presence of old secondary forests, the preferred biotope for cutting new gardens. Thus all Upper Cuao settlements are located in areas of former settlement. The presence of large tracts of secondary forest in an area is read as a sign that the site is eminently suitable for habitation. Perhaps related to their preference for the modified habitat, some residence groups practice a form of circular migration in their settlement moves, intentionally returning to former settlement sites or sites known to have been occupied by their parents or grandparents (Zent 1992:361–373).

The mobility patterns of Upper Cuao settlements are rather complex and deserve a more detailed description. The life span of Upper Cuao settlements during my fieldwork ranged from 1 to 9 years; an analysis of the movements of five traditional-style communities over the three years of fieldwork revealed that the average time interval between the building of new settlements was 2.77 years. Settlements were moved an average of 3.37 km, within a range of 460 m to 11.5 km. It is interesting to note that the residence groups with the most sedentary settlements were not necessarily the most sedentary groups, because they were also the most active travelers to communities outside the region for long-term visits. Thus it is possible to discern two general mobility patterns: (a) longer occupation of the home settlement alternating with long-distance, extended-stay visitation trips, and (b) short-lived occupations accompanied by short-distance relocations of the home settlement (Zent 1992:378–386).

Due to the frequent mobility, it is normal for a settlement group to have more than one house-garden settlement at any one time. Moves between the different settlements are decided by the house headman and are frequently associated with different tasks in the agricultural and subsistence cycle. Examples of typical inter-house migrations throughout the year include the following: (a) Movement to forest camps in the dry season to begin establishing a new settlement. (b) Movement back to the main settlement by the early-to-middle wet season. At this time, the maturing manioc plants are pruned, and bundles of manioc stem-cuttings for planting are gathered. (c) Movement to the new settlement to harvest the newly ripened maize crop. Other possible activities at this time include building the new house, finishing the manioc planting in the new gardens, and weeding the young gardens. (d) Movement back to the main settlement to prepare surpluses of dry manioc bread for the upcoming camping season. Other moves appear to be associated with the location of foraging activities that are seasonally based, such as collecting chicle fruit (*Couma macrocarpa*) in mid-to-late wet season (June–August), or frogs (e.g., *Leptodactylus knudzeni*) that congregate for mating purposes in marshy depressions by the early wet season (May–June) (Zent 1992:377–378).

In addition to the moves between *isode* belonging to a settlement group, there are temporary short-term dislocations from the current main residence by individuals or family groups, which might be called "camping out." The camping trips occur throughout the year and range from 1 to 15 km from the settlement. The main reasons given for camping out are (a) visits back to former settlement sites to harvest slow-maturing cultivated plants, such as peach palm (*Bactris gasipaes*); (b) hunting-fishing-collecting expeditions, sometimes to particular areas where seasonally abundant harvests of wild resources are expected; (c) trips to gather wild resources used in manufacturing (for domestic use or trade); and (d) separations in the service of performing ritual acts. The Piaroa camping pattern is a sedentary form of camping and therefore distinct from the nomadic trekking practiced by other Amazonian groups (Maybury-Lewis 1967; Good 1987). Piaroa campers arrive at a camping spot, quickly put up a roofed hut, and stay there for the duration of the trip (Zent 1992:374–377).

Another form of camping routinely practiced in the Upper Cuao entails a dispersal of the residential group into smaller units during the dry season and appears to be closely associated with settlement relocation. The exact timing of the dry-season dispersal varies, but the general pattern is that by mid-season (February) the main house is vacated and the residential group has divided into extended family units, each one building its own camphouse. Religious ideology supports the institutionalization of dry-season camping: it is believed that evil-doing *mæri* spirits wander more in the dry season and hence are more likely to find and enter a house and harm anyone inside. The Piaroa also choose to wander at this time and take up residence in quickly built huts under the cover of high forest. The location of the dry-season camps is usually within 5 km of the house-garden settlement. The distance between different camphouses of the same settlement group is only about 1–1.5 km, and group members continue to share meat and collaborate in foraging activities (although not as closely as when they are all under the same roof). Throughout the course of the dry season the camp groups may switch campsites once or twice, all of the camp locations being situated fairly close to one another. In anticipation of the dry-season dislocation, women produce large surpluses of dried manioc bread—but it is not uncommon for the supply to run out, so that domestic groups must make trips back to the mature gardens to harvest more; I believe that it is for this reason that the summer camp is usually within a day's round-trip from the house-garden settlement. While at the campsites, the men cut new gardens and the women clear the roots and stumps from a spot of ground in anticipation of building a more permanent house. With the onset of the rains in April–May, the settlement group reunites back at the main house. Thus do Upper Cuao settlements follow an annual rhythm of wet-season aggregation and dry-season dispersal (Zent 1992:375–377).

To summarize, traditional Upper Cuao settlements are very small and dispersed, and display a very active yet spatially limited mobility within their dispersed territories over the short term. The spatial attributes of dispersion and frequent but limited mobility articulate well with the resource dynamics imposed by the horticultural-hunting subsistence system. First, fragmentation of the population into small, scattered settlements is a response to the operating efficiency and stability of the resource system. Manioc production and small-game hunting are pursuits efficiently carried out by small groups of two or three persons. The (relatively) stable level of resource production provided by the combined horticultural-hunting resource system must be viewed as a necessary condition for the physical dispersion of small settlement groups. Meanwhile, the optimal spatial response to the thinly populated but stable wild resource base is a settlement pattern of small and equidistant residential units, because travel and transport costs are thus minimized (Wilmsen 1973; Horn 1968). Second, limited mobility appears to be a compromise solution between declining returns on game-animal capture and the implied costs of moving farther away from the gardens, the supply center and anchor of the food-energy system. Camping as well as settlement shifts help to maintain hunting return rates at low yet acceptable levels. But the distance of these moves is constrained by the extra labor needed to prepare stores of dried manioc cakes to take along, and by the considerable transport costs of hauling heavy loads of manioc cuttings from mature to juvenile garden sites. To the extent that settlement helps to manage the conflicting demands posed by the subsistence system, it allows the residential group to achieve independence in subsistence matters.

Conclusions

Traditional Piaroa settlement pattern has been characterized as interfluvial, small-sized, scattered, and semisedentary. The key spatial properties identified in the present analysis were high relative dispersion and mobility. The explanatory adequacy of the interpretation of the settlement pattern depends on the ability to properly account for these two properties, in both functional-adaptive and developmental matrices (cf. Trigger 1968).

As Overing has observed, traditional Piaroa settlement pattern may be viewed on two analytical levels: the independent settlement, and the neighborhood configuration of settlements. The distinction is important because we can note differences in the dynamic interplay of the properties of dispersion and mobility according to the level being considered. At the level of the individual settlement, the relationship between dispersion and mobility is largely complementary and conditioned by matters of defense as well as day-to-day subsistence. In the first place, dispersion and mobility are effective responses to militarily superior, hostile neighbors and to life-threatening, contagious diseases (Beckerman 1980). Secondly, the pattern of small settlements scattered over the landscape is an efficient response to a scarce, evenly dispersed wild-animal resource base. Residential dispersion is permitted by the reliability, stability, and productivity of small-scale manioc cultivation. Limited mobility of the settlement group within the home range(s) and between successive settlement sites supports the maintenance of the considerable spatial distances separating communities. Such movements also serve to counteract depressed hunting return rates, without overstepping easy access to the group's gardens. The relatively frequent yet short-distance settlement shifts signify compromise between the centripetal pull of horticulture and the centrifugal pressures of hunting. In this manner, an acceptable level of operating efficiency is maintained in both domesticated and wild subsistence spheres, which enhances overall independence of the settlement group in subsistence matters.

At the level of the intersettlement groupings, the role of subsistence diminishes and we must look to other conditioning factors of the environment: to wit, biological reproduction, health, commerce, and politics. It is in the context of these other factors that we see the functional significance of the supralocal social-spatial structures. The fluid movement of family units and individuals between settlements and neighborhood groupings indicates a level of settlement interdependence superseding the level of settlement independence. The significance of the neighborhood rests on the inherent instability of the small, separate communities. Precisely because settlement groups are so small, there are reasons for the formation of larger groupings. One reason is the importance of trade: many technological artifacts necessary to Piaroa livelihood and social life, including steel tools, are not manufactured locally and can be obtained only through trade. Demographic requirements create another kind of interdependence: the small settlements are not viable demes, and the circulation of people reflects the need—despite the stated preference for endogamy—to marry outside the natal unit. Health risk poses another problem for small settlements: in the event of the sickness or death of a key family member, people must depend on food or labor supplied by coresidents or, alternatively, relatives in neighboring communities. Similarly, when healthy people flee their community due to the risk of infection, they require support (most importantly, permission to harvest manioc) from residents of the receiving community; because of the heavy reliance on manioc cultivation for subsistence, any distant move would require setup support to establish a new house and garden. Finally, political power in this environment depends on attracting sufficient people in order to be able to produce a surplus of food. Surplus production is channeled into the ritual feasts, which are the main material vehicle for establishing, consolidating, and maintaining political power among traditional Piaroa. One of the main advantages of political prominence, in turn, is an expanded kin network in which to arrange marriages, obtain trade goods, and seek labor support when needed—i.e., to better fulfill demographic, commercial, and health needs.

At the intersettlement level of analysis, the relationship between dispersion and mobility is oppositional—in the sense that the former implies the separation and independence of individual settlements, whereas the second evinces the connections and interdependence among people residing in different settlements. The demographic, epidemiological, commercial, and political factors mentioned above constitute powerful integrative forces, which are expressed spatially in the form of the neighborhood groupings of proximate, relatively closely spaced settlements. But consolidation of these connected-yet-still-separate small settlements into larger settlements is apparently prevented by defense (i.e., human predator and microorganismic pathogen avoidance) and economic pressures (i.e., dispersed natural fauna and open land areas) that lead toward centrifugation and the disintegration of population aggregations. Thus traditional settlement pattern represents a compromise between the tendency to disperse in order to optimize defensive and subsistence relationships, and the tendency to concentrate in order to fulfill other survival needs.

The countervailing ecologically motivated tendencies to separate and to nucleate settlement are mirrored in the social organization by the institutionalization of the social-symbolic oppositions of ideal versus real settlement notions, kindred versus Dravidian kinship terminologies, and endogamous versus nonendogamous marriage. The cognatic kinship system is a social structural arrangement that permits settlement endogamy and hence the persistence and identity of the settlement group through time. At the same time, cognatic kinship facilitates alliances with other settlement groups and the eventual disintegration of the settlement group. Theoretical and comparative analyses of cognatic kinship systems have identified a resource dynamic of abundant land and scarce labor as the crucial material conditions that give rise to this social structural type (Murphy 1979; Price 1984). In a similar vein, cognatic kinship among the Piaroa, as elsewhere in native Guiana, reflects a social system designed to effect the inclusion rather than the exclusion of personnel, an orientation that derives from the fact that people themselves are the strategic resource in this environment (Rivière 1984; Zent 1993b). Cognatic kinship also permits the flexible organization and movement of small groups of people, an aspect of the settlement pattern associated with the availability of land.

This brings us back to the significance of the historical component. The resource dynamic of abundant land and scarce labor is directly attributable to the historical-ecological process of the easement of population pressure on resources, brought about by direct demographic decline and increased technological capacitation. My explanation of traditional Piaroa settlement pattern becomes intelligible in this light. A historical-ecological perspective is thus necessary in order to see that the dispersed, mobile settlement pattern embodies a spatial adaptation to a poor but underpopulated environment, rather than the strained response of a population at the point of pressing the resource envelope.

Acknowledgments

I wish to thank everyone who attended the Tulane Mini-Conference on Historical Ecology for their valuable feedback after the reading of my paper, especially Anna Roosevelt and Neil Whitehead. The critical editorial job by Bill Balée on earlier manuscript versions was most helpful. Thanks are also due to my dissertation committee, Barbara Price in particular, for earlier discussion of some of the ideas presented herein. Carlos Quintero drew the map. Fieldwork was supported by the IIE Fulbright program, Explorer's Club, and the Instituto Venezolano de Investigaciones Científicas (IVIC).

Notes

1. Considerable debate has surrounded the application and interpretation of Dravidian kinship systems, and it seems that this concept cannot be used without raising objections from someone. However, the present usage follows Shapiro's (1984) general characterization of most lowland South American systems as Dravidian, and Overing's (1984) particular characterization of Piaroa kinship as such. The point to be emphasized here is that the parallel/cross distinction consistently specified by Dravidian terminology translates to a formal contrast between consanguine and affine in Piaroa.

2. Although the exact mechanisms that might explain the megalethal impact of Old World diseases on New World aboriginals are not known, it is generally accepted that the latter's immunogenetic systems were highly vulnerable to the foreign pathogenic organisms (Crosby 1972; Neel 1977; Black 1992).

3. Average hunting and fishing efficiency by men in a sample of seventeen societies was .75 kg/hr, with a range of .11–2.34 kg/hr. The mean hunting and fishing efficiency by all adults (men and women) in a smaller sample of seven societies was .50 kg/hr (Hames 1989:64–65). See Zent (1992:319) for a more detailed illustration of the unfavorable comparison of Upper Cuao animal harvest rates with those of other Amazonian societies.

References

Acosta Saignes, Miguel. 1954. *Estudios de etnología antigua de Venezuela*. Publicaciones de la Facultad de Humanidades y Educación, Universidad Central de Venezuela. Caracas: Tipografía Vargas.

Aguirre Elorriaga, M. 1941. Misión del Orinoco. Pp. 1–91 in J. del Rey Fajardo (ed.), *La Compañía de Jesús en Venezuela*. Caracas: Biblioteca de la Academia Nacional de la Historia.

Alvarado, A. Eugenio de. 1941. Informe Sobre el manejó y conducta que yuvieron los padres èsuítas con la Expedición de la Línea Divisoria entre España y Portugal en la Península Australmi Orillas del Orinoco. Pp. 215–233 in J. del Rey Fajardo (ed.), *Decumentos jesuíticos relativos a la historia de la Compañía de Jesús en Venezuela*. Biblioteca de la Academia Nacional de la Historia, vol. 79.

Anduze, Pablo. 1963. Deyaruwa. *Boletín Indigenista Venezolana* 7:155–167.

———. 1974. *Los Dearuwa: Dueños de la selva*. Biblioteca de la Academia de Ciencias Físicas, Matemáticas y Naturales, vol. 13. Caracas: Biblioteca de la Academia Nacional de la Historia.

Arvelo-Jiménez, Nelly, F. Morales Méndez, and Horacio Biord Castillo. 1989. Repensando la historia del Orinoco. *Revista de Antropología* 5(1–2): 155–173.

Ayres, Pedro J. 1967. Las misiones del Alto Orinoco y Río Negro en 1824. Pp. 313–329 in *Las misiones de Píritu*, vol. 2. Caracas: Biblioteca de la Academia Nacional de la Historia.

Balée, William. 1992. People of the fallow: A historical ecology of foraging in lowland South America. Pp. 35–57 in K. H. Redford and C. Padoch (eds.), *Conservation of Neotropical Forests: Working from Traditional Resource Use*. New York: Columbia University Press.

Baumgartner, Juan. 1954. Apuntes de un médico-indigenista sobre los Piaroa de Venezuela. *Boletín Indigenista Venezolano* 2(1–4): 111–125.

Beckerman, Stephen. 1980. On ecology, choice, and Amazonian settlement patterns. *Current Anthropology* 21(5): 704–705.

Biord Castillo, Horacio. 1985. El contexto multilingüe del sistema de interdependencia regional del Orinoco. *Antropológica* 63–64:83–102.

Black, Francis L. 1992. Why did they die? *Science* 258:1739–1740.

Boglar, Lajos. 1969. Nota sobre la cultura de los Piaroa. *Revista Venezolana de Folklore* 2:63–67.

———. 1971. Chieftainship and the religious leader: A Venezuelan example. *Ethnographica Academiae Scientiarum Hungaricae* 20(3–4): 331–337.

Bourlière, F. 1975. Mammals, small and large: Ecological implications of size. Pp. 1–8 in F. B. Golley, K. Petrusewicz, and L. Ryszkowski (eds.), *Small Mammals: Their Productivity and Population Dynamics*. Cambridge: Cambridge University Press.

Bueno, Ramón. 1933. *Apuntes sobre la provincia misionera del Orinoco e indígenas de su territorio, con algunas otras particularidades*. Caracas: Tipografía Americana.

Butt Colson, Audrey J. 1983–84. The spatial component in the political structure of the Carib speakers of the Guiana Highlands: Kapon and Pemon. *Antropológica* 59–62: 73–124.

Carneiro, Robert L. 1964. Shifting cultivation among the Amahuaca of eastern Peru. *Völkerkundliche Abhandlungen* 1:9–18.

———. 1979. Forest clearance among the Yanomamo: Observations and implications. *Antropológica* 52:39–76.

Caulín, Fr. Antonio. 1966. [1779]. *Historia coro-gráphica natural y evangélica de la Nueva Andalucía, Provincias de Cumaná, Guayana y Vertientes del Río Orinoco*. Caracas: Biblioteca de la Academia Nacional de la Historia.

Chaffanjon, Jean. 1986. [1889]. *El Orinoco y el Caura*. Trans. Joelle Lecoin. Caracas: Fundación Cultural Orinoco.

Codazzi, Agustin. 1940. *Resumen de la geografía de Venezuela (Venezuela en 1841)*. 3 vols. Caracas: Biblioteca Venezolana de la Cultura.

Colchester, Marcus. 1984. Rethinking Stone Age economics: Some speculations concerning the pre-Columbian Yanoama economy. *Human Ecology* 12(3): 291–314.

Crevaux, Jules Nicolas. 1883. *Voyages dans l'Amérique du Sud*. Paris: Hachette.

Crosby, Alfred W. 1972. *The Colombian Exchange: Biological and Cultural Consequences of 1492*. Westport, Conn.: Greenwood Press.

Cruxent, José María. 1947. Algunas actividades explotativas de los indios Piaroa del Río Parguaza (Guayana venezolana). *El Agricultor Venezolano* 11(120): 12–15.

Cruxent, José María and Mauricio Kamen-Kaye. 1950. Reconocimiento del área del Alto Orinoco, ríos Sipapo y Autana, en el Territorio Federal Amazonas, Venezuela. *Memoria de la Sociedad de Ciencias Naturales La Salle* 10(26): 11–23.

Dalton, Leonard V. 1912. *Venezuela*. London: T. Fisher Unwin.

Deferrari, Enrique. 1945. *Tribus indígenas de la Prefectura Apostólica del Alto Orinoco*. Tercera Conferencia Interamericana de Agricultura. Cuadernos Verdes, 40. Caracas: Escuelas Gráficas Salesianas.

Del Rey Fajardo, José. 1977. *Misiones Jesuíticas en la Orinoquía: Aspectos fundacionales*, vol. 1. Colección Manoa. Caracas: Universidad Católica Andrés Bello.

Denevan, William, ed. 1976. *The Native Population of the Americas in 1492*. Madison: University of Wisconsin Press.

———. 1992. Stone vs. metal axes: The ambiguity of shifting cultivation in prehistoric Amazonia. *Journal of the Steward Anthropological Society* 20(1–2): 153–165.

Dickey, Herbert Spencer. 1932. *My Jungle Book*. Boston: Little, Brown.

Dobyns, Henry F. 1966. Estimating aboriginal American population: An appraisal of techniques with a new hemisphere estimate. *Current Anthropology* 7:395–416.

———. 1983. *Their Number Become Thinned: Native American Population Dynamics in Eastern North America*. Knoxville: University of Tennessee Press.

Dupouy, Walter. 1952. El Piaroa, hombre de la selva. *Tierra Firme* 1953 (May): 17–20.

Eden, Michael J. 1974. Ecological aspects of development among Piaroa and Guahibo Indians of the Upper Orinoco basin. *Antropológica* 39:25–56.

Ferguson, R. Brian. 1989. Ecological consequences of Amazonian warfare. *Ethnology* 28:249–264.

Gilij, Felipe Salvador. 1965. [1783]. *Ensayo de historia Americana, o sea historia natural, civil y sacra de los reinos y de las provincias españolas de tierra firme en la América meridional*. 3 vols. Caracas: Biblioteca de la Academia Nacional de la Historia.

Good, Kenneth R. 1987. Limiting factors in Amazonian ecology. Pp. 407–421 in M. Harris and E. Ross (eds.), *Food and Evolution: Toward a Theory of Human Food Habits*. Philadelphia: Temple University Press.

Grelier, Joseph. 1953. Los indios Piaroa de la región de Puerto Ayacucho. *Boletín Indigenista Venezolano* 1(2): 253–263.

———. 1955. Habitat, types d'habitation et genres de vie chez les aborigènes du Bassin de l'Orénoque. *Ethnographie*, n.s., 50:42–59.

———. 1959. *La route du poison*. Paris: Editions de la Table Ronde.

Gross, Daniel. 1975. Protein capture and cultural development in the Amazon Basin. *American Anthropologist* 77:526–549.

Gumilla, José. 1963. [1741]. *El Orinoco ilustrado y defendido*. Fuentes para la Historia Colonial de Venezuela, vol. 68. Caracas: Biblioteca de la Academia Nacional de la Historia.

Hames, Raymond B. 1983. The settlement pattern of a Yanomamö population bloc: A behavioral ecological interpretation. Pp. 393–427 in R. Hames and W. Vickers (eds.), *Adaptive Responses of Native Amazonians*. New York: Academic Press.

————. 1989. Time, efficiency, and fitness in the Amazonian protein quest. *Research in Economic Anthropology* 11:43–85.

Hames, Raymond B. and William T. Vickers. 1982. Optimal diet breadth theory as a model to explain variability in Amazonian hunting. *American Ethnologist* 9(2): 358–378.

Hanson, Earl. 1933. Social regression in the Orinoco and Amazon Basins: Notes on a journey in 1931 and 1932. *Geographical Review* 23:578–598.

Henley, Paul. 1983. Los Wánai (Mapoyo). Pp. 217–241 in W. Coppens (ed.), *Los aborígenes de Venezuela*, vol. 2. Fundación La Salle de Ciencias Naturales, Instituto Caribe de Antropología y Sociología, monograph 26. Caracas: Editorial Texto.

Holmberg, Allan. 1969. *Nomads of the Long Bow*. New York: Natural History Press.

Horn, H. S. 1968. The adaptive significance of colonial nesting in the Brewers blackbird (*Euphagus cyanocephalus*). *Ecology* 49:682–694.

Humboldt, Alexander von and Aimé Bonpland. 1876. *Personal Narrative of Travels to the Equinoctial Regions of America During the Years 1799–1804*. London: Bell.

Iribertegui, Ramón. 1987. *Amazonas: El hombre y el caucho*. Monografía N° 4. Puerto Ayacucho: Ediciones del Vicariato Apostólico de Puerto Ayacucho.

Isaac, Barry. 1977. The Siriono of eastern Bolivia: A reexamination. *Human Ecology* 5(2): 137–154.

Koch-Grünberg, Theodor. 1979. [1923]. *Del Roraima al Orinoco*. (Spanish trans. of the original by Dr. Federica de Ritter, *Vom Roraima zum Orinoco: Ergebnisse einer Reise in Nordbrasilien und Venezuela in den Jahren 1911–1913*, 3 vols.) Caracas: Ediciones del Banco Central de Venezuela.

Lathrap, Donald. 1968. The hunting economies of the tropical forest zone of South America: An attempt at historical perspective. Pp. 23–29 in R. B. Lee and I. DeVore (eds.), *Man the Hunter*. Chicago: Aldine.

Mansutti Rodríguez, Alexander. 1986. Hierro, barro cocido, curare y cerbatanas: El comercio intra e interétnico entre los Uwotjuja. *Antropológica* 65:3–75.

————. 1988. Pueblos, comunidades y fondos: Los patrones de asentamiento Uwotjuja. *Antropológica* 69:3–35.

————. 1990. *Los Piaroa y su territorio*. CEVIAP Documento de Trabajo No. 8. Caracas: CEVIAP.

————. 1994. Tres momentos, tres modelos: Los sistemas de poblamiento Piaroa. Paper presented at the Symposium "Settlement, Subsistence, and the Conquest of South America: Ongoing Threats to Indigenous Cultures," 48th International Congress of Americanists, Uppsala, Sweden, 4–9 July.

MARNR-ORSTOM. 1988. *Atlas del inventario de tierras del Territorio Federal Amazonas*. Caracas: MARNR-DGSIIA.

Martin, M. Kay. 1969. South American foragers: A case study in cultural devolution. *American Anthropologist* 71:243–260.

Maybury-Lewis, David. 1967. *Akwẽ-Shavante Society*. Oxford: Clarendon Press.

Meggers, Betty J. 1971. *Amazonia: Man and Culture in a Counterfeit Paradise*. Chicago: Aldine.

Michelena y Rojas, Francisco. 1989. [1867]. *Exploración oficial*. Nelly Arvelo-Jimenez and Horacio Biord Castillo, eds. Iquitos, Peru: IIAP-CETA.

Migliazza, Ernest C. 1980. Languages of the Orinoco-Amazon basin: Current status. *Antropológica* 53:95–162.

Monod, Jean. 1970. Los Piaroa y lo invisible: Ejercicio preliminar a un estudio sobre la región Piaroa. *Boletín Informativo de Antropología* 7:5–21.

————. 1972. *Un riche cannibale*. Paris: Union Générale d'Editions.

Morey, Nancy. 1975. Ethnohistory of the Colombian and Venezuelan llanos. Ph.D. diss. Ann Arbor: University Microfilms International.

Morey, Nancy and Robert Morey. 1973. Foragers and farmers: Differential consequences of Spanish contact. *Ethnohistory* 20(3): 229–246.

Morey, Robert V. and Nancy C. Morey. 1975. Relaciones comerciales en el pasado en los llanos de Colombia y Venezuela. *Montalbán* 4:533–564.

Murphy, Robert F. 1979. Lineage and lineality in lowland South America. Pp. 217–224 in M. L. Margolis and W. E. Carter (eds.), *Brazil: Anthropological Perspectives. Essays in Honor of Charles Wagley*. New York: Columbia University Press.

Neel, J. V. 1977. Health and disease in unacculturated Amerindian populations. Pp. 155–177 in K. Elliott and J. Whelan (eds.), *Health and Disease in Tribal Societies*. Amsterdam: Elsevier.

OCEI (Oficina Central de Estadística e Informática). 1985. *Censo indígena de Venezuela: Nomenclador de comunidades y colectividades*. Caracas: Taller Gráfico de la OCEI.

————. 1993. *Censo indígena de Venezuela 1992*, vol. 1. Caracas: Taller Gráfico de la OCEI.

Overing, Joanna. 1975. *The Piaroa, a People of the Orinoco Basin: A Study of Kinship and Marriage*. Oxford: Clarendon Press.

————. 1981. Review article: Amazonian anthropology. *Journal of Latin American Studies* 13(1): 151–164.

————. 1983–84. Elementary structures of reciprocity: A comparative note on Guianese, central Brazilian, and north-west Amazon sociopolitical thought. *Antropológica* 59–62:331–348.

————. 1984. Dualism as an expression of differencs and danger: Marriage exchange and reciprocity among the Piaroa of Venezuela. Pp. 127–155 in K. Kensinger (ed.), *Marriage Practices in Lowland South America*. Illinois Studies in Anthropology, no. 14. Urbana: University of Illinois Press.

Overing, Joanna and M. R. Kaplan. 1988. Los Wóthuha (Piaroa). Pp. 307–411 in Jacques Lizot (ed.), *Los aborígenes de Venezuela: Etnología contemporánea II*. Fundación La Salle de Ciencias Naturales, Instituto Caribe de Antropología y Sociología, vol. 3. Caracas: Monte Avila Editores.

Perera, Miguel A. 1982. Patrones de asentamiento y actividades de subsistencia en el Territorio Federal Amazonas, Venezuela. Spanish trans. of Ph.D. diss., University of Bristol.

Price, Barbara. 1984. Competition, productive intensification, and ranked society: Speculations from evolutionary theory. Pp. 209–240 in R. B. Ferguson (ed.), *Warfare, Culture, and Environment*. Orlando: Academic Press.

Pruneti, Marcello. 1968. I Piaroa. *Archivio per l'Antropologia e l'Etnologia* 98(1–2): 1–15.

Pulliam, H. R. 1988. Sources, sinks, and population regulation. *American Naturalist* 132:652–661.

Ramos Pérez, Demetrio. 1946. *El Tratado de Límites de 1750 y la expedición de Iturriaga al Orinoco*. Madrid: Consejo Superior de Investigaciones Científicas.

Rivero, Juan. 1956. [1736]. *Historia de las misiones de los llanos de Casanare y los rios Orinoco y Meta*. Bogotá: N.p.

Rivière, Peter. 1984. *Individual and Society in Guiana*. Cambridge: Cambridge University Press.

Roosevelt, Anna Curtenius. 1980. *Parmana: Prehistoric Maize and Manioc Subsistence Along the Amazon and Orinoco*. New York: Academic Press.

Rottmayr, Luis. 1949. Noticias sobre los Piaroa. *Revista de la Misión del Alto Orinoco* 1(2): 15–17.

Rouse, Irving and José María Cruxent. 1963. *Venezuelan Archaeology*. New Haven: Yale University Press.

Scaramelli, Franz G. 1993. Las pinturas rupestres de las cuevas del Parguaza, estado Bolívar, Venezuela: Mito y representación. *El Guácharo* 31:1–96.

Schusky, Ernest L. 1972. *Manual for Kinship Analysis*. 2d ed. New York: Holt, Rinehart and Winston.

Shapiro, Judith. 1984. Marriage rules, marriage exchange, and the definition of marriage in lowland South American societies. Pp. 1–30 in K. M. Kensinger (ed.), *Marriage Practices in Lowland South America*. Illinois Studies in Anthropology, no. 14. Urbana: University of Illinois Press.

Smole, William. 1966. Utilización de los recursos por los indios Piaroa y Guaica de Venezuela. Pp. 116–129 in *Unión Geográfica Internacional, Conferencia Regional Latinoamericana*, vol. 1. México City: Unión Geográfica Internacional.

———. 1976. *The Yanoama Indians: A Cultural Geography*. Austin: University of Texas Press.

Steward, Julian H. and Louis Faron. 1959. *Native Peoples of South America*. New York: McGraw-Hill.

Tavera-Acosta, Bartolomé. 1907. *En el sur (dialectos indígenas de Venezuela)*. Ciudad Bolivar: Benito Jimeno Castro.

———. 1927. *Río Negro: Reseña etnográfica, histórica, y geográfica del Territorio Amazonas*. 2d ed. Maracay, Aragua, Venezuela.

Tesch, Heinz K. 1953. Viajando con los indios Piaroa. *Revista Shell* 2(8): 67–71.

Trigger, Bruce G. 1968. The determinants of settlement patterns. Pp. 53–78 in K. C. Chang (ed.), *Settlement Archaeology*. Palo Alto: National Press Books.

Urbina, Luis. 1983–84. Some aspects of the Pemon system of social relationships. *Antropológica* 59–62:183–198.

Velez Boza, Fermín and Juan Baumgartner. 1962. Estudio general, clínico y nutricional en tribus indígenas del Territorio Federal Amazonas de Venezuela. *Archivos Venezolanos de Nutrición* 12(2): 143–225.

Wavrin, M. de. 1948. *Les indiens sauvages de l'Amérique du Sud*. Paris: Payot.

Whitehead, Neil L. 1988. *Lords of the Tiger Spirit: A History of the Caribs in Colonial Venezuela and Guyana, 1498–1820*. Royal Institute of Linguistics and Anthropology, Caribbean Studies Series 10. Dordrecht/Leiden: Foris/Koninklijk voor taal-land-en Volkenkunde.

Wilbert, Johannes. 1966. *Indios de la región Orinoco-Ventuari*. Instituto Caribe de Antropología y Sociología, Fundación La Salle de Ciencias Naturales, monograph 8. Caracas: Editorial Sucre.

Wilmsen, E. N. 1973. Interaction, spacing behavior, and the organization of hunting bands. *Journal of Anthropological Research* 29:1–31.

Zent, Stanford. 1992. Historical and ethnographic ecology of the Upper Cuao River Wõthihã: Clues for an interpretation of native Guianese social organization. Ph.D. diss., Columbia University.

———. 1993a. Discriminación cultural de la tecnología biomédica occidental y extinción cultural entre los indígenas Piaroa, Estado Amazonas, Venezuela. Pp. 227–242 *Salud y Población Indígena de la Amazonia: Memorias del I Simposio Salud y Población Indígena de la Amazonia*. N.d., n.p.

————. 1993b. Ethnic fluidity and open ecological spaces in native Guiana. Paper presented in the session "Land Reform, Territoriality, and Ecology in Latin America," 92nd Annual Meeting of the American Anthropological Association, Washington, D.C., 17–21 November.

————. In press. Donde no hay médico: Las consecuencias culturales y demográficas de la distribución desigual de los servicios médicos modernos entre los Piaroa. *Antropológica*.

Zucchi, Alberta, Kay Tarble, and J. Eduardo Vaz. 1984. The ceramic sequence and new TL and C-14 dates for the Agüerito site of the Middle Orinoco, Venezuela. *Journal of Field Archaeology* 11:155–180.

CHAPTER 13

Whatever Happened to the Stone Age?
Steel Tools and Yanomami Historical Ecology

R. BRIAN FERGUSON

Ecology and History

In a time when ecological projections run from gloomy to disastrous (Ferguson 1995b), we need to learn better how to get along with the earth. The contributors to this volume seek a new understanding of the dynamics of human interaction with the physical world. Anthropologists have long investigated this interaction and now realize that it is not, as we once imagined, a case of people adapting to a natural environment as if it were a static, external object. The environment changes, in the short term and the long, and such change is an integral part of the processes of human ecology. As William Balée puts it, human communities, their cultures, and their landscapes must be seen as integrated historical totalities.

Balée's own research (1989, 1993) on anthropogenic Amazonian forests shows how even "minimal impact" subsistence systems can, over millennia, transform vast environments. Amazonian "garden hunting"—taking game animals that are attracted by recently overgrown gardens (Linares 1976; Ross 1978:10)—illustrates ecological modification at the other end of the timescale. Many other examples will be encountered by readers of this book (also see Crumley 1994). However, this should not be taken to invalidate the basic perspective of earlier anthropological ecology, that the nonhuman environment plays a major role in shaping the contours of culture. Environmental mutability itself has limits, and for any people at any point in time, the environment is a reality that constrains in multiple and often highly specific ways what people do. Again as Balée puts it, there is a dialectic at work between nature and culture, an evolving relationship in which the present adapts to the results of past interactions.

Papers in this volume show that different kinds of societies can have different kinds of ecological interactions, with very different degrees of impact on the physical world. Modern states can bring about massive and rapid environmental transformations. The impact of tribal peoples is more limited: their actions also may bring about major change, but usually as a gradual process. Where sudden, dramatic shifts occur in tribal ecologies, it is usually because of contact between previously isolated biotic communities. Of course, the biggest set of cases in this category comes from the expansion of Europe over the past five hundred years, with its massive ecological transfers (Cronon 1983; Crosby 1986), epidemic-driven demographic restructurings (Dobyns 1983; Ramenofsky 1987), and pervasive reorientations of extractive labor (Wolf 1982).

From the perspective of an evolved nature/culture interaction in any given locality, the impact of European contact may seem unpredictable, a historical contingency. But one analyst's contingency is part of another's pattern: from a global system perspective, the encounter will appear highly regular, with expectable local variations. Thus, as Neil Whitehead suggests (chapter 2), developing a generalizing historical ecology may be dependent on developing a generally better historical anthropology, one that situates all peoples in an interconnected world.

On the other hand, the big fault of world systems perspectives in the past has been failure to grasp how "the local" reacts to and shapes "the larger." Close attention to ecological relationships provides a solid grounding for understanding how local systems interact with exogenous introductions, a grounding that can be followed out through social and symbolic organization to frame the arenas of agency in historical change. Teasing apart these connections—showing the history in ecology and the ecology in history—is the proper task of historical ecology.

In the conference that preceded this volume, we debated what "*historical* ecology" is, or should be. One issue was identifying the object of study. Several participants saw it as the process of landscape or biosphere transformation by human action, which is the primary focus of Crumley's (1994) edited volume, *Historical Ecology*. Of Balée's four postulates of historical ecology, three concern human impact on the environment. But Whitehead, in particular, questioned whether such transformative processes should be called "history." Along the same lines, my view of historical ecology is a concern with human interaction with the physical world in a specific time and place, with the process of interaction understood as conditioned *by* prior history and frequently changing *as* history. There is no reason for these different interests to be seen as contradictory: one can and should inform the other.[1]

Some of the stickiest points in the conference discussions appear to me more semantic than substantive. A matter of particular dispute was the meaning of "evolution" in relation to "history." Clearly, participants understood many different things by these terms. Some thought the two were incompatible as ways of seeing process. I do not. I see "cultural evolution" as regularity in social change, and its study as an "effort to determine similar form and process in parallel developments" (Steward 1976:15). Within all the particularities of any specific historical sequence are commonalities that can be discovered by comparing situations from different times, places, and cultures. Sometimes these regularities involve the development of more centralized and hierarchical polities—what is often meant by "political evolution"—

but that is just one possible trajectory. From this perspective, understanding evolution requires better histories, but making sense of history requires a sense of evolution (see Fried 1967).

The search for underlying regularities within history raises questions of causality and determinism. In my view, a good place to start that search is the "infrastructure"—the mutually constraining system of ecological relations, demographic patterns, work techniques, and technology (Ferguson 1995a). This chapter focuses on technology—not in order to claim that it is some sort of prime mover, or to advocate any form of technological reductionism, but to redress a long history of neglect. In particular, what has been neglected by anthropology is the cultural and ecological significance of steel tools.

The Cutting Edge

The culture of professional anthropology is as peculiar as any other, and, as with any culture, we take our peculiarities for granted. One big peculiarity is highlighted by Lauriston Sharp's article, "Steel axes for Stone Age Australians" (1993 [orig. 1952]). The introduction of steel axes to Arnhem Land in northern Australia, and the way in which they were traded, led to changes that were nothing short of revolutionary. The Yir Yiront social order based on age, gender, and kinship was subverted; relationships between trade partners changed drastically; even the logic of the totemic system was eaten away. Happily, it turned out that Sharp was wrong in predicting their cultural destruction, for the Yir Yiront creatively adjusted their understanding of totemism and "the old ways" to give meaning to their new and radically different way of life (Taylor 1988). But in retrospect, Sharp's broader point, that the introduction of steel axes set in motion massive social change, is completely confirmed.

There is no reason to consider this an unusual case. Thus it is curious that although Sharp's article is one of most reprinted in anthropology, there has been little follow-up regarding its implications for our conception of the ethnographic universe (R. F. Salisbury's *From Stone to Steel* [1962] being the most visible exception). Although by our best estimates forest clearing is three to ten times more efficient in time and effort with steel axes than with stone (Carneiro 1979a,b; Coles 1979; Harding and Young 1979; Hill and Kaplan 1989; Saraydar and Shimada 1971; Townsend 1969), and although, across cultures and history, it is frequently reported that indigenous people make great efforts to obtain metal tools, the implications of this technological revolution have been ignored.

Curious. Anthropology divides the history of humankind into ages of stone and metal, but the discipline appears simply uninterested in the fact that countless cultural groups have gone from stone to steel sometime before the anthropologists began to study them. (As Carneiro [1979b:21] observes, the "felling of a tree with a stone ax . . . is an event that has rarely been witnessed by ethnologists.") In a recent introduction to Sharp's article (1993:24), the editor notes the "startling impact of this seemingly minor innovation."[2] I think it *appears* startling only because so little has been done to investigate a change that—far from being minor—may be second only to the introduction of new epidemic diseases in its expectable impact.

The disregard of steel is part of a much broader disinterest in the impact of Western contact. For most of this century, anthropology has been in thrall to "the ethnographic present," the imagined period between initial contact by Western observers and the beginning of culture change. That was the proper object of anthropological discourse. Acculturation studies were out of the theoretical mainstream, even stigmatized. (Lauriston Sharp's article originally appeared in the applied anthropology journal *Human Organization*.) That is changing, but most studies of culture contact still focus on loftier topics than steel tools, such as cosmology and values (Turner 1993).

Among tree-felling horticulturists, including most known peoples of Amazonia, it seems likely that the impact of steel would have been even greater than for the foraging Yir Yiront. Indeed, soon after Sharp, Alfred Métraux (1959) pulled together some of the copious evidence that indigenous peoples of the Amazon endured war, disease, missionization, and subjugation—all to get their hands on these tools. He argued that their introduction created nothing less than a "revolution." Yet Métraux's article is rarely cited, and ecological studies of Amazonia have taken little notice of metal tools (cf. Harner 1963). As a result, what we have taken to be "indigenous subsistence practices" may in fact be strongly conditioned by exogenous factors.[3]

Portrayals of the Yanomami provide a good illustration of the general disregard for the difference between stone- and steel-based economies, despite the existence, for the Yanomami, of unusually good research on precisely that topic. Recent articles in the *New York Times* by a reporter with extensive anthropological contacts regularly refer to the Yanomami as a "Stone Age" people (e.g., Brooke 1991)—but no Yanomami today uses stone tools, few if any have used them in recent decades, and those who have, found their stone axes in ancient gardens of extinct peoples (Barandiaran 1967:29–31; Chagnon 1977:24; Cocco 1972:193; Wilbert 1972:30–31). There is no indication that Yanomami ever made stone tools themselves.

This chapter will demonstrate how different the Yanomami appear when seen in relation to historical changes in their basic technology. A number of established debates about the Yanomami will be considered—including whether or not they were exclusively hunters and gatherers in the recent past; the role of protein scarcity as a limiting factor; and the reasons for their warfare. Beyond that, I hope to show that several widely accepted characterizations of the Yanomami are highly questionable when seen through a "lens of steel," and to suggest how these foreign items contributed to a broad transformation of their sociocultural organization.

Some important qualifications are needed before proceeding. While the technological advantages of steel tools[4] are clearly established, how much of an advantage they represent may be exaggerated by the artificial nature of some experiments (e.g., Carneiro 1979b). Even with the limited observations available, it is clear that tree hardness and diameter strongly affect relative efficiency (Carneiro 1979a;

Townsend 1969). Obviously, the significance of cutting tools would vary by the total amount of cutting that a people might do. Thus the advantage of steel blades will expectably vary by ecology and culture, and the practical significance of their introduction merits further investigation by experiment, and via archaeology and ethnohistory.

The demand for steel tools—how many are needed—is also affected by a combination of factors besides efficiency, and these are difficult to assess. Among the Yanomami specifically, demand is affected by the requirements of different types of men's and women's labor, the possibility that one tool (a machete) can be made into several usable edges, the durability of metal tools,[5] and of course, the number of "consumers" and the exchange relationships between them (Ferguson 1995c:24). No doubt other concerns could be added.

Tools are only one of the Western trade goods sought. As basic tools become more abundant, interest shifts to other goods (see Salisbury 1962; Wike 1951). Those being drawn into the lower levels of national societies may acquire a spectrum of mass-consumption items. Yanomami today need medicines, clothes, shotguns, radios, outboard motors, and much more (Ferguson 1990:244). Even from the start of contact, nonutilitarian items, such as beads and red cloth, are coveted. That is no surprise: we know of the importance of prestige items in indigenous trade systems (Bishop 1983; Friedman and Rowlands 1977)—why not with Western contact? As Whitehead (pers. comm.) notes, tokens of connections to powerful intruders may be especially important political symbols.

Finally, trade is only one aspect of contact situations, which often include hugely negative consequences for indigenous peoples. For that reason, it is quite expectable that some peoples, familiar with the horrid details, will avoid or flee from contact (Ferguson 1990:246). But important though these exceptions and countercurrents are for understanding the role of indigenous agency, they remain exceptional. The general pattern is that those without steel—Yanomami, and nearly everybody else around the world—make great efforts to get it. Neither these complicating factors, nor the current anthropological fashion of exoticizing others and spotlighting cultural difference, should blind us to basic human commonalities: given a choice, people will not choose to work three to ten times as hard to accomplish the same tasks.

The Hunter-Gatherer Debate

Through the late 1950s, when precious little was known about the Yanomami, they were considered to be "essentially hunting, fishing, and gathering nomads" (Steward and Faron 1959:434). Fieldworkers among the Yanomami in the 1950s and early 1960s adopted the position that Yanomami began gardening only recently (Peña Vargas 1981:25; Wilbert 1963:187–188; 1972:14; Zerries 1955:73). Their "marginal" position on the fringes of the tropical forest culture pattern was explained in two ways: as survivors of ancient hunters and gatherers, or as devolved horticulturalists pushed from the rivers into less-productive zones (Steward and Faron 1959:374–378; and see Albert 1985:33–35).

Napoleon Chagnon (1966:46–49) first challenged the idea of a recent shift to agriculture, pointing out the existence of gardens in early reports, and the fragile evidentiary basis of the assumption of hunter-gatherer status. Jacques Lizot (1977: 498–499; 1980:3–7; 1988:506) has argued repeatedly and heatedly for the antiquity of agriculture. Both assert that the reliance on agriculture they observed around the juncture of the Orinoco and Mavaca Rivers (hereafter the Orinoco-Mavaca area) in the 1960s, where groups of 130 or even more people lived in one location for years with only limited use of wild foods other than game, represented Yanomami subsistence as it had been practiced for hundreds of years. And so the question: Did the Yanomami only recently "discover" agriculture, or have they been settled cultivators for centuries? Steel may provide part of the answer, but the question itself is not that simple.

Historical sources provide limited and inconclusive evidence. Five reports from 1767 to 1911 include some information on Yanomami subsistence practices. In 1767, the missionary de Jerez reported that the Yanomami "nation maintain[ed] itself" with wild cacao, which they preserved for consumption later in the year—something no twentieth-century Yanomami is reported to do (Cocco 1972:45). Around 1804 the priest Ramón Bueno (1965:135, 145) came to rather fantastic conclusions—that Yanomami wore skins, slept in trees, and subsisted entirely on gathered roots. But we can give more weight to his lament on how difficult it was to capture Yanomami because, he said, they never stayed in one place.

The botanical explorer Robert Schomburgk (1841:221), in 1838, found a northern Yanomami group along the Uraricoera River to have had a small garden with manioc (of what type is not indicated) that they visited between treks. Their location along a very active trade route would have put these Yanomami in direct contact with many horticulturalists, but that was highly unusual for Yanomami of that time. Also in the 1830s, Agustín Codazzi (1940, 2:48) produced the first ethnographic census of the Upper Orinoco region, in which the Yanomami are noted as living off fishing and hunting, without any reference to cultivation.

Theodor Koch-Grünberg (1979, 3:250, 257–258), traveling the Uraricoera route in 1911–1912, again found Yanomami with manioc gardens, but was told by them that this was a practice recently learned from their neighbors, from whom they had also obtained steel tools. In the past, they said, they had not had gardens, and even then other Yanomami in the highlands lived by hunting, fishing, and gathering. But Koch-Grünberg himself notes Schomburgk's observation of gardens in this area more than seventy years earlier.

These passages clearly indicate that Yanomami from the eighteenth century to the start of the twentieth were much more mobile and reliant on forest products than the Orinoco-Mavaca people later observed by Chagnon and Lizot. Koch-Grünberg's report indicates *some* pure foraging, but *some* Yanomami had gardens in 1838 and 1911. There is no reason to believe that the practice of gardening was restricted to them alone. Whether Yanomami who were not exposed to other horticulturalists had small gardens, or lived by hunting and gathering alone, will not be answered by historical sources.

Another kind of evidence concerns recent Yanomami subsistence patterns. Yanomami are skillful gardeners, with a highly diverse range of crops for this region (Harris 1971:478–482; Lizot 1980), suggesting a knowledge of horticulture well antedating this century. beyond that, their primary crops offer clues, although no hard answers, about past agriculture.

The Yanomami are unusual among Amazonian slash-and-burn farmers in that the banana, not bitter manioc, is their staple crop (although settled groups exposed to outsiders take up manioc quickly [Barandiaran 1967:37; Colchester 1984:300–302; Montgomery 1970:105; Ramos 1995:25]). If, as is commonly believed, bananas are a post-Columbian introduction to South America, that fact alone would strongly support the idea that Yanomami became gardeners sometime after 1492. But while the banana is not native to the New World (Holoway 1956; Simmonds 1962), the possibility of pre-Columbian banana cultivation is real (Patiño 1958).[6] And as Daniel de Barandiaran (1967:35) points out, a banana soup is an integral part of the mortuary rituals of all Yanomami, including divisions that appear to have separated centuries before Columbus. Thus, while there is reason to suspect that the ancient Yanomami were gardenless foragers, it is not definite that they actually were.[7]

The Yanomami's recent agricultural organization indicates a past—whenever that past began—in which agriculture was integrated with a great deal of mobile foraging. Bitter manioc requires a cumbersome technology—grater boards, basket presses, and griddles (see Ferguson 1988:144)—for processing the toxic tubers into food, and this burden is inconsistent with high mobility. Preparing bananas, of course, requires no such tools, and reliance on them is consistent with a pattern of trekking between scattered gardens. So is Yanomami reliance on the protein-rich peach palm, a cultivated tree that bears fruit for thirty years or so (Barandiaran 1967:43–47; Lizot 1980:34). Among some twentieth-century groups—most notably those who held the captive Helena Valero (Valero 1984)—seasons of peach palm ripening are times of visiting and trekking between old gardens.[8]

But could they live as pure hunter-gatherers? The Yanomami are superb at living off nature, and frequently opt to do so for weeks or even months at a time (Biocca 1971; Good 1983, 1989, 1991; Valero 1984). The antiquity of their trekking seems beyond question. Yet in those few reported instances when Yanomami were compelled to live for extended periods with no, or unusually small, gardens, they suffered from hunger, even in areas of comparatively rich resources (Biocca 1971:34, 39, 43; Lizot 1974:7; Valero 1984:170). Moreover, the pre-twentieth-century restriction of Yanomami to high country (see Ferguson 1995c)—which, besides having poor soil and a tendency to drought (Huber et al. 1984:108–109), is characterized by wild resources that are scarce and unreliable (Colchester 1984:294; Chagnon 1992:83)—would have exacerbated the difficulties of exclusive foraging.

Yes, Yanomami are able to survive as pure hunter-gatherers, but with difficulty. (As argued below, minimal gardening in the past may have been associated with long-term population decline.) No instance has been reported where they have freely chosen to live without gardens. Probably they would find the notion ridiculous. My conclusion: If any Yanomami had the opportunity to establish gardens in the past, they had the motivation and the knowledge to do so. But what determined their ability to develop gardens? That question brings us back to stone and steel.

It may be possible to make a garden using clearings from natural tree falls (Laura Rival, pers. comm.), or just by girdling and burning trees that are left standing (Tristram Kidder, pers. comm.). But even in such cases, ground clearing, not to mention tree girdling, would be most difficult without some cutting tool. Despite exceptions, the general rule is: to create a garden, you have to fell some trees.

As noted earlier, for tree felling a steel axe is three to ten times more efficient than one of stone. Significantly, the high measure is from Robert Carneiro's (1979a,b) experiment among the Yanomami. Carneiro concludes that garden clearing with stone alone would require more work, and would rely more on burning and domino-effect tree falls. One account also indicates that Yanomami possessing little steel use liana ropes to pull down weakened trees (Valero 1984:183).

Marcus Colchester (1984) considers Yanomami gardening at the reduced efficiency of stone axes, and concludes that its higher costs and lower productivity would tip the scales toward greater reliance on hunting and gathering than among any recently observed Yanomami. That conclusion is supported by Raymond Hames (1989:73) and Stanford Zent (1992), and is consistent with the historical ecology reconstructed above.

But it is not clear that all Yanomami had stone axes. I found no indication anywhere that any Yanomami ever knew how to make a stone axe. On the contrary, those who remember them being used recall that they were found in the surface remains of extinct peoples (Barandiaran 1967:29–31; Chagnon 1977:24; Cocco 1972:193; Steinvorth de Goetz 1969:30; Wilbert 1972:30–31)—not an unusual situation in Amazonia (Balée, pers. comm.). But these remains and axes are not found all over. They are abundant along the far Upper Orinoco (Anduze 1960:96, 210; Grelier 1957:129–130, 139, 144), where there may have been an axe-manufacturing center, probably among people wiped out in the terrible slave-trade wars of the mid-eighteenth century (Ferguson 1995c:79–82). To the north in Sanema country, stone axes are rare (Barandiaran 1967), and Yanomami encountered by Koch-Grünberg along the Uraricoera in 1911 were completely unfamiliar with one he showed to them. Thus the availability of found stone axes for the Yanomami was variable.

Also variable was the availability of steel axes. There is no way of knowing whether Yanomami received any steel through indigenous trade networks prior to the mid-eighteenth century. Comparative studies of Western contact make that seem likely, although it was probably in extremely limited amounts. By the late 1760s, however, Spanish soldiers on the Upper Orinoco were visiting Yanomami directly (Cocco 1972:45), and Portuguese did the same along the Rio Negro by the late 1770s, if not earlier (Hemming 1987:30–32). These Europeans wanted to "reduce" the Yanomami and other Indians—i.e., entice them to come down to settle in villages where they could be used and controlled; standard *reducción* procedure was to offer gifts, but how much steel was introduced in this way is completely un-known. By 1800, Upper Orinoco Yanomami were engaged in routine trade at the Spaniards' farthest outpost, La Esmeralda (Humboldt 1889:460–463). From then until today, some steel has come to some Yanomami—in trickles or in torrents, in frequently shifting geographic patterns, and with varying amounts of associated risk.

Considering the point of the availability of axes in light of this discussion, three basic subsistence patterns appear possible: (1) no axes—hunting and gathering with no, or only very limited, gardening; (2) stone axes—extensive hunting and gathering between short visits to small gardens; (3) steel axes—regular trekking be-tween longer stays at larger gardens. But still one more historical factor must be considered.

Garden clearing requires burning, more so with stone axes. Burning creates smoke. In the times when Yanomami were sought to feed captive labor markets—which occurred in many periods and places, from the mid-eighteenth century or earlier, and probably up to the early twentieth century—raiders located targets by climbing trees and looking for smoke (Gilij 1965, 3:96–97). Garden makers would be sending up a beacon for knowledgeable indigenous scouts who knew when and where to look. A strong presence of these social predators would greatly encourage maximum mobility among their intended victims.

The question that began this section, whether the Yanomami of the eighteenth and nineteenth centuries were hunter-gatherers *or* agriculturalists, may propose a false dichotomy. Taking into account all considerations—early historical reports, the implications of different aspects of Yanomami subsistence, reconstructions of subsistence options in the absence of steel, and variations in the availability of stone and steel axes and the danger of raiders—one answer seems to fit all: that pro-tohistoric Yanomami subsistence was not one pattern, but a range of patterns, in-cluding all three options noted above. This postulated range went from exclusive reliance on hunting and gathering all the way to manioc gardening, with the posi-tion along this range strongly conditioned by the availability of different kinds of axes and the danger of raiders.

But if greater reliance on foraging in the past was a more efficient use of time and effort, it did not necessarily guarantee a secure existence, especially in the rel-atively resource-poor Parima highlands. Elsewhere I have suggested that some areas of Amazonia may operate as "population sinks," places of long-term demo-graphic decline, and that the Yanomami's highland homes were such locations (Fer-guson 1989a:255; 1995c:74). At any rate, those Yanomami did not increase at any-thing like the rate of the past hundred years or so, during which Yanomami settlement area expanded dramatically, and single villages gave rise to whole clus-ters of local groups (see figure 13.1). Marvin Harris (1977:50–51) hypothesized that this growth and expansion resulted from the introduction of bananas and steel tools—a view that has since gained important support (Albert 1989:637; Colchester 1984: 292–293; cf. Lizot 1988:496–497).

To recast this hypothesis in terms of previous discussion, population growth fol-lowed a historical shift to pattern (3), regular trekking between longer stays at

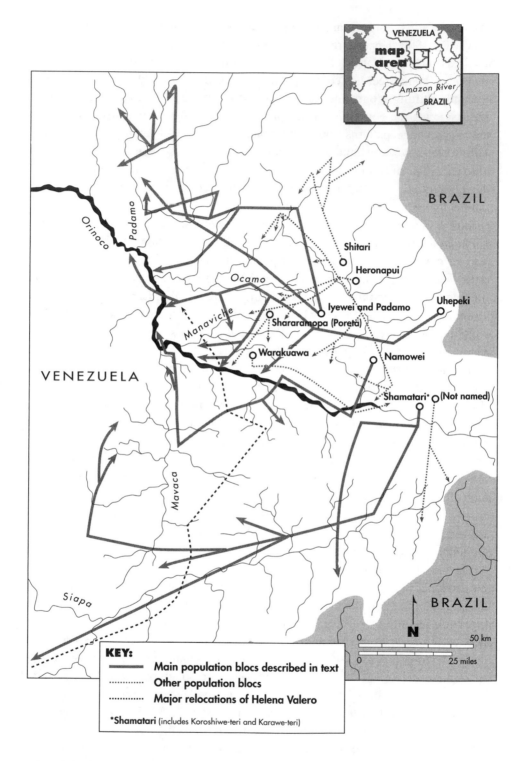

Fig. 13.1 *Main Movement of Western Yanomamo Population Blocs.*(From Ferguson 1995c: map 4.)

larger gardens. The rubber boom of the late 1900s brought major quantities of steel into the region, some of which went to Yanomami. Although some local rubber barons were terribly brutal, there is no indication of major captive raids deep into Yanomami highlands (which is not to say that none at all occurred). With more steel, and less danger, the Yanomami could develop the semisettled villages observed (beyond or before the missions) in the twentieth century, of the kind discussed in the next section.

In examining evidence related to ancient subsistence technology and patterns, facts were noted that compel a reevaluation of the widely accepted image that the Yanomami were isolated from and unaffected by the outside world until the mid-twentieth century. This brief review has noted relations of violence and exchange that directly and indirectly linked Yanomami to Europeans from the mid-eighteenth century at least, and that had major consequences for the Yanomami way of life. The rest of this chapter will follow the thread of steel through other changes that radically affected twentieth-century Yanomami, both before and during the time that anthropologists arrived on the scene.

Twentieth-Century Yanomami Ecology

Yanomami adaptation to their environment has given rise to one of the most contentious issues in recent anthropology, the "Great Protein Debate" (discussed in Ferguson 1989a,b). This debate developed on the basis of portrayals by Chagnon, and to a lesser extent on those by Lizot. As noted previously, both assert that the fairly large and relatively sedentary groups they studied are typical of Yanomami subsistence for centuries. The originators of the protein theory of Yanomami warfare—Jane Bennett Ross (1971), William Divale (1970), and Marvin Harris (Divale and Harris 1976; Harris 1984)—all accepted that this was a long-established pattern (although Harris, as noted, gave it only a century).

Previously, many ethnographers had observed the lack of protein in Amazonian cultigens, and had suggested that for people who relied on hunting to flesh out their diets, game availability set a limit on the size and duration of settlements: relatively large and settled villages would run through the fauna. Chagnon (1977:33), initially at least, indicated game scarcity in the Orinoco-Mavaca area. There seemed good reason to expect that the Yanomami—inland hunters, even though they had recently moved to large rivers—might have problems with protein. The warfare theory flowed from this reasoning. War was argued to be an evolved adaptive complex, maintaining diet by dispersing settlements and reducing population growth.

But we have already seen that the pronounced sedentism of Orinoco-Mavaca people ca. 1964—Chagnon's "Fierce People"—was not characteristic of Yanomami in the nineteenth century and earlier. If that sedentism is a recent development, how and why did it come about? Environmental changes cannot explain the shift. To understand this and earlier changes in subsistence orientation that affect the applicability of the protein hypothesis, we must turn to Upper Orinoco history (Ferguson 1992; 1995: chs. 10–13).

Yanomami around the Upper Orinoco received many steel tools during the rubber boom of the late nineteenth and early twentieth centuries. They also faced the danger of a brutal local governor, Tomás Funes, who killed hundreds of Indians who were not sufficiently productive. His death (in 1921), following the collapse of rubber prices, led to wholesale Creole abandonment of the area. From then on, local Yanomami were not in any danger from outside raiders. They still had steel tools obtained during the rubber boom, although wear brought these to extreme scarcity during the period of maximum Western retraction, in the 1920s and 1930s. From the mid-1930s onward, woodsmen slowly reentered Yanomami lands, offering a new and relatively nonthreatening source of steel, which the Yanomami obtained through work, theft, and assault. How did they subsist in this time of some steel and no raiders?

Fortunately, we have an excellent source for much of this period: Helena Valero, a Brazilian girl captured by Yanomami in 1932 or 1933, who lived with them in the Orinoco-Mavaca area until 1956. Two independent tapings of her life story (Biocca 1971; Valero 1984), and especially the more complete Spanish-language version, show a subsistence pattern of leaving main gardens for extended hunting and gathering expeditions, often combined with visits to old and new gardens, or to the villages (and gardens) of allies. The average time in different modes cannot be estimated, but a typical pattern would be to spend one to three months at one garden, and then several weeks in the forest, before arriving back home or at another garden.

Missionaries who established stations among the Yanomami in the 1950s and 1960s report that at first the surrounding groups would periodically abandon the mission to spend time in the forest (Jank 1977:85–86, 177; Vareschi 1959:169). Recently Kenneth Good (1989; 1991), who has spent years among a Yanomami group that has relatively abundant steel but is not linked to any Western settlement, has provided detailed evidence on subsistence mobility. On the basis of his and other reports, we can estimate that Yanomami with steel tools but not exposed to the danger of raiders spend *up to* 40 percent of the year on trek, and also spend time at other gardens, so that less than half of the year is spent at their primary garden. This is more time in one place than indicated for earlier periods, but much less than that reported for the center of the Orinoco-Mavaca contact zone, where villages are never left empty and forest products are much less important (see Ferguson 1992:205).

The question of why these Yanomami are not *more* sedentary raises old issues of adaptation, bringing ecology into history. Plainly, the mobile pattern is more conducive to maintaining adequate game intake than the sedentary one. But as Good has shown, game scarcity is only one reason for trekking: cultigen conditions may be just as important. The exhaustion of ripe bananas at one garden, or the ripening of forest fruits and peach palms at old gardens, are incentives to trek, as are a variety of other considerations.

Whatever triggers the movement, this mobility pattern spreads hunting across a much wider area of forest. Thus the significance of game depletion for explaining war seems less relevant to these trekking Yanomami, and the whole debate would have proceeded very differently if this subsistence pattern had been better known earlier. But the protein controversy as it did develop concerned wars involving the more settled peoples described by Chagnon and Lizot, which brings us back to the question, How do we explain that unusually sedentary pattern of the Orinoco-Mavaca area in the mid-1960s?

To understand the sedentary character of those villages, we again must consider steel—not its utility, but rather the problems of getting and protecting access to sources of it and other Western goods. A pervasive theme in twentieth-century Yanomami history is that of exploration, trade expeditions, and garden relocation to, and then down, streams that led to Western settlements—or to other indigenous people such as the Yecuana who had a source of steel tools (see figure 13.1). Equally pervasive is the observation that those who settle alongside a Westerner, or get one to stop among them, zealously attempt to monopolize the intruder and his goods (Albert 1988:102–103; Chagnon 1974:7–15, 163–171; 1977:79, 152–153; 1983:18; Peña Vargas 1981:30; Shapiro 1972:33, 43). In these and many other reports, machetes and axes are specifically noted as the objects of greatest desire.

Since the 1940s especially, the lure of steel has pulled Yanomami out of highland forest into new locations. Within the highlands, they have gone into resource-poor savannas where outsiders have built their airstrips (see Huber et al. 1984; Jank 1977; Smole 1976), but little is reported about ecological adjustments there. Somewhat better reported is what happens when Yanomami move down to Western outposts along major rivers, such as the Orinoco.

This is not a move that the Yanomami would undertake for subsistence purposes. There is little or no fertile flood plain along the Upper Orinoco (Lizot 1980:47), and flooding is a problem for lower gardens (Cocco 1972:176, 419); thus, getting better agricultural land is not an incentive here. The rivers are rich in fish, but the Yanomami who moved toward them were equipped to exploit only small streams (Biocca 1971; Valero 1984). Mosquitoes and other insects are a severe problem along the rivers (Chagnon 1977:161; Smole 1976:47), and the rivers themselves were an obstacle for canoeless Yanomami (Chagnon 1992:83; Valero 1984:503, 505, 514). At the same time, Yanomami have long understood that coming near to Westerners increases the possibility of disease (see Albert 1988; Saffirio and Hames 1983:5), along with other, less lethal aggravations.

What prompted Yanomami to move down to main rivers—investigators from all over Yanomami-land agree—was their desire for steel tools and other Western goods (Albert 1988:98; Cocco 1972:18, 32, 114; Colchester 1985:10–11; Peters 1973:62–63; Saffirio 1985:24, 91–93; Shapiro 1972:25–29; Smole 1976: 51–52). But if they came out to get access to these products, why not continue their mobile, trekking pattern from riverine bases, as they were initially reported to have done?

The answer is that obtaining Western goods involves, not merely being near to a Westerner once in a while, but also controlling access to the source. As I will explain shortly, there is competition both to get these favored positions, and to prevent others from gaining direct access to them. Thus Yanomami not only came to live close to Westerners, but stayed at home once they were there. (Missionaries actively discouraged trekking [Jank 1977:85–86, 177; Lizot 1976:12; Vareschi 1959:169], but their exhortations seem much less significant than the Yanomami's own trade interests.) Furthermore, individuals and families were allowed by mission Yanomami to move in with them, thereby adding to the size of those villages (Cocco 1972:212–213; Early and Peters 1990:66; Eguillor García 1984:54–56; Ramos 1972:41). Thus the relatively large and sedentary groups of the Orinoco-Mavaca area of the mid-1960s, formerly portrayed as typical of long-term Yanomami adaptation, are better understood as the result of recent efforts to establish and control access to sources of steel tools and other Western goods (Ferguson 1995c: chs. 11–13).

This new pattern of large and sedentary villages, well stocked with Western goods but also exposed to all the hazards of Western contact, set in motion fundamental and far-reaching changes in these groups' interaction with the natural environment (see Ferguson 1992:203–208). For one thing, women, whose gathering was critical while on trek and starting new gardens in the forest, were reduced to being little more than bearers of water and firewood—one of several changes that underwrote the low status of women in the Orinoco-Mavaca area (Ferguson 1995c:357–358). Sedentism also puts more pressure on game resources, which brings us to the protein debate, as seen in the context of all the subsistence changes entailed by moving into contact with Western providers of steel.

Despite all the argumentation, data about game availability and intake in the Orinoco-Mavaca area are limited and most arguable, especially through the middle 1960s—which encompasses the wars described by Chagnon and Lizot. Those authors vigorously dispute the existence of game scarcity until after shotguns became widespread in the later 1960s, although both agree that the entire region suffered drastic game depletion by 1975 (Chagnon 1977:148; Lizot 1976:13). Nevertheless, evidence from many other Amazonian peoples (Ferguson 1989b), from other Yanomami, and anecdotal information from the Orinoco-Mavaca area itself (Chagnon 1977:148; Colchester and Semba 1985:17; Comité 1983:27; Saffirio and Hames 1983:37–38; Saffirio and Scaglion 1982; Salazar 1967:92; Smole 1976: 163–167, 175–176) all suggest that local hunting was substantially less productive by the mid-1960s.

I have challenged (Ferguson 1989b) the game-depletion explanation of Yanomami warfare. Although several of its main postulates are supported, it is theoretically inadequate to explain war, and unnecessary because another, stronger explanation is available (below). But, following Good (1989:135–140), I also argue (Ferguson 1992:206) that the declining amount of game in the Orinoco-Mavaca contact zone was accompanied by a decrease to the point of elimination of the village-wide sharing of meat. This along with other contact-related factors led to a breakdown in reciprocity and social ties, paving the way for greater individualization of production and consumption.

The nutritional challenge posed by a declining availability of game could be off-set by an increased reliance on fish and other river resources. Local Yanomami quickly took to canoes, and one mission post gave out almost a million fishhooks between 1957 and 1978 (Peña Vargas 1981:37). Missions also provided food; the quantities are entirely unknown, although missionaries claim that their food distrib-utions prevented starvation in the crisis following drought, flood, and a malaria epi-demic in 1964, just before Napoleon Chagnon arrived (Cocco 1972:176). Individ-ual Yanomami also produced food for trade or sale in Creole towns downriver, and some developed extensive banana fields for this purpose (Chagnon 1977:144). By the beginning of the 1980s, some Yanomami near the juncture of the Mavaca and Orinoco were starting to look like peasants, selling crafts and buying food at a local cooperative and even living in single-family homes (Chagnon 1992:221–222).

Put in historical context, the large sedentary village pattern observed by Chagnon and Lizot in the middle 1960s may be seen as one phase in a long subsis-tence transformation, into which the Yanomami were drawn by the glittering prize of steel. Ironically, since Harris (1984:112) states that the Divale-Harris model is not intended to apply to peoples strongly affected by Western contact, one could say that the theory is not applicable to the main case upon which it was developed.

The Social Ramifications of Steel

Up to this point, I have kept my focus on Yanomami ecology, arguing that to under-stand their modes of interaction with the environment, one has to consider both the efficiency and the availability of cutting tools. In this section I shall discuss some of the sociocultural ramifications of the introduction of steel tools, and the other new consumer demands that follow in their wake.

The Orinoco-Mavaca area up to the mid-1960s offers the following picture (Fer-guson 1992): Greater access to Western goods brought exposure to new and deadly diseases. High rates of mortality were associated with contact, and these deaths tore at the fabric of kinship. This, along with the decline of meat sharing, led to the at-omization of society, and fostered the instrumental use of force in interpersonal rela-tionships. At the same time, the anchoring effect of controlling access to Westerners diminished the traditional option for individuals or groups in conflict situations—i.e., to move away from the problem (see Ferguson 1989b:195–196). These and other changes combined to lower the threshold for violence, to make the people seem fierce.

On top of that, entirely new antagonisms were created through competition for Western goods. The principal argument of *Yanomami Warfare* (Ferguson 1995c) is that the actual occurrence of reported wars among the Yanomami can be explained as the result of antagonisms created by competitive interests in steel tools and other Western manufactures. With the introduction of these goods from a few limited sources, intervillage trade was reoriented toward missions or other sources of

Western goods. As noted, those close to such a source made every effort to monopolize access to the provider. Mission groups with exceptional supplies of steel and other items ceased making local labor-intensive manufactures, and instead obtained them through trade with other groups (Cocco 1972:205, 376–378; Colchester 1984:298; Eguillor García 1984:126; Peters 1973:167–168). The limited information available suggests that this new trade pattern was accompanied by a shift from balanced to exploitative exchange, with realized labor time flowing toward the outpost villages. (This is something that could be measured.)

Bringing steel and other Western goods into the picture at this point allows the correction of another widespread misunderstanding about the Yanomami: that their trading is a social bond without any material basis. In a frequently cited passage, Chagnon (1977:100–101) asserts that Yanomami pretend to forget how to manufacture some item they could easily produce, in order to create a pretext for trading that might lead to a broader alliance without seeming to ask for help. The supposed "forgetting" described by Chagnon takes on an entirely different cast when one learns that it was the mission Yanomami who stopped making everything, and who were supplied with local manufactures by trade partners dependent upon them for Western goods.

The introduction of Western goods brings a parallel transformation in marriage patterns. There is a sharp increase in village exogamy, as primary interests shift from one's coresidents to those who can supply the coveted steel. This new intermarriage is unequal: as men cede wives to gain access to outpost largesse, there is a pronounced flow of women into the mission villages (Albert 1988:102–103; Chagnon 1966:57–58; 1977:80; Cocco 1972:210–213; Peters 1973:127–129; Shapiro 1972:210–213; Smole 1976:72). Bride-service patterns are similarly skewed—heavy for in-marrying males, light or nonexistent for those who can provide a wife's family with Western goods (Cocco 1972:211; Chagnon 1977:79; Early and Peters 1990:67; Peters 1973:122–129). All together, groups that can monopolize a Westerner or otherwise act as middlemen in the Western trade can receive great material and social benefits for doing so—hence the great interest in monopolization.

In this regard too I would offer an adjustment to a widespread perception about the Yanomami. Everyone knows that some Yanomami men contend over the marital disposition of women (although this is far more unusual than is generally imagined). But few realize that the most pronounced pattern in exogamic unions of all types is that women are given to men who can reciprocate with steel and other Western manufactures. In my perspective, if conflict over women does trigger war, it is because the contested women are elements within a total exchange and alliance relationship that has gone bad. Whether alliances are happy or strained is to a large degree determined by the distribution of Western goods. That brings me to political relations and war itself.

My hypothesis, stated briefly, is that Yanomami go to war to protect or enhance their access to Western goods. Sometimes this is done very straightforwardly, by plundering small but well-equipped parties or settlements (Arvelo-Jiménez 1971:42,

93; Barandiaran and Walalam 1983:98, 102–103, 191; Cocco 1972:53–54, 64–65, 67, 70, 74, 374, 376; Colchester 1985:47; Peters 1973:155). These raids, though telling much about the motivational significance of steel, constitute only a small part of the Yanomami war record. My argument is *not* that Yanomami warfare, generally, is aimed at plunder, but rather that the goal of improving access to Western goods can be discerned in the patterning of other wars. Antagonisms based on the distribution of these goods historically structure political relationships, relationships that can turn into wars ostensibly fought over revenge, sorcery, women, or prestige.

This logic, my book seeks to show, is substantiated by the spatial and temporal patterning of all reported Yanomami wars—who attacks whom and when. With some simplification, military antagonists occupy different positions in the existing Western trade, and the aggressors would benefit in that trade by driving their enemy away. Political relations go from peace to war after a major change in the Western presence. Similarly, alliances incorporate the trade partnerships and marriage relationships structured by and structuring the flow of Western goods. The use of force within alliances also affects the direction, rate, and (hypothetically) terms of trade of those items.

There is more to it than that, of course. Along with the availability of Western goods, political relations are partly determined by the ability to apply force, something that is also affected by ties to Westerners. And political relations are not talked about in the material-interest terms argued here; rather, aggressors invoke moral principles in entirely personalistic terms, so that their adversaries in war—sometimes close kin—always "have it coming." The particular logics of Yanomami culture will thus enter into decisions, as with any people anywhere, but in my view these logics are brought into accord with the practicalities of material circumstances.[9]

That, in brief, is my explanation of Yanomami politics and war, an explanation that begins with the critical significance of steel tools for Yanomami. Two other studies illuminate other aspects of the influence of steel.

John Fred Peters's thesis (1973) describes "the effect of Western material goods upon the social structure of the family" among an eastern Yanomami mission group. Along with changes in marriage patterns as already indicated, Peters describes broader transformations. The possession of Western goods became a key to both individual and village status. As with the Yir Yiront, the traditional social order ascribed by sex and age was upset in favor of one oriented to achievement, as measured in new merchandise. The most Western goods went to those younger men, and sometimes women, who were quickest to grasp what outsiders wanted. One young man with exceptional ties to the missions and Brazilians was recognized as politically precocious, already telling people what to do. New inequalities developed, with different members of one village earning from $2 to $80 worth of goods in one year.

Bruce Albert's article "La fumée du métal" (1988) demonstrates how steel tools come to occupy a central place in Yanomami symbolism. In his reconstruction of the changing interactions of Yanomami with outsiders over the past 150 years, a key element in their own understanding of the process is the idea that metal objects give off a smoke, which causes epidemics. This is just one point in a much more extensive and sophisticated analysis, an analysis that shows the need for more research on the Yanomami's own conceptualization of Western contact (also see Hill 1988). For now, it is enough to conclude that every aspect of Yanomami culture, from subsistence to worldview, is strongly affected by the introduction of steel tools.

Summary and Discussion

Following the thread of steel has led us through many corners of Yanomami ethnography. In regard to the hunter-gatherer issue, consideration of the efficiency and availability of stone and steel axes, along with other factors, provides a way to understand the mixed evidence on ancient subsistence practices, suggesting the existence of a variable range of subsistence orientations. Along the way, that discussion highlighted the illusory character of the Yanomami's presumed isolation from the outside world.

From the late nineteenth century, the chosen pattern for most Yanomami who possessed steel has been to spend most of their time at different gardens, interspersed with extended trekking. The relatively large and very sedentary villages that became the ethnographic basis for the protein-and-warfare debate are more recent innovations. In those cases, Yanomami were drawn to new environments and encouraged to stay in place in an effort to obtain and control a flow of steel tools and other Western goods. This new existence led to a major overhaul of subsistence orientations, with important social consequences. The Orinoco-Mavaca pattern of the mid-1960s was a stage in a process of transformation that may ultimately lead to their becoming "peasants" (Chagnon 1992:221–222). Seen this way, there is no reason to expect that warfare in the Orinoco-Mavaca area in the mid-1960s should or even could be an evolved complex adapting people to local ecology.

In regard to more general social ramifications, the allure of steel and other Western goods drew some Yanomami into direct and prolonged contact with Westerners—an exposure that in some cases produced social atomization and anomie, adding to an image of "fierceness." I argue that actual warfare arises out of antagonistic interests involving differential access to sources of Western goods. That argument draws on consideration of economic, marital, and political relations that are shaped by access to steel, and suggests correctives to widespread misunderstandings about intervillage trade and marriage. Other research shows additional social and symbolic transformations brought on by the introduction of Western goods (along with other things).

These are the principal findings that come out of greater attention to steel tools among the Yanomami. As with the Australian Yir Yiront, there is no reason to presume that the Yanomami are unusual. This suggests broad new avenues for research in historical ecology. Zent (1992; and chapter 12, this volume), for instance, provides a wide-ranging exploration the impact of steel tools on the subsistence, settlement, and society of the nearby Piaroa. Whitehead and I (Ferguson and Whitehead 1992) demonstrate the huge comparative base that exists for studying these and other contact-related transformations.

On the subject of war, I have shown (Ferguson 1990) how many cases of Amazonian warfare can be clearly linked to interests in Western goods. With the historicization of Amazonian studies (Roosevelt 1994), highland New Guinea has emerged as a crucial locale for formulating and testing theories about indigenous warfare relatively unaffected by Western contact (Ferguson 1995c:408; Knauft 1993:1186). But even there, outside influences have not yet been given sufficient attention. Salisbury (1962:118–119), for instance, records a major intensification of war just as steel axes began filtering into the highlands during World War II. Robert Crittenden and Edward Shieffelin (1991:132–138) have recently described the spread of new tools and diseases up valleys from the coast, starting in the late nineteenth century; their discussion of efforts to monopolize this trade, and of the unusually intensive warfare associated with the situation, has both strong parallels and intriguing contrasts with the Yanomami tribal zone (e.g., in the efflorescence of prestige economies). The political history of highland New Guinea is ripe for reappraisal from a tribal zone perspective. Once that is done, we can be more certain about what is purely local in the warfare that anthropologists have reconstructed, observed, and filmed.

William Denevan (1992) and William Doolittle (1992) take the significance of steel so far as to suggest that shifting agriculture was rare or nonexistent in the pre-Columbian New World, and developed only with the introduction of metal cutting tools. Both argue that ancient cultivators relied primarily on intensive cultivation and repeated field use—evidence of which has been accumulating throughout the Americas.[10] Indeed, the ancient inhabitants of the Upper Orinoco—not Yanomamo—apparently left their mark in what are today square patches of bamboo forest, suggesting intensive and repeated cultivation (Lizot 1980:41).

These topics also frame new areas for archaeological research (see Bamforth 1993; Rogers 1993). Archaeologists, better than anyone, may be able to trace the dissemination of Western tools and other goods, and their possible impact on subsistence. They may be able to reconstruct trade networks and the other forms of integration that typically accompany them, and the war patterns that often do the same. In that, archaeology may make a crucial contribution to our understanding of tribal warfare, and help explain the great disparity observed in the violence of prehistoric and historic peoples (see Ferguson in press).

Another reason to pay more attention to steel and other aspects of Western contact is suggested by Alcida Ramos (1990), who notes that North American anthropologists generally study topics that direct their attention away from the impact of Western contact. By recasting our research orientation to encompass the topics raised here, it would be possible to merge established North American theoretical concerns with the *indigenista* focus on the continuing problems of contact (Turner 1993).

In countless situations around the world, steel tools and the goods that follow them are the principal "means of seduction" whereby indigenous people are drawn into cooperation with intrusive agents of the state. John Hemming (1978:9) calls this the "fatal fascination" of Amazonian Indians. Anthropologists have devoted much attention to studying "cultures of resistance"—but in situations of resistance, there are many who cooperate with the invaders. Understanding such cooperation would seem a necessary counterpoint to understanding resistance; and in many cases, the introduction of steel tools can provide one starting point.

Notes

1. These two foci—how human activity produces historical change in the environment, and how historical process shapes human interaction with the environment—suggest a third potential area of research for historical ecology: the long-term effect of ecological relations on the broad processes of social history. How different, for example, would France be if its past four hundred years had occurred in a different environment? If the answer is "that would not be France," the point is made.

2. In another collection, the editors James Spradley and David McCurdy (Sharp 1990:410) refer to the "introduction of an apparently insignificant, hatchet-sized steel axe."

3. Roosevelt (1989:30–34) makes this point based on the emerging archaeology of Amazonia. But even she—in her survey of post-Columbian changes—mentions new technologies only in passing, and steel tools not at all.

4. It would be more accurate to say "iron and/or steel," but the distinction is rarely noted in available sources.

5. Thomas Headland (pers. comm.) found laughable my estimate (Ferguson 1995c:389), based on one single report, that tools could last for twenty-five or thirty years. I willingly concede the point, and hope that some fieldworker will provide a better estimate of durability in the future.

6. Patiño (1958) describes debate from the sixteenth century about whether Spaniards introduced bananas from the Canary Islands, or whether the fruit was already here. The main argument for the latter position is the extraordinary distribution of large-scale banana cultivation encountered in the subsequent decades of conquest. Then again, that may be just another demonstration of the rapidity of pre-observer tribal zone transformations.

7. Whatever the situation of the ancestral Yanomami, they were probably connected in important ways to the more advanced riverine polities of the region. Those polities were obliterated during the eighteenth century, if not before (see Whitehead 1988, 1994). Thus any complementarity of subsistence orientation of highland and river people, or any trade in technology (e.g. axes), may be recoverable only archaeologically.

8. The pattern can be so marked that, a year before *The Fierce People* was published, Barandiaran (1967:47) suggested calling the Yanomami the "people of the peach palm." One can only wonder how the debate over Yanomami warfare would have gone if that name had stuck.

9. A long-standing problem in understanding war has been a failure to distinguish emics from etics. Anthropologists who report the participants' stated reasons for war almost always produce a grab bag of motives, some obviously economic, some reflecting particular cultural values. But does this mélange of motivation truly explain the actual occurrence of war? In the perspective argued here (Ferguson 1995c:364–367), the practicalities of daily life structure social relationships, including those of politics and war—but the underlying structure

will rarely be apparent in elicited rationales. Local discussions about war will be framed in terms appropriate to local moral idioms. Explanations provided to outsiders will reflect this, plus all the vagaries of informant personality and memory, and considerations of what might "play well" with powerful outsider interviewers. Beyond that, the intergroup relationships involved in political process are multidimensional, encompassing trade, marriage, alliance, prestige, and more. Any disputed element may stand for the whole in a strained situation, and thus a seemingly trivial slight may lead to violence. For all these reasons, the emics of war will always produce a variety of explanations. One object of anthropological analysis has always been—at least until recently—to discover underlying order beneath surface explanations.

10. Roosevelt (1989) also stresses the existence of intensive cultivation regimes preceding the more mobile patterns that developed after contact. However, she also finds archaeological evidence for an earlier (ca. 3000–2000 B.C.), less intensive, horticulture based on root crops, although it is not apparent how fixed or shifting the garden plots may have been.

References

Albert, B. 1985. Temps du sang, temps des cendres: Représentation de la maladie, système rituel et espace politique chez les Yanomami du sud-est (Amazonie Brésilienne). Doctoral thesis, Université de Paris X, Naterre, France.

———. 1988. La fumée du métal: Histoire et représentations du contact chez les Yanomami (Brésil). *L'Homme* 28(2–3): 87–119.

———. 1989. Yanomami "violence": Inclusive fitness or ethnographer's representation? *Current Anthropology* 30:637–640.

Anduze, Pablo. 1960. *Shailili-Ko: Relato de un naturalista que tambien llegó a las fuentes del río Orinoco*. Caracas: Talleres Gráficos Ilustraciones.

Arvelo-Jiménez, Nelly. 1971. *Political Relations in a Tribal Society: A Study of Ye'cuana Indians of Venezuela*. Cornell University Latin American Studies Program Dissertation Series, no. 31. Ithaca, N.Y.: Cornell University.

Balée, William. 1989. The culture of Amazonian forests. Pp. 1–21 in Darrell A. Posey and William Balée (eds.), *Resource Management in Amazonia: Indigenous and Folk Strategies*. Advances in Economic Botany, vol. 7. Bronx: New York Botanical Garden.

———. 1993. Indigenous transformation of Amazonian forests. *L'Homme* 33(2–4): 231–254.

Bamforth, Douglas. 1993. Stone tools, steel tools: Contact period household technology at Helo. Pp. 49–72 in Rogers and Wilson (eds.), *Ethnohistory and Archaeology*.

Barandiaran, Daniel de. 1967. Agricultura y recolección entre los Indios Sanema-Yanoama, o el hacha de piedra y la psicología paleolítica de los mismos. *Antropológica* 19:24–50.

Barandiaran, Daniel de and Aushi Walalam. 1983. *Hijos de la luna: Monografía antropológica sobre los Indios Sanema-Yanoama*. Caracas: Editorial Arte.

Bennett Ross, Jane. 1971. Aggression as adaptation: The Yanomamo case. Unpublished MS.

Biocca, Ettore, ed. 1971. *Yanoama: The Narrative of a White Girl Kidnapped by Amazonian Indians*. New York: Dutton.

Bishop, Charles. 1983. Limiting access to limited goods: The origins of stratification in interior British Columbia. Pp. 148–161 in E. Tooker and M. Fried (eds.), *The Development of Political Organization in Native North America*. Washington, D.C.: American Ethnological Society.

Brooke, James. 1991. Venezuela befriends tribe, but what's Venezuela? *New York Times*, 11 September, p. A4.

Bueno, P. Ramón. 1965. *Tratado histórico*. Ed. F. de Lejarza. Caracas: Biblioteca de la Academia Nacional de la Historia.

Carneiro, Robert. 1979a. Forest clearance among the Yanomamo: Observations and implications. *Antropológica* 52:39–76.

———. 1979b. Tree felling with stone ax: An experiment carried out among the Yanomamo Indians of southern Venezuela. Pp. 21–58 in C. Kramer (ed.), *Ethnoarchaeology*. New York: Columbia University Press.

Chagnon, Napoleon. 1966. Yanomamo warfare, social organization, and marriage alliances. Ph.D. diss., University of Michigan.

———. 1974. *Studying the Yanomamo*. New York: Holt, Rinehart and Winston.

———. 1977. *Yanomamo: The Fierce People*. 2d ed. New York: Holt, Rinehart and Winston.

———. 1983. *Yanomamo: The Fierce People*. 3d ed. New York: Holt, Rinehart and Winston.

———. 1992. *Yanomamo*. 4th ed. Fort Worth: Harcourt Brace Jovanovich.

Cocco, P. Luis. 1972. *Iyewi-teri: Quince años entre los Yanomamos*. Caracas: Librería Editorial Salesiana.

Codazzi, Agustín. 1940. *Resumen de la geografía de Venezuela*. 3 vols. Caracas: Biblioteca Venezolana de Cultura.

Colchester, Marcus. 1984. Rethinking Stone Age economics: Some speculations concerning the pre-Columbian Yanoama economy. *Human Ecology* 12:291–314.

———, ed. 1985. *The Health and Survival of the Venezuelan Yanoama*. Copenhagen: Anthropology Resource Center/International Work Group for Indigenous Affairs, document no. 53.

Colchester, Marcus and Richard Semba. 1985. Health and survival among the Yanoama Indians. Pp. 13–30 in Colchester (ed.), *The Health and Survival of the Venezuelan Yanoama*.

Coles, J. 1979. *Experimental Archaeology*. London: Academic Press.

Comité para la Creación de la Reserva Indígena Yanomami. 1983. *Los Yanomami Venezolanos*. Vollmer Foundation.

Crittenden, Robert and Edward Shieffelin. 1991. The back door to the Purari. Pp. 125–146 in E. Schieffelin and R. Crittenden (eds.), *Like People You See in a Dream: First Contact in Six Papuan Societies*. Stanford: Stanford University Press.

Cronon, William. 1983. *Changes in the Land: Indians, Colonists, and the Ecology of New England*. New York: Hill and Wang.

Crosby, Alfred. 1986. *Ecological Imperialism: The Biological Expansion of Europe, 900–1900*. Cambridge: Cambridge University Press.

Crumley, Carole, ed. 1994. *Historical Ecology: Cultural Knowledge and Changing Landscapes*. Santa Fe: School of American Research Press.

Denevan, William M. 1992. Stone vs. metal axes: The ambiguity of shifting cultivation in prehistoric Amazonia. *Journal of the Steward Anthropological Society* 20:153–165.

Divale, William. 1970. An explanation of tribal warfare: Population control and the significance of primitive sex ratios. *New Scholar* 2:173–192.

Divale, William and Marvin Harris. 1976. Population, warfare, and the male supremacist complex. *American Anthropologist* 78:521–538.

Dobyns, Henry. 1983. *Their Number Become Thinned: Native American Population Dynamics in Eastern North America*. Knoxville: University of Tennessee Press.

Doolittle, William E. 1992. Agriculture in North America on the eve of contact: A reassessment. *Annals of the Association of American Geographers* 82:386–401.

Early, John and John Peters. 1990. *The Population Dynamics of the Mucajai Yanomama*. San Diego: Academic Press.

Eguillor García, Maria Isabel. 1984. *Yopo, shamenes y hekura: Aspectos fenomenológicos del mundo sagrado Yanomami*. Vicariato Apostólico de Puerto Ayuacucho: Librería Editorial Salesiana.

Ferguson, R. Brian. 1988. War and the sexes in Amazonia. Pp. 136–154 in R. Randolph, D. Schneider, and M. Diaz (eds.), *Dialectics and Gender: Anthropological Approaches*. Boulder, Colo.: Westview Press.

———. 1989a. Ecological consequences of Amazonian warfare. *Ethnology* 28:249–264.

———. 1989b. Game Wars? Ecology and conflict in Amazonia. *Journal of Anthropological Research* 45:179–206.

———. 1990. Blood of the Leviathan: Western contact and warfare in Amazonia. *American Ethnologist* 17:237–257.

———. 1992. A savage encounter: Western contact and the Yanomami war complex. Pp. 199–227 in Ferguson and Whitehead (eds.), *War in the Tribal Zone*.

———. 1995a. Infrastructural determinism. Pp. 21–38 in M. Murphy and M. Margolis (eds.), *Science, Materialism, and the Study of Culture*. Gainesville: University of Florida Press.

———. 1995b. (Mis)understanding resource scarcity and cultural difference. *Anthropology Newsletter*, November, p. 37.

———. 1995c. *Yanomami Warfare: A Political History*. Santa Fe: School of American Research Press.

———. In press. Violence and war in prehistory. In D. Martin and D. Frayer (eds.), *Troubled Times: Violence and Warfare in the Past*. Langhorne, Pa.: Gordon and Breach.

Ferguson, R. Brian and Ncil L. Whitehead, eds. 1992. *War in the Tribal Zone: Expanding States and Indigenous Warfare*. Santa Fe: School of American Research Press.

Fried, Morton. 1967. *The Evolution of Political Society: An Essay in Political Anthropology*. New York: Random House.

Friedman, Jonathan and M. J. Rowlands. 1977. Notes towards an epigenetic model of the evolution of "civilization." Pp. 201–276 in J. Friedman and M. J. Rowlands (eds.), *The Evolution of Social Systems*. London: Duckworth.

Gilij, Felipe Salvador. 1965. [1783]. *Ensayo de historia Americana*. 3 vols. Trans. and preliminary study by A. Tovar. Caracas: Biblioteca de la Academia Nacional de la Historia.

Good, Kenneth. 1983. Limiting factors in Amazonian ecology. Paper prepared for the symposium "Food Preferences and Aversions," Cedar Key, Fla., October.

———. 1989. Yanomami hunting patterns: Trekking and garden relocation as an adaptation to game availability in Amazonia, Venezuela. Ph.D. diss., University of Florida.

———. 1991. (With David Chanoff.) *Into the Heart: One Man's Pursuit of Love and Knowledge Among the Yanomama*. New York: Simon and Schuster.

Grelier, Joseph. 1957. *To the Source of the Orinoco*. Trans. H. Schumckler. London: Herbert Jenkins.

Hames, Raymond. 1989. Time, efficiency, and fitness in the Amazonian protein quest. *Research in Economic Anthropology* 11:43–85.

Harding, A. and R. Young. 1979. Reconstruction of the hafting methods and functions of stone implements. Pp. 102–105 in T. Clough and W. Cummins (eds.), *Stone Axe Studies*. Council for British Archaeology Research Council, Report no. 23.

Harner, Michael. 1963. Machetes, shotguns, and society: An inquiry into the social impact of technological change among the Jivaro Indians. Ph.D. diss., University of California, Berkeley.

Harris, David. 1971. The ecology of swidden cultivation in the Upper Orinoco rain forest, Venezuela. *Geographical Review* 51:475–495.

Harris, Marvin. 1977. *Cannibals and Kings: The Origins of Culture*. New York: Random House.

———. 1984. A cultural materialist theory of band and village warfare: The Yanomamo test. Pp. 11–140 in R. B. Ferguson (ed.), *Warfare, Culture, and Environment*. Orlando: Academic Press.

Hemming, John. 1978. *Red Gold: The Conquest of the Brazilian Indians, 1500–1760*. Cambridge: Harvard University Press.

———. 1987. *Amazon Frontier: The Defeat of the Brazilian Indians*. Cambridge: Harvard University Press.

Hill, Jonathan, ed. 1988. *Rethinking History and Myth: Indigenous South American Perspectives on the Past*. Urbana: University of Illinois Press.

Hill, Kim and Hillard Kaplan. 1989. Population and dry-season subsistence strategies of the recently contacted Yora of Peru. *National Geographic Research* 5:317–334.

Holoway, M. 1956. *Bananas*. Washington, D.C.: Pan American Union.

Huber, Otto, Julian Streyermark, Ghillean Prance, and Catherine Ales. 1984. The vegetation of the Sierra Parima, Venezuela-Brazil: Some results of recent exploration. *Brittonia* 36(2): 104–139.

Humboldt, Alexander von. 1889. *Personal Narrative of the Travels to the Equinoctial Regions of America, during the Years 1799–1804*, vol. 2. Trans. and ed. T. Ross. London: George Bell.

Jank, Margaret. 1977. *Culture Shock*. Chicago: Moody Press.

Knauft, Bruce. 1993. Review of R. B. Ferguson, *War in the Tribal Zone*. *Science* 260:1184–1186.

Koch-Grünberg, Theodor. 1979. [1923]. *Del Roraima al Orinoco*. 3 vols. Ed. E. Armitano. Caracas: El Banco Central de Venezuela.

Linares, O. 1976. "Garden hunting" in the American tropics. *Human Ecology* 4:331–349.

Lizot, Jacques. 1974. El río de los Periquitos: Breve relato de un viaje entre los Yanomami del alto Siapa. *Antropológica* 37:3–23.

———. 1976. *The Yanomami in the Face of Ethnocide*. Copenhagen: International Work Group for Indigenous Affairs, document no. 22.

———. 1977. Population, resources, and warfare among the Yanomami. *Man* 12:497–571.

———. 1980. La agricultura Yanomami. *Antropológica* 53:3–93.

———. 1988. Los Yanomami. Pp. 479–583 in W. Coppens with B. Escalante (eds.), *Los aborígenes de Venezuela*, vol. 3. Caracas: Fundación La Salle de Ciencias Naturales.

Métraux, Alfred. 1959. The revolution of the ax. *Diogenes* 25 (Spring): 28–40.

Montgomery, Evelyn Ina. 1970. *With the Shiriana in Brazil*. Dubuque, Iowa: Kendall/Hunt.

Patiño, Victor Manuel. 1958. Plátanos y bananos en América equinoccial. *Revista Colombiana de Antropología* 7:295–337.

Peña Vargas, Camila. 1981. *El P. Luis Cocco: Ejemplo de evangelización Salesiana en Venezuela*. Caracas: Librería Editorial Salesiana.

Peters, John Fred. 1973. The effect of Western material goods upon the social structure of the family among the Shirishana. Ph.D. diss., Western Michigan University.

Ramenofsky, Ann. 1987. *Vectors of Death: The Archaeology of European Contact*. Albuquerque: University of New Mexico Press.

Ramos, Alcida. 1972. The social system of the Sanuma of northern Brazil. Ph.D. diss., University of Wisconsin.

———. 1990. *Ethnology Brazilian Style*. Série Antropologia, no. 89. Brasília: Universidade de Brasília, Instituto de Ciências Humanas.

———. 1995. *Sanumá Memories: Yanomami Ethnography in Times of Crisis*. Madison: University of Wisconsin Press.

Rogers, J. Daniel. 1993. The social and material implications of culture contact on the Northern Plains. Pp. 73–88 in J. Daniel Rogers and Samuel M. Wilson (eds.), *Ethnohistory and Archaeology: Approaches to Postcontact Change in the Americas*. New York: Plenum.

Roosevelt, Anna. 1989. Resource management in Amazonia before the conquest: Beyond ethnographic projection. Pp. 30–62 in Darrell A. Posey and William Balée (eds.), *Resource Management in Amazonia: Indigenous and Folk Strategies*. Advances in Economic Botany, vol. 7. Bronx: New York Botanical Garden.

———, ed. 1994. *Amazonian Indians from Prehistory to the Present: Anthropological Perspectives*. Tucson: University of Arizona Press.

Ross, Eric. 1978. Food taboos, diet, and hunting strategy: The adaptation to animals in Amazon cultural ecology. *Current Anthropology* 19:1–36.

Saffirio, Giovanni. 1985. Ideal and actual kinship terminology among the Yanomama Indians of the Catrimani River basin (Brazil). Ph.D. diss., University of Pittsburgh.

Saffirio, Giovanni and Richard Scaglion. 1982. Hunting efficiencies in acculturated and unacculturated Yanomama villages. *Journal of Anthropological Research* 38:315–327.

Saffirio, John and Raymond Hames. 1983. The forest and the highway. Pp. 1–52 in *The Impact of Contact: Two Yanomamo Case Studies*. Cambridge, Mass.: Cultural Survival Report no. 11; Bennington, Vt.: Bennington College Working Papers on South American Indians, no. 6.

Salazar, Fred with Jack Herschlag. 1967. *The Innocent Assassins*. New York: Dutton.

Salisbury, R. F. 1962. *From Stone to Steel: Economic Consequences of a Technological Change in New Guinea*. London: Cambridge University Press.

Saraydar, Stephen and Izumi Shimada. 1971. A quantitative comparison of efficiency between a stone axe and a steel axe. *American Antiquity* 36(2): 216–217.

Schomburgk, Robert. 1841. Report of the third expedition into the interior of Guyana. . . . *Journal of the Royal Geographical Society of London* 10:159–267.

Shapiro, Judith Rae. 1972. Sex roles and social structure among the Yanomama Indians of northern Brazil. Ph.D. diss., Columbia University.

Sharp, Lauriston. 1990. Steel axes for Stone Age Australians. Pp. 410–424 in James Spradley and David McCurdy (eds.), *Conformity and Conflict: Readings in Cultural Anthropology*, 7th ed. Glenview, Ill.: Scott, Foresman.

———. 1993. Steel axes for Stone Age Australians. Pp. 24–31 in Arthur Lehmann and James Myers (eds.), *Magic, Witchcraft, and Religion: An Anthropological Study of the Supernatural*, 3d ed. Mountain View, Calif.: Mayfield.

Simmonds, N. W. 1962. *The Evolution of Bananas*. London: Longmans.

Smole, William. 1976. *The Yanoama Indians: A Cultural Geography*. Austin: University of Texas Press.

Steinvorth de Goetz, Inga. 1969. *Uriji Jami! Impresiones de viajes Orinoquenses por aire, agua y tierra*. Caracas: Asociación Cultural Humboldt.

Steward, Julian. 1976. *Theory of Culture Change: The Methodology of Multilinear Evolution*. Urbana: University of Illinois Press.

Steward, Julian and Luis Faron. 1959. *Native Peoples of South America*. New York: McGraw-Hill.

Taylor, John. 1988. Goods and gods: A follow-up of "Steel axes for Stone-Age Australians." Pp. 438–451 in Tony Swain and Deborah Bird Rose (eds.), *Aboriginal Australians and Christian Missionaries*. Bedford Park, South Australia: Association for the Study of Religions.

Townsend, William. 1969. Stone and steel tool use in a New Guinea society. *Ethnology* 8:199–205.

Turner, Terence, ed. 1993. *Cosmology, Values, and Inter-Ethnic Contact in South America*. South American Indian Studies, no. 2. Bennington, Vt.: Bennington College.

Valero, Helena. 1984. *Yo soy Napeyoma: Relato de una mujer raptada por los indígenas Yanomami*. Caracas: Fundación La Salle de Ciencias Naturales, Monografía no. 35.

Vareschi, Volkmar. 1959. *Orinoco arriba: A través de Venezuela siguiendo a Humboldt*. Caracas: Lectura.

Whitehead, Neil. 1988. *Lords of the Tiger Spirit: A History of the Caribs in Colonial Venezuela and Guyana, 1498–1820*. Providence, R.I.: Foris.

———. 1994.

Whitehead, Neil L. 1994. The ancient Amerindian polities of the Amazon, Orinoco, and Atlantic Coast: A preliminary analysis of their passage from antiquity to extinction. Pp. 33–54 in Roosevelt (ed.), *Amazonian Indians from Prehistory to the Present*.

Wike, Joyce. 1951. The effect of the maritime fur trade on Northwest Coast Indian society. Ph.D. diss., Columbia University.

Wilbert, Johannes. 1963, *Indios de la región Orinoco-Ventuari*. Caracas: Fundación La Salle de Ciencias Naturales.

———. 1972. *Survivors of Eldorado: Four Indian Cultures of South America*. New York: Praeger.

Wilmsen, Edwin N. and James R. Denbow. 1990. Paradigmatic history of San-speaking peoples and current attempts at revision. *Current Anthropology* 31:489–524.

Wolf, Eric. 1982. *Europe and the People Without History*. Berkeley: University of California Press.

Zent, Stanford. 1992. Historical and ethnographic ecology of the upper Cuao River Wothiha: Clues for an interpretation of native Guianese social organization. Ph.D. diss., Columbia University.

Zerries, Otto. 1955. Some aspects of Waica culture. *Proceedings of the 31st International Congress of Americanists*, pp. 73–88. São Paulo.

CHAPTER 14

Missionary Activity and Indian Labor in the Upper Rio Negro of Brazil, 1680–1980: A Historical-Ecological Approach

JANET M. CHERNELA

The role of missions in postcolonial Brazil remains a significant gap in contemporary models of Latin American social and economic history. The prevailing view among historians holds that the mission village system, which flourished in Brazil between 1680 and 1750, was succeeded by an irreversible shift toward secularization (MacLachlan 1973; Boxer 1962), severing permanently the practical and ideological ties between church and state in Brazil. Taking as its case the mission village enterprises of the Upper Rio Negro in Brazil (see figure 14.1), this chapter shows the persistence of the mission village system in the north-central Amazon into the last quarter of the twentieth century. I discuss the environmental, economic, and political factors that together have contributed to a "history apart" for the native American populations inhabiting the northern frontiers of Brazil, and the role of the mission village as economic and political agent in the region through a number of significantly different governmental regimes.

The mission villages of the northern Amazon received the attention of modern reformers in 1980 when an international tribunal on human rights found the Order of Salesians of the Upper Rio Negro in violation of laws and agreements on ethnocide and racial discrimination against the 20,000 Indians of the Rio Negro watershed. The allegations against the mission, charged by both urban reformers and indigenous witnesses, focused on its active participation in the supply of indigenous labor to urban centers. The case raised a number of questions regarding the continued role of the frontier ecclesiastical mission in modern Brazilian society.

Rather than treat the events of 1980 as isolated from the parameters of political and economic contexts, I emphasize the historic embeddedness of the role of the frontier mission as mediator in the organization, distribution, and control of Indian

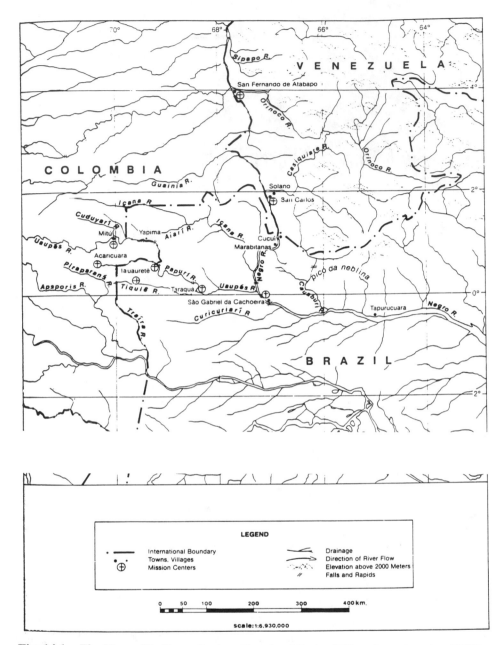

Fig. 14.1 *The Upper Rio Negro Region, Showing Mission Village Locations* (ca. 1980).

labor over centuries. In order to do so I will review the structure of labor in the mission villages of north central Brazil between 1680 and 1980.

I begin with a brief summary of the 1980 decision by the Russell Tribunal on the rights of indigenous peoples. I then turn to the seventeenth century to trace the foundations of structures and conditions that contribute to the 1980 findings. The years 1680–1682 mark the establishment of a Portuguese colonial policy in which indigenous peoples were placed in labor pools under the agency and administration of European missionaries. I briefly review events in the Rio Negro in the eighteenth and nineteenth centuries before focusing on the Mission Village of 1980, the period in which I conducted fieldwork in the region. I point to the continuity in social relations over four centuries of Indian-state interaction despite major differences in contrasting state polities and the increasing modernization of the Brazilian economy. I argue that only an approach that is at once historical, ecological, and political—a "historical ecology"—permits an understanding of the interactive factors and their outcomes that contribute to the processes constituting that which we call "history."

Introduction: The Russell Tribunal

The 1980 Russell Tribunal on indigenous human rights, held in the Netherlands, brought to public light labor practices by the Salesian Mission of the Upper Rio Negro in Brazil that were alleged to be infractions of international agreements on ethnocide. At the center of the case against the Salesian Mission was the role it played in transporting Tukanoan women from the indigenous area along the northern affluents of the Rio Negro at the Colombian-Brazilian border to distant urban centers, such as the city of Manaus, to labor as domestic servants.

Álvaro Sampaio, a Tukanoan student who testified at the Tribunal, made this statement to the press:

> The Tukano Indian population is suffering a nightmare. . . . The signs are in the number of Indian women who have been abandoned with their children. They are the reminders of so-called "progress." . . . Some young girls go to Manaus to work in schools run by Salesian nuns. The work is hard and the hours long; they do not have Sundays or holidays off, and they do not receive the minimum wage. . . . Dozens of Indian women . . . are now working for low salaries in the homes of Brazilian Air Force officials. Many of the women employed as domestics are not well-treated and are frequently humiliated. . . . In general, these women have been sent to [work in] cities by the Salesian sisters of the Upper Rio Negro . . . [where] they . . . can only take care of the children of the lieutenants, captains and brigadiers [whom they serve]. For the Indian child there is nothing. . . . I call this ethnocide. (Wright and Ismaelillo 1982:17)

The Brazilian Catholic Church, with a strong record of advocacy in issues concerning indigenous peoples, published this commentary on the Tribunal's conclusions in its missionary journal:

> The Salesians of the Rio Negro . . . have attempted to defend themselves against these accusations . . . [but] have never denied the actual "cultural massacre" of the Indians of the region. Many of these Indians are now migrating to Venezuela or Colombia, or have sought refuge in the outskirts of the city of Manaus, where they are ashamed to be recognized as Indians. The Upper Rio

Negro today has been transformed into a kind of "Salesian feudalism." Thousands of Indians in the region, considered to be the largest Indian area in the country, are losing their culture, traditions, customs, identity and even their languages. For Bishop Dom Miguel Alagna, this is called "the integration of the Indian into the national community." For the authors of the Tribunal case, it is ethnocide. (Conselho Indigenista Misionaria, *Folhetim*, 24 May 1981)[1]

The discovery of underpaid and unpaid Indian laborers in Manaus culminates a long-standing history of Indian/white relations characterized by harsh exploitation of the Amerindian labor force. To better understand why this pattern has prevailed, I begin by reviewing the relationship between missions and indigenous people of the Upper Rio Negro as it was established in the Portuguese Amazon in the last quarter of the seventeenth century. I broadly compare the mission villages of the Rio Negro between 1680 and 1980, briefly discussing the intervening periods.

Indian Labor Laws in the Portuguese Colony: 1680

Iberian powers saw indigenous Indian labor as the key to financial success in the Portuguese Amazon of the 1600s. While lucrative plantations flourished in the Lusitanian colonies of the Atlantic coast where soils were well suited to agricultural production, the interior of the colony had few profitable exports. Principal among these were extractive forest resources that served as New World substitutes for East Indian spices, found only in the farthest reaches of the rainforest. Given the poor soils, impenetrable forests, and widely dispersed flora, the instrument of greatest economic value in the Amazon was, ultimately, the Indian (Chernela 1993). European powers competed for dominion over the inaccessible wild resources and the Indian laborers required to collect them.

Until the 1680s slaving expeditions went relatively unimpeded. It is estimated that hundreds of thousands of Indians were forcibly removed from the interior to the downriver colonial settlement at Belém, where they performed both public and private services (Sweet 1974; Wright 1981). Men were utilized as oarsmen, gatherers, and builders. Women were placed in colonial homes, where they carried out domestic chores including cleaning and childcare (Hemming 1978).

Native populations downriver were decimated by the introduction of diseases and the abuses of slavery. Pressure for Indian labor by colonists at the mouth of the Amazon resulted in scarcities in the areas closest to colonial centers, driving expeditions ever deeper into the forest in search of captive Indian labor. Epidemics of smallpox and measles devastated the banks of the Amazon in 1665 (Hemming 1978:410). The Jesuit missionary Antônio Vieira, writing in 1673, reports that mistreatment by the Portuguese resulted in the deaths of more than two million Indians within a forty-year period (Vieira 1925–28). So thorough was the depopulation of the river edge caused by slaving, disease, and flight, that in the last quarter of the seventeenth century, it was necessary to travel as much as two months up the Amazon in order to capture Indian slaves (Hemming 1978:411).

In response to the extreme competition for Indian labor, the Crown in the 1680s instituted an official system of labor recruitment controlled by the Crown, to be administered through the clergy. Indians from diverse backgrounds and languages were removed from hinterland regions and resettled in colonial mission centers, known as *aldeias*. (The Portuguese word *aldeia* is derived from the Moorish administrative unit best translated as township or village. It is used here to refer to farms and estates, as well as villages, administered by state-authorized missions.) By securing Indians into labor villages under authority of the Crown, the Portuguese could establish their presence in contested areas, such as the northern headwater regions of the Rio Negro, an area threatened by European competitors. In the controlled settlement, Indians would serve as a labor reservoir for the Portuguese Crown and colonists, and, at the same time, secure the region for Portugal through their sedentarized and nationalized presence.

It was in the mission village that relations between clergy and the Indian labor force of the colony were forged. Regulations divided Indians into three labor pools: those residing and working in mission villages; those assisting missionaries in recruiting and resettling Indians into mission villages; and finally, those who were, theoretically at least, available to meet the needs of government and private employers (MacLachlan 1973; Kiemen 1973). This last group of Indians, constituting only 20 percent of the total Indian labor force, was parceled out from the mission village to European settlers for contractual periods of two to six months. Missionaries thus controlled between 80 and 100 percent of the total Indian labor force at any given time.

Indian women were regularly assigned to the households of public officials, where they were required to perform domestic tasks, such as wet-nursing and maid service (Hemming 1978:413). They were excluded from participating in husbandry and forest collection, as well as agricultural work, although these had long been among women's principal traditional responsibilities.

Life in the mission territory "followed the European model" (Burns 1980:39) with its well-defined hierarchy: its regular and secular clergy, its students, and its laborers. According to Bradford Burns (1980:39), "Churchmen . . . rigidly controlled the lives of their charges. . . . Under their guidance, the Indians contributed to the imperial economy . . . , dressed like Europeans, mastered European trades, and paid homage to the king in Lisbon. Thus, those touched by the *aldeia* system were brought by the determined hand of the missionaries within the pale of empire."

As required by the Crown, the missionaries documented and registered all births, baptisms, marriages, and deaths (Hemming 1978:412–424). In the institutionalized relationship of the mission community, assimilation was the long-term goal. Indians living in mission villages were given Portuguese surnames, taught the Portuguese language and European values, trained and encouraged in techniques of intensive agriculture, policed, and generally educated in the ways of whites. In short, missionaries attempted to substitute one cultural system for another, imposing civilization through regulation and reward, and through the strict control of daily routine, ritual and festivities, political leadership, and all productive activity.

Wages for Indian labor in the mission village were held to a minimum—a policy favored both by mission administrators and by the Crown. In 1686, payment for a month's labor was set at two yards of cotton cloth. The artificial level of Indian wages and the substitution of goods for wages did not contradict the "essentially communal philosophy of the mission system" (MacLachlan 1973:206); indeed, the missionaries themselves were not permitted to receive monetary compensation (Hemming 1978:413). Whatever proceeds were gained from the sale of agricultural surpluses in regional markets were used to purchase tools and other materials needed for the upkeep of the village church, school, or hospital.

At the close of the seventeenth century (1693–1695), the Crown apportioned the large, uncharted region of the Amazon to the religious orders. The Jesuits, Franciscans, Mercedarians, and Carmelites were each assigned a parcel of land over which they would exercise exclusive authority. By 1750, sixty-three mission villages had been established by the four religious orders in the Bishopric of Belém; they varied in size from 150 to more than 800 inhabitants, the average being 475. In the year 1730, there were approximately 50,000 Indians registered in the mission villages of Amazonia (Boxer 1962:290; MacLachlan 1973:208).

The Directorate System: 1751–1780

The control of Indian labor by the clergy was strongly opposed by the colonists, who increasingly applied pressure on the Crown to secularize this key source of labor (Alden 1969). By the mid-eighteenth century, popular sentiment opposing the clergy charged that the religious orders had profited from forest collection, where the determining factor was the control of Indian labor.

The intent of the Directorate system, installed by the Marquês de Pombal in 1757, was to transfer temporal authority from the religious orders to secular administrators responsible only to the governor. Pombal was an anticleric who expelled the Jesuits from Portugal and its colonies and redrafted the property laws to prevent the accumulation of great wealth by the Church. The governing powers of the religious orders, as authorized by the regulations of 1680 and 1686, were abolished. The Marquês de Pombal conceived of a vast program to develop Brazil through increased production in agriculture and industry. In recognition of the growing importance of the Amazon, the captaincy of São José do Rio Negro (present-day Amazonas) was created in 1755, subordinate to the captaincy of Grão Pará Maranhão. The newly apointed governor, Francisco Xavier de Mendonça Furtado, brother of the marquês, accused the missionaries of directing Indian labor to their own advantage and thereby accumulating financial gain to the detriment of private colonists and the state (MacLachlan 1973). Under the new system, the role of missionaries as mediators between Indian villages and the government was to be removed. The Indian, in turn, was to assume greater responsibility and "acquire an appreciation of the material benefits of labor" (MacLachlan 1973:209). Through this policy the Crown hoped to better absorb the Indian into Portuguese culture and into an ever-widening economic world order.

The *aldeia* system, however, had not disappeared (Burns 1980:107). In the farthest reaches of the Amazon River basin, the transfer to secular directors was never accomplished. There, mission administrators continued to serve the state in implementing the new developmental directives of the marquês. In contrast to the lower reaches of the basin, where state efforts at secularization were successfully applied during the same period, mission village organization and production in the north central region actually increased and intensified. In fact, the Pombaline government entrusted the goals of the integration and civilization of the Indian to the regional administrators who continued to be, in the northern Amazon, missionaries. The Pombal regime, did, however, set forth the limits of their charge. For example, missionaries could no longer buy or sell Indians or engage in private commerce of any kind (Sweet 1974:317). Missionaries were to register, make available for royal service, and deliver any Indians requested by the governor; those missionaries who did not cooperate with the Crown could be sent to Pará for punishment (Sweet 1974).

Renewed threats of Spanish incursion into the northern frontier areas in the 1760s inspired the Pombalines to strengthen the Portuguese presence in the Upper Rio Negro valley by creating a number of fortified mission villages. Indians from the Içana, Uaupés, and Xié regions were removed and settled into large mission forts, where they were employed to demarcate borders, manufacture goods, and grow crops (Wright 1981:145, 169, 173).

It was during the Pombal era, in 1761, that one of the principal mission-fort centers in the Uaupés region was established at the confluence of the Uaupés River with the Rio Negro at São Gabriel da Cachoeira (Brüzzi 1977:17–18). A large parish, integrating clerical powers with secular administrative functions, was formed with the Franciscan José de Santa Ursula appointed as its vicar. The location functioned as both a military outpost and a center for religious conversion. Tukanoan and Arawakan Indians of diverse language groups were brought downriver from the Uaupés, Papurí, Tiquié, and Içana Rivers to build and occupy the mission-fort complex (see figure 14.1).

The Pombaline state invested large amounts of capital and effort into the economic development of the north central Amazon and, within it, the Rio Negro valley. Agricultural experimentation, introduced into the area at the beginning of the eighteenth century (Reis 1943:27), was further intensified under the Pombal administration. The promotion of exportable perennial crops was the centerpiece of the Pombaline economic strategy for the Amazon basin. The program was based principally on the tree crops coffee and cacao, but it also promoted the lucrative export annuals sugarcane, indigo, and cotton (Reis 1943:33). The government provided a wide range of fiscal incentives for the intensification of agriculture, supplying each mission village with seed and technical assistance. In addition to agriculture, the government also promoted ranching through direct subsidies to mission villages: in exchange for government-supplied cattle, mission administrators were expected to institute breeding programs and increase livestock in the villages (Reis 1943:28). Administrative units staffed by missionaries who represented the state, oversaw Indian production and ensured that work quotas were met (Reis 1943; Galvão 1979). By 1785 there were an estimated 220,000 coffee and 200,000 cacao trees reported for plantations in the Rio Negro basin, with Indians the principal agricultural producers (Reis 1943; Galvão 1979).

New agricultural and ranching enterprises increased at the expense of traditional hunting, fishing, and horticulture. Extensive cultivation methods of long fallow and short cultivation periods customary to indigenous agriculture were replaced by intensive techniques. To the native crops of the region, such as manioc, which required little labor, were added labor-intensive, nutrient-costly regimes. Annual crops quickly exhausted the nutrient loads of the senile white sand soils of the Rio Negro region, requiring that land cleared for cash monocropping be abandoned or transformed into pasturage after a few harvests. In areas nearest the mission centers, the rapid rate of land clearance, coupled with increased population density, resulted in the accelerated depletion of the forest cover and loss of the faunal wildlife associated with it. The result was a drastic reduction in yields from subsistence gardening and food gathering, rendering the mission village and its methods incompatible with traditional means of food procurement within the limits of the local ecosystem.

Faced with the growing difficulty of carrying out traditional subsistence methods, native populations adjacent to the mission village had no choice but to seek labor there in exchange for sustenance. Surrounding populations were forced to move into new areas or to enter the mission's sphere of influence in order merely to survive. By eliminating the possibilities of traditional subsistence means, the mission enterprise, whether intentionally or not, had created the conditions of its own necessity.

In spite of its stated goals of secularization, then, the Directorate government further situated the hinterland missionary in the role of mediator and charge of Indian life and labor. The missions, although closely regulated, had provided the Pombalines with the most effective means of mobilizing and organizing Indian manpower. Even in the latter half of the eighteenth century, a period of public backlash aimed at clerics, the mission system was still recognized as the most efficient and least costly means of enforcing the policies of the government in frontier regions.

But despite the intensity of effort, the high expectations of agriculture and ranching were never met. By 1779, the economic focus had shifted away from the Rio Negro basin in the central Amazon to the Solimões basin in the western Amazon. The administrative center of the captaincy of the Rio Negro was dissolved and the region placed under the jurisdiction of Pará. For the next sixty years, the economy of the Rio Negro valley proceeded to decline, and with it, missionization efforts (Reis 1943; Spix and von Martius 1824).

One prominent explanation attributes this period of decline to the overexploitation of resources near the principal collection centers (MacLachlan 1973). Seventeenth-century mission centers had focused on collection, and the wealth of the Jesuits was attributed in part to the collection of forest spices.[2] According to Colin MacLachlan (1973:206), "Overeager collectors had depleted the sources of supply by stripping trees bare, often before the fruit matured." To this explanation must be added the failed attempts at intensive agriculture at the close of the eighteenth century.

Whatever scant profits were earned from the extraction of forest products or from agricultural enterprises, the mission villages of the Upper Rio Negro were never truly profitable. Instead, both extraction practices and land clearance resulted in the depletion of resources, with increasingly intensive land-use regimes producing increasingly poor input-output ratios. By exhausting local resources and furthering a transition to export-oriented, labor-intensive crops, such as coffee, the land-use policies of the Pombal regime had contributed to the region's economic decline. By the end of the eighteenth century, faced with the demise of its collection and agricultural programs, and finding the indigenous population dramatically reduced (Lopes de Sousa 1959:205), the Crown relaxed its hold on the Rio Negro, temporarily abandoning the area until the mid-nineteenth century.

Expansion: Nineteenth Century

Mission centers of the north central Amazon that were left to languish during the economic decline of the early nineteenth century experienced a resurgence by the middle of that century. Spurred by investments in new industries, and by advances in communications (the steamship, railway, and telegraph), Europe in the mid-nineteenth century entered a period of prosperity and worldwide expansion that would last four decades.

In Brazil, coffee exports provided the economy with a renewed impulse. By 1850, Brazilian coffee accounted for 50 percent of the world's coffee market. Coffee production continued to rise steadily, and by 1880 it accounted for three-fifths of all Brazilian exports (Schneider 1991:43).

The sudden growth in world markets, coupled with general prosperity elsewhere in Brazil, stimulated new demand for northern products. Prices per commodity rose, as did the total number of items exported. The expanding market for Amazon products brought with it a stimulus for products and labor from the region, thereby expanding the export sector, and, in turn, stimulating the production of foodstuffs consumed by and exchanged for that sector's labor force. In the Amazon, massive transportation projects linked commercial zones with the previously inaccessible interior (Reis 1943). In the Rio Negro alone, four new shipping companies were formed during the decade of the 1850s. By the 1880s commercial activity in the Amazon basin was thriving. In the three-year period between 1880 and 1883, for example, the port of Manaus was visited by no fewer than 457 steamships with a total cargo of 187,481 tons; of these steamships, 68 were foreign and 389 Brazilian national (Reis 1943:81).[3]

As industrial development accelerated, mission activities resumed worldwide with renewed zeal (Oliver 1952). In the independent Brazilian Empire of 1845, missionaries were charged with the "catechization and civilization, education and governing" of all Amazonian indigenous peoples, as decreed by the governor of Pará (Regulamento no. 426, 1845, in Wright 1981:288). The impact of this renewed missionary activity was felt acutely in the Upper Rio Negro. In 1849, a large parcel of land encompassing the two major tributaries of the Upper Rio Negro, the Uaupés and the Içana, was designated a single parish. The governor of Amazonas appointed the Capuchin friar Gregorio José Maria de Bené "vicar in charge of population and missions on the Uaupés and Içana Rivers" (Brüzzi 1977:18).[4] Between 1852 and 1854 de Bené baptized 550 people—an estimated one-fourth of the total indigenous population of the Brazilian portion of the Uaupés river basin. Two other mission centers were established along the Uaupés River in 1852. Between 1881 and 1883, three Franciscan missionaries zealously concentrated and organized Indians into large mission centers on the Uaupés, Tiquié, and Papurí Rivers.[5] Their string of posts, spanning more than 800 km, allowed river contact with some 3,000 inhabitants (Koch-Grünberg 1909, 2:7–9).

The largest settlement by far was that of Ipanorê (later known as Taraquá), at the confluence of the Tiquié and Uaupés Rivers. Founded in 1790, Ipanorê had soon after been abandoned for a half-century. As economic growth reached into the hinterland regions, Ipanorê became the new center of mission enterprise (Chernela 1988, 1989; Chernela and Leed in press). Campaigns were carried out to attract Indians to the labor settlement. The writer Henri Coudreau called the Ipanorê of 1883 "the capital of all Uaupés and Papori villages. . . . [Ipanorê] possessed the largest number of houses and inhabitants, besides the church, which was, in architecture and proportion, the largest and best of any on this river" (Coudreau 1887; Aranha 1907:33). Coudreau continues, "The Ipanorê of 1883 had a cemetery, school, missionary residence, a uniformed police force, and a jail with separate quarters for men and women" (Coudreau 1887; Aranha 1907:58).

State and Church in the Twentieth-Century Upper Rio Negro

Within the three years between 1888 and 1890 the Brazilian government had abolished slavery (1888); established the republic (1889); and separated the powers of church and state (1890). Even so, these major events did little to advance the rights of the Indians in fact, for the rubber boom era had arrived in the Amazon.

Even during the decade of 1880, years of heightened economic expansion, the portion of exports from the Rio Negro had reached no more than 2 percent of total Amazonian export value. This changed with the surge in world demand for Amazonian rubber. The drive for rubber reached the Upper Rio Negro in the early 1870s, and by the close of the century rubber traders had penetrated into the farthest reaches of the Uaupés basin. Rubber camps were established along the Rio Negro to which workers from the Uaupés and Içana Rivers were enticed by promises, or brought by force (Wright 1981:324). In accounts of his visit to the Upper Rio Negro in 1904, the German ethnographer Koch-Grünberg describes Colombian rubber traders who captured young boys and raped and carried off Tukanoan women (Koch-Grünberg 1909).

When the anthropologist Curt Nimuendajú arrived in the Uaupés basin in 1927, he found rubber collectors forcibly removing Indians from the lower Uaupés in Brazil and transporting them to upriver camps in Colombia, where, according to Sutti Ortiz, they "were enslaved and tortured to increase their productivity in rubber collection" (Ortiz 1984:208). Abuses in living and working conditions, work hours and compensation, were common occurrences in the Upper Rio Negro rubber camps. Nimuendajú recorded the names of whites he found engaged in debt servitude; one of the names listed was that of the mayor of the Brazilian Uaupés (Nimuendajú 1950:144).

It was in just this atmosphere of extreme abuse that the Salesian Mission of the Upper Rio Negro was founded in 1914. Arriving in the wake of rubber camp atrocities, the missionaries of the Salesian Order offered sanctuary and protection from forced recruitment. Following directly on the abolition of slavery, the rubber years represented to a habitually exploited work force a model of labor as "free." The mission settlement was once again the only refuge for the Indian, a favorable alternative to outright slavery or the so-called free labor conditions of the rubber camps. The mission continued its expansion, with Salesians assuming administration of their predecessors' mission sites at São Gabriel (1915), Taraquá/Ipanorê (1924), and Iauaretê (1929). In 1945 a fourth mission town was established in the Uaupés basin at Pari Cachoeira on the Tiquié River (see figure 14.1).

The Salesian Order of Missions

The Society of St. Francis de Sales, founded in Turin, Italy, in 1845, is one of many similar brotherhoods founded in the mid-nineteenth century as part of an expanded effort to convert remote populations to Catholicism. Initially, the order distinguished itself from other Roman Catholic orders through its dedication to the educational and vocational training of poor urban youth. The Salesian commitment to vocational training, bred in the industrial centers of Europe, was transferred, by the turn of the century, to those populations most distant from the reach of European influence. In 1915, Pope Pius X assigned the Prefeitura Apostólica do Rio Negro, Amazonas, Brazil, to the Salesian Order. In the same year Pius X also assigned adjacent portions of southern Venezuela to the Salesians. By 1941, the prefecture had been elevated to a prelacy whose size exceeded 400,000 square kilometers.

Salesian texts show the Order to have perceived its task as an arduous one of educating, professionally training, catechizing, and generally civilizing a remote population through the training of its youth. One text, written in 1950, summarized this view: The task of the Salesians in the Upper Rio Negro was to provide

> religious assistance . . . , elementary and agro-professional schooling, sanitation, hospitalization, and attachment of the Indian to the soil [*fixação dos índios ao solo*]. . . . The greatest task of all before the missionary . . . [is] the

challenge of dedicating oneself to the farthest outreaches of Brazilian territory, a region enveloped in virgin forest; characterized by treacherous, wild, and black watercourses; and inhabited by one of the most [ethnically] "mixed" populations on earth, . . . two steps from Colombia and Venezuela: the Frontier itself. (Missões Salesianas do Amazonas 1950:17; see also Azevedo 1933 and Giacone 1949)

At the time of my visit to the region in the years 1978 through 1981, some 10,000 Indians in the Upper Rio Negro were living in or associated with three mission villages, then subsidized at least partially by the Brazilian state. Along the Uaupés River, population was distributed into three parishes: 4,531 in Iauaretê, 3,238 in Pari Cachoeira, and 960 in Taraquá/Ipanorê.

The Mission Village of 1980

The mission village of 1980 bore a strong resemblance to the mission villages of the colonial period. When I visited Iauaretê, at the confluence of the Uaupés and Papurí Rivers, in 1980, the mission village consisted of a large residential and production complex with barns, stables, poultry houses, hospitals, garages, warehouses, machine shops, carpentry shops, crafts workshops, greenhouses, laundries, and dispensaries. There were dormitories for students and workers, as well as large and well-guarded living quarters for clerics and visitors. All buildings and public works were constructed and maintained by Indian labor.

Contact with the world beyond the mission was minimal; the occasional river trader to barter with, or air force pilot delivering supplies, were the only exceptions to an otherwise wholly Indian and missionary population. Roads linked sectors of the complex but did not extend beyond it. The mission village also encompassed a port, with three motor launches and a mooring area for riverboats and canoes. All buying and selling took place within mission grounds and under mission authority. Health care was provided through a mission hospital and trained medical missionaries. The high portals surrounding the grounds were closed daily at sundown.

In many ways the labor relations of the mission village of 1980 recalled those of the *aldeia* system of three centuries before. The clerics still controlled all means of production and commerce. The economy, like that of the colonial period, was based on a system of *aviamento*, in which an employer advances food and other consumer goods to individuals in exchange for the future delivery of labor. The mission maintained a trading store where the missionaries sold Indians finished goods on credit and collected payment in the form of products or labor. The mission determined both the price of the finished goods it sold and the value of the labor and products it purchased.

As far as possible the mission village was economically self-sufficient. Food was grown in gardens on the mission farm. Corn, beans, and rice were grown for consumption. In addition, cows, sheep, horses, pigs, and chickens were raised. Grasses to maintain pasturelands, however, were imported at high cost and replaced with frequency. All agricultural production was turned over to the mission, to be returned, in theory, to the Indians in the form of goods and services.

Among the services provided by the mission was the training of young men and women in skilled and semiskilled occupations. In eight "agro-industrial learning centers," Indians were trained and employed as ranchers, farmers, masons, locksmiths, carpenters, bakers, ranchhands, weavers, seamstresses, sweepers, cooks, laundresses, candlemakers, groundskeepers, and a host of other occupations. To the girls was given the task of cleaning the large and elaborately furnished interiors: kitchens, classrooms, meeting rooms, dormitories, offices, chapels, and grounds. The Salesian ideology of work had as its central concern the conversion of the "savage" into a productive citizen. Women's work was critical in the creation of a domestic and "civilized" interior space distinct and separate from the wild and "uncivilized" exterior.

While religion and devotion were stressed in the educations of both sexes, religion for girls emphasized the concept of devotion as practice—particularly in the selfless service to the religious body, which meant, in this case, to the mission community. Girls, excluded from the agricultural tasks that were the principal activity of Tukanoan women in outlying villages (Chernela 1985), were now trained in domestic services and crafts. This constitutes a significant change from traditional gender roles. Two crafts traditionally performed by men, weaving and featherwork, were now made female specializations and recast for marketability. Their products were sold by the missionaries in Manaus to support the work of the mission. Compensation for Indian labor was held to a minimum, in keeping with the nonmonetary values of mission ideology and the view of work as apprenticeship.[6] Indians were compensated through goods, such as clothing, combs, notebooks, and pencils.

Government programs were propagated through mission contact with outlying villages. In a program that encouraged export agriculture, the mission supplied seed to nearby villages for cash-cropping and then appointed and trained "animators" in each village to oversee production. Those who reported successful projects were rewarded.

The infertile lands that had been cleared and subjected to intensive agriculture were abandoned to pasturage with frequency. Ranching itself proved costly, requiring the importation of exotic grasses at great expense, with government cattle subsidies and international church donations eventually providing the financial support for these enterprises. Cash crops performed below expectations. The formerly lucrative cacao production proved vulnerable to the viral disease "witches'-broom," and its value on the world market proved unstable. Forest spices such as the neotropical substitutes for cloves and cinnamon—as well as pepper, sarsaparilla, and dyestuffs, once major contributors to wealth in the seventeenth century—were either exhausted or no longer in demand.

The intensification of agriculture, with the ensuing erosion and the closure of vast tracts of land to Indian foraging, left the Indians nearest the mission centers unable to continue traditional forms of subsistence. Hunting, fishing, and forest cultivation were replaced by varying forms of dependence upon the mission. On occasion, outright predation took place. During my stay at the mission center at Iauaretê a case of "fishing-for-fowl" was reported: in what were said to be common occurrences, young men hurled hook and line over the mission wall in order to "fish" for chickens. This form of preying upon the mission estate along its peripheries is a direct consequence of the depletion of free-ranging food sources, such as fish and game, within the vicinity of the mission.

From a different Tukanoan perspective, however, the Salesian missionaries were a reliable source of trade medicines, goods, and protection. In 1965 the mission listed eight hospitals for the Rio Negro region. The mission hospital at Iauaretê, for example, treated countless indigenous patients who suffered, in 1980, from tuberculosis, influenza, measles, whooping cough, and malaria. Families accompanied patients interned in the hospital, where space was allotted for hearths and hammocks much as in a village household. The hospital's director made river rounds to outlying villages within a 100 km radius to attend to health needs. (The importance of the hospital and of itinerant healthcare to the region underscores the need of Western medicines to treat new illnesses.) In other words, in a context of disease and violence, the mission center served an important function to Tukanoans who chose to ally themselves with it.

The mission hospital director described represents the many compassionate Salesian missionaries I encountered in my visit in 1980. It was not individual missionaries who necessarily created the oppressive conditions of the mission center; many, in fact, opposed them. Rather, it was the policies issued from outside, not inside, the indigenous region.

State-Mission Alliance

A succession of governments from the 1930s to the 1950s sought to modernize the north, but the first to do so effectively was the military government of Castelo Branco, which seized power in 1964. It established the program PIN (Plan for National Integration), aimed at industrializing the north through the creation of a free-trade sector and increased state presence.

The advantages of the mission system to the Brazilian state were as apparent to the military leaders of the republic of the 1970s and 1980s as they had been to the rulers of the independent Brazilian Empire of 1880, and to the colonial governors of two centuries before. Recognizing as indispensable to national security both the strategic placement of the frontier missions and their role in assimilating Indians into the Brazilian nation, the military government entered into a mutually beneficial alliance with the missions of the north central Amazon, proffering financial aid and other forms of assistance in return for mission allegiance and services in securing national borders.

Unlike the Jesuits of the seventeenth century, the Salesians aligned themselves clearly with the state through a relationship that served the needs of both entities. An alliance of mutual assistance, known as the "Triangle of Integration," was established between the Salesian Mission of the North Central Amazon and two federal agencies, the Brazilian Air Force (FAB) and the Bureau of Indian Affairs (FUNAI).[7] The missions were to act on behalf of the military in defense of the frontier. Between 1964 and 1980, the military government of Brazil, interested in maintaining mission outposts on its northern frontiers and in furthering the integration of the Indian into Brazilian society, supplied the mission with food, fuel, construction materials and labor, educational services and supplies, medical supplies, and much-needed cost-free air transportation. In return for goods and services granted by the air force, the mission conferred favors on air force personnel. One of these was the supply of young girls as domestic servants.

Modernization: Manaus

For three centuries the city of Manaus, founded in 1669 at the confluence of the Rio Negro and the Solimões River, served as a military outpost for the Portuguese and as a center for trade in forest products from the Rio Negro Valley. With the exception of the brief period from 1896 through 1912, when a single forest product, rubber, was in great demand in overseas markets, the extractive industries of Manaus brought little wealth or economic activity into the region.

In 1967, the new military government of Castelo Branco designated the city of Manaus as a "pole," or site, for industrial development in the Amazon region, and established within it a Free Trade Zone in order to attract capital to the north. The modernization scheme included powerful fiscal incentives and investments amounting to more than $47 million U.S. dollars (Despres 1988). In the subsequent years, approximately two hundred new industrial firms relocated to the Free Zone. The majority of these industries were labor-intensive assembly plants for electrical exports, and by 1988 they had in their employ approximately 60 percent of the industrial workforce.

The new economic activity in the Free Zone profoundly affected the city. The number of economically active persons in metropolitan Manaus soared from 39,000 in 1960 to 216,000 in 1980, reflecting a 500 percent increase (Despres 1988). In 1960 Manaus accounted for only one-fourth of the state's population; twenty years later, it accounted for nearly half. In the five years between 1975 and 1980, the number of commercial establishments in Manaus increased from roughly 3,000 to nearly 8,000 (Despres 1988; see also Despres 1991). The new industrial metropolis was maintained and extended through the combination of military force and free trade. In less than two decades, nearly every facet of social, cultural, and economic life of the city was altered as capitalist relations replaced former patrimonial ones.

The new professional middle class, drawn from large southern cities, including Rio de Janeiro and São Paulo, sought the services of household laborers. With the rapidly inflating economy of the free port, domestic services were in scarce supply, and a premium was placed on agencies that could supply this labor and mediate the transactions necessary to secure it. Missionaries in remote Indian regions were opportunely situated to fill this need.

Contact between the little-known indigenous area of the Upper Rio Negro and the city of Manaus, 1,200 km to the south, dramatically increased following the 1967 establishment of the Free Zone and the great demand for domestic service. Between 1967 and 1985, the mission provided young Tukanoan women to serve as household domestics in at least one hundred homes. The majority of these households were those of military personnel; on the departure of a military family from Manaus,[8] the woman (or women) hired by them entered the urban labor market of domestic workers.

Discussion

In the early 1980s a small group of intellectuals in Manaus, discovering the role of the mission in providing domestic workers to private households—the "maid trade," as it came to be known—protested through the human rights forum of the Russell Tribunal. The moral problem addressed by these urban intellectuals, however, reflects a more general conflict: that between two political economies and ideological systems at odds with one another.

The contemporary findings of Indian domestic laborers working for little or no pay with the agency of the missionaries can now be interpreted in a new way. The mission village had constituted a single economic and social formation, based upon monopolist principles that served the interests of both church and state, in which regional or traditional interests were subordinated to a central authority. The salient features of such patrimonial arrangements are the effective monopoly over resources and the consolidation of power so as to limit the options of participants, forcing them to seek power within a single framework. As the only source of trade goods, and the only purchaser of resources and labor, the mission controlled all economic opportunity and offered a sphere of exchange limited to the extreme.

With the establishment of a commercial metropolis at Manaus, however, modern capitalism, as a system that relies on wage labor, was brought into contact with the traditional patrimonial systems of the frontier, precipitating a crisis for the mission system of the Upper Rio Negro Valley in the middle twentieth century. The "maid trade" is but one indicator of a long-term crisis for the mission village as a social and economic institution.

The combined state/mission alliance had monopolized key resources and labor in the outlying areas, hindering the emergence of free markets in labor, commodities or land, by means of the centuries-old patrimonial mission village system. With the establishment of a capitalist metropole in Manaus—made possible both by state initiative and by the accumulation of southern capital—a new industrial labor market exceeded the capability of a regime based upon exclusionary, patrimonial principles.

Conclusions

Although mission villages were found throughout Portuguese America—from the Orinoco River in the north to the Plata River in the south—they had their longest duration at the periphery of the colony. The extreme persistence of this social and economic form in the north central Amazon may best be explained by its ongoing utility to nation states and ruling elites.

Established to provide labor for the extractive industry of the sixteenth century, the mission village persisted long after it had proved unprofitable. As markets for extractive industries declined, or as these products were harvested out, control over Indian labor was continued for benefits other than economic ones. From the seventeenth through the twentieth centuries, mission village enterprises were maintained by states not for the accumulation of capital, but for political reasons alone. So concerned was the state with the occupation and fortification of the frontier zones that it financially supported mission villages regardless of profitability, with income opportunities distributed or withdrawn for political rather than economic gain.

Central to a succession of governments of Brazil, from the Portuguese Crown to the modern Brazilian state, have been the dual concerns for national security and the value of the Indian as a source of labor and revenue. It was indigenous labor, rather than forest or farm products, that emerged historically as the single most important economic resource in the northwest Amazon.

As slaving proved too costly and risky a means of obtaining needed Indian labor, a more efficient method of colonial labor recruitment was instituted in the form of the consolidated mission village, or *aldeia*. Through the mission village, the Portuguese Crown and later the Brazilian state could draw the Indian into a world system and secure their own outlying border regions. In granting the missions access to remote Indian populations for the purposes of religious conversion and social and national integration, the state ensured its own ends of border surveillance, and the mobilization and organization of Indian labor. Landscape and individual were simultaneously "converted" and "domesticated."

In the farthest reaches of the Amazon basin, the mission system has remained relatively intact until the present day. Secular takeover, presumed to have transpired throughout Brazil, was impeded in the border areas where national considerations of security continued to dominate policy. In these regions, the missions continued to serve the national governments by organizing Indians into productive units, instilling national identities, claiming and cultivating lands, and providing a permanent Brazilian presence in an otherwise underpopulated and potentially disputed region.

The approach I have taken here differs from an exclusively historical one in which individual achievement and interest are given primacy. Instead, this chapter stresses the impact of cultural factors and environmental constraints on individual and institutional roles. The efforts of individuals or institutions might account for the presence of the mission village in the Upper Rio Negro for a short duration of time, but cannot account for patterns of social structure that persist over four centuries in the context of extreme changes in policy-making within both church and state.

Interest is itself shaped by historic circumstance. One might argue, for example, as did the urban intellectuals of the Russell Tribunal, that it is within the interests of the missionaries to act as agents in the labor transactions described. However, the agency and positioning of Salesian missionaries in the Upper Rio Negro is the exception, rather than the rule. Elsewhere in Latin America, including central Brazil, missionaries of the Salesian Order have taken the lead in the advocacy of indigenous people. Indeed, numbers of individual Salesian missionaries within the Upper Rio Negro system attempted, over time, to change policy. This argument underscores the importance of political and economic factors over and above those of individual contribution or design.

Cultural ecology, then, as a processual approach, would seem to provide greater explanatory force. In the mission village, factors such as low investment in technology, overexploitation of basic natural resources, and lack of elasticity in internal markets resulted in low productivity and inhibited capital accumulation—impeding, ultimately, the potential for generating new productive enterprises. It may be said that the mission village system persisted in the north central region because environmental constraints resulted in its late incorporation into the Brazilian economy. The private entrepreneurial sector, prodigiously expanding in the southern states during the nineteenth and twentieth centuries, took little interest in the Upper Rio Negro with its poor soils, scarce population, and primitive technological infrastructure. The low potential for profitability in the region reduced economic competition, leaving the area accessible to mission enterprises that were growth-limited, subsidized by state or international organizations for motives other than profit alone.

Environmental constraints, however, do not solely account for political and economic events in the Upper Rio Negro. The interest of the state in demarcating and defending its borders through continuous occupation of a Brazilianized citizenry accounts, too, for the state's interest and investment in the maintenance of mission centers in the farthest reaches of its territories. The goals of the hinterland missionaries fit well within the particular goals of the state in securing borders. Through increased capital and labor investment, the combined mission-state venture attempted to extract whatever exportable resources might be produced. The investment of greatest yield, however, was the occupational training of the population of the region—the formation of a labor force prepared to serve in an increasingly modernized economy.

An approach that is both historical and ecological—namely, historical ecology—should therefore provide a more powerful instrument for understanding the interactions of society and environment over time. Insofar as it is a synthetic approach, historical ecology provides insight into the mutual influences of state and mission interests and the environment as these forces impact upon one another. It is these relationships, and their interactions, that form the core of this synthetic approach.

The exceptionally long duration of the mission villages of the Upper Rio Negro provides a case with which to examine the explanatory power of the historical ecology approach. The ebb and flow of missions in this region advanced and retreated following the pulse of national economic expansion. Each period of growth experienced by the Brazilian economy was accompanied on the frontiers by increased missionary presence. The mission village itself constitutes a microcosm of historical ecology in the Amazon, at once the agent of social and political change and the first to feel its impact. The changes observable in the mission formations of a single river basin clearly reflect the coinciding economic, environmental, and nationalistic circumstances of the entire region within a widening world market.

Acknowledgments

Versions of this paper were presented in the Latin American Studies Program, University of Arizona, Tucson, 31 March 1993; the Columbia University School of International and Public Affairs, New York, 1 March 1994; and Tulane University, New Orleans, 9 June 1994.

Notes

1. All translations from the Portuguese to English in this paper are mine.

2. Sugar holdings in the northeast of Brazil account for the majority of the wealth attributed to the Jesuits.

3. Within this context of thriving Amazonian commerce, the portion of export from the Rio Negro basin represented a mere 2 percent of the total value (Reis 1943).

4. Koch-Grünberg (1909, 2:8) calls Frei de Bené a Carmelite Padre—unlike Brüzzi and Wright, who describe him as Capuchin.

5. The three missionaries were Frei Venancio Zilochi, on the Tiquié; Frei Iluminato José Coppi, on the Uaupés and Papurí at Ipanorê; and Frei Matheus Camioni, on the lower Uaupés at Taraquá.

6. In reports, for example, trainees were referred to as "aprendizados" (Figoli 1982).

7. The basis for this plan was laid down by Getúlio Vargas in 1940. It was further strengthened by the government of Castelo Branco.

8. Most military families were assigned to the frontier regions for a period of two years.

References

Alden, Dauril. 1969. Economics of the expulsion of the Jesuits. Pp. 25–65 in Henry H. Keith and S. F. Edwards (eds.), *Conflict and Continuity in Brazilian Society*. Columbia: University of South Carolina Press.

———. 1973. *Colonial Roots of Modern Brazil: Papers of the Newberry Library Conference*. Berkeley: University of California Press.

Aranha, Bento de Figueiredo Tenreiro. 1907. *Archivo do Amazonas, Revista Destinada a Vulgarisação de Documentos Geográficos e Históricos do Estado do Amazonas* 1(3).

Azevedo, Soares de. 1933. *Pelo Rio Mar*. (Report compiled by the Salesian Prelacy of the Upper Rio Negro, under the coordination and supervision of Dom Pedro Massa.) Rio de Janeiro: C. Medes.

Boxer, Charles. 1962. *The Golden Age of Brazil*. Berkeley: University of California Press.

Brüzzi Alves da Silva, Pe. Aucionílio. 1977. *A civilização indígena do Uaupés*. 2d ed. Missão Salesiana do Rio Negro, Amazonas, Brazil. São Paulo: Centro de Pesquisas de Iauaretê.

Burns, E. Bradford. 1980. *A History of Brazil*. New York: Columbia University Press.

Chernela, Janet M. 1985. Os cultivares de mandioca (tucano). Pp. 151–158 in Berta Ribeiro (ed.), *SUMA: Etnológica Brasileira*, vol. 1, *Etnobiologia*. Rio de Janeiro: Vozes.

———. 1988. Righting history in the Northwest Amazon. Pp. 35–49 in Jonathan Hill (ed.), *Rethinking History and Myth: Indigenous South American Perspectives on the Past*. Urbana: University of Illinois Press.

———. 1989. Marriage, language, and history among eastern-speaking peoples of the Northwest Amazon. *Latin American Anthropology Review* 1(2): 36–42.

———. 1993. *The Wanano Indians of the Brazilian Amazon: A Sense of Space*. Austin: University of Texas Press.

Chernela, Janet and Eric Leed. In press. The deficits of history: The whiteman in a myth cycle from the Northwest Amazon. in Bruce Albert and Alcida Ramos (eds.), *Pacificando o Branco*.

Coudreau, Henri. 1887. *La France équinoxiale*. Vol. 2, *Voyage à travers les Guyanes et l'Amazonie*. Paris.

Despres, Leo A. 1988. Macrotheories, microcontexts, and the informal sector: Self-employment in three Brazilian cities. Kellogg Working Paper #110.

———. 1991. *Manaus: Social Life and Work in Brazil's Free Trade Zone*. Albany: State University of New York Press.

Figoli, Leonardo. 1982. Identidad étnica y regional: Trayecto constitutivo de una identidad social. Master's thesis, University of Brasilia.

Galvão, Eduardo. 1979. The encounter of tribal and national societies in the Brazilian Amazon. Pp. 25–38 in Maxine L. Margolis and William E. Carter (eds.), *Brazil: Anthropological Perspectives: Essays in Honor of Charles Wagley*. New York: Columbia University Press.

Giacone, Antonio. 1949. *Os Tucanos e outras tribus do Rio Uaupés afluente do Negro-Amazonas: Notas etnográficas e folclóricas de um missionario salesiano*. Associação Brasileira dos Amerindianistas. São Paulo: Impr. Oficial do Estado.

Hemming, John. 1978. *Red Gold: The Conquest of the Brazilian Indians, 1500–1760*. Cambridge: Harvard University Press.

Kiemen, Mathias. 1973. *The Indian Policy of Portugal in the Amazon Region, 1614–1693*. New York: Farrar, Straus and Giroux, Octagon Press.

Koch-Grünberg, Theodor. 1909. *Zwei Jahre unter den Indianern: Reisen in Nordwestbrasilien 1903/1905*. 2 vols. Berlin: Ernst Wasmuth.

Lopes de Sousa, Marechal Boanerges. 1959. *Do Rio Negro ao Orinoco*. Rio de Janeiro: Conselho Nacional de Proteção aos Indios.

MacLachlan, Colin M. 1973. The Indian labor structure in the Portuguese Amazon, 1700–1800. Pp. 199–230 in Alden (ed.), *Colonial Roots of Modern Brazil*.

Missões Salesianas do Amazonas. 1950. *Nas Fronteiras do Brasil*. Rio de Janeiro.

Nimuendajú, Curt. 1950. Reconhecimento dos Rios Içana, Aiarí, e Uaupés: Relatorio apresentado ao Serviço de Proteção aos Indios do Amazonas e Acre, 1927. *Journal de la Société des Américanistes* 39:125–182.

Oliver, Roland. 1952. *The Missionary Factor in East Africa*. London: Longmans, Green.

Ortiz, Sutti. 1984. Colonization in the Colombian Amazon. Pp. 204–230 in Marianne Schmink and Charles H. Wood (eds.), *Frontier Expansion in Amazonia*. Gainesville: University of Florida Press.

Reis, Arthur Cezar Ferreira. 1943. *O processo histórico da economia Amazonense*. Rio de Janeiro: Editora Paralelo.

Schneider, Ronald M. 1991. *Order and Progress: A Political History of Brazil*. Boulder, Colo.: Westview Press.

Spix, Johann B. von and Karl F. P. von Martius. 1824. *Travels in Brazil, 1817–1820*. 2 vols. London: H. E. Lloyd.

Sweet, David G. 1974. A rich realm of nature destroyed: The Middle Amazon Valley, 1640–1750. Ph.D. diss., University of Wisconsin.

Vieira, Pe. Antônio. 1925–28. *Cartas do Padre Antônio Vieira*, collected and annotated by J. Lúcio de Azevedo. Coimbra: Imprensa da Universidade.

Wolf, Eric. 1982. *Europe and the People Without History*. Berkeley: University of California Press.

Wright, Robin. 1981. History and religion of the Baniwa peoples of the Upper Rio Negro Valley. Ph.D. diss., Stanford University.

Wright, Robin, and Ismaelillo. 1982. *Native Peoples in Struggle: Cases from the Fourth Russell Tribunal and Other International Forums*. New York: Anthropology Resource Center and E.R.I.N. Publications.

CHAPTER 15

Cultural Persistence and Environmental Change: The Otomí of the Valle del Mezquital, Mexico

Elinor G. K. Melville

Ethnohistorians ascribe the persistence of distinct ethnic groups in colonial Mexico to "the simple degree of contact, measured in distance, frequency, or hours spent" (Lockhart 1992:4–5). It is true that groups distant from the center of colonial society, and with little to attract colonial entrepreneurs, tended to remain ethnically and culturally distinct, whereas groups in the central regions exhibited the greatest degree of change. But there are exceptions to this rule: one such is the Otomí of the Valle del Mezquital, located a scant sixty miles north of Mexico City.

The Otomí have been in continuous and close contact with first Spaniards and then Mexicans for nearly five hundred years, and yet they have remained linguistically, culturally, and ethnically distinct from the neighboring ladinos. Indeed, Charles Gibson (1967:10) notes that at contact and throughout the colonial era (1521–1810) they spoke the only non-Nahuatl-based language in the Valley of Mexico—and they continue to do so. The region they inhabit, the Valle del Mezquital, has also achieved a distinction of sorts in the twentieth century: as the archetype of the barren regions of Mexico, and as a vast, informal treatment plant for the waste waters of Mexico City.

The Valle del Mezquital comprises the catchment area and drainage basin of the Tula River (see figure 15.1), and in the 1930s an irrigation district was created that channeled the flood waters that regularly inundated Mexico City into the Tula River system. These waters were used to irrigate lands in the Tula River valley, and as the effluent from the city increased in volume the irrigation district grew in importance and extent. Despite its current status as a major source of vegetables for the city's markets, however, the Valle del Mezquital continues to be perceived as a traditional, backward, and poor "indigenous" region by scholars and lay people alike. Indeed, it is an indication of the power of this perception to shape policy that the chemical-laden waters of Mexico City's effluent could be thought to "improve" the soils of the region.[1]

334

Fig. 15.1 *The Valle del Mezquital, Showing Sub-Areas and* Cabeceras.(From Melville 1994b:26.)

The perceived poverty of the Valle del Mezquital is invariably identified with the land-management systems of the Otomí. Geographer Kirsten Johnson's study (1977) comparing the attitudes of the Otomí toward the land and its use with the technocratic approaches of government experts sent to prepare Otomí villagers for the expansion of irrigation system #3 demonstrates attitudes toward this region and toward the Otomí that persist until the present. Not too surprisingly, perhaps, her study demonstrates profound differences in goals and approaches between the two groups: the technocratic approach to land use in Mexico's arid regions is based on the idea that scientific knowledge rationally applied could correct the perceived poverty of both the soils and the people of these regions; the Otomí approach is summed up in the saying "Do as the land bids."

For the technocrats, the poverty of a region such as the Valle del Mezquital is a consequence of outmoded technology and a lack of access to scientific knowledge; they argue that formal education would teach new ways to use the land, which would provide the means to correct the poverty of both the Otomí and their region (Johnson 1977:161). For the Otomí, their poverty is a consequence of their lack of access to water—ultimately a lack of power rather than education. They argue that access to irrigation water would allow them to grow more and different types of plants. While they appreciate modern material technology—tractors, for example—they express no need to change radically their approach to the land or their land-management systems (Johnson 1977:147–150).

Drawing on her own ethnographic research and the secondary literature dealing with the Otomí, Kirsten Johnson developed a synthesis of Otomí land use and attitudes to the land. (Since her study was completed, irrigation system #3 has spread to include extensive areas of the Valle del Mezquital; the practices she discusses are still carried out in areas where the national irrigation system does not reach.) The Otomí apply what might be termed an "ecosystem" approach to agriculture: they use their detailed knowledge of local conditions to select niches for particular species according to the type of soil and probability of rain and/or floodwater. Yet they do not simply rely on the bounty or lack thereof from the land. Rather, they also practice various methods to increase the productivity of the soils: canal irrigation, hillside terracing, floodwater dams, and wetlands management (drainage mostly) are used to obtain optimal soil/water ratios; rotting vegetation and animal manure are used to improve soil structure and fertility (Johnson 1977:110–147). The technocrats interviewed by Johnson viewed these techniques as wholly inadequate, however, since they provide only a bare subsistence for the Otomí themselves and contribute little or nothing to the national economy.

The modern landscape is assumed—certainly by the technocrats, but also by the majority of social scientists—to reflect the continuity of ecosystems as well as culture from the pre-Hispanic era (Johnson 1977:161).[2] But when the history of the Valle del Mezquital is examined more closely, one finds that far from reflecting a traditional indigenous past, the modern environments and landscapes of this region reflect profound environmental changes set in train by the Spanish invasion; and that although certain Otomí agricultural techniques are almost exact replicas of those found in the archaeological record—for example, their methods for managing soil-water regimes (Sanders, Parsons, and Santley 1979)—they are applied to very different environmental conditions.

In the pages that follow I will explore the hypothesis that the transformation of the environment associated with the introduction of sheep by the Spaniards in the sixteenth century brought about the effective marginalization of the Otomí, so that while they were not distant from the center of Spanish society they were structurally isolated and hence were forced/allowed to remain distinct. I will argue that the mode by which sheep-grazing was introduced into the Valle del Mezquital allowed for its adoption by the Otomí, while yet allowing for the retention of traditional land-management systems in the same space.[3]

The Introduction of Sheep

The Valle del Mezquital was, at the time of initial contact between the Otomí and the Spaniards (ca. 1524), a densely populated agricultural region where extensive areas of land had been cleared of trees to make way for agricultural fields. Despite the extent of forest clearance, however, sufficient water was generated within the catchment area of the Valle del Mezquital to sustain the flow of springs and streams and to support quite extensive irrigated systems; only the hillside towns and villages in the very arid northeast were without irrigated fields. The extent of irrigation, together with the presence of perennial streams and springs, indicates a healthy catchment area until the middle of the sixteenth century. By the end of that century, however, the catchment area had deteriorated, and there was no longer sufficient water available to support both Otomí agriculture and growing Spanish enterprises such as agricultural holdings, flour mills, and mines (Melville 1994b: 31–39).

The deterioration of the catchment area of the Valle del Mezquital over the last quarter of the sixteenth century resulted from the expansion of sheep-grazing into this region following the Spanish defeat of the Mexica in 1521. As the Spaniards expanded into the Valle del Mezquital, regional production shifted away from intensive indigenous irrigation agriculture to small-scale intensive sheep-grazing that was characterized by high densities of animals (reaching ca. 785 head of sheep/km^2 in the Tula River Valley in the 1570s) and large numbers of flock owners who had title to (or squatted on) small areas of land.[4] The initially abundant vegetation and the custom of grazing in common obviated the need to own large areas of land in order to secure forage for flocks that numbered ten to fifteen thousand (and at times twenty thousand) head in the 1570s.[5] By the 1570s sheep-grazing dominated the region.

At the same time, the Spaniards accelerated wood cutting for mining and for lime and charcoal manufacture, cleared forests for grazing, and fired the range for pasture regeneration; by the 1580s, the hills were cleared of trees. With increased soil temperatures and the development of more arid microenvironments, the vegetative cover shifted from a complex mosaic of cultivated species such as maize, beans, squash, tomatoes, and maguey; shrubs such as mesquite (growing as single-stemmed, massive trees); isolated dicots such as willows, cedars, and native cherries; forests of pines and oaks; and open woodlands of mixed species, to a more homogeneous desert scrub dominated by small, multistemmed mesquite. Water was no longer detained for sufficient time to replenish groundwater; instead, it flowed rapidly over the surface, carrying the soils down from the high, steep-sided hills onto the lower slopes. By the 1590s extensive gullying had begun to cut down through slope wash and the underlying soils; at the same time, water tables dropped, and springs failed (Melville 1994b:39–59).

The transformation of the physical environment of the Valle del Mezquital enabled the Spaniards to gain control over this region. By the end of the sixteenth century, two-thirds of the surface area had been transferred into the Spanish system of land tenure; the Otomí villages and their much-reduced agricultural lands were surrounded by extensively grazed latifundia; and the Otomí were marginalized within the new political economy (Melville 1994b:116–50).

This collapse of indigenous agroforestry was not simply the result of Spanish expansion, however, at the expense of indigenous communities, nor was it a straight-

forward case of ranching competing with agriculture, for many Otomí adopted
sheep-grazing on a large scale themselves during the sixteenth century. The sheep
stations that surrounded Otomí villages in the sixteenth century may well have
been the property of the villages themselves, or of Otomí nobles; and the sheep
that transformed the biological regime were owned by Otomí as well as Spaniards.
The Otomí had become agents of the profound transformation occurring in their
landscape.

The Adoption of Sheep-Grazing by the Otomí

Otomí were grazing sheep as early as the 1550s, apparently without license to do
so. In the 1560s, Viceroy Velasco began to grant licenses for sheep-grazing to individual Otomí nobles and to Otomí communities; during the six-year period
1560–1565 they received 78.9% of the *mercedes* for *estancias de ganado menor*
(grants of land for corrals and shepherds' huts, and license to graze sheep or goats
in common in the surrounding area), and by 1565 they owned 35.8% (99) of the
total 276 sheep stations in the region. In the years 1566–1569, 5 of 11 grants went
to Otomí, almost as many as to Spaniards; and in 1570–1579 Otomí received 26.5
of 44.5 grants—considerably more than the Spaniards. And although Spaniards acquired more stations by squatting during these years, the Otomí still held slightly
more than one-third of the lands given over to sheep-grazing by 1579. The acquisition of sheep stations by the Otomí declined in the decade of the 1580s, and they received only 18.5% (13) of a total of 70 grants. But in the 1590s, Otomí nobles and
villages again received more grants than the Spaniards: 69.8% (94) of the 154.5
grants awarded. By 1600, 37% of the grazing lands were in Otomí hands (Melville
1994b:116–150). Until the end of the century, when severe restrictions were placed
on Indian participation in the pastoral sector of the political economy of New
Spain, many Otomí seem to have grazed animals far in excess of the legal maximum of 2,000 head per station—as did at least one "mulatto" and many Spanish
pastoralists (Melville 1994b:49–54).[6]

The process of land acquisition through the Spanish system of land tenure does
not negate the fact of the eventual economic marginalization of the Otomí, however: by the seventeenth century Otomí villagers were transformed from the dominant economic force in the Valle del Mezquital into marginalized peasants. Nor
does it negate the fact that Otomí communities had no room for future expansion:
so much land had been transferred into the Spanish system of land tenure by 1600
that, despite the demographic collapse of the Otomí population, which had reached
90–92 percent by this date, there was insufficient land for the *fondo legal* (townsite
to which every community was entitled) of the villages in the fertile southern half
of the region (Melville 1994b:146–150). But it does put a rather different com-
plexion on Otomí survival and persistence—making it difficult, at least at first
sight, to argue structural isolation or cultural involution as their means of cultural
survival.

Parallel Systems of Resource Exploitation

The enthusiastic adoption of sheep-grazing by the Otomí, the profound transformation of the vegetative and soil cover of the Valle del Mezquital under sheep-grazing
and the consequent loss of traditional resources, the proximity of the region to the
center of Spanish power, and the intensity of contact over several centuries are the
same forces that in many other regions transformed "Indians" into "Ladinos."[7] That
transformation invites questions with regard to the Otomí. Specifically, how did
they remain so distinct? How were they able to adopt a cultural element characteristic of the conquerors—one that was, moreover, a major sector of the evolving
colonial political economy—and yet not become ladinos? How were they able to
maintain traditional systems of land management when their physical world was
transformed literally under their feet?

In reply to these questions, I would propose a mechanism that may have enabled
the Otomí of the Valle del Mezquital to adopt pastoralism while at the same time
maintaining agricultural practices and approaches to the land that were distinct from
Spanish or mestizo systems of land management. This mechanism concerns the
formation of ranching as a *parallel system* of production *outside* Indian village/
community production. Historically, environmental degradation brought about
Otomí marginalization, destroyed the basis of intensive pastoralism itself, and undermined irrigation agriculture using internally generated water. The combination of
political and economic marginalization on the one hand, and physical isolation in a
transformed landscape on the other, I suggest, produced a situation that almost guar-
anteed Otomí ethnic persistence. It should be noted, however, that I am not suggesting that the Otomí remained the same, nor that they did not undergo change; simply,
that they remained distinct from Spaniards and mestizos, and, later, Mexicans.

I originally constructed the model of parallel systems of resource exploitation as
a means to conceptualize the processes by which Spanish pastoralists and their animals were able to expand into densely populated agricultural regions, such as the
Valle del Mezquital, without the use of organized force—though with violence, as
will be seen. I suggested that the processes by which pastoralism evolved in New
Spain were essentially different from those by which Spanish colonial agriculture
evolved. Whereas Spanish agriculture was, from the first, embedded in Indian systems of land tenure and use together with village systems of production, pastoralism evolved "outside"—or better, parallel to—village production (Melville 1990).

The initial development of Spanish agriculture occurred within the context of the
encomienda system set up immediately following the Spanish defeat of the Triple
Alliance, as a system of rewards given to individual conquistadores. As an *encomendero*, a conquistador was to be supplied with labor and tribute from a specified community of Indians in return for the defense of their customary rights after
the collapse of any centralized indigenous authority, within the extremely unsettled
conditions of the early conquest era when Spanish authority was insecure. ("Conquest" is here equated with the process by which the colonial regime evolved.)

While the encomienda system clearly served to siphon off tribute to the conquerors, it also functioned as a decentralized system of government. And because it resembled, in broad terms, indigenous tribute systems, it was imposed with remarkably little difficulty.[8]

Although encomenderos had a virtual monopoly on production within the Spanish world up to 1550, primarily by virtue of their access to indigenous labor, they were ultimately dependent on indigenous land-use and land-tenure systems for their agricultural enterprises. They could state the tribute they required and the crops they wanted cultivated, and they could direct their laborers into a variety of entrepreneurial pursuits; but they did not have rights to the lands encompassed by their encomienda communities.[9] Crops were grown on Indian lands using Indian labor, very often according to indigenous management systems and schedules. We find wheat being cultivated in the mounds typical of maize agriculture in the early years, for example (Gibson 1964:322). And later, when it became clear that wheat production on a commercial scale competed with Indian work schedules, most especially maize cultivation, and that the best grains for bread could not be successfully grown during the summer rainy season, the Spaniards changed their scheduling: from spring sowing and fall harvest of rain-fed wheat, to fall sowing and spring harvest of irrigated wheat. As Arnold Bauer (1986) has demonstrated, the shift to a spring harvest meant that threshing was postponed to the following fall, when traditional methods of threshing with the *trilla*, or with large numbers of mares on a flat area of hardened ground, were not threatened by early rains and did not use animals needed for preparing the ground for maize.

Encomenderos bought or otherwise acquired land (by squatting, grants, gifts, or force) within their encomiendas (Gibson 1964:272–274)—thereby appearing to remove their enterprises from dependence on Indian production systems. But a case can still be made for the dependence of the new agricultural enterprises on the villages by examining developments following the imposition of the *repartimiento* in 1550. In a move to reduce the power of the encomenderos the Crown took to itself the right to Indian labor and set up government-controlled work gangs: *repartimiento* in New Spain, *mita* in the Andes (Gibson 1964:224–236). The shift to the repartimiento system of labor recruitment took the control of Indian labor out of the hands of the encomenderos and made Indian labor available to non-encomenderos, but it did not eliminate the Spanish agriculturalists' dependence on Indian villages; rather, it consolidated this dependence, producing a land-labor system that was carried into the era of the hacienda. Indeed, not only is management of a grain or mixed hacienda inconceivable without Indian villages to supply the work gangs needed to weed and harvest, but the relations between haciendas and Indian villages are treated as a defining characteristic of the hacienda system (Van Young 1983; Gibson 1964:292).

In contrast to the development of Spanish agriculture, the introduction and expansion of pastoralism were not dependent on Indian labor, land tenure, or land-management systems. Pastoralists did not need title to a specific area of land, since grass was a "fruit of nature" in Spanish custom, and animals could graze wherever

grass grew (Vassberg 1984:6; Chevalier 1975:12). Nor did they have to rely on Indian labor to care for their animals: they seem often to have relied on black slaves
in the early years.[10] There was really no reason why anyone who could afford the
cost of a few breeding animals and of someone to care for them could not set up a
herd, request a grazing license, and be in business. Thus pastoralism evolved, in
structural and organizational terms, in parallel to the village-encomienda system of
production; it was not embedded within it.

That most early pastoralists were encomenderos, however, is not surprising,
since they had the means for capital accumulation and thus the means to acquire
breeding stock and develop herds (Miranda 1965; McLeod 1973:46). Nor is it surprising that they often did not bother to obtain licenses, but simply squatted on the
lands they needed for corrals and shepherds' huts and grazed their animals in common on the surrounding lands (Gibson 1964:275; Simpson 1952:6; Chevalier 1975:
131). But it is interesting to note that in the Valle del Mezquital, encomenderos
grazed their animals in others' encomiendas in the early decades, and rarely in their
own: of the fourteen local encomenderos grazing herds of sheep and cattle in the
Valle del Mezquital during the 1530s, only three grazed their animals in their own
encomienda (Melville 1994b:128–129 n. 25).

Perhaps it was simple pragmatism that kept the encomenderos from grazing in
their own encomienda lands. Relations between the herdsmen and the villagers was
notoriously violent; indeed, violence seems to have functioned as a means to expand into this densely populated agricultural region and to gain access to forage.[11]
In such circumstances, it would make more sense to annoy someone else's tributaries rather than one's own. But the independence of the encomendero-pastoralists
from their own village-encomiendas in the Valle del Mezquital nicely illustrates my
point that pastoralism was not structurally and organizationally embedded in the
village-encomienda system of production.[12]

As well as serving as a conceptual model for the Spanish invasion of densely
populated regions, the model of pastoralism as a parallel system of resource exploitation also serves to clarify the development of the striking differences between
the labor regimes of grain and livestock haciendas, and further illustrates the development of pastoralism in New Spain. The marked independence of laborers on livestock haciendas is often ascribed to the low labor needs and less-intensive land-
management systems of traditional pastoralism (J. Riley 1984:261–262; Zeitlin
1989; Gutiérrez-Brockington 1989:103). I suggest, however, that such differences
were not simply the result of the different labor needs of pastoralism and agriculture; rather, they arose in the formative pre-hacienda era as a result of the different
developmental processes by which grazing and Spanish agriculture evolved. I base
my argument on differences between the labor regimes of agriculture and pastoralism that appeared at an early stage in the Valle del Mezquital. Various categories of
wage laborers (*sirvientes*, "*gente*," *naboríos*, *gañanes*) were resident on the small-
scale agricultural and pastoral units of production prior to the appearance of the hacienda in this region.[13] But while Indian sheep-shearers living in villages hired
themselves out to *estancia* (sheep station) owners, agricultural specialists did not

live in villages and hire themselves out to *labor* (agricultural units of ca. 200 ha) owners; rather, they lived on the *labores* in a manner reminiscent of the later hacienda system.[14]

The structural and organizational separation of pastoralism from village production provided Otomí commoners with an opportunity to learn new skills and earn money outside the community—and perhaps to escape increasingly onerous obligations (Zeitlin 1989:47; Hoekstra 1993). It also provided a means for the Otomí elite, and for entire Otomí communities, to engage actively in the new political economy (Melville 1994a). The benefits of raising domestic livestock were many: it was a relatively easy and secure way to obtain needed items such as fat, meat, fiber, traction, manure, and leather. The Otomí evidently preferred sheep over cattle. Several factors shaped their choice of these animals. First, sheep have one great advantage over the other domesticates brought by the Spaniards: their wool; wool is a much warmer fiber than cotton, eminently suitable for the cool highlands, and this would probably have influenced Otomí acceptance of these animals. Second, sheep thrived particularly well in the cool highlands. Third, mutton was preferred over beef in the sixteenth century (Chevalier 1975:142–143). Finally, the development of the Otomí as sheepherders rather than as, say, cattle ranchers was probably shaped by governmental decrees: *ganado mayor* (cattle and horses) were officially expelled from the central highland regions by viceregal order in the 1550s (except those animals destined for use on agricultural lands), and the region was thus destined to be a sheep-raising region—at least during the sixteenth century (Chevalier 1975:133–135).

The resistance to or acceptance of Old World species was not only based on perceived need, ecosystem compatibility, and government order, however, it also depended on the process by which the new cultural element entered the society. Sheep-grazing was new, it did not replace a traditional system of resource exploitation; and the processes by which pastoralism developed parallel to the village-encomienda system of production meant that it did not unduly disturb the organization of traditional agriculture, even if it helped condense the space for that endeavor. It is an intriguing fact that there was really very little opposition to the initial introduction of livestock, and this despite the fact that cattle and horses, goats and sheep were all grazed within the same space used by the Otomí for agriculture and were corralled next to their houses at night. Livestock raising did not (theoretically) compete with agriculture because grazing animals exploited a previously unexploited (at least from the point of view of the Spanish pastoralists) resource, grass, and pastoralists occupied previously unknown socioeconomic niches.

When opposition to the expansion of grazing did appear, it developed in the context of access to vegetation grown in one particular place: the planted agricultural field. As long as the numbers of domestic livestock remained low, and as long as they ate only grass, there was little difficulty. But domesticated livestock did not distinguish between cultigens and other forage, and Spanish pastoralists were not terribly interested in maintaining indigenous agriculture at the expense—as they saw it—of their rapidly increasing herds, unless forced to do so by viceregal order.[15] Furthermore, the fluid power relations inherent in the conquest situation meant that, wherever possible, the Spaniards ignored both Indian needs and vicere-gal orders and let their animals forage in planted fields.[16] The Otomí agricultural-ists, at least at first, had no knowledge of their rights vis-à-vis the pastoralists, and could not effectively defend themselves against the invading animals. Viceregal or-ders, such as those of Viceroy Velasco in the 1550s, attempted to codify the timing and use of agricultural fields by grazing animals (Chevalier 1975:133–135). But it was really only with time and experience and many lawsuits—and, most impor-tantly, with the involvement of indigenous peoples such as the Otomí in livestock raising—that the relations between agriculture and pastoralism were codified by local custom and use.

Far more threatening to Otomí agriculture than the actual adoption of pastoral-ism was the process of environmental transformation that accompanied the expan-sion of sheep into the Valle del Mezquital.[17] Forest clearance, burning, and over-grazing degraded the catchment area of the Valle del Mezquital and destroyed the basis of initial Otomí success in this arid region. As the water table dropped and springs failed, landowners and villagers fought each other in and out of court over the reduced water sources, drawing up agreements that usually spelled out the weakening power of villages to control this essential resource. The extent to which the catchment value had deteriorated in the Valle del Mezquital by the end of the sixteenth century is indicated by the fact that the Otomí had to fight for water to ir-rigate fields that were vastly reduced in extent as a result of the demographic col-lapse (ca. 92 percent by 1600).[18] But, to the considerable chagrin of the Spaniards, the Otomí, though reduced in number and with reduced agricultural fields, man-aged to retain control over the best lands in the region: the humid bottomlands.[19]

The environmental degradation that accompanied the development of intensive sheep-grazing in the Valle del Mezquital resulted in the deterioration of the carrying capacity of the range and the collapse of intensive pastoralism. Small-scale inten-sive sheep-grazing, with large numbers of small landowners grazing huge numbers of animals in common, developed in the 1550s and 1560s, reached its apogee in the 1570s, and began to decline rapidly thereafter; by the last two decades of the six-teenth century, much more land was needed to maintain a single head of sheep. The only way to maintain profitable flocks was to acquire several stations and to enforce exclusive access to the pasture. Those who could afford to buy land did so, and by 1600 regional production was dominated by extensively grazed latifundia. Pas-toralists who had been known as *señores de ganado* (livestock owners on a grand scale) in the period of expansion, were now known as *terratenientes* (landowners).

Another consequence of environmental transformation was a shift in regional production from sheep to cattle and goats in the seventeenth century (Melville 1994b). But the Otomí did not take part in either cattle-raising or goat-herding on the same scale as they had sheep-grazing in the era of expansion. Indeed, their mar-ginalization can be ascribed, in part, to the depression of sheep-grazing in this re-gion after the sixteenth century. Restrictions on the number of animals Indians could graze, and on the number their abbatoirs could butcher, also limited the ability of the Otomí to actively engage in the pastoral sector of the economy by the end of the century (Dusenberry 1963:111; Chevalier 1975:138).

In the early stages of contact with the Spanish, the Otomí had taken advantage of opportunities provided by a rapidly changing political economy in order to maintain their economic and political status in relation to the newcomers, and perhaps to leave old obligations behind. The evolution of pastoralism as a parallel system of resource exploitation allowed them to adopt livestock-raising while maintaining traditional agricultural systems. The importance of this is very clear when one considers the problems that the Spaniards faced when they tried to introduce wheat production. The Indians of New Spain persistently and successfully resisted the efforts of the Spaniards to force them to grow wheat. It was even difficult to have wheat grown as an item of tribute. Although wheat was as crucial an element of Spanish culture as sheep- or hog-raising, and was as profitable a sector of the colonial economy, it was not readily accepted. Wheat was not an add-on in the way livestock was—it was a replacement. As a replacement it required extensive changes in many areas of social and cultural life, so that in most places it was rejected (Gibson 1964:322–323).[20] But livestock could not simply be added to the culture and landscape of the Valle del Mezquital: pastoralism brought with it profound changes in the environment that, in the end, destroyed the basis of small-scale intensive pastoralism and, as well, the basis for Otomí participation in the colonial political economy, and produced a landscape within which the Otomí villages became islands in a sea of mesquite. Finally, then, one may argue that the persistence of Otomí land-management techniques after the sixteenth century resulted from cultural involution and structural—and physical—isolation.

Acknowledgments

A preliminary draft of this paper was presented at the Annual Meeting of the Conference on Latin American History in Chicago, 5 January 1995, in the panel on "Ranching Labor and the Environment in the Americas: A Comparative Perspective." I would like to thank José Cuello and Peter Herlihy for their careful reading and critique of that earlier draft, and William Balée for his editing of the final draft and for his encouragement. I also benefited enormously from discussions with Elizabeth Graham, Michael Levin, Cynthia Radding, Donald Worster, and the participants in the 1994–95 Seminar in Environmental History at the University of Kansas. I am grateful for Máximo Alvarado Guerrero's help with terminology. All errors and omissions are, of course, entirely my responsibility. Research for the material contained in this chapter was funded by the Social Sciences and Humanities Research Council of Canada: Canada Research Fellowship (1998–1992), Post Doctoral Fellowship (1996–1998), and Doctoral Fellowships (1976–1978).

Notes

1. Over the short term, irrigation by untreated human and chemical waste stimulated increases in productivity. Kaja Finkler (1973:105), for example, noted a ninefold increase in

productivity in some areas. It is only in the last fifteen years that environmentalists have called attention to the deleterious side effects of these waters (see, for example, Pérez Urbe 1991a,b). The ambivalent attitude of the Mexicans to the use of untreated waste in this region is nicely captured in the title of an article by Anselmo Estrada Albuquerque (1983), "Valle del Mezquital: Las aguas matan la fauna y dan vida agrícola." More commonly, however, the Valle del Mezquital is viewed as a vast treatment plant for Mexico City's waste.

2. By far the best-known scholar of the Valle del Mezquital is Miguel Othón de Mendizábal. An entire volume of his collected works (1946–47) was devoted to a comprehensive and detailed history of the Valle del Mezquital from the preconquest era to the 1940s. Mendizábal's belief that the natural resources of the region were inherently poor has done much to provide a historical basis for the region's image as a backward place. And his promotion of the idea, current at the time, that traditional ways should be eradicated in order to fully integrate backward indigenous regions such as the Valle del Mezquital into the national economy, reinforced the perception that the land-management systems of the Otomí were at the base of the region's poverty.

During the 1960s and 1970s the Valle del Mezquital was considered the prime example of Mexico's socially, economically, and politically disadvantaged regions. During this period the region was studied as internal colony by a large interdisciplinary group of social scientists under the auspices of the Patrimonio Indígena del Valle del Mezquital, the Instituto de Investigaciones Sociales, the Instituto de Investigaciones Económicas, and the Instituto de Investigaciones Históricos (which included a group of anthropologists). For examples of the publications that resulted from this research see Medina and Quesada 1975, and Cristiani Canabal and Martínez Assad 1973.

3. The discussion set out below is based in large part on primary documentary research carried out in the Archivo General de Indias (Seville) and the Archivo General de la Nación (Mexico City) and published in Melville 1990, 1994a, and 1994b. Since the data base is of considerable size, I have referred to these publications rather than setting out the individual citations.

4. Land grants for sheep stations were 780.5 hectares; while in absolute terms this is quite large, they were small in comparison with the often immense holdings found in this region by the end of the sixteenth century, some of which were 500 km^2 in extent.

5. Archivo General de la Nación (AGN), General de Parte, vol. 1, fol. 181r. See Melville 1994b:49–55 for a discussion of changes in stocking rates and sources, and pp. 78–84 for a discussion of changes in grazing rates in this region over the period 1530–1600 (and for sources).

6. The major source for this discussion of Indian land holding was the collection of documents having to do with land grants and titles in the colonial era: the Ramo de Mercedes of the Archivo General de la Nación (AGN) in Mexico City. This source was augmented by material found in AGN, Tierras, General de Parte, and Indios, and the Justicia, Escribanía de cámara, and the Audiencia de México of the Archivo General de Indias (AGI) in Seville, Spain. See Melville 1994b, appendix C, for a complete listing of the sources for land holding and land use.

7. See, for example, Lockhart 1992 for a discussion of the historiography of Indian-Spanish relations, and Zeitlin 1989 for a discussion of model building and criteria used to define "Indian."

8. Lockhart suggests that coincidences between Spanish culture and the indigenous cultures of the densely populated central highlands allowed "the quick, large-scale implantation among indigenous people of European forms, or what appeared on the face of it to be such,"

and he proposes that "large-scale lucrative encomiendas" were possible only in those regions where such coincidences occurred (1992:5).

9. For a detailed discussion of the activities of the encomenderos and the role they played in the formation of the colonial political economy, see Miranda 1965 and G. M. Riley 1971.

10. My evidence for the use of black slaves as herdsmen comes primarily from complaints filed by Indians about the violent behavior of the slaves; see Melville 1994b:50 n. 92 for documentary citations of these complaints and their frequency during the sixteenth century.

11. Recent ethnohistorical research stresses the fact that the small Spanish population had to negotiate their way into power during the first sixty years, because they did not have at their command either a standing army sufficient to force their way into all the lands they desired, or even an organized police force (Hoekstra 1993). While each village had a constable, there was no "police force" in the modern sense during the colonial era, and a formal state militia was not created until the eighteenth century. The distinction being made here between "violence" and "organized force," then, is between calculated and nonrandom violence, which was quite clearly present during the conquest (1521–1580) and colonial (1580s–1810) eras, and the organized force implicit in military or paramilitary institutions such as police forces, which was not present until the eighteenth century.

12. It is tempting to generalize from this region, but there is evidence of encomenderos grazing livestock on the lands of their own encomiendas in other regions. Cortes, for example, grazed livestock on the lands of his encomienda in Morelos (G. M. Riley 1971:241).

13. *Ganañes*: AGI Audiencia de México, leg. 111, ramo 2, doc. 12; AGN Tierras, vol. 2105, exp. 1, fol. 10v; vol. 2812, exp. 13, fol. 411; AGN General de Parte, vol. 5, fols. 26v, 65r–v. *Navorios, "gente"*: AGI Audiencia de México, leg. 111, ramo 2, doc. 12. *Sirvientes*: AGN General de Parte, vol. 3, fol. 202r.

14. Evidence of Indian sheep-shearers selling their labor outside the repartimiento: AGN Civil, vol. 809, exp. 6; AGN Tierras, vol. 2713, exp. 18, fol. 8v. Evidence of agricultural specialists living on *labores*: AGI Audiencia de México, leg. 111, ramo 2, doc. 12.

15. Evidence of the rapid increase of the introduced grazing animals is extensive: contemporaries exclaimed in wonder at the speed of increase, and the incredible numbers of animals; viceroys and other royal officials complained equally of the difficulty in protecting Indian plantings in the face of the desire of the Spanish pastoralists to achieve renown as *señores de ganado* (livestock owners on a grand scale). See Simpson 1952:2–6 for contemporary witness to the increase of grazing animals, and Chevalier 1975:133, 135 for a discussion of the problems facing the viceroy in his efforts to protect Indian land.

16. Virtually all students of rural New Spain in the conquest era note the problems that developed between pastoralists and villagers. See, for example, Gibson 1964:280–281. For a listing of the complaints filed by the Otomí of the Valle del Mezquital, see Melville 1994a:50 n. 25.

17. The major sources for indications of changes in the physical environment over the last half of the sixteenth century were the geographic descriptions carried out by order of the crown or the church, land suits, grants for grazing rights, and land surveys carried out for various reasons—population censuses, for example, or court evidence. This information was augmented by material taken from a wide range of other sources, such as claims to encomienda rights and descriptions of villages and their resources for *congregaciones* (the reduction of several villages into one). The major sources for the physical environment prior to

the changes associated with the irrupting ungulate populations include the geographic descriptions carried out ca. 1548 by order of the king (the so-called *Sume de visitas*), land suits, grants for grazing rights, and land surveys. (See Melville 1994a:31–39 for a more detailed description of the Otomí landscapes.)

In addition to the self-conscious descriptions of landscapes at specific moments, which when repeated in different years indicate broad changes over time, I was able to make use of "throw-away" data—i.e., pieces of information gleaned from many different documents that were not necessarily written to describe or explain the environment. When hundreds of these bits and pieces were ordered chronologically and geographically, they provided evidence of the processes of environmental change. (See Melville 1994a:84–87 for a full discussion of these sources and the problems in their use.)

18. The case between Catalina Méndez and San José Atlan over access to water from a spring illustrates the problems facing the inhabitants of the Valle del Mezquital at the end of the century: in 1600 spring flow was no longer sufficient to supply the town and the 4,000 grapevines belonging to Méndez, as it had done up to then (AGN Tierras, vol. 3, exp. 1).

19. AGN Tierras, vol. 1525, exp. 1; vol. 2697, exp. 11; AGN Mercedes, vol. 4, fols. 330–332; vol. 7, fol. 87.

20. Ruvulcaba (1984:424), however, argues that new cultigens such as wheat could be added to the indigenous cultivation systems without drastic changes in the temporal or spatial distribution of native cultigens.

References

Bauer, Arnold J. 1986. La cultura Mediterranea en las condiciones del Nuevo Mundo. *Historia* 21:31–53.

Chevalier, François. 1975. *La formación de los grandes latifundios en México*. Mexico City: Fondo de Cultura Económica.

Cristiani Canabal, Beatriz and Carlos R. Martínez Assad. 1973. *Explotación y dominio en el Mezquital*. Mexico City: UNAM, Centro de Estudios del Desarrollo.

Dusenberry, William H. 1963. *The Mexican Mesta: The Administration of Ranching in Colonial Mexico*. Urbana, Ill.: University of Illinois Press.

Estrada Albuquerque, Anselmo. 1983. Valle del Mezquital: Las aguas mata la fauna y dan vida agrícola. *Unomásuno* (Mexico City), 15 August, p. 22.

Finkler, Kaja. 1973. A comparative study of the economy of two village communities in Mexico, with special reference to the role of irrigation. Ph.D. diss., City University of New York.

Gibson, Charles. 1964. *The Aztecs Under Spanish Rule: A History of the Indians of the Valley of Mexico, 1519–1800*. Stanford: Stanford University Press.

Gutiérrez Brockington, Lolita. 1989. *The Leverage of Labor: Managing the Cortés Haciendas in Tehuantepec, 1588–1688*. Durham: Duke University Press.

Hoekstra, Rik. 1993. *Two Worlds Merging: The Transformation of Society in the Valley of Puebla, 1570–1640*. Amsterdam: Centrum voor Studie en Documentatie van Latijns Amerika.

Johnson, Kirsten. 1977. "Do as the land bids": A study of Otomí resource-use on the eve of irrigation. Ph.D. diss., Clark University.

Lockhart, James. 1992. *The Nahuas After the Conquest: A Social and Cultural History of the Indians of Central Mexico, Sixteenth through Eighteenth Centuries.* Stanford: Stanford University Press.

McLeod, Murdo J. 1973. *Spanish Central America: A Socioeconomic History, 1520–1720.* Berkeley and Los Angeles: University of California Press.

Medina, Andrés and Noemi Quesada. 1975. *Panorama de los Otomies del Valle del Mezquital: Ensayo metodológico.* Mexico City: UNAM, Instituto de Investigaciones Antropológicas.

Melville, Elinor G. K. 1994a. Land-labor relations in sixteenth-century Mexico: The formation of grazing haciendas. *Slavery and Abolition* 15(2): 26–35.

———. 1994b. *A Plague of Sheep: Environmental Consequences of the Conquest of Mexico.* New York: Cambridge University Press.

———. 1990. Environmental and social change in the Valle del Mezquital, Mexico, 1521–1600. *Comparative Studies in Society and History* 32(1): 24–53.

Mendizábal, Miguel Othón de. 1946–47. *Obras completas*, vol. 6. Mexico City.

Miranda, José. 1965. *La función económica del encomendero en los orígenes del régimen colonial.* Mexico City: Universidad Nacional Autónoma de México.

Pérez Urbe, Matilde. 1991a. Rechazan suspender el uso de aguas negras en Hidalgo. *La Jornada* (Mexico City), 23 September, pp. 1, 14.

———. 1991b. Se reducirá el uso de aguas negras en el Mezquital: CNA. *La Jornada* (Mexico City), 24 September, pp. 1, 16.

Riley, G. Michael. 1971. Land in Spanish enterprise: Colonial Morelos, 1522–1547. *Americas* 27(3): 233–251.

Riley, James D. 1984. Crown law and rural labor in New Spain: The status of the gañanes during the eighteenth century. *Hispanic American Historical Review* 64(2): 259–285.

Ruvulcaba Mercado, Jesús. 1984. Agricultura colonial temprana y transformación social en Tepeapulco y Tulancingo (1521–1600). *Historia Mexicana* 33(4): 424–444.

Sanders, W. J., J. R. Parsons, and R. Santley. 1979. *The Basin of Mexico.* New York: Academic Press.

Simpson, Lesley Byrd. 1952. *Exploitation of Land in Sixteenth-Century Mexico.* Berkeley and Los Angeles: University of California Press.

Van Young, Eric. 1983. The historiography of the colonial hacienda. *Latin American Research Review* 18:5–61.

Vassberg, David E. 1984. *Land and Society in Golden Age Castile.* Cambridge: Cambridge University Press.

Zeitlin, Judith Francis. 1989. Ranchers and Indians on the southern isthmus of Tehuantepec: Economic change and indigenous survival in colonial Mexico. *Hispanic American Historical Review* 69(1): 23–60.

CHAPTER 16

The Great Cow Explosion in Rajasthan

CAROL HENDERSON

This chapter examines the cattle population boom and crash that occurred in India's northwestern state of Rajasthan between 1951 and 1991. During the first thirty years of this period, the population of cattle soared from 10.8 million to 13.5 million head of animals (Rajasthan Government 1979:60–62). In the next five years the cattle population dropped, reaching 10.9 million in 1988 (Prabhakar 1992:198); subsequently this trend appears to have continued. Ironically, the reversal from growth to decline of the cow population appears at a time of rising Hindu militancy in the state, in the form of both support for political parties espousing a Hindu fundamentalist line, such as the Bharatya Janata Party (BJP), and the adoption in public discourse of specific symbols differentiating Hindus from the state's Muslims.

Rajasthan, a state whose largely arid nature—60 percent, or about 213,800 km^2, lies in the Thar desert—defines agricultural possibilities, has long depended on livestock production as a principal component of the rural economy (Acharya, Patnayak, and Ahuja 1977; Jodha 1985:247; Malhotra 1977:310). The impact of livestock on the environment comprises a leitmotif in discussions of economic development in Rajasthan, which ranks as one of India's poorest and least-developed states in terms of its literacy, child mortality, provision of social services, and modern infrastructure—such as tube wells, paved roads, electrification, and the use of tractors (Dandia 1981; Rajasthan's Backwardness 1983). Furthermore, Rajasthan's twelve desert districts, its "arid zone," lag behind the rest of the state in terms of these same variables.

These factors are sometimes regarded as the result of beliefs and knowledge characterized as "superstitious . . . traditionalist and laden with innumerable taboos," in the words of a report on the Luni Development Block, located on the margins of the Thar Desert, that was prepared for the United Nations Conference on Desertification (UNCOD 1977a:51). They have also been viewed as the consequence of the marginal environmental conditions for agricultural production in the Thar Desert (Mann 1982). Neither explanation, however, appears to account for recent shifts in

349

the livestock population of the state. The hypothesis of shifts based on religious af-filiation proves counterintuitive, because forms of Hindu religious sentiment that in-clude the veneration of cows grew during this period. A Malthusian hypothesis based on environmental degradation and the droughts that occurred during the 1980s also is inadequate, because it removes from the analysis historical elements that sig-nificantly affected producers' access to resources and the costs and benefits of di-verse strategies.

On a more global scale than Rajasthan, such shifts in livestock populations are not unknown: drought, war, and epidemic disease may decimate livestock within a brief period. Wartime "scorched earth" policies and peasant resistance to collec-tivization can lead to the mass slaughter of herds and flocks. Mechanization pro-motes livestock population reductions, as owners replace their animals with ma-chines. Clearly, political, technological, and economic factors may result in shifts in livestock populations, as may epidemics or conditions leading to the lack of fod-der. In the last example, livestock producers' ability to move their animals out of hazardous zones suggests that in this instance, cases of starvation may reflect changes in the political economy of former host regions, which no longer welcome drought migrants.

In this presentation, I hypothesize that the livestock population boom and crash reflects institutional changes associated with shifting costs and benefits related to access to resources during the mid-to-late twentieth century. Principal components include a dialectical relationship between people and the environment, and between groups of people within the apparatus of the Indian state.

The data for this paper derive from a field study of adaptations and adjustments to drought in the Thar Desert in 1981–1982, and field research on pastoralism in this region in 1988–1989 (Henderson 1989, 1992, 1993). Interviews and participant observation, carried out during two years of drought, provided data on people's re-sponses to shifting conditions in the Thar Desert. The research initially focused on factors that promoted livestock population growth, since most sources identified this as the preeminent cause of environmental degradation in this region.

Although early results suggested that livestock population reverses affected pro-ducers who were based in villages (Henderson 1987, 1989), economic variables such as price increases also motivated producers to attempt to expand their live-stock holdings (Henderson 1992). By the end of the decade, the specter of villages bereft of their cows, sheep, and other livestock compared to earlier observations, along with rural informants' comments, made it clear that larger-scale trends were afoot. The present chapter thus represents an effort to integrate this material into a historical-ecological framework and to provide a preliminary synthesis of recent complex events in western India.

To begin with, I review the ways in which a historical-ecological approach might be useful for this enterprise, and I summarize recent approaches to livestock man-agement in India that have profoundly affected debate and criticism in this area. Second, I review the characteristics of the arid zone and the production strategies of its residents in relation to ecological, economic, and political events, seen in histor-ical perspective. In this, my focus will be on Marwar, a traditional cultural region and the largest of the princely states in the Thar Desert prior to their reorganization into the modern state of Rajasthan. Third, I attempt to demonstrate specific relationships between shifting strategies and changes in the region's political and economic relationships (particularly during the late 1980s), and the transformation of the costs and benefits to producers that occurred at this time, in order to identify the proximate causes of this region's cattle population boom and crash.

On Historical Ecology and Sacred Cows

The premise that historical events impel the principal changes in relationships linking humans and their immediate environment separates historical ccology from other ecological models that regard evolutionary events as the source of shifts (Balée 1994:2). In its approach, historical ecology argues that the diverse calculus of interests manifested in human behavior may instigate a trajectory, here defined as a bundle of relationships characterizing the human/environment nexus, that may persist or expand relative to alternatives.

Historical ecology's focus emphasizes political, economic, cultural, and social units—and, within these, actors who construct meaningful behavior. Their actions present a panorama of responses to the anticipated and unanticipated consequences of the intertwined environmental and human behavioral relationships that comprise the arena for action. In this, as William Marquardt comments, the analysis examines "a long succession of human decisions, human relationships with each other, and human relationships within the physical surroundings" (1994:219). These units and the events with which they are concerned, in the historical ecology perspective, will always be subjected to selection pressure based on regularities attached to the exchange and flow of energy among components of this trajectory, but the proximate causes of changes in human/environmental relationships may be at a remove from these pressures. Thus, within the historical ecology framework, any specific human/environment relationship, so identified, is a historically contingent entity (Patterson 1994:230).

The assumption that actors are goal-oriented and seek to optimize certain desired outcomes comprises a starting point in the historical-ecological approach: resource management reflects the relative costs and benefits of different actions within a specific context. For livestock producers, this might mean expanding herds when the demand for their product rises, reducing herds when it drops; managing herds to increase outputs, relative to inputs (efficiency); and selective breeding to enhance certain desired characteristics, such as speed, strength, heat tolerance, and ability to produce on poor forage and saline water sources. Social goals also enter into these calculations, particularly if a household's actions (such as wedding gifts of cattle) provide a ground for future claims to assistance, for one's household or livestock.

Many anthropologists working in India have not found this perspective congenial when the subject is the management of zebu cattle (*Bos indicus*). These anthropologists, by and large, question whether the vast majority of cattle-owners (who are Hindus) optimize resources to achieve a mix of desired goals, saying instead that they pose an exception to what has become a basic assumption regarding resource-use strategies by other groups (Cancian 1989:165–166). In fact, the issue of the management of cattle in India informs one of the most acrimonious debates ever to grace the pages of anthropology journals from the 1960s to the early 1980s, starting with Marvin Harris's paper on "The Cultural Ecology of India's Sacred Cattle" (1966a). The debate generated by Harris's model was unusually transnational in its scope. It also persisted well beyond the five-to-ten-year period within which most academic brouhahas seem to run their course. In fact, one participant in the debate published an article on this in December 1994 (Lodrick 1994)—almost thirty years after Harris presented portions of his model at a Columbia University seminar (Harris 1964).

Harris (1966a) hypothesized that the belief in the sacredness of cattle in India could be understood by utilizing a cultural-ecological framework. His paper demonstrated that cattle supplied essential products to their owners: milk, dung, traction, and replacement heifers and oxen. Thus, cattle-keeping supported agricultural production. In subsequent publications, supporters of a cultural ecology position argued that there was a fairly close fit between the characteristics of the bovine population of a region and that region's ecological, demographic, technological, and economic characteristics (MacLachlan 1982; Odend'hal 1972, 1988; Vaidyanathan 1988; Vaidyanathan, Nair, and Harris 1982). The focus of discussion moved from general questions about "cattle" (i.e., bovines), to animals identified as members of specific age, sex, and species groups, and particularly, to adult female zebu cattle ("cows"), seen in relation to these categories.

Harris proposed that the "positive functions" of cattle-keeping (for their owners) should be considered when interpreting the belief in the sacred cow. In his theory, the "sacred cow complex"—that is, the set of beliefs and practices associated with the veneration of cattle—helped to motivate the conservation of cattle, because otherwise owners might be inclined to cull their least-productive animals (Harris 1966b:64; 1977:220–222). To state this another way, the belief in the sacred cow suggests a cost-benefit equation that trades off short-term gains obtained from reducing individual household outlays to maintain less-productive animals, against household long-term lowered risk of failure if these animals later have value (Harris 1966a:56). This suboptimizing strategy, again, looking comparatively at other cultural groups, frequently features in accounts of subsistence producers (Scott 1976:2–7). Harris then proposed that cattle owners who adopted this perspective might enjoy a slight competitive advantage over those who did not, because the sacred cow complex generally supported the conservation of farm capital, in the form of cows (Harris 1977).

Opponents claimed that the most important functional relationship connecting cows and Indians stemmed from beliefs held by the Indians (Diener, Nonini, and Robkin 1978; Freed and Freed 1981; Lodrick 1979, 1994; Simoons 1979). Arguments against the cultural ecology position were varied and complex; in general, however, they contrasted two belief systems, "Hindu" and "non-Hindu." Hindus were committed to *ahimsa*, the principle of the protection of life, as this is set forth in the second-century laws of Manu and diverse ancient *shastras*, or texts. "Non-

Hindus" were not bound by these rules. Hindus venerated the cow in her adult, lac-tating, guise, whereas non-Hindus did not. Hindus protected cows, whereas non-Hindus were free to pursue other management strategies. Consequently, Hindus did not efficiently manage their cattle in order to achieve economic goals (because this might involve culling), even if it cost them in terms of productivity (Dandekar 1969:1564; 1970:531; Heston 1971:196–197; Simoons 1979:471, 473). In Ra-jasthan, for example, where the population is about 90 percent Hindu (Lodrick 1994:20), this perspective implies that for the majority of the population, adult fe-male zebu cattle were an unmanaged—albeit, as will be discussed, extraordinarily important—resource.

These writers argued that one could best account for the sex-, age-, and species-mix of bovines (cattle and water buffalo) in a particular region by examining the proportion of Hindus and non-Hindus. A region with a preponderance of Hindus should favor zebu cattle, even if this contradicted its ecological and economic char-acteristics. In a study of dairy owners in Ajmer District of Rajasthan, for example, Deryck Lodrick (1979) concluded that Hindus tended to own cows, even though it cost more to produce a unit of milk with a cow than with a buffalo; while Muslim owners, not bound by a religion-sanctioned preference, owned buffalo. Stanley A. Freed and Ruth S. Freed (1981) argued for a similar relationship, based on data col-lected in a village quite near to New Delhi, the nation's capital.

In rejecting the cultural-ecological hypothesis that protecting cows served the long-term interests of individual owners, proponents of the ideological perspective focused instead on the aggregate consequences of "Hindu" cow management, which supposedly maximized the number of cows. Cows were identified in this ar-gument as the main source of environmental problems in India (Diener, Nonini, and Robkin 1978; Simoons 1979:473, 478), and evidence for environmental degrada-tion was featured as evidence that "Hindu" standards ruled the management of cat-tle. Some of the most compelling data that were cited to support this perspective (Diener, Nonini, and Robkin 1978:225–226; Simoons 1979:468) derived from grassland studies conducted in the Thar Desert in the mid-1950s and early 1960s, popularized by Reid Bryson (Bryson and Baerreis 1967; Bryson 1972; Bryson and Murray 1977). Bryson noted that the air over the Thar Desert was full of moisture during the monsoon period, yet little rain fell, and he hypothesized that dust kicked up by cattle and other animals created subsidence (sinking air) that physically blocked rainfall (1972:139, 142). The outcome was an escalating cycle of dust, drought, and vegetational failure, each event triggering and intensifying the next (Bryson and Murray 1977:113). Bryson suggested that the natural state of the Thar Desert was not sandy desert, but rather savanna, and in support he cited a study car-ried out at the Central Arid Zone Research Institute (CAZRI), a few miles to the south of Jodhpur City, capital of the district of that name: when a plot of land at CAZRI was fenced to keep out animals, it produced lush grass "except within one goat-neck distance from the fence" (Bryson 1972:142).

While Bryson seemingly had strong evidence for his position, his data could not account for the positive correlation between the frequency of dust storms and the amount of monsoon rain noted in the Luni Development Block study (UNCOD

1977a:8). That study suggests that dust storms and rainfall reflect environmental processes that underlie both. Indeed, the U.N. Conference on Desertification, in its review of desertification pressures (1977b:70), concluded that atmospheric dust played a minor role in reducing rainfall.

Ironically, an influential paper at this time—but one that was not, so far as I know, cited in the sacred cow debate—was Garrett Hardin's (1968) work on the "tragedy of the commons." Hardin persuasively argued that the economic decision-making of individuals could, when aggregated, lead to negative environmental consequences. The existence of negative environmental conditions was not necessarily evidence of a lack of economic rationality, but its reverse. Nevertheless, opponents of the cultural ecology position focused on poor environmental quality as crucial evidence that Hindu cow managers did not behave in an economic manner.

It should be noted that the view that Indian livestock owners were environmentally destructive had much support from other researchers working in India. For example, scientists conducting research on the Thar Desert energetically subscribed to the position that its "traditional" occupants caused environmental degradation (Dhir 1982; Kumar and Bhandari 1992; Prakash 1987). In this context—and in the setting of the sacred cow debate—these positions flourished despite the shortage of data on how people actually managed their resources, or on the effects of the land reforms that were then in progress.

Most scientists in the sixties and seventies working in this area either overlooked indigenous technologies or regarded them as "superstitions"—and thus ignored potential approaches to understanding decision-making in local contexts (Jodha 1991:A98; and see Jain 1993). Anthropologists conducting field research in India focused on religious belief systems, caste, and kinship; there was little empirical testing of the assumption that beliefs motivated behavior. Even when ecological data were collected by ethnographic field workers they were seldom published, relative to other research interests. For example, the Freeds had been working on the impact of urbanization on Shanti Nagar village when the sacred cow debate stimulated them to work up their field notes on this subject and to collect additional data on cattle (1981:489). As a number of commentators note, the early debate on the sacred cow utilized a surprisingly limited set of data on how people actually managed their livestock (Adams 1982:373; Mencher 1971:203; Nonini 1982; Rao 1969:A225; Schwartzberg 1979). In this case, how were researchers to assess their material?

In examining the question of cow-management strategies it is evident that the sacred cow debate lacked clarity regarding the difference between individual and aggregate levels of analysis. Research results were seldom reported in ways that placed current events in a historical context. Thus, the actual set of goals sought by cattle owners (who might be, among many possibilities, rich or poor; Hindu or non-Hindu; urban or rural; subsistence- or entrepreneurially oriented) failed to emerge comprehensively in the debate.

Authors, particularly those who espoused the "Hindu" cow-management model, conflated individual actions with systemic responses and effects. My reading of the work of Harris (1966a,b, 1977) and of Vaidyanathan, Nair, and Harris (1982) is that they did not argue that the Indian environment was improved through the actions of farmers; they argued, rather, that farmers simply achieved individual, personal benefits—at whatever cost to the total system of energetic relationships.[1] Nevertheless, cultural ecology's then-prominent position that human adaptation to the environment manifested itself in the achievement of balanced energy equilibria (Orlove 1980:240–243) contributed to this misunderstanding.

Historical ecology offers a response to these claims, for it separates individuals from underlying selective pressures. William Balée's call (1994) for studying human/environment relationships as a dialectic is one way of acknowledging the complexity of mediated cultural behavior in specific political, technological, ecological, and economic settings. Individual and group interests may drive specific initiatives that create environmental impacts; these, in turn, are part of processes leading to ongoing problems, options, and pressures that rebound on individual and group interests. Historical ecology focuses on this relationship, for it generates the proximate behavior involved in shifting the bundles of energetic relationships comprised in these complex behavioral assemblages.

In epistemological terms, historical ecology analytically separates energetic outcomes (seen in evolutionary perspective) from their antecedent processes and conditions. In contrast to cultural ecology, historical ecology's quest appears not to be an evolutionary one: for example, as Balée argues (1994:3), there is no necessary connection between creating a "more habitable biosphere" and degrading the non-human biosphere. Evolutionary events are emergent properties of the relationships with which historical ecology concerns itself, and are analytically separate from historical ecology's focus.

This approach examines actors in specific contexts; the regularity is to assume that these actors seek goals, authoring themselves and, dialectically, environmental changes. Institutional factors related to resource control and access provide settings within which specific strategies may have greater or lesser utility, given actors' goals. The framework thus seeks to comprehend the mediations that link humans and environmental processes, and to explore the shifting historical and ecological contexts of action. In the case of the cattle population boom and crash in Rajasthan, understanding this context is the first step toward comprehending the complex ways in which diverse interests press upon people's options and influence outcomes, seen as an unfolding process.

Ecology and Landscape in the Arid Zone

The Thar Desert spreads between Pakistan and India, where it includes parts of four states: Gujarat, Haryana, Punjab, and Rajasthan (see figure 16.1). Although, strictly speaking, it lies in the zone where annual rainfall is less than 100 mm (Le Houérou 1977:18), it is surrounded by arid and semiarid zones, known in Rajasthan as the Marusthal, or "desert of death," where annual rainfall is less than 400 mm—below the lower limits to regularly produce crops without irrigation. In India, this arid region is bordered on the south by the Aravalli range, and stretches to the Pakistan border. Its northern limits lie in the southernmost portions of Punjab state, and it is bordered in the south, in Gujarat, by the sea (Bharadwaj 1961:143). Rajasthan's arid region belongs to the west Asian desert zone that stretches from India to the west across Pakistan and the Iranian plateau.

Rajasthan's arid region is relatively small, compared to Africa's Sahel zone—itself being about one-fifth the size of each of the Sahel's arid countries. Rajasthan distinguishes itself with a high human population density: population densities in

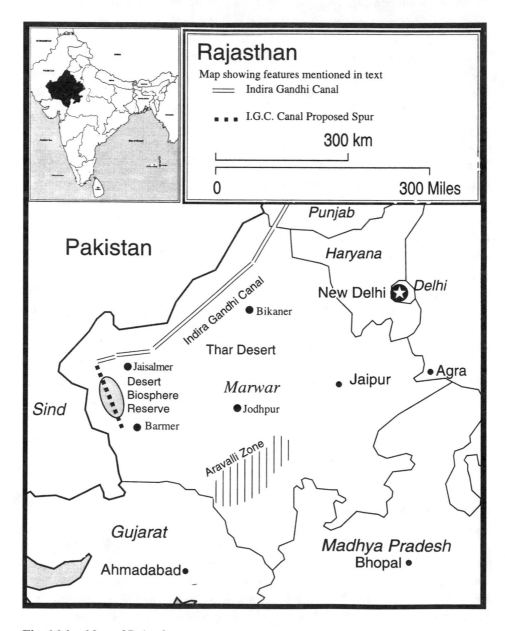

Fig. 16.1 *Map of Rajasthan.*

Chad, Mali, Niger, and Algeria are estimated at between four and ten persons per km^2 (World Bank 1991:204–205)—whereas in Rajasthan's arid zone, the rural population density as calculated from the 1991 census figures is approximately sixty-six persons per km^2 (*Census of India* 1991a,b).[2] Rajasthan's arid region, like others in the world, is prone to desertification; its residents pursue mixed cropping and livestock-production strategies, and rely on spatial mobility to survive frequent droughts.

In Rajasthan, the arid zone, as scientists dubbed the region, includes twelve districts. Jalor, Pali, Nagaur, Sikar, Jhunjhunu, and Sirohi Districts are located along the western slopes of the Aravallis; through them falls the 400 mm isohyet of average annual rainfall (they are the most humid districts). Barmer, Jodhpur, and Churu stretch toward the north and west; the 300 mm isohyet passes through them, although their northernmost portions, like the very arid Jaisalmer, Bikaner, and Ganganagar Districts, are crossed by the 200 mm isohyet. Rainfall drops to less than 100 mm in western Jaisalmer District, near the Pakistan border (Pisharoty 1980:41).

Not only are low average annual rainfall totals problematic for agricultural producers, the lower these annual averages, the greater is the chance that the monsoon will fail: unreliability and variability increase with aridity. Fluctuations within the desert are much greater than in its border areas (Mann, Malhotra, and Kalla 1974: 108). For the arid zone as a whole, the probability is that agricultural drought— defined as rainfall less than 80 percent of normal—will occur about two years out of five (UNCOD 1977a:11). High-velocity winds, heat, and very bright solar radiation also characterize the region.

Sandy gray-brown desert soil types, saline water sources, and a lack of permanent water courses or springs mark the region. Rocky knolls and outcrops punctuate the horizon. Dunes and alluvial depressions where water settles after rain comprise most of the land where people graze their animals and try to grow crops. Differences in slope and geomorphology play an important role in determining the flow of surface and subsurface moisture: these runoff patterns channel water and concentrate it at sites that form ponds, which may be surrounded by silt and salt pans.

The vegetation in the region is drought-adapted, heat-tolerant, and resistant to salinity. Prickly shrubs and trees predominate, especially individuals of *Prosopis cineraria*, locally termed *khejri*, which have important economic uses. Grass cover in the desert consists of *Lasiurus*, *Cenchrus*, and *Dichanthum* species; these are also used by people for food (particularly *Cenchrus*), as well as fodder for livestock. L. D. Ahuja and H. S. Mann (1975:30–31) note that these grasslands are extremely heterogeneous. Climax stands occur where a lack of drinking water precludes grazing by livestock. Plants respond rapidly to opportunities for growth during the brief monsoon season, which begins in July and ends in late September. In general almost all annual rainfall occurs during this three-month season, and it is far from a steady supply: after a spell of three or four days of afternoon showers announces the monsoon's arrival, there may be several weeks of drought, followed by one or two additional brief spells of light showers. This is enough to add a bright green tinge to the area's usual brown, tan, or gray aspect, as the plants carry out their reproductive cycles. By the end of October, the flush of green disappears.

This environment is characterized by "patchiness"—that is, small areas within it that are ecologically distinct (Winterhalder 1994:33). The Thar Desert by no means consists of uniform sandy plains or dunes: patches of different soils, plant communities, and moisture availability create different opportunities for humans to identify and use resources within a locality (defined as the area within which people ordinarily perform their daily productive activities). Localities thus emerge in relationship to a particular community's patterned interactions. These intervene in environmental and ecosystem processes to create landscapes, "the material manifestation of the relation between humans and the environment" (Crumley 1994:6).

Two considerations are evident in examining human interactions within the Thar Desert. One: human settlements appear where patches converge; very frequently, settlements perch on the edge of a slope or a hill overlooking a depression that receives runoff sufficient to provide drinking water for the settlement and to water, through underground percolation, patches of alluvial soil. Two: humans create both positive and negative environmental impacts; although the latter receive most attention in the literature, because of the severe risk of desertification (Jaiswal 1977; Kumar and Bhandari 1992), people attempt to tilt ecosystem processes in their favor.

For example, R. Thomas Rosin shows (1993) that villagers create farmland through a combination of land- and livestock-management practices. One such practice is the construction of low levees ("bunds") that impound water runoff from local catchment systems and increase soil moisture, thereby improving opportunities for plant growth and the formation of humus. Another is the herding of animals into fields, where the manure they deposit will help to improve the soil. The location of aboveground tanks and ponds can affect the subsurface percolation of water and downstream agricultural productivity (Rosin 1993). In the high-water-table zones lying along the Luni river basin, which drains the Aravalli mountains, animal-powered lifts were used to irrigate cultivated fields prior to the adoption of diesel and electric pumps for this purpose. Most recently, along the Pakistan border, construction of the Indira Gandhi Canal has transformed the landscape from desert grasslands to intensively cultivated plots.

People selectively encourage trees to grow within fields, because the tree canopy and root system help to concentrate moisture and to shelter the crops. People in Jodhpur District classified *khejri* as a "shrub-cum-tree": something that became a tree if properly pruned—a process that requires patience and the ability to forgo lopping branches for livestock feed in drought years. "Sacred groves," known as *oraans*, where users prohibited plowing and the cutting of trees, conserved grazing for livestock and probably functioned to improve local water and grass stocks, since the shallow root systems typical of many of the desert species were not disturbed. Other regions contained dedicated pastures (*gochar:* literally, "cow-fodder" in the Rajasthani language).

Human impacts also transform resources in less positive fashion, such as the grazing of livestock, which reduces the proportion of palatable species. A prominent theme of research by scientists is the anthropogenic sources of environmental degradation (Mann 1982; Malhotra 1988). The outcome is a mix of impacts that, over time, can be seen as the interaction of dynamic processes.

Production and Access to Resources

Production strategies in the arid zone rest on a symbiosis among livestock, cultivation, and wage-labor migration or entrepreneurship. Households attempt to diversify their activities, rather than focusing exclusively on only one productive activity; this reduces the risk of failure if, for instance, drought or a locust infestation destroys crops, or an epidemic kills a household's livestock. Entrepreneurship and wage work may provide cash inputs to support households in times of failure.

In varying combinations of these activities, the general production strategies of individual households may be characterized as (a) pastoral nomadism; (b) agropastoralism; (c) cultivation; and (d) wage labor or commercial activities. (a) Pastoral nomadism, based on specialized animal husbandry requiring periodic movements, today is rarely followed even by members of groups such as Rabaris and Raikas, who claim this as their traditional occupation. (b) Agropastoralism is probably the most common activity. In this, producers combine cultivation with animal husbandry that, depending on grazing and water conditions, usually requires transhumance. Part of a household—that is, an independent managerial unit comprised of family members and unrelated dependents—remains at the crop-production site, while others migrate with animals. Occasionally, members trade off these roles. (c) Cultivators focus on crops, augmented by livestock production. These households move their animals rarely, usually only in the case of drought. Finally, (d) wage labor is common among the landless or those with poor and inadequate holdings of land and livestock. These are joined by a small group of professionals (teachers and social workers, for instance) and entrepreneurial households.

Until the middle of the twentieth century, up to two-thirds of rural employment in the arid zone depended on livestock production (Acharya, Patnayak, and Ahuja 1977:275). Pastoral nomads and agropastoralists shared the landscape with one another; where feasible, people grew small patches of millet, sorghum, and drought-adapted pulses in favored spots. In the desert interior, seasonal stands of hardy grasses provided fodder to animals, which moved in circuits following the annual rains that came with the monsoon. People foraged for wild plants and animals to add vegetables, fruits, and protein sources to their diet, which depended on millet, pulses, and the milk of camels, cattle, sheep, and goats.

Seasonal migrations followed the retreating monsoon in September southward along its track. Other destinations included irrigated regions of Sind and Punjab, and the hill tracts of the central Indian plateau. This reduced grazing pressure in the home regions, for the migrants would not return until the onset of the expected following monsoon. The majority of the important seasonal migration routes crossed the Aravalli mountains, which also served as important grazing grounds for traditional sheep- and camel-herding pastoral nomads (Agrawal 1993; Prasad 1984).

During droughts, the entire population of a district or village might migrate across the desert to Sind, or move south into Gujarat and central India. Later, they might return to their home village or opportunistically resettle in another village, with relatives or at the invitation of village leaders or friends.

Not all production was subsistence-oriented. Producers bred dairy and traction cattle and exported them to areas outside the Thar Desert. An 1874 source notes: "The main wealth of the desert lands of Marwar and Bikanir consists of the vast herds of camels, horned cattle, and sheep which roam over their sandy wastes and thrive admirably in the dry climate. Camels and horned cattle are bred in such numbers that they supply the neighboring provinces" (Blair 1874:78). Military agents sought the horses produced in the western sections of Marwar. Weavers transformed wool into blankets, which were exported to west Asia. Throughout the nineteenth century, Marwari traders fanned out into western and eastern India and remitted earnings that could be invested in homes, farm and livestock capital, and philanthropic ventures (Timberg 1978). Many elements of these productive activities were still visible in the 1980s.

A significant factor within the rural economy was producers' reliance on common-property regimes, communal systems of regulated access to shared resources (McCay and Acheson 1987). A patchiness of resource availability and a high dependence on pastoral resources supply the ecological conditions under which common-property regimes may be more efficient in distributing grazing resources among producers than will privately owned pastures, since owners cannot predict whether a particular field will receive rainfall. Common-property regimes provide fodder buffers that are especially important for the poorer members of a community.

Subject to local rules of access, livestock herders could move flexibly across the landscape. The extensive form of agropastoralism practiced in the region involved sharing grazing resources among the residents of many villages, even districts. Some of the different categories of lands that were subject to common rules of access include community grazing lands, fallows, *oraans* and *gochar*, ponds, migration routes, and just-harvested croplands (Jodha 1985:248). R. R. Prasad (1984:17) reports that the demand for manure was so strong in some areas that farmers paid migrants to fold their animals in their fields at night, in return for *jaggery* (unrefined sugar), food, and in some cases, a little cash.

Persons grazing livestock needed permission from village leaders to bring their animals onto communal grazing land, or from individual landowners. The basic rules about access to grazing and water established that herders must avoid currently cultivated land; they must move on in a day or two; and they must ask permission to camp on village lands. Users defended their lands. Violators could be expelled by violence. Relatively little is known about these institutions, for scholars have paid little attention to non-elite mechanisms of control. An additional complicating factor is that social and cultural changes during the past forty years have eroded many of the institutional features that support common-property regimes.

Until India became independent of British colonial rule in August 1947, Rajasthan, then known as Rajputana, was home to a number of principalities with traditional rulers. In the Thar Desert the most important of these princely states, as they were known, were Marwar (also called Jodhpur state), Bikaner, and Jaisalmer. During the sixteenth century there was sporadic warfare with the Mughal rulers, which was complicated by the latters' supply problems of sustaining armies across the desert; this ended in the last four decades of the sixteenth century, when the empire incorporated these states through conquest and marital alliance (Bhargava 1966). In securing the Rajput states, the Mughals obtained desert-capable military forces, which could be used for strikes into contemporary northern Pakistan and Afghanistan; strategic access to war camels and horses; and control over the important trade routes linking Delhi to coastal Surat and Bombay (Tod 1914, 2:127–128; Ziegler 1994). The significance of overland trade declined with the emergence of Calcutta as India's preeminent center during the period of British rule. During the first two decades of the nineteenth century, the British negotiated treaties subordinating the desert states' international political relations to the government in Calcutta. Marwar retained control of its internal affairs, however—particularly over land and agricultural production.

These relationships changed greatly with India's independence and the creation of the modern state of Rajasthan. In 1947, the rulers of the princely states discussed whether to seek their own independence, join Pakistan, or become part of the Indian Union. Three of these states shared a border with the then-forming new Pakistan state. In Marwar, Maharajah Hanuwant Singh, who succeeded to the throne in June 1947—virtually on the eve of independence—was "intractable" from the Indian nationalist viewpoint and reportedly "on the verge" of acceding to Pakistan (Prabhakar 1992:23–24). This threat formed a bargaining chip in discussions of how a Rajasthan state might be formulated: would Marwar's regime continue as it was, exist in the form of a constitutional monarchy, or adopt a parliamentary type of government and the direct franchise? The stakes were high, from both sides' viewpoints.

Approximately 60 percent of landholdings—and 80 percent of villages—in Marwar were directly controlled by the ruler under what was known as *khaslo* tenures (D. Singh 1964; H. Singh 1979:102). Hereditary right-holders (*jagirdars*) controlled most of the remainder. Local people obtained lands to cultivate at the will of village landlords, who were themselves either hereditary or transferred in at the will of the ruler. Control of the ruler's share of the produce was organized on a village-by-village basis. Possession of a village (or shares in one) meant the rights to keep a proportion of the harvest, to personally cultivate lands, and to collect such other tributes in the form of goods or labor as could be wrested from the residents. About the only recourse that people had against a rapacious landlord was desertion to a region outside his control.

Almost 97 percent of arid Shergarh subdistrict in Jodhpur District (where I conducted field research) belonged to *jagirdars*. Essentially all of the remainder was *khaslo* (Purohit 1980:335). As they had since the early nineteenth century, landlords—and through them, Marwar state—collected taxes on livestock, grass, well water, wool, sales of livestock at seasonal fairs, official stamps on weights, game, slaughterhouses, tanneries, and houses. This list is far from exhaustive, since rulers also collected taxes on weddings, religious fairs, tools, liquor, gambling, prostitution, guilds, and mineral extraction, along with special "one-time" taxes (Sharma 1979:99). To say that rural non-elites regarded these as a burden is a vast understatement. Calls for agrarian and democratic reforms stimulated protests, some violent, in Marwar from the early to mid-twentieth century.

Independence and Institutional Change

At the end of March 1949 Marwar and the other desert states signed on as part of the new Rajasthan Union. One of the promises kept by the interim government—shortly before the first general elections in 1953—was passage of the Rajasthan Land Reforms and Resumption of Jagir Act of 1952. Revenue laws enacted at this time created a new legal structure for regulating access to land, based on whether or not an individual personally plowed a plot of land. Subsequently, land ceiling laws established in 1963 attempted to reduce the embarrassingly large number of landlords who retained control of the lion's share of their former holdings (Purohit 1980:338).

The new government of Rajasthan also abolished forced labor (*begar*), eliminated most of the taxes formerly levied against the desert residents, and lowered land taxes to a fraction of the previous total (Jodha 1985:256). Marwar, which had been the largest of the desert princely states, was divided into five new districts of Rajasthan: Barmer, Jalor, Jodhpur, Nagaur, and Pali—thus effectively dismembering the ruler's seat (Jodhpur) from its richest Luni basin holdings (Jalor, Nagaur, and Pali) and from the Pakistan border, which fronts on Barmer District.

One of the biggest changes that occurred was the transformation of land-use patterns. Edward S. Haynes, who traces shifts in land use in Rajasthan between 1660 and 1950, estimates that between these dates the portion of land that was cultivated ranged from 10 to 26 percent (Haynes 1984:49). (Haynes includes both net cultivated lands and fallows in the category of "cultivated land.") From 1950 to 1960, the proportion of cultivated land shot up to 45 percent. Since many of the former rulers held on to their better landholdings, the new lands for cultivation had to come from somewhere—and they did.

Croplands were carved out of previously uncultivated lands—particularly pastures, and land that fell into the category of "waste land" or "cultivable waste."[3] (In the Thar Desert, lands classified as "cultivable waste" functioned as grazing lands; the term is meaningless from the local perspective.) "Grasslands" declined from 18 percent to 12 percent, and the category of "cultivable waste" dropped from 40 to 28 percent between 1950 and 1960 (Haynes 1984:49). This masks a variability among districts during the most intense period of reforms. Using as his basis the period from 1953/54 to 1963/64, N. S. Jodha (1985:249) estimates that the proportion of "cultivable waste" in Jodhpur District dropped from 15 percent to 7 percent of total land; in Jaisalmer, it dropped from 26 percent to 12 percent.

In tandem with its land reforms, the new government embarked on a series of campaigns to develop agriculture in the Thar Desert. These programs promoted sedentary village life, and attempted to create the infrastructural conditions through which the desert could be made to produce high-yielding varieties of wheat, barley, and other crops. The desert was to be made to bloom; its nomadic and seminomadic peoples were to be settled. A belt of sedentary farmers was to be established along the Pakistan border in the Indira Gandhi Canal command area—first paralleling Rajasthan's northern frontier, from Punjab to Jaisalmer, and later with a Barmer spur extending south, along the western international line.

By the 1970s, motor vehicles and tractors were replacing animals as sources of traction power. Rubber tires replaced wooden wheels on ox- and camel-carts, increasing their efficiency and range in the sandy desert. The "green revolution" in Punjab and Haryana states (the term coined by Francine R. Frankel [1972] to describe the adoption of high-yielding varieties of wheat and other crops, industrially produced fertilizers, and heavy machinery by farmers in these states) vastly expanded harvests in these regions, and attracted farm laborers from the Thar Desert.

The human population of the arid zone, estimated at 6.1 million in 1951, grew steadily during the following decades: 7.9 million in 1961, 10.2 million in 1971, 13.5 million in 1981, and 18 million in 1991 (Prabhakar 1992:45–46; Prakash 1987:1; UNCOD 1977a:15). This growth represents a remarkable change from the past. B. L. Bhadani estimates that the population of Marwar state hovered around approximately 2 million between 1654 and the end of the nineteenth century (Bhadani 1979:426). It is likely that in the other arid zone territories the population processes were similar.

The numbers of livestock also rose. Not only did the total number of livestock of all species grow, their aggregate impact on the resource base also expanded. This is evident when different species of livestock are converted to "animal units," a calculation based on their caloric intake. In this system, adult cattle rank as 1.0 animal unit, buffalo as 1.3, sheep and goats as 0.15, camels and horses as 1.0, and donkeys as 0.75 (National Council of Applied Economic Research 1964:5). Calculating the livestock population of Rajasthan state on this basis, the total number of animal units steadily increased from 16.9 million in 1951 to 26.4 million in 1983, a 53 percent increase (Prabhakar 1992:198)—less than the increase in the human population, roughly 70 percent, during the same time period.[4] Thereafter, the total number of animal units declined slightly to 23.2 million in 1988 (Prabhakar 1992:198).

Historical and Ecological Processes in the Thar Desert

The foregoing description outlines a number of the most important transformations in the Thar Desert and its surrounding arid zone during the past several decades. In this section I will attempt to provide a synthesis of these shifts in terms of three interconnected issues. I will examine, first, how individual and group strategies shifted to respond to new opportunities between 1950 and 1990; and second, the ways in which these transformations interact with and affect the general utility of systems of access to resources, particularly common-property regimes. This leads to considering a political ecology of the ways in which interactions between institutional and ecological variables heightened the risk of failure, leading to the cattle population collapse.

Institutional changes that promoted sedentary cultivation stimulated numerous changes in the relationships among individuals and toward the environment. The outcome was a new trajectory of resource use. Before this, a number of factors had promoted pastoral and agropastoral activities in the arid zone.

Although rulers of the traditional princely states in the Thar Desert were, on occasion, keen to stimulate agricultural settlement and cultivation, the relationship between rulers and the ruled reduced interest in this as a goal. Extra inputs of labor were required to create cropland. The risk of failure was substantial, because of the high frequency of drought. Investing in new croplands bound cultivators to the many demands for cash, in-kind payments, forced labor, and services demanded by the landlord. The perspective of potential cultivators suggests few incentives to invest in cropland-creating activities.

Indeed, conflicts with landlords appear as a theme in family oral histories and village-origin myths. One frequent answer to the question, "Why did your ancestor so-and-so move to this village?" evoked stories about disputes with a landlord. One version of the founding myth of the village in which I studied adjustments to drought asserted that the ancestor of the founding lineage was expelled from his former lands for insulting the Maharajah. According to a report on the districts of Marwar (*Marwar ra Pargana ri Vigat*, cited in Bhati 1968:320) collected shortly before the end of the seventeenth century, this was in 1668. In the villagers' contemporary account, the new lands were inferior to the remembered old village—which might reflect the synergy between people and land required to create crop fields.

The rulers also may have valued desert residents as more than cultivators: they comprised a work force that could be mobilized to build forts and waterworks, serve in the military, and provide services that ranged from those that were hereditary to those required at the ruler's whim. Haynes provides evidence that cultivation competed with other possible land uses, such as grazing for the camels, cattle, and horses required by the ruler's military forces, or maintaining private hunting and wood reserves. Finally, the rulers faced the problem of attracting and keeping a labor force. The population might decamp suddenly and thoroughly in order to avoid death by starvation, if the monsoon failed—and not all might return to their original villages. In 1868, for example, reportedly 1 million out of Marwar's 1.5 million residents migrated to escape famine conditions in the region (Maloo 1987:35). After the great famine at the end of the nineteenth century, called *chappaniya kal*, some villagers in Jodhpur District told me, their grandfathers and great-grandfathers abandoned both their old lands and old landlords. A ruler's grazing reserves, which could be opened to local residents' cattle during famines, presumably encouraged people to return.

From the standpoint of both ruler and ruled, therefore, prior to 1950, incentives to expand the amount of cultivated land beyond about one-quarter of the total seem to have been absent. The most important mechanism for bringing new areas into cultivation used by the rulers—creating a new village unit and granting rights to a new landlord—was infrequently practiced. B. L. Bhadani, who has studied the seventeenth-century population of Marwar, found the list of names of villages existing in the 1930s to be almost the same as that compiled during the seventeenth century by Muhta Nainsi in his *Marwar ra Pargana ri Vigat* (Bhadani 1979:425–427).

The "traditional" landscape appears, in this perspective, to reflect the relationships between the interlinked, yet divergent, interests of rulers and non-elites in the Thar Desert region. The data generally suggest that these relationships were far from static, prior to 1950: the landscape, as this existed on the eve of reforms, itself resulted from historical and ecological processes.

After independence, land and political reforms profoundly transformed these relationships. Significantly, agrarian reforms made it economically viable for poor households to cultivate submarginal lands (Jodha 1985:253). Land reform probably stimulated a demand for plow oxen and—with lowered costs to producers—also made it economically feasible for these new farmers to own cows. Cow ownership in the arid zone significantly functions as a status marker for most groups bent on upward social and economic mobility. Further, this period was characterized by expanding markets for livestock and livestock products (Henderson 1992), stemming from Rajasthan's comparative advantage in livestock production (Jodha 1985).

The biggest increase in the state's cattle population occurred between 1951 and 1961, when the number of animals shot up by 21 percent, according to Manohar Prabhakar (1992:198). Thereafter, according to this source, it declined by 5 percent in 1971, then increased modestly (3 percent) from 1972 to 1977, and then registered a 5 percent increase between 1977 and 1983. Yet the growth in the number of cattle in the three decades prior to 1981, 47 percent (Kumar and Bhandari 1992), was actually outstripped by the expansion of other species: buffalo increased by 138 percent, sheep by 163 percent, and goats by 238 percent; even the number of camels, which breed more slowly than these other species, grew by 170 percent during this period (Kumar and Bhandari 1992).

The reforms freed people from reliance on potentially capricious landlords and patrons and made it feasible for members of stigmatized occupational groups to abandon such work. These new landowners proudly identified their occupation as "*kheti*"-farming. The traditional patron-client system of relationships (*jajmani*) crumbled, mourned by few at the bottom of the caste hierarchy; by the early 1980s, many people denied that they, or their families, had ever been involved in such relationships.

These transformations were not without costs. One of the most striking is the impact of land reform on resource-access systems. Jodha (1985) and Haynes (1984) argue that the cessation of the rulers' control over grazing preserves, forests, and fodder-producing areas resulted in a "free-for-all" of uncontrolled access and subsequent resource depletion. Jodha suggests that while the older system was exploitative, because it limited user access, the new system of land management lacked controls in areas where community institutions were weak (Jodha 1985: 254). The new, democratically constituted village *panchayats* (councils) found that it was unpopular and difficult to impose taxes on newly minted farmers. Further, they lacked the authoritarian stance of the previous rulers and they were unable to enforce decrees, limiting their ability both to invest in improvements and to prevent overutilization.

The new land-tenure system, drawn up to meet the demands of peasant-led agrarian reform movements, did not recognize common-property resources: land-use titles, under the law, depended upon individuals' demonstration of continuous cultivation of a plot of land. Communities of users who managed property might continue this under the new regime as long as they were able to prevent encroachments—but it was difficult to hold back determined claims, for common-property user communities lacked legal standing (Henderson 1993).

An example of the ways in which the law was blind to the arid zone's ecological characteristics appears in the Rajasthan Agricultural Lands Utilisation Act of 1954 (section 4), which, as Jodha puts the matter, "completely ignoring the traditional wisdom of the desert farmer, empowers the district collector [the executive in charge of the district] to prohibit the fallowing of crop lands" (Jodha 1977:339). Claiming that certain lands were "village pastures," if such a claim was recorded, placed them under the management of the state bureaucracy, and potentially opened them to redistribution to landless and poor households—a Catch-22 for anyone attempting to protect pastures from cultivation, for at this time the state's priority was to distribute lands for cultivation.

Toward a Political Ecology of Cattle

By the middle of the 1970s, it was clear that the expansion of cultivated land simultaneously with livestock population growth was leading to a catastrophic collision of interests. Institutional change and ecological factors contributed to these problems. Pressures existed for intense competition: between cattle- and sheep-herders, between people remaining in their village sites and livestock migrants, and between rich and poor resource-users (Henderson 1993; Jain 1993:60–61; Prasad 1984:20). Migration, however, remained a central strategy to reduce pressures on the arid zone environment during its lengthy dry season, as well as to survive drought.

While the decade of the 1950s was not noted for a severe famine, scarcities occurred throughout these years, according to O. P. Kachhawaha (1992:349–355). The more difficult famine years lay ahead—particularly 1963/64, 1965/66, and 1968/69, which caused intense suffering because each year was preceded by two or three scarcity years. Similar phenomena characterized the 1970s, with 1972 and 1974 being particularly difficult. There were a few good years in the middle of the 1970s, and then persistent monsoon failures began—a situation that lasted for most of the ensuing decade. In the arid zone, there was inadequate rainfall to produce "normal" crop outputs from 1977 to 1987 (Kachhawaha 1992; Dandia 1981:5; Goyal 1991:100–102). Thus, at the outset, from 1983 to the end of the decade, ecological conditions were not favorable to continue the expansion of livestock, such as occurred during the previous three decades. The only monsoon during the decade that did *not* fail was in 1988, and even then scarcities were declared in districts of western Rajasthan because water supplies remained problematic.

As in the past, livestock owners attempted to leave the famine zone—but now they migrated in a profoundly altered landscape. India's government identified the Pakistan border as a security problem, both because Pakistani invasions were feared and because it teemed with herders who attempted to sneak across with their animals, with smugglers, and with other undesirables (who often were indistinguishable from the herders). The northern border was now occupied by the Indira Gandhi Canal command area, which was stretching to occupy 5,420 km^2 (Prabhakar 1992:204). Livestock migrants who had formerly used the grass stands in this area found their pastures turned into cropland or destroyed by canal overflow and salinization. Although this area represents just 2.5 percent of the total arid zone, its strategic location and the transformations of social and economic relationships that came about with intensive irrigation and the immigration of cultivators from Punjab and Haryana curtailed its previous uses. Along the southern migration routes that led across the Aravalli mountains to Gujarat and central India, irrigation development reduced the fall and spring fallows upon which herders had grazed their animals. Now migrants found that previously hospitable villagers had become hostile. Armed skirmishes were reported as taking place between migrants and villagers (Workshop on Drought 1988:15–16).

Many areas outside Rajasthan's arid zone that were still open to grazing were declining in quality. In the Banni grasslands of Gujarat state, not only did droughts compromise grass regeneration, the construction of dams reduced the recharge of sweet-water systems in the area (Bharara 1993:3). Other regions were closed to herders as a part of efforts instigated by the government to improve environmental quality.

The Desert Biosphere Reserve, established in the heart of sheep country in western Barmer and Jaisalmer Districts—and adjacent to a military zone that abuts the border—removed approximately 3,100 km^2 from local communities (Ministry of Environment and Forests 1988:17). This represents approximately 1.5 percent of the total arid zone, or about 5 percent of its lands that are classified as "cultivable waste" and "pastures."[5] As with the Indira Gandhi Canal command area, the percentage of the total lands in the arid zone is slight, but it encompasses a local region that is significant for sheep-grazing and other activities, which are connected to the nearby Pakistan border. The Ministry of the Environment and Forests report (1988:40) acknowledges this factor in noting that closing the region might create "considerable public outcry" by the residents, who formerly (and not insignificantly to the local economy) enjoyed access to the border. The government planned to install barbed-wire fences to eliminate "indiscriminate" grazing and permit "controlled" access on about two-thirds of the preserve—a solution that is highly likely to be vulnerable to political decisions, if previous experiments are any indication (see Jain 1993:60–61). The status of this region as a protected preserve, however, might be considered somewhat endangered, since maps of the preserve show a projected Indira Gandhi Canal spur extending southward through the center of this designated zone.

The government also declared large sections of the Aravalli mountains "off-limits" to pastoralists and agropastoral migrants in order to undertake environmental restoration. The protected Aravalli zone was bitterly resented by livestock herders. Some observers have suggested that the situation in the Thar Desert comprises a "tragedy of the commons" (Hardin 1968). If so, it is a tragedy that reflects the institutional changes and transformations in the production system of the arid zone. The management vacuum, which lowered costs to producers at the same time that demand boomed for their product, stimulated livestock producers to expand their herds. Simultaneously, the real-world conditions for herders increasingly mimicked the assumptions in Hardin's model, one of which is that resource users are not allowed to leave a given territory.

To put the situation another way, a set of relationships characterized by a relatively low degree of risk, from the standpoint of individuals, was superseded by one with a higher degree of collective risk. If pastoralists and agropastoralists switched to, or increased their reliance on, cultivation, the risk of failure grew. While initially this may have seemed a good bet from the perspective of individuals, the costs to them and the risks of failure steadily rose with the erosion of the resource base and the failure of crops during the 1980s. The patches in the environment got patchier; there were fewer of them. The costs of migration grew, since animals starved along routes where formerly they grazed, and herders found themselves fighting for grass. People who relied on former common-property resources to support their livestock found production compromised.

Only those individuals and households who were rich enough to maintain livestock on purchased or produced fodder—those not dependent on patchy resources—maintained their stock. In Rajasthan, sheep declined by 26 percent; cattle by 19 percent; goats by 18 percent; and even the number of camels fell by 12 percent. But buffalo, the favorite animal of well-off dairy producers, increased by 6 percent (Prabhakar 1992:189).

At this point, it may be useful to add a further historical perspective to the issue of livestock population declines in Rajasthan. The cattle population crash between 1983 and 1988 represents a smaller decline than occurred after the great famine of 1899–1900 in western India. Kamala Maloo (1987:184) estimates that the cattle population of the princely states in Rajputana (not including Ajmer-Merwara, which was administered by the British) was 13 million before the famine; subsequently, it dropped by 56 percent.[6] However, he reports, losses were substantially higher in the desert states: Marwar estimates losses of 90 percent; Bikaner, 75 percent; and in Jaisalmer, contemporary sources claimed that 67 percent of cattle were lost. Looking to the 1980s, these patterns still hold: arid Jaisalmer and Barmer Districts reportedly lost 50 percent of their cattle between 1983 and 1988, while the losses in Jodhpur and Pali were 34 and 36 percent.[7]

These data suggest that the rise and fall of the cattle population in Rajasthan during the past forty years is not a unique event, though it represents a substantively different combination of institutional, economic, technological, political, and ecological factors than existed at the turn of the century. An analysis of the dynamics at that time might explore—as the present discussion has done—the interests that individuals have in their livestock. How does this fit into the broader series of goals, circumstances, and options or constraints that characterize different resource-users' groups? What is the aggregate impact of these actions on the relationships linking humans and the environment?

One might look at the events in 1899–1900 for clues to factors that might, or might not, lead to a shift in the trajectory of energetic relationships encompassed within the human/environment nexus at that time. Although in the 1980s *chappaniya kal* (the famine of 1899–1900) was passing out of living memory, it was an event that burned with great clarity in people's thoughts as a benchmark against which to judge current events, and as a model for assessing how one might respond. It is possible that the events of the 1980s will one day be recalled in a similar light.

Returning to issues raised in the sacred cow debate discussed in this paper, it is clear that the historical and ecological context of cattle management in Rajasthan places these questions in an entirely different dimension. First, the data that featured in the debate cannot be understood without reference to historical and ecological processes. The land reforms in Rajasthan profoundly affected the demand for livestock and stimulated massive cattle population growth at that time; the data collected during the 1950s and 1960s (often then utilized in the debate) reflect a period that possesses some unique demographic, technological, economic, and ecological characteristics relative to earlier periods. Subsequent technological and economic shifts also have promoted a number of general trends that were not necessarily noted by participants in the debate. For example, Lodrick's efforts (1979, 1994) to demonstrate that Hindu urban dairy owners in Ajmer preferred less-efficient zebu cows, and that Muslim urban dairy owners featured more-efficient buffalo, ignore the big trend—and incontrovertible fact—that during the 1970s and 1980s both Hindu and non-Hindu dairy producers consistently replaced zebu cows with buffalo. The documented growth of buffalo and decline of zebu cattle during the 1980s cited in this paper support observations made at the local level that commercial dairying promoted buffalo ownership.

Second, the debate's prevailing reliance on decontextualized scraps of information obscured recognition of important distinctions between individual and aggregate levels of response. When the scientists at CAZRI fenced off a section of ground and grass grew on it, they created a landscape—not an environment. The important fact is that the Thar Desert, for its individual users, consists of landscapes; responses are to this, which is the result of intertwined human behavioral and environmental processes. The general historical context and the policy to sedentarize the desert population lead me to the conclusion that what was really at stake in the big picture was not fencing out the animals, as seen in the case of the CAZRI grass plot, but fencing in the herders. As the foregoing discussion has shown, this helped to trigger a complex interaction of events and interests, as these unfolded over time.

Conclusion

Historical ecology studies the intertwined relationships of humans and the environment, seeking thus to comprehend the proximate causes of changes in these relationships. A primary focus of analysis is the landscape: its creation and modification over time. Within a specific landscape, regarded as a historically contingent set of relationships, actors and interest groups interact to achieve diverse ends. This approach emphasizes identifying separate and possibly divergent interests, and questioning the representations of interests against their behavioral and material consequences set in short- and long-term perspective.

Recovering landscape as dialectically created breaks down the dualities between humans and nature that have bedeviled efforts to account for specific relationships. Insights into the case of cattle management in Rajasthan, India, may be gained from examining the impact of institutional change, the prospects of establishing new linkages between subaltern groups and the state, and the properties of individual and group responses to alterations in the landscape and the options or constraints that operate therein.

Acknowledgments

Research for this paper was supported by grants from the American Institute of Indian Studies and the U.S. Education Foundation in India, as well as by logistical support from Rupayan Sansthan and the School of Desert Sciences in Jodhpur, Rajasthan. I thank Marvin Harris for stimulating my interest in livestock issues in this region, and for his generous comments on this work. William Balée encouraged the writing of this paper and its presentation at the Historical Ecology conference, and made useful suggestions throughout the process. I am grateful to Padmashri Komal Kothari, Edward S. Haynes, and Joan L. Erdman, who incisively commented on events in Rajasthan, and to Brian Ferguson, Barbara J. Price, and Maria Lagos, along with participants in the Conference on Historical Ecology, for their numerous helpful suggestions.

Notes

1. In fact, the theme that the Indian environment is degraded flows through these works. The closest Harris comes to indicating that an environmental equilibrium may result is when

he states that his intention is "to urge that the explanation of taboos, customs, and rituals associated with management of Indian cattle be sought in 'positive-functioned' and probably 'adaptive' processes of the ecological system of which they are a part" (1966a:51). The ambivalent language regarding the "adaptive processes" and the fact that the article does not set out to demonstrate that India has a homeostatic energetic system suggest that a key problem resides in the then-current conception of analytic terms such as *adaptation*.

2. Rural population density is calculated on the basis of figures provided in the table "Ranking of Districts by Area" (*Census of India* 1991a:20), and of rural population figures in Statement 7, "Progress in Rural Population of State/District, 1901–1991" (*Census of India* 1991b:12).

3. Another term for this seen in the literature is *culturable waste*. Both terms stem from British categories.

4. Calculated from Prabhakar's table (1992:198) showing livestock in Rajasthan from 1951 to 1988. The figures are conservative, since the category of "other" livestock (which includes horses and donkeys) is not included in this estimate, as Prabhakar did not provide a breakdown of the numbers of "other" species. The number of "other" livestock tallied was estimated as constant at 400,000 for the entire period in question.

5. This is calculated using data on these categories for the arid zone provided in the report of the Workshop on Drought (1988), p. 34.

6. Maloo does not explain whether this estimate refers to *Bos indicus* only, or to all livestock; contemporary authors sometimes referred to "horned cattle" to distinguish the former from other species.

7. Calculated from data provided in the *Statistical Abstract of Rajasthan, 1983* (Rajasthan Government 1986:114–115), table 15.1, "Livestock (1983)," and from the 1988 *Statistical Abstract* (Rajasthan Government 1991:115–119), table 15.1, "Livestock (1988)."

References

Acharya, R. M., B. C. Patnayak, and L. D. Ahuja. 1977. Livestock production: Problems and prospects. Pp. 275–280 in Jaiswal (ed.), *Desertification and Its Control*.

Adams, Richard N. 1982. Comment on: Bovine sex and species ratios in India, by A. Vaidyanathan, K. N. Nair, and Marvin Harris. *Current Anthropology* 23(4): 365–383.

Agrawal, Arun. 1993. Mobility and cooperation among nomadic shepherds: The case of the Raikas. *Human Ecology* 21(3): 261–279.

Ahuja, L. D. and H. S. Mann. 1975. Rangeland development management in Western Rajasthan. *Annals of Arid Zone* 14(1): 29–44.

Balée, William. 1994. Historical ecology: Premises and postulates. Paper presented at the Conference on Historical Ecology, Tulane University, New Orleans, 9–11 June.

Bhadani, B. L. 1979. Population of Marwar in the middle of the seventeenth century. *Indian Economic and Social History Review* 16(4): 415–427.

Bharadwaj, O. P. 1961. The arid zone of India and Pakistan. Pp. 143–174 in L. Dudley Stamp (ed.), *A History of Land Use in Arid Regions*. Paris: UNESCO.

Bharara, Laj Paul. 1993. Socio-economic aspects of Banni pastoralists and factors effecting changes in their perceptions towards the future. Pp. 1–7 in Richard P. Cincotta and Ganesh Pangare (eds.), *Pastoralism and Pastoral Migration in Gujarat*. (Proceedings of the Workshop on Transhumant Pastoralism in Gujarat, 24–25 July 1992.) Anand, Gujarat: Institute of Rural Management.

Bhargava, Visheswar Sarup. 1966. *Marwar and the Mughal Emperors*. Delhi: Munshiram Manohar Lal.

Bhati, Narayan Singh, ed. 1968. *Marwar ra Pargana ri Vigat* (Report on the districts of Marwar), part 1. Jodhpur: Rajasthan Oriental Research Institute.

Blair, Charles. 1874. *Indian Famines*. Edinburgh: William Blackwood.

Bryson, Reid. 1972. Climatic modification by air pollution. Pp. 133–155 in N. Polunin (ed.), *The Environmental Future*. (Proceedings of the First International Conference on the Environmental Future, Finland, 27 June–3 July 1971.) New York: Barnes and Noble.

Bryson, Reid and D. A. Baerreis. 1967. Possibilities for major climatic modification and their implications: Northwest India, a case for study. *Bulletin of the American Meteorological Society* 48(3): 136–142.

Bryson, Reid A. and Thomas J. Murray. 1977. *Climates of Hunger*. Madison: University of Wisconsin Press.

Cancian, Frank. 1989. Economic behavior in peasant communities. Pp. 127–170 in Stuart Plattner (ed.), *Economic Anthropology*. Stanford, Calif.: Stanford University Press.

Census of India, 1991. 1991a. Series 21, Rajasthan: Provisional population totals. Paper 1 of 1991. Jaipur: Government Printing Office.

Census of India, 1991. 1991b. Series 21, Rajasthan: Provisional population totals: Rural-urban distribution. Paper 2 of 1991. Jaipur: Government Printing Office.

Crumley, Carole L. 1994. Historical ecology: A multidimensional ecological orientation. Pp. 1–16 in C. L. Crumley (ed.), *Historical Ecology: Cultural Knowledge and Changing Landscapes*. Santa Fe: School of American Research.

Dandekar, V. M. 1969. India's sacred cattle and cultural ecology. *Economic and Political Weekly* (Bombay) 4:1559–1566.

———. 1970. Sacred cattle and more sacred production functions. *Economic and Political Weekly* (Bombay) 5:527–531.

Dandia, Milap Chand. 1981. Why is Rajasthan backward? *Yojana* 25(22): 4–7, 26.

Dhir, R. P. 1982. The human factor in ecological history. Pp. 311–331 in Brian Spooner and H. S. Mann (eds.), *Desertification and Its Control: Dryland Ecology in Social Perspective*. New York: Academic Press.

Diener, Paul, Donald Nonini, and Eugene Robkin. 1978. The dynamics of the sacred cow: Ecological adaptation vs. political appropriation in the origins of the sacred cow complex. *Dialectical Anthropology* 3(3): 221–241.

Frankel, Francine R. 1972. *India's Green Revolution: Economic Gains and Political Costs*. Princeton: Princeton University Press.

Freed, Stanley A. and Ruth S. Freed. 1981. Sacred cows and water buffalo in India: The uses of ethnography. *Current Anthropology* 22(5): 483–502.

Goyal, Suresh. 1991. *Management of Scarcity*. Jaipur: RBSA Publishers.

Hardin, Garrett. 1968. The tragedy of the commons. *Science* 162(859): 1243–1246.

Harris, Marvin. 1964. The sacred cow in India. Paper presented at the Columbia University Seminar on Ecological Systems and Cultural Evolution, 2 November, New York.

———. 1966a. The cultural ecology of India's sacred cattle. *Current Anthropology* 7:51–66.

———. 1966b. Reply to comments on: The cultural ecology of India's sacred cattle, by M. Harris. *Current Anthropology* 7:63–64.

———. 1977. *Cannibals and Kings: The Origins of Culture*. New York: Vintage Books.

Haynes, Edward S. 1984. Land use and land use ethic in Rajasthan, 1850–1980. Paper presented at the 13th Wisconsin Conference on South Asia, 4 November, Madison, Wis.

Henderson, Carol. 1987. Famines, droughts, and the "norm" in arid western Rajasthan: Problems of modeling environmental variability. *Research in Economic Anthropology* 9:251–280.

———. 1989. Life in the land of death: Famine and drought in arid western Rajasthan. Ph.D. diss., Columbia University.

———. 1992. What is a sheep: Cultural concepts, economic realities. Pp. 68–77 in Elizabeth Hansen (ed.), *Proceedings of the Annual Conference of the Virginia Consortium for Asian Studies*. Occasional Papers of the Virginia Consortium, vol. 9. Arlington: Virginia Consortium for Asian Studies.

———. 1993. Policy, property, and pastoralists in the Thar Desert: Dilemmas in understanding the commons. Paper presented at the International Workshop on Common Property Resources and the Crisis of Pastoralism in the Thar Desert, Jodhpur, India, 17–19 March.

Heston, Alan. 1971. An approach to the sacred cow of India. *Current Anthropology* 12(2): 191–209.

Jain, H. K. 1993. The institutional environment and its relationship to sustainable development. Pp. 59–63 in Richard P. Cincotta and Ganesh Pangare (eds.), *Pastoralism and Pastoral Migration in Gujarat*. (Proceedings of the Workshop on Transhumant Pastoralism in Gujarat, 24–25 July 1992.) Anand, Gujarat: Institute of Rural Management.

Jaiswal, P. L., ed. 1977. *Desertification and Its Control*. New Delhi: Indian Council of Agricultural Research.

Jodha, N. S. 1977. Land-tenure problems and policy in the arid region of Rajasthan. Pp. 335–347 in Jaiswal (ed.), *Desertification and Its Control*.

———. 1985. Population growth and the decline of common property resources in Rajasthan, India. *Population and Development Review* 11(2): 247–264.

———. 1991. Drought management: Farmers' strategies and their policy implications. *Economic and Political Weekly* (Bombay) 26(39): A98–A104.

Kachhawaha, O. P. 1992. *History of Famines in Rajasthan, 1900 A.D.–1990 A.D.* Jodhpur: Research Publishers.

Kumar, M. and M. M. Bhandari. 1992. Human use of the sand dune ecosystem in the semi-arid zone of the Rajasthan Desert, India. *Land Degradation and Rehabilitation*, vol. 4.

Le Houérou, H. N. 1977. The nature and causes of desertization. Pp. 17–38 in Michael H. Glantz (ed.), *Desertification: Environmental Degradation in and Around Arid Lands*. Boulder, Colo.: Westview Press.

Lodrick, Deryck O. 1979. On religion and milk bovines in an urban Indian setting. *Current Anthropology* 20:241–242.

———. 1994. Rajasthan as a region. Pp. 1–44 in K. Schomer, J. Erdman, and D. Lodrick (eds.), *The Idea of Rajasthan: Explorations in Regional Identity*, vol. 1. New Delhi: Manohar - American Institute of Indian Studies.

MacLachlan, Morgan D. 1982. Comment on: Bovine sex and species ratios in India, by A. Vaidyanathan, K. N. Nair, and Marvin Harris. *Current Anthropology* 23(4): 376–377.

Malhotra, S. P. 1977. Socio-demographic factors and nomadism in the arid zone. Pp. 310–323 in Jaiswal (ed.), *Desertification and Its Control*.

———. 1988. Human resources assessment in the arid zone of Rajasthan. Pp. 116–129 in A. K. Tewari (ed.), *Desertification: Monitoring and Control*. Jodhpur: Scientific Publishers.

Maloo, Kamala. 1987. *The History of Famines in Rajputana*. New Delhi: Himanshu Publications.

Mann, H. S. 1982. The Central Arid Zone Research Institute (India). Pp. 293–311 in B. Spooner and H. S. Mann (eds.), *Desertification and Its Control: Dryland Ecology in Social Perspective*. New York: Academic Press.

Mann, H. S., S. P. Malhotra, and J. C. Kalla. 1974. Desert spread: A quantitative analysis in the arid zone of Rajasthan. *Annals of Arid Zone* 13(2): 103–113.

Marquardt, William H. 1994. The role of archaeology in raising environmental consciousness: An example from southwest Florida. Pp. 203–222 in Crumley (ed.), *Historical Ecology*.

McCay, Bonnie J. and James M. Acheson, eds. 1987. *The Question of the Commons*. Tucson: University of Arizona Press.

Mencher, Joan. 1971. Comment on: An approach to the sacred cow of India, by Alan Heston. *Current Anthropology* 12(2): 202–204.

Ministry of Environment and Forests (India). 1988. *Thar Desert Biosphere Reserve*. Project Document 7. New Delhi: Kapoor Art Press.

National Council of Applied Economic Research (NCAER). 1964. *Agriculture and Livestock in Rajasthan*. New Delhi: NCAER.

Nonini, Donald. 1982. Comment on: Bovine sex and species ratios in India, by A. Vaidyanathan, K. N. Nair, and M. Harris. *Current Anthropology* 23(4): 337–338.

Odend'hal, Stewart. 1972. Energetics of Indian cattle in their environment. *Human Ecology* 1(1): 3–22.

———. 1988. Human and cattle population changes in deltaic West Bengal, India, between 1977 and 1987. *Human Ecology* 16(1): 23–33.

Orlove, Benjamin S. 1980. Ecological anthropology. *Annual Review of Anthropology* 9:235–273.

Patterson, Thomas C. 1994. Toward a properly historical ecology. Pp. 223–238 in Crumley (ed.), *Historical Ecology*.

Pisharoty, P. R. 1980. Semi-arid and arid areas of Rajasthan: A physicist's view. Pp. 41–46 in H. S. Mann (ed.), *Arid Zone Research and Development*. Jodhpur: Scientific Publishers.

Prabhakar, Manohar. 1992. *Reference Rajasthan*. Jaipur: Creative Communicators.

Prakash, Ishwar. 1987. Livestock as a biotic factor causing desertification. Paper presented at the First International Conference on Rajasthan Studies, Jaipur, Rajasthan, 12–16 December.

Prasad, R. R. 1984. Aspects of pastoral nomadism in rural Rajasthan. *Journal of Social Research* 27(2): 11–22.

Purohit, M. L. 1980. The chronology of socio-economic dynamics of land ownership in arid zone of western Rajasthan. *Annals of Arid Zone* 19(3): 335–343.

Rajasthan Government, Directorate of Economics and Statistics. 1979. *Statistical Abstract of Rajasthan*. Jaipur: Directorate of Economics and Statistics.

———. 1986. *Statistical Abstract of Rajasthan, 1983*. Jaipur: Directorate of Economics and Statistics.

———. 1991. *Statistical Abstract of Rajasthan, 1988*. Jaipur: Directorate of Economics and Statistics.

Rajasthan's backwardness. 1983. *Hindustan Times*, 20 June.

Rao, C. H. H. 1969. India's "surplus" cattle: Some empirical results. *Economic and Political Weekly* (Bombay), 21 March, 4:A225–A227.

Rosin, R. Thomas. 1993. The tradition of groundwater irrigation in northwestern India. *Human Ecology* 21(1): 51–86.

Schwartzberg, Joseph E. 1979. Comment on: Questions in the sacred-cow controversy, by Frederick J. Simoons. *Current Anthropology* 20(3): 489.

Scott, James C. 1976. *The Moral Economy of the Peasant*. New Haven: Yale University Press.

Sharma, G. C. 1979. *Administrative System of the Rajputs*. New Delhi: Rajesh Publications.

Simoons, Frederick J. 1979. Questions in the sacred-cow controversy. *Current Anthropology* 20(3): 467–493.

Singh, Dool. 1964. *A Study of Land Reforms in Rajasthan*. Alwar, India: Sharma Brothers.

Singh, Hira. 1979. Kin, caste, and Kisan movement in Marwar: Some questions in the conventional sociology of caste. *Journal of Peasant Studies* 7(1): 101–118.

Timberg, Thomas A. 1978. *The Marwaris: From Traders to Industrialists*. New Delhi: Vikas.

Tod, James. 1914. [1829 and 1832]. *The Annals and Antiquities of Rajasthan*. 2 vols. London: Routledge.

United Nations Conference on Desertification (UNCOD). 1977a. *Case Study on Desertification: Luni Development Block, India*. Nairobi: UNCOD. A Conf. 74/11.

———. 1977b. *Climate and Desertification*. Prepared by F. Kenneth Hare. Nairobi: UNCOD. A Conf. 74/5.

Vaidyanathan, A. 1988. *Bovine Economy in India*. Center for Development Studies Monograph Series. New Delhi: IBF Publishing.

Vaidyanathan, A., K. N. Nair, and Marvin Harris. 1982. Bovine sex and species ratios in India. *Current Anthropology* 23(4): 365–383.

Winterhalder, Bruce. 1994. Concepts in historical ecology: The view from evolutionary ecology. Pp. 17–41 in Crumley (ed.), *Historical Ecology*.

Workshop on Drought. 1988. *Report of the Workshop on Drought, Desertification, and Wasteland Development Processes in the Thar Desert; Jodhpur, 26–29 February 1988*. New Delhi: Centre for Science and Environment.

World Bank. 1991. *World Development Report 1991: The Challenge of Development*. Washington, D.C.: International Bank for Reconstruction and Development.

Ziegler, Norman P. 1994. Evolution of the Rathor state of Marvar: Horses, structural change, and warfare. Pp. 192–216 in K. Schomer, J. Erdman, and D. Lodrick (eds.), *The Idea of Rajasthan: Explorations in Regional Identity*, vol. 2. New Delhi: Manohar - American Institute of Indian Studies.

CHAPTER 17

The Historical Ecology of Thailand: Increasing Thresholds of Human Environmental Impact from Prehistory to the Present

Leslie E. Sponsel

Although some would argue that historical ecology is quite distinctive from environmental history, I think it is appropriate to acknowledge some of our intellectual roots and affinities. If any individual can be singled out as the principal founder of modern environmental history it is Donald Worster (see Bailes 1985; Cronon 1993; Crumley 1994; Marsh 1965; White 1985; Worster 1985, 1988a,b, 1990, 1993). In a seminal paper entitled "History as Natural History: An Essay on Theory and Method," Worster (1990:16) writes:

> First, let us be clear about one thing: there is no special new theory that ecological *history* can or should be expected to add to the anthropological models. To believe otherwise is to suppose that history is a self-contained discipline, with its own models of society and its own peculiar epistemology. It is not. History is more a clustering of interests than a discipline, and it has never had a unique, discrete paradigm to work with.

Worster appeals to ecological anthropology, and especially Marvin Harris's research strategy of cultural materialism, as a source for general theory. While I have great respect for Worster, and in general I find substantial validity and utility in Harris's strategy (Harris 1979, 1987, 1994), John Bennett's diachronic and comparative approach to human systems ecology as well as his concept of the ecological *transition* is more appealing to me for exploring historical ecology and ecological history (Bennett 1969, 1976, 1993). In the words of his students, Sheldon Smith and Ed Reeves (1989:14):

> The "ecological transition" is Bennett's term to refer to the historical trend which has accelerated in the past century, away from human communities which adapt to nature toward communities which incorporate nature and adapt to one another. In *The Ecological Transition* Bennett sought to construct a paradigm that would bridge the arbitrary distinction between culture and nature.

The purpose of this chapter is to provide one test of the validity and utility of Bennett's concept of the ecological transition against what is known or can be surmised about human environmental impacts from prehistory to the present in Thailand. This is also simultaneously an illustrative exercise in description (environmental history) and explanation (historical ecology).

The general scheme of cultural evolution, with an emphasis on the *dynamic* interplay of environmental, demographic, technological, and economic components and their changes, is adopted to organize the discussion chronologically. Precedents for this framework are provided in Bellwood (1985), Dunn (1970), Higham (1989), Hutterer (1976, 1988), McNeely and Sochaczewski (1995), Pelzer (1968), Padoch and Vayda (1983), and Watson and Watson (1969), although I have made some modifications. In particular, Jeffrey McNeely and Paul Sochaczewski (1995:10) identify what they call four major ecocultural revolutions in Southeast Asia: control of fire, domestication of plants and animals, evolution of irrigation agriculture, and development of the world marketplace. Each of these marks a new *threshold* of increasing human environmental impact—or, in Bennett's terms, ecological *disequilibrium*. After a brief explanation of the ecological transition, these ecocultural revolutions provide a convenient framework for this brief survey of the environmental history of Thailand: foraging, farming, commerce, and finally, population growth and the impact of modernization during the twentieth century will be considered separately.

Ecological Transition

The concept of the ecological transition as developed by Bennett (1976, 1980, 1993) is diagrammed in table 17.1. Societies may be ranked along a continuum from equilibrium to disequilibrium, which are opposite tendencies. At the outset it should be clearly understood that this is not a dualistic typology or a simplistic either-or proposition—namely, that either societies are in equilibrium or they are in disequilbrium. Rather, equilibrium and disequilibrium are a matter of degree; that is, they are *relative* rather than absolute conditions. Moreover, this is not a simple unilinear scheme: a particular society may shift its position through time along this *continuum* in *either* direction. However, it appears that the more common direction along this continuum is toward greater disequilibrium, the direction of the ecological transition—this in accord with entropy, the second law of thermodynamics (Rifkin 1990, 1993).

The essence of an equilibrium society is reflected in biologist Daniel Kozlovsky's fundamental rule of human ecology (1974:106): "Live as simply and as naturally and as close to the earth as possible, inhibiting only two aspects of your

Table 17.1   The Ecological Transition

Component	Equilibrium	Disequilibrium
Population	Small, controlled	Large, expanding, weakly controlled
Needs and wants	Minimized, limited mostly to satisfying physiological requirements	Maximized to include culturally defined and promoted wants, which are expanding
Resources	Self-sufficient use of local environment	Increasing dependence on import from distant places through extensive trade
Technological capacity	Limited, based on human energy	Highly developed, industrial, based on multiple sources of energy, including fossil fuels
Environmental contact	Large percent of society	Small percent of society
Environmental impact	Low	High, including resource depletion, species extinction, environmental transformation, pollution

Source: Bennett 1976:13.

unlimited self: your capacity to reproduce and your desire for material things."
Many relatively traditional small-scale foraging, swiddening, and herding societies
studied by anthropologists tended to be equilibrium societies to some degree; this
does *not* mean that all these societies, including all groups or individuals within
them, were always in perfect balance and harmony with nature. But by virtue of
being small and mobile populations at low density with limited technological capacity, a focus on subsistence needs, and (usually) a viable environmental ethic,
such societies commonly possess the attributes that Bennett describes for equilibrium. Accordingly, equilibrium refers to the maintenance of a stable population,
well-being at the individual and population levels, and a sustainable use of resources and ecosystems (Bodley 1994:259).

In contrast, a disequilibrium society is characterized by uncontrolled growth in
both population and the consumption of resources, with marked resource depletion
and environmental degradation. Such societies exploit natural resources from distant ecosystems through an extensive network of trade and commerce. This allows
them the temporary luxury of depleting, degrading, and even destroying distant
ecosystems—but of course, ultimately it is ecocidal, since the biosphere contains a
finite number of ecosystems and they are interconnected. During recent centuries,
through European colonization and industrialization, and particularly with economic development and modernization since World War II, most societies of the
world have increasingly undergone the ecological transition toward greater disequilibrium (Bennett 1976; Bodley 1994; Dwivedi 1988; Marsh 1965; Myers 1993;
Rifkin 1990, 1993; Simmons 1989; Turner et al. 1990; Turner and Meyer 1993;

Watt et al. 1977). The best analogy for this recent spiral of disequilibrium is cancer, which by definition is uncontrolled growth that eventually destroys its host (e.g., Hern 1993).

Unfortunately, in table 17.1 Bennett omits a very important influence on human/ environment interactions, namely, cultural knowledge, beliefs, values, and attitudes (Callicott 1994; Hamilton 1993; Kearney 1984). For instance, indigenous foragers often view nature variously as alive, feminine, sacred, mythical, and/or cyclical (Oelschlaeger 1991).

It is important to realize that ecological disequilibrium and environmental impact are not synonymous, since *all* human societies, including those in some degree of equilibrium, cause some environmental impact. (Indeed, many animals do so as well, such as leaf-cutter ants, prairie dogs, beavers, and elephants). However, environmental impact becomes destructive when it is *irreversible* (i.e., resources and ecosystem can no longer regenerate), and/or when it threatens the sustainability of the society. Thus, some impacts are clearly negative or maladaptive, such as reducing biodiversity through deforestation; whereas others may be positive or adaptive, such as enhancing biodiversity through traditional swiddening, which creates a mosaic of plant and associated animal communities at different stages of succession (Sponsel 1992).

Of course, ideas such as equilibrium, balance, and harmony have become not only controversial but somewhat unfashionable in recent years with the development of chaos theory. Perhaps this reflects the apparent increased disorder in society and the environment, and also the rise of postmodernism and revisionism (Botkin 1990; Glacken 1956; F. Edgerton 1993; R. B. Edgerton 1992; Hastings et al. 1993; Headland 1997; McDonnell and Pickett 1993; Pimm 1991; Simberloff 1982; Soule and Lease 1995; Worster 1990). Nevertheless, I have yet to find a better heuristic model than Bennett's ecological transition for viewing the cultural ecology, historical ecology, and recent environmental crises of Thailand, although criticisms and alternatives are most welcome. Others continue to find this conceptual framework useful as well, whether or not they derive it from Bennett (Bodley 1994:16, 161, 403; Drengson 1989:163; Simmons 1989:84, 87, 96–97, 102, 227, 259–260, 342; Turner and Meyer 1993:41–44; Worster 1988b). A knowledge of the history of anthropology reveals that intellectual fads and fashions come and go, but also that some elements of previous paradigms endure. I join others who still find the concept of the ecological transition useful, albeit with qualifications and recognition of its limitations.

In theory, equilibrium and disequilibrium remain logical possibilities. If they are considered as the poles of a continuum along which dynamic societies interacting with dynamic environments (biophysical and sociopolitical) may shift through time in *multiple ways* and at *varying rates*, then the concept has some utility as a heuristic model—but the ecological transition and the factors that Bennett itemizes under the poles of equilibrium and disequilibrium must all be subjected to *empirical testing* in specific temporal, spatial, ecological, and cultural contexts. However, the concept of the ecological transition is not useful if equilibrium and disequilibrium are asserted as simply two contrasting types of societies that are monolithic, fixed, and absolute.

In this paper it will not be argued that any societies in Thailand were ever in perfect equilibrium; rather, I will discuss *successively higher thresholds of disequilibrium* through the principal stages of what McNeely and Sochaczewski (1995:10) referred to as the four major ecocultural revolutions. A main goal here is to simultaneously test and illustrate the applicability of this heuristic model in the case of Thailand. Although for some readers it may appear to be too broad and ambitious an undertaking to consider the entire human prehistory and history of all of Thailand given the limited data available, this is attempted precisely to identify possibilities and needs for further research, as well to test the ecological transition as a primary hypothesis. Such an overview should also provide useful background for better understanding recent environmental problems and issues.

Thailand

Thailand is a useful unit for analysis here not so much because it is a nation-state (Winichakul 1996), but because it constitutes a "region" in the historical-ecological sense (Crumley 1994). Thailand is a relatively natural regional unit of mainland Southeast Asia, being bounded by mountains in the west and north, the Mekong River in the northeast, and the Gulf of Thailand and Andaman Sea in the south. It is strategically located at the crossroads of East, South, and Southeast Asia in many respects, including the perspectives of physical, biological, cultural, religious, and historical geography.

Today Thailand remains a Buddhist kingdom, although it is also a rapidly modernizing nation-state (Keyes 1987). It is far from homogeneous, but is naturally divided into four main regions (Donner 1982; Holdridge et al. 1971; Takaya 1975). The *north region* is a hilly-to-mountainous area with deciduous forests and narrow alluvial valleys. This is the border zone with Laos and Burma (now Myanmar), or the infamous Golden Triangle of the opium trade. The north includes the habitat of half a dozen ethnically and linguistically distinct animistic societies, usually referred to as hill tribes, who practice various kinds of swidden horticulture as well as foraging. The Phi Tong Luang ("spirits of the yellow leaves"), a mainly foraging society, lived in part of the northern forests, but have been almost entirely assimilated.

The *central region* is focused on the Chao Phraya River Basin, a broad valley with very fertile soils. It is where the most typical Thai culture is found among the descendants of the kingdoms of Sukhothai, Ayutthaya, and Siam. It is also the most productive rice-growing portion of the country, often called the "rice bowl" of Asia.

The *northeast region* is formed by the Khorat Plateau. It is the poorest region of the country today, an arid area with low and erratic rainfall, recurrent droughts, poor and often saline soils, and frequent crop failures. Perhaps not by coincidence, the Khorat Plateau appears to have been occupied by one of the oldest civilizations, the Khmer, which also extended into Cambodia. This civilization may have had a heavy environmental impact and have left as part of its legacy the difficult environment of the northeast.

The *south region* is a very long and narrow peninsula, with the Gulf of Thailand on the east and the Andaman Sea on the west. Its forests range from mangroves along the coasts to rain forests in the hilly-to-mountainous interior. About 70 percent of the people in the south are Muslim, who retain close affinity to Malay culture, language, and history; traditionally they are mostly subsistence fishers along the coast of the Gulf of Thailand. Two foraging societies are also found in the region: the Chay Lao (sea nomads) along the coast of the Andaman Sea, and the Sakai or Semang in the rain forests of the mountains near the border with Malaysia.

It should be recognized that each of these regions encompasses a wide variety of ecosystems and microenvironments. For instance, the very long and narrow southern peninsula has a remarkable diversity of environments in close proximity—forested limestone mountains, caves, rivers, streams, estuaries, swamps, mangroves, intertidal zones, beach forests, sand dunes, islands, bays, coral reefs, and deep waters, among others.

Thailand has the following types of forest: open forest (34.7%), lowland evergreen rain forest (1.9%), semi-evergreen rain forest (4.3%), dry evergreen (21.0%), hill evergreen (1.9%), swamp, mangrove (1.7%), beach, mixed deciduous (22.6%), bamboo (4.3%), coniferous (1.3), and scrub (0.8%) (Hirsch 1993:54). There is even finer variation in these forests, such as in the different combinations of tree and shrub species found in mangroves depending on factors such as the frequency of inundation (Somboon 1984).

Ideally, the ecological history of each region and each society should be considered separately in depth, but neither space nor available data allow this here. Instead, with the caution that Thailand is far from homogeneous, this chapter will proceed through a panoramic exploration of the historical ecology of Thailand as a whole region.

Natural Environmental Changes

The environment cannot be treated as uniform in either space or time. Ecosystems are dynamic, and they change in response to internal as well as external influences—ranging from local natural treefalls in the forest to widespread climatic oscillations triggered by volcanic eruptions, El Niño, and periods of glacial advance and recession. Human responses to natural changes are an important aspect of research in ecological history and historical ecology. Thus, before examining anthropogenic influences, the natural changes should be outlined briefly.

During much of prehistory the environment (via climatic change) exerted more influence on humans than they did on it (although with agriculture and increased population density and sedentariness, the influence of humans grew as their relationship to the environment became increasingly active). Between 150,000 and

12,000 B.P. there were no less than five substantial climatic oscillations from warmer to colder periods and back again in the region now called Thailand. There were also major changes in sea levels during the Pleistocene and Holocene; for example, between 22,000 and 16,000 B.P. a lowered sea level united all of Malaysia, as well as most of Indonesia and the Philippines, with mainland Southeast Asia. During this period woodlands covered northern Thailand and savannas covered central and southern Thailand, while, supposedly, rain forests did not even exist in the region. However, during the end of the Pleistocene, around 12,000 B.P. climate changed rapidly. By 10,000 to 9,000 B.P. temperatures and rainfall were well above the present levels in most areas of Southeast Asia; savannas and most woodlands disappeared, while rain forests expanded beyond their present natural limits. Probably by about 8,000 B.P. climatic conditions similar to those of today had developed. By the Middle Holocene, 6,000 to 4,000 B.P. sea level had increased above the present level by about three meters, inundating most of the Chao Phraya River Basin up to the site of Ayutthaya. Seas finally lowered to present levels as recently as around 4,000 B.P. (Anderson 1989; Flenley 1979; Sinsakul 1992).

Archaeologists and other students of prehistory and history in Thailand seem to have paid most attention to the influence of such natural environmental changes on human ecology, and much less attention to the influence of human societies in changing the environment. The latter is the emphasis of the remainder of this paper, in accord with one of the main concerns of historical ecology.

Foraging

So far the oldest proposed archaeological site within Thailand is Mae Tha in Lampang Province in the north, where artifacts were found in Pleistocene alluvium capped by a basaltic lava flow dated between 800,000 and 600,000 B.P. However, the archaeological discoveries have yet to be fully reported (Anderson 1989:103).

The oldest accepted radiocarbon-dated site in Thailand is Lang Rongrian, a stratified rock shelter in Krabi Province of the southern peninsula, which is probably older than 37,000 B.P. The Pleistocene artifacts are mainly small, unifacially retouched flaked tools. The early Holocene artifacts are large, heavy, unifacially and bifacially flaked core tools identical to Hoabinhian finds from Malaysia. The middle Holocene deposits contain burials, pedestal pots, and small-cord-marked vessels, similar to other Neolithic sites in Thailand and Malaysia. Marine shells were found in the Holocene deposits, indicating foraging along the river estuary and adjacent coast. Lang Rongrian overlaps the period when hominids may have been evolving into *Homo sapiens sapiens*, which was also a time of lower sea levels (Anderson 1987).

At the end of the Pleistocene, human populations increased markedly, perhaps because they took advantage of a much wider range of food resources during the Holocene than previously. Gastropods, particularly land snails and freshwater mollusks, became an important component of the diet of inland societies (Anderson

1989). It can be hypothesized that this shift to a wider range of animal species for food reflects a depletion of species previously favored and a new threshold of disequilibrium. However, the Pleistocene megafaunal extinction spasm that has been found elsewhere does not appear to have occurred in Southeast Asia—perhaps because there, as in Africa, hominid predators and their prey coevolved over hundreds of thousands of years or more, unlike the regions of the Americas, Australia, and New Zealand where such extinctions were important phenomena that some scholars associate with human colonization.

The archaeological remains of these early hunter-gatherer societies—called Hoabinhian, after the type site in the Hoa Binh district of northern Vietnam—are found in an area that reaches from south China, throughout mainland Southeast Asia, to north Sumatra, ranging in date from 12,000 to 5,000 B.P. (Pookajorn 1985; Reynolds 1990). The Hoabinhian was a broad-spectrum foraging pattern exploiting a very extensive and diverse range of animal and plant species. Some of the prey included monkeys and apes; thus, some of the foragers were exploiting the rain forest and/or its edge (Hutterer 1988:65). Early humans in Thailand apparently adapted to a wide diversity of local environments including coastal, riverine, and forest habitats, as well as various microenvironments.

While in historic times there have been at least three distinct cultures focusing on foraging in Thailand, only with caution and qualification can they be taken to reflect to any degree the foragers of prehistory (Engelhardt 1989; Pookajorn 1985, 1988). The main reason is that these cultures are involved (to varying degrees) in exchange with farming and other societies in Thailand. The conditions and adaptations of foraging must have been very different prior to the evolution of agriculture. However, some groups of foragers in the forests, like the Phi Tong Luang, may have survived exclusively on wild foods without any recourse to farming themselves or to trade with farmers (Pookajorn 1985:215; cf. Headland and Bailey 1991). Another problem is that research on foragers in Thailand is very limited; by this time, they have been largely assimilated by Thai society.

In the historic period, the three distinct foraging cultures in Thailand—the Mlabri, the "Sakai" or Semang (Orang Asli in Malaysia), and the Chao Lay, Urak Lawoi, and other so-called sea nomads along the Andaman Sea side of the southern peninsula—are found, respectively, in three distinct biomes: the deciduous forests of the north, the rain forests of the south, and the coastal zone, including mangrove forests, of the south (Bernatzik and Bernatzik 1958; Engelhardt 1989; Hogan 1972; Sauer 1965a). In prehistory, prior to farming, other foraging societies with somewhat different adaptations must have lived in other parts of Thailand that were probably more favorable habitats than the other biomes—such as the environments of central Thailand, which included savannas occasionally broken by gallery or riverine forests.

Here space does not allow me, nor is it necessary, to enter into a detailed description of each of these three cultures. The main points to be made are that, like most traditional foragers, those in Thailand probably were small bands of a few dozen individuals who were able to extract their subsistence from a wide diversity of natural environments and wildlife resources because they were highly mobile, with a low population density, and with a wealth of ecological knowledge and acute sensitivity to nature. These are the societies in Thailand most likely to approximate the characteristics that Bennett lists for equilibrium societies.

The sea nomads provide an interesting example of environmental sensitivity:

> The Urak Lawoi monitor the size of oysters taken and quantity of oysters gathered, by depositing the refuse in carefully stratified middens. When the daily take falls below the usual harvest, as shown by a visual comparison of midden lenses, the Urak Lawoi interpret this to mean that the carrying capacity of the oyster population has been exceeded and the band makes preparations to move to a new camp site. (Engelhardt 1989:138)

As another illustration of the ecological sensitivity of foragers, the Phi Tong Luang, which in Thai means "spirits of the yellow leaves," get their name from the temporary lean-to shelters they construct. The leaves used as roofing are from banana plants, rattan, or other palms; when they turn yellow, the Phi Tong Luang know it is time to move to a new campsite. This custom regularly relieves the exploitation pressures on the resources of the local environment, and thereby may avoid severe or irreversible depletion. These foragers harvest only what they can eat the same day; they do not accumulate a surplus or store food (Pookajorn 1985).

Although some might classify such foragers as equilibrium societies, this does not mean that they did not manage and transform aspects of their environment and resources to some degree (Heizer 1955). For example, forest foragers may have owned and manipulated fruit trees such as durian, constructed fish weirs in streams, and influenced animal species through selective predation. The Phi Tong Luang reinserted the tops of yam tubers into the ground to regenerate and remembered their location for future harvesting (Pookajorn 1985). While still remaining primarily foragers, in recent history some of the Sakai tended tiny plots of bananas, cassava, and wild yams (Engelhardt 1989). The sea nomads even altered coral reefs by cutting channels for easier access by their boats (Engelhardt 1989). Further, the selective use of certain species of animals and plants over the long term could substantially alter ecosystems.

Fire

Certainly one of the most powerful influences on the environment that foragers applied was fire (see chapter 4). The antiquity of hominid use of fire remains equivocal, although there are enticing claims from the Lower and Middle Pleistocene in Asia, including one in association with *Homo erectus* remains at Zhoukoudian cave in China (James 1989). Whatever the oldest periods for the use of natural fires, and the subsequent control of fire by hominids, surely these must be on the order of many thousands of years in Southeast Asia. Less contentious are the long-recognized multiple uses and adaptive advantages of fire (Clark and Harris 1985; Sauer 1965b; Stewart 1956). For instance, cooking would have rendered digestible plant foods that were high in cellulose and/or starch, as well as those containing secondary compounds or toxins (Stahl 1984). In my opinion, the use of fire for cooking is a key factor in the adaptive success of humans in tropical rain-forest ecosystems where toxic compounds are so common in plants (Sponsel in press). However, here we are more concerned with the environmental impact of fire.

While several of the previously mentioned attributes of foragers would appear to limit their environmental impact, it is important to realize that the *cumulative* effect of foragers over long periods of time could be quite substantial (see Pelzer 1968:275–276; Pyne 1991). Many thousands of hunter-gatherers setting patches of vegetation on fire over many thousands of years would create a mosaic of plant communities in different stages of succession. In some situations, the result is even a pyrosere: a stage of plant succession maintained by recurrent burning, as found in aboriginal Australia and elsewhere (e.g., Pyne 1991). Thus, already with prehistoric foragers, perhaps hundreds of thousands of years ago, we have the beginning of the transformation of the natural landscape into an anthropogenic landscape. By the time of farming, the impact would have been even greater (Bartlett 1956). Indeed, J. E. Spencer (1966) concluded that there is little if any forest left in Southeast Asia that has not been cut and burned for swiddens at one time or another in prehistory and history. However, in most areas the forest would have regenerated.

The immediate effect of the fire would be to stimulate the fresh growth of grasses and other vegetation that would attract grazing animals, which in turn could be hunted. Hunters in the recent history of Thailand clearly recognized this relationship and searched for game by looking for smoke or recently burned areas (McNeely and Sochaczewski 1995:29–30). Over time, this burning could also provide the foundation for the domestication of some species of herbivores. Charles Wharton (1968:162) goes even further: "We may conclude that the living wild cattle of southeast Asia appear intimately dependent on an environment which is, if not entirely created by man and fire, certainly maintained by these agencies."

Farming

It is not clear whether Hoabinhians practiced any kind of cultivation (Solheim 1972; Yen 1977). However, beans, gourds, peppers, and other plants that flourish in disturbed soils would have been naturally selected for, regardless of whether the intention was merely gathering—thus foreshadowing domestication (Anderson 1989). By 9,000 to 7,500 B.P. the Hoabinhians in northwestern Thailand were manufacturing ground and polished stone ax and adze blades (Gorman 1970). These may have been employed in clearing small areas of forest for cultivating crops (Anderson 1989); however, it seems likely that foragers would first have taken advantage of natural tree gaps for semicultivation, prior to regular clearing with an ax. The development of metal axes with their much greater efficiency over stone axes would probably have increased the pressure on forests (Denevan 1992).

Holocene sites such as Non Nok Tha, Ban Chiang, and Ban Kao contain animal bones of chickens, pigs, and cattle that later, perhaps as early as the middle Holocene, became domesticated (Anderson 1989).

By at least 7,000 B.P. pottery was widespread in Thailand, suggesting an increase in both food storage and sedentariness, and thus the evolution of a very different relationship with the natural environment (Anderson 1989; Pookajorn 1985). The Hoabinhians, unlike their predecessors, used caves and rock shelters for prolonged residence rather than just temporary camping, and sedentary village sites were established in coastal and other areas prior to agriculture (Anderson 1989; Higham 1989). Of course, increased sedentariness and other practices would put more pressure on the local environment and its resources, including forests.

Archaeological sites along the coast of the Gulf of Thailand range over a period of some 2,000 years. Sites such as Khok Phanom Di indicate that by 5,000 B.P. there were sedentary settlements up to five hectares in size, some of which persisted for as long as six centuries. Their subsistence economy was based on a varied mixture of hunting and gathering in inland forests, and on riverine and marine fishing. They raised domesticated pigs and cattle. Wild rice was harvested from the naturally flooded marshlands along rivers and streams where it still grows to this day. They may have also domesticated rice, or it may have spread into the area from the Yangtze Valley of China between 7,000 and 4,000 B.P. There is clear evidence for domesticated rice at Khok Phanom Di around 4,000 B.P. (Higham and Thosarat 1994:10, 15, 142–146); however, analysis of pollen cores from that site indicates that humans were modifying the plant communities in the area through burning and other practices from as early as 8,000 B.P. onward. (Palynological studies reveal that in Western Malesia, human forest clearance with the use of fire may have occurred as early as 17,800 B.P. and certainly by 2,500 B.P. [Maloney 1985].) Shell jewelry and pottery were traded inland. The first intensification of agriculture likely accompanied the development of chiefdoms, between 2,500 and 2,000 B.P. (Higham and Maloney 1989).

The relative chronology of swiddening and wet rice cultivation in Thailand is not clear. Swiddening may well have developed first in the forested hills, while the seasonally flooded lowlands were used mainly for hunting (McNeely and Sochaczewski 1995:38). On the other hand, because important Southeast Asian cultigens like wet rice and taro are dependent on inundation for growth, farming may have started in lowland swamps, with swiddening of dry rice and other rainfed crops developing in upland forests only later as population pressure forced people to colonize them (Hutterer 1988). In either case, by about 3,000 B.P. there is evidence of bunded fields and water buffalo in northeast Thailand (Higham and Kijngam 1979), whereas the central plains of Thailand seem to have been settled later (Anderson 1989). Eventually, wet rice allowed the conversion of the savannas and marshes of the lowlands. This also meant the destruction of wildlife habitat, particularly for the larger mammals. But the floodplain environments were more productive, with more-fertile soils, because of their annual rejuvenation through sedimentation from the floodwaters of major rivers.

In comparison with swiddening, wet rice paddy agriculture requires a much higher level of technology, as well as social and political organization for the construction and management of irrigation networks, as has long been recognized (Wittfogel 1956; cf. Cohen 1993:176–179). While swiddening has some environmental impact, especially cumulatively over the long term, wet rice paddy farming is usually a relatively permanent transformation of the natural landscape—what Lucien Hanks (1972) referred to as holding capacity, or Stephen Lansing (1987, 1991) as an engineered landscape. Unlike swiddens, wet rice paddies can be sustained for centuries or more, thanks to nutrients from river sediments and the nitrogen-fixing blue-green algae. However, whereas swiddening under traditional circumstances probably enhances biodiversity by creating a mosaic of biotic communities at different stages of succession, clearly wet rice paddies decrease biodiversity, even though they may still contain hundreds of species of plants and animals (Heckman 1979; and see Flenley 1988).

State

Rice is not just a crop, but a whole way of life in Thailand. It is also the agricultural foundation for the great civilizations of Southeast Asia, although states did not evolve in the region until relatively late in prehistory, around 2500–2000 B.P. (Higham 1989:307). In Thailand the first large-scale complex societies included cities with monumental architecture and fine art, Hindu-Buddhist state religion, centralized government based on semidivine kings, and military institutions. Of course, not all of these developments were indigenous; most were influenced to some degree from India. The Mon civilization called Dvaravati flourished in what is now central and western Thailand during the ninth to eleventh centuries A.D. It was an Indianized culture dominated by Therevada Buddhist religion. Many of the kings were devout Buddhists and encouraged that religion among the populace, and temples traditionally served as centers for religious, educational, ceremonial, and social functions. (See also Bayard 1984; Coedes 1968; Cohen 1993; Hall 1985; Winzeler 1976; and Wyatt 1984.)

It can be hypothesized that Buddhism tempered human impact on the natural environment in Thailand, given the principles for a sustainable environmental philosophy and ethics latent in this religion (Batchelor and Brown 1992; Brockelman 1989; Kunstadter 1989:548–550). As discussed later, both reasoned arguments and circumstantial evidence suggest that this was probably the case until recent decades (see table 17.2, below). However, this is not the place to discuss the matter in detail (see Sponsel and Natadecha-Sponsel 1988, 1993, 1995, in press). Two examples will have to suffice. Traditionally, most males in Thailand spent some time in the monkhood (the period could be months, years, or even a lifetime), and it seems reasonable to assume that the celibacy of the monks suppressed population growth to some degree. Also, there is a mutualistic association between Buddhism and forests, reflecting the life of the Buddha, and to this day forests are the optimum context for meditation (Roscoe 1994). This tradition continues in various ways. (See figure 17.1.) Indeed, Buddhist monks are prominent among the environmental activists campaigning against the rampant deforestation in Thailand in recent years (Leungaramsri and Rajesh 1992). Unfortunately, direct evidence for any role of Buddhism in the environmental history of Thailand is lacking. For instance, pollen analysis does not yet have fine enough resolution to detect any impact of Buddhism on the vegetation ecology of Thailand (Bernard Maloney, pers. comm.). It is noteworthy that Buddhism initially developed in the Ganges Plain of central India in the sixth century B.C. during a period of marked population growth, agricultural expansion, and deforestation (Ling 1973:42–43).

It can also be hypothesized that there are various beliefs and taboos that to some degree serve a conservation function, and that these are often related to the animism and Hinduism that underly the Buddhist and (in the south) Muslim strata in Thai society (Prime 1992; Sponsel 1995). For instance, to this day in rural areas most Thai would not dare to cut down a large tree for fear of offending the resident spirit (Kriengkraipetch 1989:205–206; Rajadhon 1988:336).

Trade

One of the major characteristics of disequilibrium societies is the exploitation of resources from distant ecosystems through extensive trade on a large scale. This developed very gradually throughout Thailand, but it can be hypothesized that it only reached destructive levels in recent history in connection with Westernization (see Hall 1985; Hutterer 1977; Parnwell and Bryant 1996:4–5; Reid 1988).

Early in prehistory, appropriate kinds of stone for tools were exploited and/or traded over a large area in Thailand (Anderson 1989). Foragers and farmers have long been involved in exchange systems (Headland and Reid 1989). For instance, in historic times the Phi Tong Luang in the northern forests employed silent barter to trade forest products to neighboring farming villages; traditionally the trade items included honey, beeswax, rattan baskets and mats, rhinoceros horn, bear galls, wild animals, bamboo shoots, herbal plants, and wild yams, in return for salt, steel, clothes, blankets, tobacco, and matches (Pookajorn 1985). Trade may have developed in part as a measure to relieve growing competition; that is, through niche differentiation, the potential competitors specialize in providing complementary resources for each other in a cooperative exchange system (Engelhardt 1989).

For many centuries forest products have also been traded over very long distances, from Southeast Asian areas to as far away as India and China (Dunn 1975; Glover 1989; Reid 1995). For example, in historic times the Sakai in southern Thailand extracted natural resources such as beeswax, resins, and exotic woods from their forest habitat for exchange with Indian, Chinese, and (later) European traders

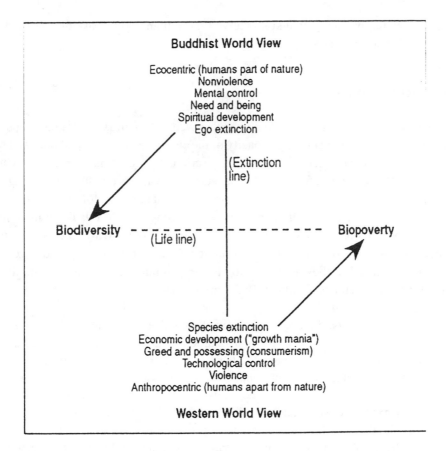

Fig. 17.1 *Worldview and Biodiversity.* (After Sponsel and Natadecha-Sponsel 1993:87.)

(Engelhardt 1989). In the thirteenth to fifteenth centuries glazed ceramic wares were exported in great quantities from the kingdom of Sukhothai to other regions around the South China Sea, and even as far as Indonesia and the Philippines. The kingdom of Ayutthaya became a center for maritime trade, including forest and wildlife products such as sapanwood, eaglewood, benzoin, gumlac, deerhides, elephant tusks, rhinoceros horn, and dried fish. Also during this period farmers produced a sufficient surplus of rice for export to other regions of Asia.

The "age of commerce" between the fifteenth and seventeenth centuries in Southeast Asia (Reid 1988, 1990) was made possible by the development of large trading ships ("junks") of 200–500 tons, which for the first time permitted the overseas transport of bulk cargoes (Evers 1987:755). Trade subsequently developed with the Portuguese and Spanish in the sixteenth century, and with the British, Dutch, and French in the seventeenth century. (Also see Kathirithamby-Wells 1995 and Lieberman 1990.)

It should be noted that local natural resources and goods were also part of the taxes paid by commoners to the Thai court during the nineteenth century. For instance, as tax payment Bangkok received more than 3,000 teak logs and 2,400 logs of other species each year (Wilson n.d.).

Twentieth Century

For Thailand a common view is that until the twentieth century the human popula- tion was well below the carrying capacity of the environment, plenty of land and re- sources remained, and there were few if any serious environmental problems. Thus Constance Wilson (n.d.: 25) concludes her survey of various documents bearing on the environmental history of nineteenth-century Thailand as follows: "Looking back, it is surprising how well the country was holding up in 1900. The human im- pact on the environment was, with but few exceptions—teak forests and the Phuket tin mines—still limited. This situation, however, was not to survive the twentieth century." In the conclusion to his compendium of travelers' accounts of early-nineteenth-century Thailand, B. J. Terwiel (1989:256–257) observes:

> Prior to the massive influx of European advisers that came with the Bowring Treaty, and independently of the Industrial Revolution, the Thais developed a system that grew increasingly effective at drawing on natural resources. . . . During the first half of the nineteenth century there was seemingly still infinite wealth to be tapped. Much of the landscape was as yet untouched by the human hand. The Thais had only just begun their "conquest" of the lower Chao Phraya Delta. Our travellers saw pioneering farming settlers who feared being trampled by herds of wild elephants, who could be attacked by croco- diles and who had to battle with large flocks of wild birds in order to grow their crops. Such battles continued throughout the nineteenth and into the twentieth century, and gradually they were all "won." The rhinoceros, the crocodile, the tiger, monkeys, birds and elephants all had to make way. It is one aim of this book to make people realize how recently these battles have been fought . . . how in only a few generations the countryside was altered, first to make way for a seemingly endless series of rice fields, and more recently through a much more ominous series of changes that seem to have turned the Gulf from one of the earth's richest fishing grounds into an industrial cesspool.

Similarly, Juree Vichit-Vadakan (1989:429) asserts that, "prior to the great popula- tion increase of the twentieth century, population pressure on land was not consid- erable, natural resources were plentiful, and virgin forests and frontierland were seemingly unlimited" (also see Kunstadter 1989:543). Such *impressions* require much more research to be either confirmed, revised, or rejected, but they appear plausible following the previous considerations in this essay.

Thailand managed to remain independent during the colonial era by sacrificing some of its territory to colonials, by astute diplomacy, and because the British and French agreed to maintain it as a buffer zone between their colonies. It can be hy- pothesized that this relieved some of the pressure on forests and other natural re-

sources that would otherwise have come with colonization (Poffenberger 1990)—
but the impact of the Western powers was still felt, at least indirectly, through trade.

The introduction of the steamboat in the 1830s to 1840s, and, toward the end of
the nineteenth century, of the telegraph, electricity, railroads, and roads, all contributed to much higher levels of wood consumption (Wilson n.d.). From 1850 to
1984 the area under rice production was markedly increased, from 9,000 to 70,000
km^2, mainly to earn foreign exchange through rice exports. Unfortunately, this emphasis on agricultural expansion rather than intensification was accomplished
mostly through forest conversion: between 1913 and 1986, the area under forest decreased from 75 percent to 15 percent (Chunkao 1987; Hirsch 1993:26–27). In recent years Thailand has had the highest annual rate of deforestation in all of Southeast Asia at 2.7 percent—more than twice that of the countries with the next-highest
rate, Laos and Malaysia (Dinerstein and Wikramanayake 1993). A major contributing factor to deforestation has been the alienation of local communities from their
land and resources by the central government, a process that began in 1896 when
Bangkok took the control of northern teak forests away from local princely states by
establishing the Royal Forestry Department (Ekachai 1990; Hirsch 1993:53; Hafner
1990; Pragtong and Thomas 1990). By the early 1970s Thailand had changed from a
net exporter of timber to a net importer (Hirsch 1993:149).

The last major resource frontiers in Thailand were the deciduous forests of the
north and the rain forests of the south. In the northern highlands wildlife was still
rich in the forests as late as the 1950s, but in recent decades this region has turned
into a faunal desert as the result of various factors mostly related to economic development. In many areas wildlife has been depleted through overhunting; at the
same time, its habitat has been degraded or destroyed through swiddening. In general the population of the hill peoples has increased markedly, while the subsistence
economy has added a market component, and fallow periods for swiddens have declined. National and international economic development programs did not take
into account the importance of nontimber forest products in the local economy of
the hill peoples. For instance, 34 percent of the annual income for the Hmong is derived from forest products. Also with economic development came the decline of
community cohesion, as well as a deterioration in beliefs, values, and practices that
appear formerly to have promoted the sustainable use of wildlife. The land base has
become privatized for individual households who compete for remaining resources
for short-term gain. Simultaneously, the situation of the hill peoples has been complicated and aggravated by a massive invasion of impoverished and landless Thais
from the lowlands (Dearden 1995). In the north, from 1950 to 1984, the area under
cash crops increased thirtyfold with new roads, access to markets, machinery, and
new crops (McNeely and Sochaczewski 1995:55–58). This serious environmental
deterioration of recent decades has transpired in an area where at least some of the
hill peoples had apparently lived in sustainable societies for centuries. (Also see
Aiken and Leigh 1985; Hirsch 1993; Ramitanondh 1989.)

In the south, similarly, during the early twentieth century and up to this day
rain forests have been converted to rubber and oil-palm plantations for the export

industry. The area covered by forest in the south declined from about 42 percent in 1961 to 22 percent in 1982 (Chunkao 1987:81). In recent years the mangroves have been targeted for conversion to shrimp ponds and other projects for economic development. Mangrove destruction is impoverishing local subsistence and commercial fisheries, accelerating coastal erosion, and having other detrimental environmental and socioeconomic impacts.

The wildlife trade remains important. For instance, a survey for twenty-five weekends at the Bangkok Weekend Market revealed a total of 68,654 native birds of 225 species; 3,132 exotic birds of 51 species; and 3,418 native mammals of 24 species. Most were being openly sold illegally. Profits reach at least U.S.$20,544 each weekend (Round 1990). This is just one among many ways in which wildlife continues to be exploited commercially in Thailand (see Robinson 1994; Sponsel and Natadecha-Sponsel 1991).

Despite the long history of trade within and beyond the region of Thailand, from the foregoing it can be hypothesized that only within the twentieth century have the level of consumption, modern transportation, and the global marketplace put unsustainable pressure on the environment for resource extraction. Thus, for instance, forest products can now be extracted from remote areas and reach cities within a few days—unlike in antiquity, when human backs, ox carts, elephants, and boats were the primary modes of transport. The difference that roads make, and their recency in many areas of Thailand, is illustrated by Michael Moerman's successive field trips to the same study village in 1959 and 1979: his first trip to the village took more than two and a half days by horseback, but after the construction of a road into the area it took only ninety minutes by car (Moerman and Miller 1989:306). Most of the road network in Thailand has been constructed in recent decades; from 1950 to 1980, the total length of roads in Thailand increased almost fivefold, from 5,831 to 28,151 km (Hirsch 1993:33). Much of the development of roads in the northeast took place during the Vietnam War period when this region was of special concern to the United States. Currently under construction are a series of superhighways designed to crisscross Thailand and link it with neighboring countries, a development bound to have far-reaching environmental effects.

In recent decades, especially in the northeast and north, development programs have encouraged farmers to convert forest to gardens for new commercial crops including maize, sugarcane, and manioc for export to Japan, Europe, and North America. Manioc is now second only to rice among the crops exported from Thailand. Ironically, maize is a Mesoamerican domesticate, and manioc may be a northwest Amazon domesticate. These and many other introduced cultigens emphasize the global network to which Thailand has become progressively linked in the last two centuries, as it has moved further along the ecological transition to higher levels of disequilibrium. Some exotic species that have been introduced to Thailand have had a significant environmental impact, among them the water hyacinth, kenaf, eucalyptus, tilapia fish, and goat. The environmental and social impact of introduced species would be a fascinating and worthwhile subject for further research on the historical ecology of Thailand.

This recent threshold of disequilibrium also reflects the general acceptance by many Thai people of the capitalism, frontier mentality, materialism, consumerism, and growthmania of Western society. For example, in a recent review on resource depletion, environmental degradation, and economic development in Thailand, D. Phantumvanit and K. S. Sathirathai (1988:13) observe: "For several decades, Thailand has indulged in the abundance of its natural resources without considering their long-term sustainability. As a result there are now ample signs of ecological stresses facing the nation." In other words, there is growing recognition by the government and populace, including rural communities, that an unprecedented threshold of disequilibrium has been reached, which calls into question the long-term sustainability of modern Thai society (Arbhabhirama et al. 1987; Brookfield 1993; Ekachai 1990, 1994; Kunstadter 1989; Lohmann 1993; Rigg 1995; Schnaiberg and Gould 1994).

In recent years Thailand has exhausted its frontiers for resource extraction and agricultural expansion. Now it is turning to neighboring countries that still have frontiers to be exploited (Hirsch 1995). Thus, since the logging ban in Thailand of 1990, the country has increasingly turned to its neighbors for importing timber. It can get away with this in the short term because, except for Malaysia, these countries are desperate for foreign income after years of warfare, they still have extensive areas of forest, and local environmental movements have not developed sufficiently to protest the depletion of resources like timber together with the destruction of the forests and habitats for wildlife (Hirsch 1993; Hurst 1990; Rush 1991).

Population

Levels of resource consumption are related to population levels as well as to trade and commerce, and a few remarks on some aspects of the historic demography of Thailand are appropriate here. In recent Southeast Asia, the range of population densities (persons/km^2) under different types of economies is instructive: 0.005–0.12 for foragers in tropical rain forests; up to 60 for shifting horticulturalists, and more in areas of high soil fertility in tropical rain and dry forests; and several hundred to more than a thousand for wet rice paddy agriculturalists on the rich volcanic soils of Java (Bellwood 1985) (see table 17.2). It can be hypothesized that these different levels of population density coincide with different degrees of technological capacity, natural-resource consumption, and, accordingly, environmental impact and disequilibrium (e.g., Turner and Meyer 1993:45).

Disease has long been a limiting factor on populations in Thailand. It is of interest that the cranial-vault fragments from the burials in the upper unit of Lang Rongrian were unusually thick, similar to counterparts from the Neolithic sites at Ban Chiang and Non Nok Tha in northeastern Thailand. The thickening may indicate endemic malaria (Anderson 1987); the Phi Tong Luang foragers in the northern forests of Thailand also suffer from malaria (Pookajorn 1985). Furthermore, it is important to realize here that rates of infant mortality were much higher, while in general

Table 17.2   Increasing Environmental Impact

Economy	Time (B.P.)	Population Density (persons/km^2)
Foragers	40,000	<1
Swiddeners	11,000	10s
Wet rice farmers	6,000	100s
State	2,500	1000s

longevity was much lower, in prehistory and in much of history—conditions that would have also depressed population growth overall (Hassan 1981).

During the Middle Holocene, population was growing more rapidly—as reflected in the noticeably increased number and size of settlements in the archaeological record of Thailand. While this may reflect the emergence of agriculture, the population growth may also have been tempered by the increasing health problems associated with increased sedentarization and by narrowing of the range of food resources consumed by farmers compared to foragers, as revealed elsewhere (Anderson 1989).

A study of historical sources by Larry Sternstein (1993) indicates that around 1825, just at the beginning of the European colonization of mainland Southeast Asia, the size of most settlements in the Kingdom of Siam ranged from a few dozen up to several thousand people. There were also a few fairly large cities, such as Ayutthaya (88,883), Chiang Mai (50,600), Lamphun (40,600), Chanthaburi (16,000), Nakhon Si Thammarat (14,200), Songkhla (10,190), and Pattani (11,160). Nevertheless, at this time the total population in the region now called Thailand was surprisingly low at 930,000: this would yield a crude population density of only 1.8 persons/km^2, but actual population density in rural areas would be even lower, given the concentration of a portion of the population in urban centers. The current crude population density of Thailand is more than 111/km^2. The low level in 1825 is surprising, considering the substantial agricultural potential of Thailand and the abundance of food apparent there to this day (Bronson 1989:296; Higham and Thosarat 1994).

Sternstein (1993:24) attributes this low population in 1825 to limiting factors such as frequent warfare between the Siamese and their neighbors (especially with the Burmese in the sixteenth to eighteenth centuries), corvée labor, late marriage for commoners (rarely under twenty-one years of age for men and eighteen years for women), a high proportion of celibate Buddhist monks in the population, and recurrent epidemic diseases such as malaria, cholera, and smallpox. Famine from crop failure related to drought may have been another limiting factor in some areas like the northeast (Chai Podhisita, pers. comm.). If this figure is accurate, then it supports the hypothesis that as recently as the first quarter of the nineteenth century ecological disequilibrium was still at a moderate level at most. Indeed, in Southeast Asia as a whole, rapid population growth at 1–3 percent/year did not develop until after European colonization, from the eighteenth century onward. Before colonization population was very low, with a density estimated at 5.5/km^2—in contrast to 30/km^2 in South Asia and 37/km^2 in China (Reid 1987:36).

From the first census in 1911 to the present the population of Thailand increased from 8.2 million to more than 55.5 million, and it could reach 85 million by the year 2000 (Sermsri 1989:72). During the last half of the nineteenth century the presence of European colonial powers in adjacent regions brought an end to the frequent warfare between the Kingdom of Siam and its neighbors, the Burmese, Malays, Lao, Khmer, and Vietnamese—thus relieving one of the most important and persistent limiting factors on population growth. Also, the Thai encouraged immigration to expand the labor force, especially from China: from 1825 to 1910, the number of Chinese increased from 230,000 to 792,000, while their proportion of the national population increased from 5 to 9.5 percent (Wyatt 1984:217).

Much of the population increase during the twentieth century reflects the introduction of Western medicine, which substantially decreased infant mortality rates and increased life expectancy, while modern birth control measures were introduced only in recent decades. The recent annual growth rate of the population of Thailand is one of the highest in the world, at 3 percent in the 1950s and 1960s—although it has begun to decline with government family planning programs introduced in the 1970s (Sermsri 1989:73). Additionally, the celibacy of males in the monkhood is less important in suppressing population than previously, because their relative numbers have declined: while the population of the country has increased by 300 percent since the 1920s, during the same period the number of Buddhist novices and monks increased by less than 50 percent (Wells 1975:26–27).

Conclusions

The concept of the ecological transition simultaneously describes and explains to some extent the historical ecology of Thailand. However, the evidence for low levels of disequilibrium in the prehistoric foraging and swiddening societies of Thailand remains at best circumstantial, although reasoned and reasonable arguments can be marshalled. At the same time, the evidence and arguments for successively higher levels of disequilibrium correlated with the cultural evolutionary sequence of wet rice farming, extensive international trade, and "modernization" are certainly much stronger. The primary hypothesis of this essay is sustained—that in Thailand, *high levels of disequilibrium are remarkably recent developments*, largely confined to the twentieth century, and most pronounced during the period since World War II (also see Rigg 1995:19). They may have been foreshadowed to some degree by the development of international trade in prior centuries. The recent threshold of disequilibrium is revealed by a diversity of lines of evidence that include the interrelated phenomena of the population explosion, agricultural expansion, the growth of road and railroad networks, the development of international trade, and deforestation.

Thailand mirrors the situation of the biosphere in general, as B. L. Turner and William B. Meyer (1993:40) describe the latter:

> Virtually every corner of the biosphere . . . has been touched in some ways by human action. The long history of human occupance has left few, if any, ecosystems in a pristine or fully natural condition. So large has our overall impact been that we may speak of an Earth transformed [cf. Soule and Lease 1995].

However, the specifics of the context of Thailand must also be considered. For instance, its high levels of ecological disequilibrium appear to have been reached only during the present century, and especially in recent decades. Perhaps a greater awareness of this fundamental point in the historical ecology of Thailand allows some hope that there may still be time to stop, or at least to temper, some of these destructive trends, and to create a more sustainable society. This possibility may be reinforced by the growing environmental awareness and activism in Thailand, including the latent potential that Buddhism provides for the construction of a viable environmental ethic (Ekachai 1994; Hirsch 1993; Rush 1991; Sponsel and Natadecha-Sponsel 1993, 1995, in press). Complementing these considerations are the advances in environmental science, technology, and education, as well as the increasing recognition of the fact that an economy, and accordingly a society, can only be as healthy as its environment (Brookfield 1993; Ekachai 1994; Hughes and Thirgood 1982; Ketudat and Textor 1990; Leungaramsri and Rajesh 1992; Parnwell and Bryant 1996; Permpongsacharoen 1993).

Acknowledgments

I wish to thank William Balée, Bernard Maloney, Chai Podhisita, Wilhelm G. Solheim II, and Miriam Stark for their valuable comments on the manuscript. I remain solely responsible for any deficiencies.

References

Aiken, S. Robert and Colin H. Leigh. 1985. On the declining fauna of peninsular Malaysia in the post-colonial period. *Ambio* 14(1): 15–22.

Anderson, Douglas D. 1987. A Pleistocene–early Holocene rock shelter in peninsular Thailand. *National Geographic Research* 3(2): 184–198.

———. 1989. Prehistoric human adaptations to environments in Thailand. Pp. 101–123 in Siam Society (eds.), *Culture and Environment in Thailand*. Bangkok: Siam Society.

Arbhabhirama, Anat, Dhira Phantumvanit, John Elkington, and Phaitoon Ingkasuwan, eds. 1987. *Thailand Natural Resources Profile: Is the Resource Base for Thailand's Development Sustainable?* Bangkok: Thailand Development Research Institute.

Bailes, Kendall E., ed. 1985. *Environmental History: Critical Issues in Comparative Perspective*. Lanham, Md.: University Press of America.

Bartlett, H. H. 1956. Fire, primitive agriculture, and grazing in the tropics. Pp. 692–720 in Thomas (ed.), *Man's Role in Changing the Face of the Earth*.

Batchelor, Martine and Kerry Brown. 1992. *Buddhism and Ecology*. New York: Cassell.

Bayard, Donn T., ed. 1984. *The Origins of Agriculture, Metallurgy, and the State in Mainland Southeast Asia*. Dunedin, N.Z.: University of Otago Press.

Bellwood, Peter. 1985. *Prehistory of the Indo-Malaysian Archipelago*. New York: Academic Press.

Bennett, John W. 1969. *Northern Plainsmen: Adaptive Strategy and Agrarian Life*. Chicago: Aldine.

———. 1976. The ecological transition: From equilibrium to disequilibrium. Pp. 123–155 in John W. Bennett, *The Ecological Transition: Cultural Anthropology and Human Adaptation*. New York: Pergamon Press.

———. 1980. Human ecology as human behavior: A normative anthropology of resource use and abuse. Pp. 243–277 in Irwin Altman, Amos Rapoport, and Joachim F. Wohlwill (eds.), *Human Behavior and Environment: Advances in Theory and Research*. New York: Plenum Press.

———. 1993. Underlying ideas: Ecological transitions, socionatural systems, and adaptive behavior. Pp. 3–22 in John W. Bennett, *Human Ecology as Human Behavior: Essays in Environmental and Developmental Anthropology*. New Brunswick, N.J.: Transaction.

Bernatzik, Hugo Adolf and Emmy Bernatzik. 1958. *The Spirits of the Yellow Leaves*. London: Hale.

Bodley, John H. 1994. *Cultural Anthropology: Tribes, States, and the Global System*. Mountain View, Calif.: Mayfield.

Botkin, D. B. 1990. *Discordant Harmonies*. New York: Oxford University Press.

Brockelman, Warren Y. 1989. Differing approaches to environmental protection in Thailand. Pp. 475–493 in Siam Society (eds.), *Culture and Environment in Thailand*.

Bronson, Bennet. 1989. The extraction of natural resources in early Thailand. Pp. 291–301 in Siam Society (eds.), *Culture and Environment in Thailand*.

Brookfield, Harold. 1993. The dimensions of environmental change and management in South-East Asian region. Pp. 5–32 in Harold Brookfield and Yvonne Byron (eds.), *Southeast Asia's Environmental Future: The Search for Sustainability*. Tokyo: United Nations University Press.

Callicott, J. Baird. 1994. *Earth's Insights: A Multicultural Survey of Ecological Ethics*. Berkeley: University of California Press.

Chunkao, Kasem. 1987. Forest resources. Pp. 73–88 in Arbhabhirama et al. (eds.), *Thailand Natural Resources Profile*.

Clark, J. D. and J. W. K. Harris. 1985. Fire and its role in early hominid lifeways. *African Archaeological Review* 3:3–27.

Coedes, G. 1968. *The Indianized States of Southeast Asia*. Honolulu: University of Hawaii Press/East-West Center.

Cohen, Paul T. 1993. Order under heaven: Anthropology and the state. Pp. 175–204 in Grant Evans (ed.), *Asia's Cultural Mosaic: An Anthropological Introduction*. New York: Prentice-Hall.

Cronon, William J. 1993. Foreword: The turn toward history. Pp. vii–x in McDonnell and Pickett (eds.), *Humans as Components of Ecosystems*.

Crumley, Carole L., ed. 1994. *Historical Ecology: Cultural Knowledge and Changing Landscapes*. Santa Fe: School of American Research Press.

Dearden, Philip. 1995. Development, the environment, and social differentiation in northern Thailand. Pp. 111–130 in Rigg (ed.), *Counting the Costs*.

Denevan, William M. 1992. Stone vs. metal axes: The ambiguity of shifting cultivation in prehistoric Amazonia. *Journal of the Steward Anthropological Society* 20(1–2): 153–165.

Dinerstein, Eric and Eric D. Wikramanayake. 1993. Beyond "hot spots": How to prioritize investments to conserve biodiversity in the Indo-Pacific region. *Conservation Biology* 7(1): 53–65.

Donner, Wolf. 1982. *The Five Faces of Thailand: An Economic Geography*. St. Lucia, Queensland: University of Queensland Press.

Drengson, Alan R. 1989. *Beyond Environmental Crisis: From Technocrat to Planetary Person*. New York: Peter Lang.

Dunn, Frederick L. 1970. Cultural evolution in the late Pleistocene and Holocene of Southeast Asia. *American Anthropologist* 72(5): 1041–1054.

———. 1975. *Rainforest Collectors and Traders: A Study of Resource Utilization in Modern and Ancient Malaysia*. Monograph of the Malayan Branch of the Royal Asiatic Society, no. 5. Kuala Lumpur: Royal Asiatic Society.

Dwivedi, O. P. 1988. Man and nature: An holistic approach to a theory of ecology. *Environmental Professional* 10:8–15.

Edgerton, F. N. 1993. The history and present entanglements of some general ecological perspectives. Pp. 9–23 in McDonnell and Pickett (eds.), *Humans as Components of Ecosystems*.

Edgerton, Robert B. 1992. *Sick Societies: Challenging the Myth of Primitive Harmony*. New York: Free Press.

Ekachai, Sanitsuda. 1990. *Behind the Smile: Voices of Thailand*. Bangkok: Thai Development Support Committee.

———. 1994. *Seeds of Hope: Local Initiatives in Thailand*. Bangkok: Thai Development Support Committee.

Engelhardt, Richard A., Jr. 1989. Forest-gatherers and strand-loopers: Econiche specialization in Thailand. Pp. 125–141 in Siam Society (eds.), *Culture and Environment in Thailand*.

Evers, Hans-Dieter. 1987. Trade and state formation: Siam in the early Bangkok period. *Modern Asian Studies* 21(4): 751–771.

Flenley, J. R. 1979. *The Equatorial Rainforest: A Geological History*. London: Butterworth.

———. 1988. Palynological evidence for land use changes in South-east Asia. *Journal of Biogeography* 15:185–197.

Glacken, Clarence J. 1956. Changing ideas of the habitable world. Pp. 70–92 in Thomas (ed.), *Man's Role in Changing the Face of the Earth*.

Glover, I. C. 1989. Early trade between India and Southeast Asia: A link in the development of a world trading system. Pp. 1–57 in *Occasional Papers no. 16, University of Hull Centre for South-East Asian Studies*. Hull: University of Hull.

Gorman, C. F. 1970. Excavation at Spirit Cave, north Thailand. *Asian Perspectives* 13: 81–108.

Hafner, James A. 1990. Forest and policy issues affecting forest use in northeast Thailand, 1900–1985. Pp. 69–94 in Poffenberger (ed.), *Keepers of the Forest*.

Hall, Kenneth R. 1985. *Maritime Trade and State Development in Early Southeast Asia*. Honolulu: University of Hawaii Press.

Hamilton, Lawrence S., ed. 1993. *Ethics, Religion, and Biodiversity: Relations Between Conservation and Cultural Values*. Cambridge: White Horse Press.

Hanks, Lucien. 1972. *Rice and Man: Agricultural Ecology in Southeast Asia*. Arlington Heights, Ill: AHM Publishing.

Harris, Marvin. 1979. *Cultural Materialism: The Struggle for a Science of Culture*. New York: Random House.

———. 1987. Cultural materialism: Alarms and excursions. Pp. 107–126 in Kenneth Moore (ed.), *Waymarks*. Notre Dame: University of Notre Dame Press.

———. 1994. Cultural materialism is alive and well and won't go away until something better comes along. Pp. 62–76 in Robert Borofsky (ed.), *Assessing Cultural Anthropology*. New York: McGraw-Hill.

Hassan, Fekri A. 1981. *Demographic Archaeology*. New York: Academic Press.

Hastings, Alan, Carole L. Hom, Stephen Ellner, Peter Turchin, and H. Charles J. Godfray. 1993. Chaos in ecology: Is Mother Nature a strange attractor? *Annual Review of Ecology and Systematics* 24:1–33.

Headland, Thomas N. 1997. Revisionism in ecological anthropology. *Current Anthropology* 38(4) (in press).

Headland, Thomas N. and Robert C. Bailey. 1991. Introduction: Have hunter-gatherers ever lived in tropical rain forest independently of agriculture? *Human Ecology* 19(2): 115–122.

Headland, Thomas N. and Lawrence A. Reid. 1989. Hunter-gatherers and their neighbors from prehistory to the present. *Current Anthropology* 30:43–66.

Heckman, Charles. 1979. *Rice Field Ecology in Northeast Thailand*. The Hague: Junk.

Heizer, Robert. 1955. Primitive man as an ecological factor. *Kroeber Anthropological Society Papers* 13:1–31.

Hern, Warren M. 1993. Is human culture carcinogenic for uncontrolled population growth and ecological destruction? *BioScience* 43(11): 768–773.

Higham, Charles. 1989. *The Archaeology of Mainland Southeast Asia*. New York: Cambridge University Press.

Higham, Charles and A. Kijngam. 1979. Ban Ciang and northeast Thailand: The palaeoenvironment and economy. *Journal of Archaeological Science* 6:211–233.

Higham, Charles and Bernard Maloney. 1989. Coastal adaptation, sedentism, and domestication: A model for socio-economic intensification in prehistoric Southeast Asia. Pp. 650–666 in David R. Harris and Gordon C. Hillman (eds.), *Foraging and Farming: The Evolution of Plant Exploitation*. London: Unwin and Hyman.

Higham, Charles and Rachanie Thosarat. 1994. *Khok Phanom Di: Prehistoric Adaptation to the World's Richest Habitat*. New York: Harcourt Brace College Publishers.

Hirsch, Philip. 1993. *Political Economy of Environment in Thailand*. Manila: Journal of Contemporary Asia Publishers.

———. 1995. Thailand and the new geopolitics of Southeast Asia: Resource and environmental issues. Pp. 235–259 in Rigg (ed.), *Counting the Costs*.

Hogan, D. W. 1972. Men of the sea: Coastal tribes of South Thailand's West Coast. *Journal of the Siam Society* 60(1): 205–235.

Holdridge, L. R., W. C. Grenke, W. H. Hatheway, T. Liang, and J. A. Tosi, Jr. 1971. A trial application of the basic life zone system in Southeast Asia. Pp. 679–720 in idem, *Forest Environments in Tropical Life Zones: A Pilot Study*. New York: Pergamon Press.

Hughes, J. Donald and J. V. Thirgood. 1982. Deforestation in ancient Greece and Rome: A cause of collapse. *Ecologist* 12(5): 196–208.

Hurst, Philip. 1990. *Rainforest Politics: Ecological Destruction in South-East Asia*. Atlantic Highlands, N.J.: Zed Books.

Hutterer, Karl L. 1976. An evolutionary approach to the Southeast Asian cultural sequence. *Current Anthropology* 17(2): 221–242.

———. 1988. The prehistory of the Asian rain forests. Pp. 63–72 in Julie Sloan Denslow and Christine Padoch (eds.), *People of the Tropical Rain Forest*. Berkeley: University of California Press.

———, ed. 1977. *Economic Exchange and Social Interaction in Southeast Asia: Perspectives from Prehistory, History, and Ethnography*. Papers on South and Southeast Asia, no. 27. Ann Arbor: University of Michigan Press

James, Steven R. 1989. Hominid use of fire in the lower and middle Pleistocene. *Current Anthropology* 30(1): 1–26.

Kathirithamby-Wells, J. 1995. Socio-political structures and the Southeast Asian ecosystem: An historical perspective up to the mid-nineteenth century. Pp. 25–46 in Ole Bruun and Arne Kalland (eds.), *Asian Perceptions of Nature: A Critical Approach*. Richmond, Surrey: Curzon.

Kearney, Michael. 1984. *World View*. Novato, Calif.: Chandler and Sharp.

Ketudat, Sippanondha and Robert B. Textor. 1990. *The Middle Path for the Future of Thailand: Technology in Harmony with Culture and Environment*. Honolulu: East-West Center and Chiang Mai University.

Keyes, Charles F. 1987. *Thailand: Buddhist Kingdom as Modern Nation-State*. Boulder, Colo.: Westview Press.

Kozlovsky, Daniel G. 1974. *An Ecological and Evolutionary Ethic*. Englewood Cliffs, N.J.: Prentice-Hall.

Kriengkraipetch, Suvanna. 1989. Thai folk beliefs about animals and plants and attitudes toward nature. Pp. 195–211 in Siam Society (eds.), *Culture and Environment in Thailand*.

Kunstadter, Peter. 1989. The end of the frontier: Culture and environmental interactions in Thailand. Pp. 543–552 in Siam Society (eds.), *Culture and Environment in Thailand*.

Lansing, Stephen J. 1987. Balinese water temples and the management of irrigation. *American Anthropologist* 89:326–341.

———. 1991. *Priests and Programmers: Technologies of Power in the Engineered Landscape of Bali*. Princeton: Princeton University Press.

Leungaramsri, Pinkaew and Noel Rajesh. 1992. *The Future of People and Forests in Thailand After the Logging Ban*. Bangkok: Project for Ecological Recovery.

Lieberman, J. B. 1990. Wallerstein's systems and the international context of early modern Southeast Asian history. *Journal of Asian History* 24(1): 70–90.

Ling, Trevor. 1973. *The Buddha*. London: Maurice Temple Smith.

Lohmann, Larry. 1993. Land, power, and forest colonization in Thailand. Pp. 198–227 in Marcus Colchester and Larry Lohmann (eds.), *The Struggle for Land and the Fate of the Forests*. London: Zed Books.

Maloney, B. K. 1985. Man's impact on the rainforests of West Malesia: The palynological record. *Journal of Biogeography* 12:537–558.

Marsh, George Perkins. 1965. [1864]. *Man and Nature: Or, Physical Geography as Modified by Human Action*. Cambridge: Harvard University Press.

McDonnell, Mark J. and Steward T. A. Pickett, eds. 1993. *Humans as Components of Ecosystems: The Ecology of Subtle Human Effects and Populated Areas*. New York: Springer-Verlag.

McNeely, Jeffrey A. and Paul Spencer Sochaczewski. 1995. *Soul of the Tiger: Searching for Nature's Answers in Exotic Southeast Asia*. 2d ed. Honolulu: University of Hawaii Press.

Moerman, Michael and Patricia L. Miller. 1989. Changes in a village's relations with its environment. Pp. 303–326 in Siam Society (eds.), *Culture and Environment in Thailand*.

Myers, Norman, ed. 1993. *Atlas of Planet Management*. Garden City, N.Y.: Doubleday.

Oelschlaeger, Max. 1991. *The Idea of Wilderness: From Prehistory to the Age of Ecology*. New Haven: Yale University Press.

Padoch, Christine and Andrew P. Vayda. 1983. Patterns of resource use and human settlements in tropical forests. Pp. 301–313 in F. B. Golley (ed.), *Tropical Rain Forest Ecosystems: Structure and Function*. New York: Elsevier.

Parnwell, Michael J. G. and Raymond L. Bryant, eds. 1996. *Environmental Change in South-East Asia: People, Politics, and Sustainable Development*. London: Routledge.

Pelzer, Karl J. 1968. Man's role in changing the landscape of Southeast Asia. *Journal of Asian Studies* 27(2): 269–279.

Permpongsacharoen, Witoon. 1993. Environmental education alternatives from the Thai environmental movement. Pp. 185–197 in Hartmut Schneider, Jacoline Vinke, and Winifred Weekes-Vagliani (eds.), *Environmental Education: An Approach to Sustainable Development*. Paris: Organization for Economic Cooperation and Development.

Phantumvanit, D. and K. S. Sathirathai. 1988. Thailand: Degradation and development in a resource-rich land. *Environment* 30(1): 10–15.

Pimm, Stuart L. 1991. *The Balance of Nature? Ecological Issues in the Conservation of Species and Communities*. Chicago: University of Chicago Press.

Poffenberger, Mark, ed. 1990. *Keepers of the Forest: Land Management Alternatives in Southeast Asia*. West Hartford, Conn.: Kumarian Press.

Pookajorn, Surin. 1985. Ethnoarchaeology with the Phi Tong Luang (Mlabrai): Forest hunters of northern Thailand. *World Archaeology* 17(2): 206–221.

———. 1988. *Archaeological Research of the Hoabinhian Culture or Technocomplex and Its Comparison with Ethnoarchaeology of the Phi Tong Luang, a Hunter-Gatherer Group of Thailand*. Tübingen: Verlag Archaeologica Venatoria Institute für Urgeschichte der Universität Tübingen.

Pragtong, Mark and David E. Thomas. 1990. Evolving management systems in Thailand. Pp. 167–186 in Poffenberger (ed.), *Keepers of the Forest*.

Prime, Ranchor. 1992. *Hinduism and Ecology*. London: Cassell.

Pyne, Stephen J. 1991. *Burning Bush: A Fire History of Australia*. New York: Holt.

Rajadhon, Phya Anuman. 1988. *Essays on Thai Folklore*. Bangkok: Thai Inter-Religious Commission for Development, and Sathirakoses Nagapradipa Foundation.

Ramitanondh, Shalardchai. 1989. Forests and deforestation in Thailand: A pandisciplinary approach. Pp. 23–50 in Siam Society (eds.), *Culture and Environment in Thailand*.

Reid, Anthony. 1987. Low population growth and its causes in pre-colonial Southeast Asia. Pp. 33–47 in N. G. Owen (ed.), *Death and Diseases in Southeast Asia: Explorations in Social, Medical, and Demographic History*. New York: Oxford University Press.

———. 1988. *Southeast Asia in the Age of Commerce, 1450–1680*. New Haven: Yale University Press.

———. 1990. An "age of commerce" in Southeast Asian history. *Modern Asian Studies* 24(1): 1–30.

———. 1995. Humans and forests in pre-colonial Southeast Asia. *Environment and History* 1:93–110.

Reynolds, T. E. G. 1990. The Hoabinhian: A review. Pp. 1–30 in Gina L. Barnes (ed.), *Bibliographic Reviews of Far Eastern Archaeology*. Oxford: Oxbow Books.

Rifkin, Jeremy. 1990. *Entropy: A New World View*. New York: Bantam/Viking.

———. 1993. *Biosphere Politics: A Cultural Odyssey from the Middle Ages to the New Age*. New York: HarperCollins.

Rigg, Jonathan, ed. 1995. *Counting the Costs: Economic Growth and Environmental Change in Thailand*. Singapore: Institute of Southeast Asian Studies.

Robinson, Mark F. 1994. Observation on the wildlife trade at the daily market in Chiang Khan, northeast Thailand. *Natural History Bulletin of the Siam Society* 42:117–120.

Roscoe, Gerald. 1994. *The Triple Gem: An Introduction to Buddhism*. Chiang Mai, Thailand: Silkworm Books.

Round, Philip D. 1990. Bangkok Bird Club survey of the bird and mammal trade in the Bangkok weekend market. *Natural History Bulletin of the Siam Society* 38:1–43.

Rush, James. 1991. *The Last Tree: Reclaiming the Environment in Tropical Asia*. New York: Asia Society.

Sauer, Carl O. 1965a. Fire and early man. Pp. 288–299 in John Leighly (ed.), *Land and Life: A Selection from the Writings of Carl Ortwin Sauer*. Berkeley: University of California Press.

———. 1965b. Seashore—Primitive home of man? Pp. 300–312 in Leighly (ed.), *Land and Life*.

Schnaiberg, Allan and Kenneth Alan Gould. 1994. *Environment and Society: The Enduring Conflict*. New York: St. Martin's Press.

Sermsri, Santhat. 1989. Population growth and environmental change in Thailand. Pp. 71–91 in Siam Society (eds.), *Culture and Environment in Thailand*.

Siam Society, eds. *Culture and Environment in Thailand*. Bangkok: Siam Society.

Simberloff, D. 1982. A succession of paradigms in ecology: Essentialism to materialism and probabilism. Pp. 63–99 in E. Saarinen (ed.), *Conceptual Issues in Ecology*. Boston: Reidel.

Simmons, I. G. 1989. *Changing the Face of the Earth: Culture, Environment, History*. New York: Blackwell.

Sinsakul, Sin. 1992. Evidence of Quaternary sea level changes in the coastal areas of Thailand: A review. *Journal of Southeast Asian Earth Sciences* 7(1): 23–35.

Smith, Sheldon and Ed Reeves. 1989. Introduction. Pp. 1–18 in Sheldon Smith and Ed Reeves (eds.), *Human Systems Ecology: Studies in the Integration of Political Economy, Adaptation, and Socionatural Regions*. Boulder, Colo.: Westview Press.

Solheim, W. G. II. 1972. An earlier agricultural revolution. *Scientific American* 226:34–41.

Somboon, J. R. P. 1984. Palynological study of mangrove and marine sediments of the Gulf of Thailand. *Journal of Southeast Asian Earth Sciences* 4(2): 85–97.

Soule, Michael E. and Gary Lease, eds. 1995. *Reinventing Nature? Responses to Postmodern Deconstruction*. Washington, D.C.: Island Press.

Spencer, J. E. 1966. *Shifting Cultivation in Southeast Asia*. Berkeley: University of California Press.

Sponsel, Leslie E. 1992. The environmental history of Amazonia: Natural and human disturbances, and the ecological transition. Pp. 233–251 in Harold K. Steen and Richard P. Tucker (eds.), *Changing Tropical Forests: Historical Perspective on Today's Challenges in Central and South America*. Durham, N.C.: Forest History Society.

———. 1997. The human niche in Amazonia: Explorations in ethnoprimatology. Pp. 143–165 in Warren Kinzey (ed.), *New World Primates: Ecology, Evolution, and Behavior*. New York: Aldine de Gruyter.

Sponsel, Leslie E. and Poranee Natadecha-Sponsel. 1988. Buddhism, ecology, and forests in Thailand: Past, present, and future. Pp. 305–326 in John Dargavel, Kay Dixon, and Noel Semple (eds.), *Changing Tropical Forests: Historical Perspectives on Today's Challenges in Asia, Australasia, and Oceania*. Canberra: Centre for Resources and Environmental Studies.

———. 1991. A comparison of the cultural ecology of adjacent Muslim and Buddhist villages in southern Thailand: A preliminary field report. *Journal of the National Research Council of Thailand* 23(2): 31–42.

———. 1993. The potential contribution of Buddhism in developing an environmental ethic for the conservation of biodiversity. Pp. 75–97 in Hamilton (ed.), *Ethics, Religion, and Biodiversity*.

———. 1995. The role of Buddhism in creating a more sustainable society in Thailand. Pp. 27–46 in Rigg (ed.), *Counting the Costs*.

———. In press. The Buddhist monastic community as a green society in Thailand: Its potential role in environmental ethics, education, and action. In Mary Evelyn Tucker and Duncan Williams (eds.), *Consultation on Buddhism and Ecology*.

Stahl, Ann Bower. 1984. Hominid dietary selection before fire. *Current Anthropology* 25(2): 151–168.

Sternstein, Larry. 1993. Population of Siam on the eve of European colonization of mainland Southeast Asia. Paper delivered at the Fifth International Conference on Thai Studies, held at the School of Oriental and African Studies, University of London, July.

Stewart, Omer C. 1956. Fire as the first great force employed by man. Pp. 115–133 in Thomas (ed.), *Man's Role in Changing the Face of the Earth*.

Takaya, Yoshikazu. 1975. An ecological interpretation of Thai history. *Journal of Southeast Asian Studies* 6(2): 190–195.

Terwiel, B. J. 1989. *Through Travellers' Eyes: An Approach to Early Nineteenth Century Thai History*. Bangkok: Duang Kamol.

Thomas, W. L., Jr., ed. 1956. *Man's Role in Changing the Face of the Earth*. Chicago: University of Chicago Press.

Turner, B. L. II, W. C. Clark, R. W. Kates, J. F. Richards, J. T. Mathews, and W. B. Meyer, eds. 1990. *The Earth as Transformed by Human Action: Global and Regional Changes in the Biosphere over the Past Three Hundred Years*. New York: Cambridge University Press with Clark University.

Turner, B. L. II and William B. Meyer. 1993. Environmental change: The human factor. Pp. 40–50 in McDonnell and Pickett (eds.), *Humans as Components of Ecosystems*.

Vichit-Vadakan, Juree. 1989. Thai social structure and behavior patterns: Nature versus culture. Pp. 425–447 in Siam Society (eds.), *Culture and Environment in Thailand*.

Watson, Richard A. and Patty Jo Watson. 1969. *Man and Nature: An Anthropological Essay in Human Ecology*. New York: Harcourt, Brace and World.

Watt, Kenneth E. F., Leslie F. Molloy, C. K. Varshney, Dudley Weeks, and Soetjipto Wirosardjono. 1977. *The Unsteady State: Environmental Problems, Growth, and Culture*. Honolulu: University of Hawaii Press/East-West Center.

Wells, K. E. 1975. *Thai Buddhism: Its Rites and Activities*. Bangkok: Suriyabun.

Wharton, Charles H. 1968. Man, fire, and wild cattle in Southeast Asia. *Proceedings of the Annual Tall Timbers Fire Ecology Conference* 8:107–167.

White, Richard. 1985. Historiographical essay: American environmental history: The development of a new historical field. *Pacific Historical Review* 54:297–335.

Wilson, Constance. n.d. Environmental change in nineteenth-century Thailand. Unpublished paper.

Winichakul, Thongchai. 1996. Siam mapped: The making of Thai nationhood. *Ecologist* 26(5): 215–221.

Winzeler, Robert L. 1976. Ecology, culture, social organization, and state formation in Southeast Asia. *Current Anthropology* 17(4): 623–640.

Wittfogel, Karl A. 1956. The hydraulic civilizations. Pp. 152–164 in Thomas (ed.), *Man's Role in Changing the Face of the Earth*.

Worster, Donald. 1985. History as natural history: An essay on theory and method. *Pacific Historical Review* 54:1–19.

———. 1988a. Appendix: Doing environmental history. Pp. 289–307 in Donald Worster (ed.), *The Ends of the Earth: Perspectives on Modern Environmental History*. New York: Cambridge University Press.

———. 1988b. The vulnerable earth: Toward a planetary history. Pp. 3–20 in Worster (ed.), *The Ends of the Earth*.

———. 1990. The ecology of order and chaos. *Environmental History Review* 14(1–2): 1–18.

———. 1993. *The Wealth of Nature: Environmental History and the Ecological Imagination*. New York: Oxford University Press.

Wyatt, D. 1984. *Thailand: A Short History*. New Haven: Yale University Press.

Yen, Douglas E. 1977. Hoabinhian horticulture: The evidence and the questions from northwest Thailand. Pp. 567–600 in J. Allen, J. Golson, and R. Jones (eds.), *Sunda and Sahul: Prehistoric Studies in Southeast Asia, Melanesia, and Australia*. New York: Academic Press.

EPILOGUE

Tristram R. Kidder and William Balée

> The distinction between natural and social science is beginning to seem meaningless.
>
> —Boaventura de Sousa Santos (1992:12)

The conference that gave birth to this volume centered around defining the role and place of historical ecology: was it a new field, a new paradigm, or simply a reconstitution of an extant, albeit moribund, scholarly direction? The answers to these questions remain, to some extent, elusive. Historical ecology is more than one thing to different people. The participants in the conference generally demurred at labeling it as an emerging paradigm, since it encompassed too little and too much at the same time. Historical ecology is not a new method or a way of organizing one's thinking. Nor is it simply a set of vintage ideas rebottled. Rather, we see it as part of a broader, far more consequential transformation in science as a whole (Santos 1992).

The hallmarks of this emerging scientific transformation lie in two critical distinctions with regard to the normative scientific worldview: (1) no a priori separation between humans and nature can be empirically discerned; and (2) in regard to how we study the world, the distinction between natural and social sciences is tautological. The central theme that integrates the emergent science is a growing recognition that history and contingency are crucial concepts that need to be appreciated, and even embraced, if we are to enter into an effective scholarly dialogue within and among existing academic disciplines and comprehend the holism inherent in global ecology. The papers in this volume challenge us to acknowledge that humans are natural, and nature is human. The emergent science, as exemplified by historical ecology, reveals the local and global linkages that we, as a species, embody.

Historical ecology has an important role to play in helping us to explore and challenge the human/nature dichotomy so prevalent in normative, or modern, science. From the beginnings of the Enlightenment, nature has been objectified by science (Coveney and Highfield 1990; Crombie 1988). Nature has been the subject of study as if it were an entity separated from those who study it (or perhaps better put, we study nature as if we are separate from it). Nature in this regard has been seen as "passive, eternal, and reversible" (Santos 1992:14). Nature as the object of scientific knowledge became a mechanical entity that could be disassembled and put back together again based on eternal rules or laws. Harnessed in this fashion by knowledge, nature could be tamed and controlled. Since nature is regular and regulated by laws or rules, nature is also deterministic and thus subject to reductionist principles. Cartesian methods broke down problems into the sum of their parts; modern science used laws as the means for reconstructing these parts into a whole.

In normative science the identification of laws and rules of nature requires, on the one hand, that initial conditions can be sufficiently determined, and on the other, that the outcome will be independent of the place and time of these initial conditions. In other words, absolute time and space are not relevant initial conditions (Wigner 1970:226). The presuppositions of modern science, in fact, rest on the notion of the timelessness (and spacelessness) of natural laws (Coveney and Highfield 1990; Prigogine and Stengers 1984). Historical ecology rejects that premise (Crumley 1996).

A logical outgrowth of normative science was the realization that if it were possible to codify and understand the laws of nature then it would surely be possible to do the same for the laws of society. From this premise came the origins of the sociology of Auguste Comte, Herbert Spencer, and Emile Durkheim. By the nineteenth and early twentieth centuries two antagonistic modes of thinking about the science of social sciences had emerged: one mode consisted of applying to social sciences, so far as possible, all the epistemological and methodological principles that had dominated normative science since the Enlightenment; the second was to claim for the social sciences an epistemological and methodological independence, based on the supposition that human "nature" was distinct and radically different from the subject matter of natural history (see Santos 1992). The former claim seems to be one that would be more sympathetic to breaking down the human/nature dichotomy, whereas the other should encourage the erection of boundaries between humans and nature. Paradoxically, however, both themes of social scientific study perpetuate the distinction between humans and nature.

The first idea—viewing social science as if it were science—takes as its underlying epistemology the idea that the dominant normative model for studying natural history (as in physics, geology, and biology) is the correct and universally applicable means for achieving knowledge. Therefore, as noted by Boaventura de Sousa Santos (1992:19), "no matter how large the differences between natural and social phenomena, it is always possible to study the latter as if they were the former." Social reality must be reduced to social facts that are things (as first explicitly advocated by Durkheim [1964]) so that these "facts" are measurable, observable, and quantifiable. Of course, this kind of reductionism is challenging, to say the least, and is fraught with great complexity if social behavior is not to be reduced to irrelevance. Much of the critique of positivist archaeology and of cultural ecology and cultural materialism, as these developed in the 1960s and 1970s, focused on the hyperreductionism that allowed us to speak only of laws of such universal generalization as to be meaningless (Flannery 1982). Considerable energy was given to increasing the methodological and epistemological sophistication of the social sciences (Hempel 1965; Nagel 1961). But Ernest Nagel and Thomas Kuhn (Kuhn 1970) proposed that social sciences were "retarded" relative to the natural sciences for the reason that they were more difficult to reduce from reality to things, and because there was not a universally agreed-upon (paradigmatic) consensus that would focus the acquisition and synthesis of knowledge.

The postmodern mood presents social reality as radically subjective and autobiographical. This reality cannot be understood with the methodological tenets of science. Social and human behavior are not reducible or objectifiable, because the same external reality may have several or multiple meanings. The subjectivity of the object—an ethnographic fact—requires that the methods used be qualitative rather than quantitative, descriptive as opposed to explanatory, and intersubjective rather than objective. Although it is presented as antipositivist (or postmodern), this type of thinking shares some of the same epistemological roots as normative science. First, it embraces the human/nature dichotomy—perhaps even more so than the normative model. Second, it presupposes a timelessness and spacelessness of the subject of inquiry, since the time and space of the agent are irrelevant to the subjective locus of the interpreter. While certainly not the same as the normative mode of scientific thinking, the radical subjectivism of postmodernism is much less a *re*placement than a *dis*placement of paths to knowledge developed during the Enlightenment.

Where does historical ecology fit in relation to these claims for social science knowledge? We noted above that it cannot claim to have a distinctive methodological status, but it does resonate logically with a broader trend toward the development of a new, global view of science and nature. The first rumblings of this new standard emerged from the so-called hard sciences in the early twentieth century in the work of physicists such as Einstein and Heisenberg, to name only two (Coveney and Highfield 1990; Prigogine and Stengers 1984). These and other physical scientists achieved two remarkable findings that have profound importance for fields beyond physics and quantum mechanics. First, Einstein recognized the "relativity of simultaneity." This theoretical insight admitted that the simultaneity of distant events cannot be verified, but merely defined. It revolutionized the concept of time and space: Newton's absolute time and space, the predicates of scientific determinism from the seventeenth century on, are rendered abstractions. The second impact was the introduction of measurement uncertainty: Heisenberg's uncertainty principle states that it is not possible to observe or measure an object without interfering with it in such a way that it is no longer precisely the same as it was before. Thus, by the latter twentieth century the bedrocks of science, absolute time and space, as well as the objective role of the scientist, had been cast into doubt or uncertainty.

A final but crucial turning point in the development of the new science has been the recognition of chaos science—or, more appropriately, "far-from-equilibrium dynamics." Whereas normative science saw order as the rule, modern science is turning to disorder. The work of Ilya Prigogine (1980; Prigogine and Stengers 1984) has demonstrated that at states far from equilibrium, evolution takes place as a result of fluctuations of energy that occur at unpredictable moments and spontaneously generate reactions that in turn cause the system to respond in any number of ways. Rather than being a predictable phenomenon, these critical points, called points of bifurcation, represent the potentiality of a system to be attracted to a lesser state of entropy. This implies that open systems are both irreversible and, more critically, the product of their own history (Beyerchen 1989; Prigogine 1981). In contrast to classical physics, these new concepts suggest that "in place of eternity, we now have history; in place of determinism, unpredictability . . . in place of order, disorder; in place of necessity, creativity and contingency" (Santos 1992:25). Nonequilibrium dynamics and historical contingency are comprehended in historical ecology as part of the supersession of the paradox succinctly stated by Fernand Braudel (1993:12): "Responses to natural challenges . . . continually free humanity from its environment and at the same time subject it to the resultant solutions. We exchange one form of determinism for another." The recognition of nonequilibrium and contingency is helping to mediate the human/nature dichotomy empirically by way of the landscape.

This recognition is having a significant impact on how scientists conceptualize their units of study. Whereas ecosystems, for example, represent arbitrary and self-contained isolates of matter and energy, landscapes are subject to nonequilibrium processes, contingent historical phenomena, and stochastic disturbances that lead to global change. The growing recognition of the landscape as a focus of research represents an implicit (and sometimes explicit) awareness that global (and even local) ecology is more than the sum of its parts (Crumley 1996). The new science is emphasizing as never before the holism of ecology in the present—and, by doing so, is recognizing that the present can be understood only in the context of its history. Acknowledging these concepts leads anthropologists, especially, to rethink our often implicit functionalist focus on the role of culture as a homeostatic regulatory mechanism, and to turn our interest to understanding how humans relate to a world that is relentlessly far from equilibrium.

These ideas dovetail with a series of concepts or theories labeled by Erich Jantsch as the "self-organization" paradigm (Allen 1981; Haken 1981; Jantsch 1980, 1981; Santos 1992). The implications of these ideas are profound. First, we must challenge or at least examine critically the role of laws and the related concept of causality. Invariant laws must be time symmetrical (Coveney and Highfield 1990)—that is, they must operate outside the linear conditions of time. This condition of determinism cannot, in fact, exist, since all objects, particles, or phenomena have their own history (Prigogine and Stengers 1984). Evolution itself is a kind of history. It is precisely in this recognition of the irreversibility of open systems that historical ecology has one of its main strengths. Unlike mechanistic determinism, which ignores history in favor of laws and rules, or postmodernism, which eschews history for a radical here-and-now subjectivism, historical ecology recognizes and even demands the centrality of the notion that history counts. Simply put, we contend that no object of understanding is completely ahistorical.

A second value that historical ecology takes from and contributes to the emerging science is the realization and actualization that the distinction between humans and nature is arbitrary and no longer tenable in light of a holistic, global ecology. If history counts for humans, it counts equally for plants, other animals, and even particles. Interdependence is inherent in the idea of historical ecology. So, too, is the apprehension that by action and measurement we intrude upon the natural world while the natural world affects us. There is no subject/object, human/nature dichotomy because we and they are simultaneously both subject and object of study.

Historical ecology is perhaps more a movement than a way of seeing or doing science. The scholarship in this book follows from a broader paradigmatic shift that is advancing science away from the absolute, mechanistic, and deterministic concepts of natural and human behavior toward a different, but no less scientific, understanding of our place and role on the globe. As the world continues to shrink, science must turn from its specialization and introspection to producing a commodity that transcends disciplinary boundaries and that seeks to reintegrate humans into nature. To speak of natural history requires an understanding of human actions, and to write human history demands an appreciation of nature and its influence. Western science cannot, as Bacon once said, "command nature except by obeying her" (Anderson 1960:119). Historical ecology is one small step toward better situating ourselves in the global village. This village is one where all people and things have a history, and where that history serves as a dialogue between past and present in order to better prepare us for the future.

References

Allen, Peter M. 1981. The evolutionary paradigm of dissipative structures. Pp. 25–72 in Jantsch (ed.), *The Evolutionary Vision*.

Anderson, Fulton H., ed. 1960. *Francis Bacon: The New Organon and Related Writings*. New York: Liberal Arts Press.

Beyerchen, Alan D. 1989. Nonlinear science and the unfolding of a new intellectual vision. *Papers in Comparative Studies* 6:25–49.

Braudel, Fernand. 1993. [1963]. *A History of Civilizations*. Trans. R. Mayne. New York: Penguin.

Coveney, Peter and Roger Highfield. 1990. *The Arrow of Time*. New York: Fawcett Columbine.

Crombie, Alistair C. 1988. Designed in the mind: Western visions of science, nature, and humankind. *History of Science* 26:1–12.

Crumley, Carole L. 1996. Historical ecology. Pp. 558–560 in D. Levinson and M. Ember (eds.), *Encyclopedia of Cultural Anthropology*, vol. 2. New York: Holt.

Durkheim, Emile. 1964. [1895]. *The Rules of Sociological Method*. Trans. Sarah A. Solovay and John H. Mueller. New York: Free Press.

Flannery, Kent V. 1982. The golden Marshalltown: A parable for the archaeology of the 1980s. *American Anthropologist* 84:265–278.

Haken, Hermann. 1981. Synergetics: Is self-organization governed by universal principles? Pp. 15–23 in Jantsch (ed.), *The Evolutionary Vision*.

Hempel, Carl G. 1965. Aspects of scientific explanation. Pp. 331–496 in C. G. Hempel (ed.), *Aspects of Scientific Explanation and Other Essays in the Philosophy of Science*. New York: Free Press.

Jantsch, Erich. 1980. *The Self-Organizing Universe: Scientific and Human Implications of the Emerging Paradigm of Evolution*. Oxford: Pergamon Press.

Jantsch, Erich. 1981. Unifying principles of evolution. Pp. 83–115 in E. Jantsch (ed.), *The Evolutionary Vision*. AAAS Selected Symposium 61. Boulder, Colo.: Westview Press.

Kuhn, Thomas S. 1970. *The Structure of Scientific Revolutions*. Chicago: University of Chicago Press.

Nagel, Ernest. 1961. *The Structure of Science: Problems in the Logic of Scientific Explanation*. New York: Harcourt, Brace and World.

Prigogine, Ilya. 1980. *From Being to Becoming*. San Francisco: Freeman.

———. 1981. Time, irreversibility, and randomness. Pp. 73–82 in Jantsch (ed.), *The Evolutionary Vision*.

Prigogine, Ilya and Isabelle Stengers. 1984. *Order out of Chaos: Man's New Dialogue with Nature*. Toronto: Bantam.

Santos, Boaventura de Sousa. 1992. A discourse on the sciences. *Review Fernand Braudel Center* 15:9–47.

Wigner, Eugene. 1970. *Symmetries and Reflections: Scientific Essays*. Cambridge: Cambridge University Press.

NOTES ON THE CONTRIBUTORS

WILLIAM BALÉE is Associate Professor of Anthropology at Tulane University. His research has focused on the human ecology and ethnobotany of several groups of Amazonian Indians. He is the author of *Footprints of the Forest: Ka'apor Ethnobotany—The Historical Ecology of Plant Utilization by an Amazonian People* (1994).

ROBERT L. BETTINGER is Professor of Anthropology at the University of California, Davis. He has specialized in the study of the prehistory of the Great Basin of North America. He has carried out major comparative research on foraging peoples. He is the author of *Hunter-Gatherers: Archaeological and Evolutionary Theory* (1991).

JANET M. CHERNELA is Associate Professor in the Department of Sociology and Anthropology at Florida International University. She has carried out long-term fieldwork among Tukanoan-speaking peoples of the Northwest Amazon. Her publications include *The Wanano Indians of the Brazilian Amazon: A Sense of Space* (1993).

CAROLE L. CRUMLEY is Professor of Anthropology at the University of North Carolina, Chapel Hill. Her research has focused on the prehistory of Burgundy, France. She has published widely on historical ecology and is one of the pioneers of the movement. She edited the volume *Historical Ecology: Cultural Knowledge and Changing Landscapes* (1994).

R. BRIAN FERGUSON is Professor in the Department of Sociology, Anthropology, and Criminal Justice at Rutgers University, Newark. He has conducted long-term research on the issue of warfare. He is the author of *Yanomami Warfare: A Political History* (1995).

TED GRAGSON is Assistant Professor of Anthropology at the University of Georgia, Athens. He has published extensively on the ecology of foraging peoples and on fish ecology in Amazonia. His work has appeared in *American Anthropologist*, *Human Ecology*, *Principes*, *Journal of Ethnobiology*, and elsewhere.

ELIZABETH GRAHAM is Associate Professor of Anthropology at York University, Toronto. She has carried out long-term archaeological fieldwork in Belize. She wrote *The Highlands of the Lowlands: Environment and Archaeology in the Stann Creek District, Belize, Central America* (1994).

CAROL HENDERSON is President of Rajasthan Studies Group and Postdoctoral Research Scholar, Mellon Project on Democratization, School of International and Public Affairs, Columbia University. Her work has focused on responses to environmental disaster and disaster relief in South Asia. Some of her work has appeared in *Research in Economic Anthropology*, among other journals.

TRISTRAM R. KIDDER is Associate Professor of Anthropology at Tulane University. His research deals mainly with the prehistory of the southeastern United States. Some of his work has appeared in *Journal of Field Archaeology*, *Science*, *Geoarchaeology*, *Southeastern Archaeology*, and other journals.

ELINOR MELVILLE is Professor of History at York University, Toronto. She has worked for many years on the environmental history of central Mexico. She is the author of *A Plague of Sheep: Environmental Consequences of the Conquest of Mexico* (1994).

LINDA A. NEWSON is Professor of Geography and Head of the School of Humanities at King's College London. She has carried out research on the spread of epidemic disease in colonial Latin America. Her publications include *Life and Death in Early Colonial Ecuador* (1995).

DARRELL A. POSEY is a resident scholar of Mansfield College, University of Oxford. He has conducted long-term fieldwork on the ethnobiology of the Kayapó Indians of central Brazil. His work has appeared in *Journal of Ethnobiology*, *Advances in Economic Botany*, *Agroforestry Systems*, and elsewhere.

STEPHEN J. PYNE is Professor of American Studies at Arizona State University West, Phoenix. He has carried out extensive research, even while serving as a firefighter, on the influence of fire on world landscapes. He is the author of *Fire on the Rim* (1995 [1989]).

LAURA RIVAL is Lecturer in Social Anthropology at the University of Kent, England. Educated in France, she has conducted long-term fieldwork on how the Huaorani Indians of Amazonian Ecuador construct social reality and objectify their relationship to nature. Some of her work has appeared in *Man* (now *Journal of the Royal Anthropological Institute*), among other journals.

ANNA C. ROOSEVELT is Professor of Anthropology at the University of Illinois - Chicago Circle, and Curator of Anthropology at the Field Museum of Natural History. She has carried out archaeological fieldwork in the Amazon and Orinoco basins. She is the author of *Moundbuilders of the Amazon: Geophysical Archaeology on Marajó Island, Brazil* (1991).

LESLIE E. SPONSEL is Professor of Anthropology at the University of Hawai'i. He has carried out ethnographic research with the Sanuma, a northern subgroup of the Yanomami, Curripaco, and other native peoples. He is the editor of *Indigenous Peoples and the Future of Amazonia: An Ecological Anthropology of an Endangered World* (1995).

NEIL L. WHITEHEAD is Associate Professor of Anthropology at the University of Wisconsin-Madison. He has specialized in the historical reconstruction of prehistoric and colonial Guiana. He is the author of *Lords of the Tiger Spirit: A History of the Caribs in Colonial Venezuela and Guyana, 1498–1820* (1988).

STANFORD ZENT is a researcher at the Instituto Venezolano de Investigaciones Científicas, Caracas, Venezuela. He has focused on ethnographic work among the Piaroa and Hodi of the tropical forest of Venezuela. His dissertation (1992, Columbia University) is entitled "Historical and ethnographic ecology of the Upper Cuao River Wothiha: Clues for an interpretation of native Guianese social organization."